Applied Probability and Statistics

BAILEY · The Elements of Stochastic Processes with Applications to the Natural Sciences

BARTHOLOMEW · Stochastic Models for Social Processes, *Second Edition*

BENNETT and FRANKLIN · Statistical Analysis in Chemistry and the Chemical Industry

BHAT · Elements of Applied Stochastic Processes

BLOOMFIELD · Fourier Analysis of Time Series: An Introduction

BOX and DRAPER · Evolutionary Operation: A Statistical Method for Process Improvement

BROWNLEE · Statistical Theory and Methodology in Science and Engineering, *Second Edition*

BURY · Statistical Models in Applied Science

CHERNOFF and MOSES · Elementary Decision Theory

CHOW · Analysis and Control of Dynamic Economic Systems

CLELLAND, deCANI, BROWN, BURSK, and MURRAY · Basic Statistics with Business Applications, *Second Edition*

COCHRAN · Sampling Techniques, *Second Edition*

COCHRAN and COX · Experimental Designs, *Second Edition*

COX · Planning of Experiments

COX and MILLER · The Theory of Stochastic Processes, *Second Edition*

DANIEL · Application of Statistics to Industrial Experimentation

DANIEL and WOOD · Fitting Equations to Data

DAVID · Order Statistics

DEMING · Sample Design in Business Research

DODGE and ROMIG · Sampling Inspection Tables. *Second Edition*

DRAPER and SMITH · Applied Regression Analysis

DUNN and CLARK · Applied Statistics: Analysis of Variance and Regression

ELANDT-JOHNSON · Probability Models and Statistical Methods in Genetics

FLEISS · Statistical Methods for Rates and Proportions

GNANADESIKAN · Methods for Statistical Data Analysis of Multivariate Observations

GOLDBERGER · Econometric Theory

GROSS and CLARK · Survival Distributions

GROSS and HARRIS · Fundamentals of Queueing Theory

GUTTMAN, WILKS and HUNTER · Introductory Engineering Statistics, *Second Edition*

HAHN and SHAPIRO · Statistical Models in Engineering

HALD · Statistical Tables and Formulas

HALD · Statistical Theory with Engineering Applications

HARTIGAN · Clustering Algorithms

HILDEBRAND, LAING and ROSENTHAL · Prediction Analysis of Cross Classifications

HOEL · Elementary Statistics, *Fourth Edition*

HOLLANDER and WOLFE · Nonparametric Statistical Methods

HUANG · Regression and Econometric Methods

JAGERS · Branching Processes with Biological Applications

continued on back

Experimental
Designs

A WILEY PUBLICATION IN APPLIED STATISTICS

Experimental Designs

WILLIAM G. COCHRAN
Professor of Biostatistics
The Johns Hopkins University

and

GERTRUDE M. COX
Director, Institute of Statistics
University of North Carolina

SECOND EDITION

JOHN WILEY & SONS
New York • Chichester • Brisbane • Toronto • Singapore

SECOND EDITION

30 29 28 27 26

Library of Congress Catalog Card Number: 57–5908

PRINTED IN THE UNITED STATES OF AMERICA

Preface to the Second Edition

DURING THE SIX YEARS SINCE THE PUBLICATION OF THE FIRST edition, there has been a substantial increase in the attention paid by research workers to the principles of experimental design. Evidence of this change can be seen in the medical and social sciences, in physics and chemistry, and in industrial research. Another encouraging trend is that workers in these areas, although still willing to utilize appropriate designs taken from agricultural experimentation, have begun to examine their own problems of experimentation in a fresh light and to produce new designs better suited to their particular conditions. The recent literature also reflects a move toward greater depth and comprehensiveness in experimental work, as instanced by numerous papers devoted to experimentation with more than one factor.

In preparing a second edition we decided to leave the framework of the book unchanged. New material is included wherever necessary to fulfill our original objective of making available the types of design most likely to be useful in practice.

Two chapters have been added. Chapter 6A deals with the fractional replication of factorial experiments. This technique, initially applied in agricultural field experiments, was described briefly in the first edition. In the intervening period, fractional replication has proved so productive in exploratory research, particularly in industry, that a more complete account is justified.

Chapter 8A is also concerned with experiments of the factorial type. It presents methods and new designs for experiments in which the factors represent quantitative variables, measured on a continuous scale. In this situation it is often natural to think of the yield or response as related to the levels of the factors by some mathematical function which, if it could be found, would summarize compactly all the results of the experiment. The new methods are applicable when this function can be approximated, within the limits of the experimental region, by a polynomial of the first or second degree. The second half of this chapter gives an introductory account of experimental strategies for finding, as quickly as possible, the levels at which the factors must be set in order to obtain the maximum response. These techniques, produced for industrial research and development, have many potential applications in other areas.

The appearance of several series of new incomplete block designs during the last six years necessitated expansions in some of the later chapters. The investigator who has found these designs useful now has a considerably wider selection. In fact, the large number of new designs forced us to revise one of our original objectives. When it appeared, the first edition contained the plans of substantially all the designs in common use at that time. To include the new plans, however, would involve devoting over 200 extra pages to plans alone. Even if this were done, our set of plans would soon become incomplete as further research is published. We have tried to handle this problem by giving, at the end of Chapter 9, an index to the incomplete block designs now available. For designs for which no plan is presented, the purpose of the design and the method of analysis are explained at the appropriate place in a later chapter, with references to the sources of the plans. The principal series of designs represented in this way are some new confounded factorial designs in single replication, partially balanced incomplete block designs (with elimination of either a single or a double grouping), the chain block and generalized chain block designs and some further designs related to the latin square.

To cover other developments, a number of sections have been added. Section 2.22a presents a table for estimating the numbers of replications needed for comparing two treatments when the data are arranged in two classes, as dead or alive, sound or defective. Sections 4.15a and 4.27a give the methods of analysis with data of this type for completely randomized and randomized block designs. These sections fill a gap in the first edition, which gave methods of analysis only for continuous data.

Sections 4.61a to 4.66a contain a discussion of the use of latin squares in adjusting for residual effects which may be present when the treatments are applied in sequence to the same subject.

Topics that are presented more briefly are sequential experimentation (section 2.23a), the testing of effects suggested by the data (section 3.53), the problem of making several tests of significance in the same experiment (section 3.54a), Yates' automatic method of computing factorial effect totals (section 5.24a), additional standard error formulae for split-plot experiments (section 7.22), the effects of errors in the weights on the recovery of inter-block information (section 10.12a), and the use of balanced incomplete block designs in taste and preference testing (section 11.1a). Some of these topics deserve more space than we have given them, but the number of pages added in this edition is already greater than we should have liked.

A number of users of the book, particularly those who use it for teaching, asked us to prepare this edition so that the new material is clearly identifiable. To accomplish this, we have numbered the new chapters as 6A and 8A. Any new section, table, or plan carries the letter *a*, e.g., section 2.23a, table 11.1a. Where a minor revision or addition has been made within a section, the old section number remains unchanged.

For permission to use data as examples we are indebted to the General Foods Corporation and to Doctors J. Doull and J. N. Tolentino. Considerable help has been received from staff members at our respective institutions and from colleagues elsewhere.

The large sale of the first edition of a book that is not elementary has been highly gratifying. It testifies to the alertness of workers in the experimental sciences in seeking out any new ideas and methods that may make their investigations more efficient and productive.

W. G. Cochran
G. M. Cox

May, 1957

Preface to the First Edition

WORK ON THIS BOOK WAS STARTED WHEN BOTH OF US WERE MEMbers of the staff of Iowa State College. At that time requests were received rather frequently from research workers. Some wanted advice on the conduct of a specific experiment: others, who had decided to use one of the more complex designs that have been discovered in recent years, asked for a plan or layout that could be followed during the experimental operations. Although the logical principles governing the subject of experimentation are admirably expounded in Fisher's book *The Design of Experiments*, these requests indicated a need for a different type of book, one which would describe in some detail the most useful of the designs that have been developed, with accompanying plans and an account of the experimental situations for which each design is most suitable. Such a book is directed at the experimenter and is intended to serve as a handbook which is consulted when a new experiment is under consideration.

Mainly on account of the war, slow progress was made. In 1944 we completed a mimeographed draft of which several hundred copies were distributed. Many helpful suggestions were made by readers. Of these, the most frequent was to the effect that we should include more material dealing with the statistical analysis of the results. In the mimeographed draft, our practice was to give references to worked examples of the analysis in cases where they could be found in the literature, and to present examples for only those designs whose analysis was not available. To this it was objected that many research workers did not have easy access to our references.

This suggestion raised a difficult issue. To present a self-contained account of the analysis of variance in all its ramifications would make the book, it seemed to us, unwarrantably long and expensive. Consequently we have continued to assume that the reader has some knowledge of the principles of analysis of variance and of the computational methods involved. We have included a brief review of the basic theory and an extensive set of worked examples of the analysis for both common and less common types of design. Although strenuous efforts were made to obtain a selection of examples from diverse fields of research, a preponderance from biology, and more particularly from agriculture, was dictated by our own experience in those areas. On several occasions we rejected an example which would have made an attractive addition to the scope because we did not feel sufficiently familiar with the conditions of experimentation to give a realistic account of the problems encountered.

Since courses of lectures on the design of experiments are being introduced in many colleges and universities, some teachers may be interested in the potentialities of the book as a textbook. Several comments are prompted by our own experience in this connection. First, it will often be necessary for the teacher to provide a more systematic development of the analysis of variance than is given here. Second, interest in such a course is greatly enhanced by examples from an environment with which the listeners are familiar, and especially by examples of experiments that have been conducted by some of the listeners. Thus the teacher is well advised to build the course around his own examples, using those in this book mainly as supplementary material for the students. Finally, a selective use of the book is in order, because it contains much more material than can be covered in a typical one-quarter course, and because some difficult topics have been dealt with in the early parts of the book.

We wish to express our gratitude to the many staff members at Iowa State College and North Carolina State College who helped us by putting experimental data at our disposal, by providing painstaking descriptions of their experimental techniques, or by lending their assistance in the preparation of the manuscript and plans. For similar kindnesses in connection with certain of the examples we are indebted to G. F. Potter, D. Y. Solandt, D. B. DeLury and J. Hunter, F. M. Wadley and C. F. Rainwater, and the Wailuku Sugar Company, Honolulu. Some theoretical results used in Chapter 14 were developed from research conducted at Raleigh under a contract with the Office of Naval Research. Finally, our thanks go to George W. Snedecor, who participated in the original plans for this book and made a careful reading of the first draft.

W. G. COCHRAN
G. M. COX

January, 1950

Contents

CHAPTER 1

Introduction

1.1 The Contribution of Statistics to Experimentation

1.11 The Problem of Interpretation. Since statisticians do not usually perform experiments, their claim to attention when they write on this subject requires some explanation. It is true that on many important aspects of experimentation the statistician has no expert knowledge. Nevertheless, in recent years, research workers have turned increasingly to statisticians for help both in planning their experiments and in drawing conclusions from the results. That this has happened is convincing evidence that statistics has something to contribute.

At first, requests for assistance were nearly always concerned with the interpretation of the results. It is a common characteristic of experiments in widely diverse fields of research that, when they are repeated, the effects of the experimental treatments vary from trial to trial. This variation introduces a degree of uncertainty into any conclusions that are drawn from the results. Even after a number of repetitions, or *replications* as they are called, the investigator still does not know by how much his results would be changed if the experiment were repeated further under the same conditions. Successive trials may be so discrepant in their results that it is doubtful which of two treatments would turn out better in the long run.

As an illustration of this variation, data are given in table 1.1 from a simple experiment. The data are the times (minus 2 minutes) required to perform a routine statistical calculation, that of finding the sum of squares of 27 observations. Each sum of squares was computed separately by the same person on both of two standard computing machines. In all, 10 different sums of squares were worked, making 10 trials or replications of the experiment. It will be noted that the differences in speed range from 17 seconds in favor of machine B to 2 seconds in favor of machine A. Some experimenters may comment that the results of this experiment are remarkably well behaved, and exhibit nothing like the variation with which they have to contend. The results will, however, serve our purpose.

TABLE 1.1 Time in seconds (minus 2 minutes) required for computing sums of squares

Replication	Machine A	Machine B	Difference ($A - B$)
1	30	14	16
2	21	21	0
3	22	5	17
4	22	13	9
5	18	13	5
6	29	17	12
7	16	7	9
8	12	14	−2
9	23	8	15
10	23	24	−1
Means	21.6	13.6	8.0

The object of the experiment is, of course, to compare the speeds of the two machines for this calculation. More specifically, two objects might be stated. The first is to answer the question: is there any difference in speed? or, to put it another way, to test the hypothesis that there is no difference in speed. The second object, which is related to the first, is to estimate the size of the difference in speed. Almost all experiments are carried out for one or both of these purposes—the testing of hypotheses and the estimation of differences in the effects of different treatments.

As regards the test of the hypothesis that there is no difference in speed, we might report, as relevant evidence, that B proved faster 7 times out of 10, A twice, while once there was a tie. For the problem of estimation we might report that the average difference in speed in the experiment was 8 seconds in favor of B. These purely descriptive statements do not carry us very far. Their weakness is that they supply no information about the reliability of the figures presented. For example, have we any confidence that, if the experiment were continued for another set of 10 trials, the advantage at the end would still be close to 8 seconds in favor of B?

Because of the deficiencies in the descriptive approach, experimenters have adopted a different point of view in the summarization of their results. They tend to reason as follows. Suppose that it were feasible to continue the experiment indefinitely under the same conditions. The average difference in speed between the two machines would presumably settle down to some fixed value. This value, which will be independent of the size of the experiment that was actually carried out, may reasonably be called the *true* difference between A and B. From this point of view, the problem of summarizing the results may be restated in the question: what can we say about the true difference between A and B? This is a problem in *induction* from the part to the whole, or in statistical language, from the sample to the population. A solution to this problem has been developed by means of the theory of statistics. It is this solution that constitutes the principal contribution of statistics to the interpretation of the results.

1.12 Statistical Inference. Obviously, it cannot be expected that the solution will provide the exact value of the unknown true difference. As a less ambitious goal we might hope to be able to find 2 limits within which the exact value is certain to lie, but even this cannot quite be attained. What can be done is that for any chosen probability, say .95, two limits are found such that the probability that they enclose the true difference is .95. In other words, limits can be found that are almost certain to enclose the true difference, where the degree of certainty, as measured by the probability, can be chosen by the experimenter. Since we wish to focus attention on the type of inferential statement that can be made rather than on the method of calculating the limits, the computations will not be discussed at present. For the example in table 1.1 they will be found in section 4.42. When the probability is .95, the limits for the true difference in speed between the 2 machines turn out to be 3.3 and 12.7 seconds in favor of machine B. A statement that B is faster by an amount that lies between 3.3 and 12.7 seconds has a 1 in 20 chance of being wrong. If the degree of certainty is decreased by lowering the probability to .8, the limits are narrowed to 5.1 and 10.9 seconds. If the probability is raised to .99, the limits become 1.1 and 14.9 seconds, and as the probability is brought closer to certainty, the limits steadily become farther apart. The limits are called confidence limits, and the probabilities are called confidence probabilities.*

These probabilities are not merely academic abstractions: they can be subjected to at least a rough experimental verification. For verification, we need a situation where the true difference between the effects of two treatments is known. In toxicology, for instance, this situation can sometimes be obtained by diluting a standard poison to a known extent. The dilution is sent to the laboratory, labelled as an "unknown" poison. By experiments on animals, confidence limits are found for the

* Fisher (1.2) has developed statistical inference in terms of *fiducial* limits. The two concepts, fiducial and confidence limits, have different logical backgrounds, although in all simple applications to controlled experiments the actual values of the fiducial and confidence limits are identical. For a discussion, see Kendall (1.1).

amount of the unknown that has the same toxicity as a given amount of the standard. Since the persons originating the experiment know the true value for this amount, they can verify whether the statement that the true amount lies within the limits is correct. The practical difficulty with this type of check is the labor required. A large number of experiments would be needed to verify whether about 95% of the statements made with confidence probability .95 were in fact correct, and about 5% were wrong.

As we have seen, the statistical solution to the problem of estimation consists of a statement that the true difference lies between certain limits, plus a probability that the statement is correct. It is of interest to consider whether this type of information is sufficiently precise to permit decisions of practical importance to be made. Although a thorough discussion of this question would be rather lengthy, inferences of this type often permit definite action to be taken with confidence that the action will be fruitful. When they fail to decide the point at issue, the reason is nearly always that the data obtained are insufficient. For illustration, suppose that it is desired to discover whether the application of a dressing of some fertilizer to a crop will be profitable. The cost of the fertilizer is such that its application will be profitable only if it increases yields on the average by 2 or more bushels. A series of experiments is carried out in order to estimate the true average response to the fertilizer. If the 95% confidence limits for the increase due to the fertilizer are 4 and 11 bushels, its use can be adopted with a good deal of assurance that it will be profitable. Similarly, if the 95% confidence limits are -5 and 1 bushels, a decision not to use the fertilizer follows. A case where there is uncertainty occurs when the limits are 0 and 5 bushels. Here it is likely that there will be either a small gain or a small loss, but no recommendation can be made without considerable risk of its being wrong. If it is important to make the correct decision, further experiments must be conducted in order to narrow the distance between the confidence limits.

Thus far we have considered the problem of estimating the true difference between the effects of two treatments. In testing hypotheses, we are interested in the supposition that the true difference has some specified value, most commonly zero. As in the case of estimation, difficulty arises because of the variability that is typical of experimental data. As a result of this variability, the data are never exactly in agreement with the hypothesis, and the problem is to decide whether the discrepancy between the data and the hypothesis is to be ascribed to these variations or to the fact that the hypothesis is not true. The contribution of statistics is the operation known as a *test of significance*

Essentially, this is a rule for deciding, from examination of the data, whether or not to reject the hypothesis. Such rules are made to satisfy two conditions that are obviously desirable: (i) hypotheses that are true shall be rejected only very occasionally, and the probability of rejection may be chosen by the experimenter; (ii) hypotheses that are false shall be rejected as often as possible.

This technique enables the experimenter to test his hypotheses about the action of the treatments, with the assurance that there is little risk of erroneously rejecting a hypothesis that happens to be correct. Probabilities of .05 and .01 are most commonly used for this risk, and in these cases the tests are said to be made at the 5 and 1% significance levels respectively. These levels are just useful conventions, and a lower probability may be used if the consequences of an erroneous rejection of the hypothesis are very serious. It should be remembered, however, that in lowering this probability value we automatically diminish the chance of rejecting a hypothesis that is false.

A useful property of a test of significance is that it exerts a sobering influence on the type of experimenter who jumps to conclusions on scanty data, and who might otherwise try to make everyone excited about some sensational treatment effect that can well be ascribed to the ordinary variation in his experiment. On the whole, however, tests of significance are less frequently useful in experimental work than confidence limits. In many experiments it seems obvious that the different treatments must have produced some difference, however small, in effect. Thus the hypothesis that there is *no* difference is unrealistic: the real problem is to obtain estimates of the sizes of the differences.

The construction of confidence limits may add something to the interpretation of a test of significance. For instance, suppose that the speeds of the 2 machines in the example had not been significantly different. It is a commonplace that this result would not prove that the 2 machines were identical in speed. However, if the 95% confidence limits for the difference in speed were -2 and $+4$ seconds, we might argue that a true difference of 4 seconds, even if it existed, would be of no practical significance. Consequently, it could be said that for all practical purposes the 2 machines are identical in speed. This is much more positive and useful than the mere statement that the difference in speeds was not statistically significant. If, on the other hand, the confidence limits were -30 and $+32$ seconds, there is no justification for the conclusion that the machines can be regarded as equivalent. All that we have learned is that the data are not sufficiently accurate to show whether there is a difference in speed that is of practical importance.

To summarize: variability in results is typical in many branches of experimentation. Because of this, the problem of drawing conclusions from the results is a problem in induction from the sample to the population. The statistical theories of estimation and of testing hypotheses provide solutions to this problem in the form of definite statements that have a known and controllable probability of being correct. These statements are specific enough to be useful in deciding whether action can be taken on the basis of the results.

1.13 The Function of Randomization. As would be expected, the type of statistical inference that can be made from a body of data depends on the nature of the data. It is easy to conduct an experiment in such a way that no useful inferences can be made, and many of the experiments brought to the statistician, particularly in earlier years, have been of this type. To take a simple example, suppose that in the comparison of the calculating machines each sum of squares had been computed first on machine A and then on machine B. Now it is quite possible that increased familiarity with the data will enable the second computation to be done faster than the first. The advantage is unlikely to be great in such an easy calculation; it could be so if the computation were more difficult. (In the experiment this advantage was estimated at about 4 seconds, though the confidence limits for the true advantage are rather far apart.)

If the experiment is conducted in this way, the observed difference in speed $(B - A)$ is an estimate of the true difference, plus the unknown difference in speed between a second calculation and a first. Confidence limits set up by statistical techniques apply not to the true difference between B and A but to the true difference plus this unknown advantage. Consequently, the limits tell us nothing definite about the true difference between B and A. The interpretation of tests of significance also becomes confused. If A is found significantly faster than B by ordinary statistical tests, we can be confident that there is a real difference in favor of A, since A was handicapped in the course of the experiment. But if B is found significantly faster than A, we do not know what to conclude. In this case we are dealing with a bias whose nature can be anticipated before the experiment has started. In other experiments where less is known about the type of variability that is present, similar biases that are quite unexpected can occur from some apparently innocuous rule about the way in which different treatments are handled.

In order to avoid these biases we need some means of insuring that a treatment will not be continually favored or handicapped in successive replications by some extraneous source of variation, known or unknown. This is done by the device known as *randomization*, due to Fisher. Instead of performing every calculation first on machine A, we apply the principle of randomization by tossing a coin to determine whether A or B shall be used first in any trial. The decision is made independently in each trial. The effect is that in any trial each machine has an equal chance of being tested under the more favorable conditions. Of course, the result of any specific randomization may favor one or the other treatment. But this happens only to an extent that is allowed for in the calculations that are used for tests of significance and confidence limits. This important result has been illustrated in detail by Fisher (1.2), who has shown how tests of significance and confidence limits can be constructed, using only the fact that randomization has been properly applied in the experiment. Randomization is one of the few characteristics of modern experimental design that appears to be really modern. One can find experiments made 100 or 150 years ago that embody the principles that are now regarded as sound, with the conspicuous exception of randomization.

The occasions on which randomization is required vary with the type of experiment and must be left to the judgment of the experimenter. One occasion arises when the treatments are allotted to the experimental material. Suppose that the effects of different diets on the heights and weights of children are to be ascertained. Since different children grow at different rates, a treatment that happens to be assigned to a group of fast-growing youngsters is favored. Consequently, we allot the diets in any replication at random to the children who are to receive them, with a new random allotment in each replication. Similarly, if 4 different oven temperatures for the cooking of roasts are under comparison, the 4 temperatures are assigned at random to the 4 roasts which form the material for any replication. Sometimes this is the only randomization required, but frequently other operations that are carried out in the course of the experiment are also potential sources of bias. With a repetitious operation the order of events may be important, either because a learning process is involved which tends to make later operations better than the earlier ones, or because fatigue tends in the opposite direction. Systematic biases may be guarded against by randomizing the order in which the operation is performed on the different treatments in a replication. In other cases the equipment that is used introduces variation. For example, if 4 ovens are available to cook the 4 roasts, we would not always use the same oven for the same temperature, in case biases should be introduced because of systematic differences among the ovens. Instead, the temperatures could be assigned at random to the ovens in each replication. Thus we have *two* randomizations, one to assign the temperatures to the roasts and one to assign them to the ovens. If, however, we decide, before randomizing, which roast is to go in which oven, the two randomizations can be reduced to one. This method can always be used, if convenient, to cut down the number of randomizations that must be made.

Randomization is somewhat analogous to insurance, in that it is a precaution against disturbances that may or may not occur and that may or may not be serious if they do occur. It is generally advisable to take the trouble to randomize even when it is not expected that there will be any serious bias from failure to randomize. The experimenter is thus protected against unusual events that upset his expectations. Of course in experiments where a great number of physical operations are involved, the application of randomization to every operation becomes time consuming, and the experimenter may use his judgment in omitting randomization where there is real knowledge that the results will not be vitiated. It should be realized, however, that failure to randomize at any stage may introduce bias unless either the variation introduced in that stage is negligible or the experiment effectively randomizes itself.

1.14 Restricted Randomization. We now reconsider the randomization proposed for the experiment on computing machines, in order to discuss a criticism that presumably has occurred to the reader. When this experiment is being planned, we possess the knowledge that there is some advantage to the machine on which the second calculation of any sum of squares is made. In the light of this knowledge, it seems wise to make sure that each machine receives the advantage in five of the 10 replications, rather than to leave this to be decided by the tossing of coins. In fact, if the decision is made by tossing coins, the chances are only about 1 in 4 that each machine will be used first in five replications; they are also about 1 in 4 that one machine will be used first in 7 replications. The criticism is that the randomization is unlikely to give as accurate a comparison as the proposed alternative. The criticism is sound, and the experiment was actually planned so that each machine would be used first in half the replications.

Randomization was also applied in the design actually used, though it is a different type from that originally proposed, being subject to the restriction that each machine must appear first exactly five times. The process of randomizing therefore consisted in choosing five numbers at random from the numbers 1 to 10; these turned out to be 1, 3, 6, 8, and 9. Accordingly, machine A was used first in replications 1, 3, 6, 8, and 9. This is one of the cases where it is difficult to see why failure to randomize could have led to any serious danger of bias. It probably would have been quite satisfactory to use machine A first in the first 5 replicates, and machine B first in the remainder. Nevertheless, it *might* happen that the advantage in the second calculation of a sum of squares would diminish in later replications; and, to save the trouble of trying to guess whether biases of this type are likely, the randomization was used as a precaution.

The example represents a fairly common situation, in which knowledge of the variation that will affect the results is sufficiently detailed so that the simplest type of randomization is objectionable on the grounds of accuracy. Most of the designs in this book were constructed for this situation. By suitable restrictions on the randomization, these designs enable the experimenter to utilize any knowledge that will increase the precision of the experiment. In addition, each design allows sufficient randomization to prevent biases from sources of variation about which knowledge is less certain. For each of the principal designs described, the appropriate method of randomization is also presented.

One further point should be mentioned. We have seen that the experiment on computing speeds could have been carried out either by means of the simple randomization originally proposed or by means of the restricted randomization which ensures that each machine is used first five times. Both methods provide a valid test of significance and valid confidence limits. However, the calculations made for the test of significance and for the confidence limits are different in the 2 cases. The first experiment is of the type known as randomized blocks, and calculations are made as described in section 4.23; the second experiment is of the "cross-over" type, for which the calculations are given in section 4.42. This is an instance of the general rule that the way in which the experiment is conducted determines not only whether inferences can be made, but also the calculations required to make them. The experimenter must make sure that the calculations which he uses are appropriate for the experiment.

1.2 Initial Steps in the Planning of Experiments

1.21 Importance of the Initial Steps. It has been mentioned that statisticians are often asked for advice in making inferences from the results of experiments. Since the inferences that can be made depend on the way in which the experiment was carried out, the statistician should request a detailed description of the experiment and its objectives. It may then become evident that no inferences can be made or that those which can be made do not answer the questions to which the experimenter had hoped to find answers. In these unhappy circumstances, about all that can be done is to indicate, if possible, how to avoid this outcome in future experiments. Consequently, it has come to be realized that the time to think about statistical inference, or to seek advice, is when the experiment is being planned.

Participation in the initial stages of experiments in different areas of research leads to a strong conviction that too little time and effort is put into the planning of experiments. The statistician who expects that his contribution to the planning will involve some technical matter in statistical theory finds repeatedly that he makes a much more valuable contribution simply by getting the investigator to explain clearly why he is doing the experiment, to justify the experimental treatments whose effects he proposes to compare, and to defend his claim that the completed experiment will enable its objectives to be realized. For this reason the remainder of this chapter is devoted to some elementary comments on the subject of planning. These comments are offered with diffidence, because they concern questions on which the statistician has, or should have, no special authority, and because some of the advice is so trite that it would be unnecessary if it were not so often overlooked.

It is a good practice to make a written draft of the proposals for any experiment. This draft will in general have three parts: (i) a statement of the objectives; (ii) a description of the experiment, covering such matters as the experimental treatments, the size of the experiment, and the experimental material; and (iii) an outline of the method of analysis of the results.

1.22 The Statement of Objectives. This statement may be in the form of the questions to be answered, the hypotheses to be tested, or the effects to be estimated. The aim should be to make the statement lucid and specific. The most common faults are vagueness and excessive ambition, in the sense that a twenty-year research program would be required in order to realize the stated objectives. Often it is advisable to classify the objectives as major and minor, because some types of experiment give high precision for certain treatment comparisons but low precision for others. When the experiment represents a cooperation among people of different interests, this classification is particularly helpful in that it makes clear which objectives take priority and helps to avoid an unhappy compromise that is adopted in the hope of pleasing everybody.

The statement should include an account of the area over which generalizations are to be made, or in other words, of the population about which it is hoped to make inferences. If an experiment is to be conducted on persons suffering from some disease, are the results presumed to apply to the patients in a specific hospital, to patients in all hospitals, or to all sufferers whether in hospital or not? This sort of question is crucial in applied research, where often the experimenter has some definite population in mind to which he would like to apply the results. It is obvious that worth-while inferences about an extensive population cannot be made from a single experiment. For instance, inferences made from the experiment on the two computing machines are restricted to the person who made the calculations and to the type of calculation made. There is no guarantee that the results would be the same for a different type of calculation or with different operators. Consequently, if the object were to find out which machine is the speedier in general use in a computing laboratory, the experiment that was made has only scratched the surface of the problem.

1.23 The Experimental Treatments. We have used the term *treatments* to denote the different procedures whose effects are to be measured and compared. In the selection of treatments it is important to define clearly each treatment and to understand the role that each treatment will play in reaching the objectives of the experiment. Some issues that arise in particular cases are as follows.

1. Confusion is sometimes caused by failure to distinguish whether the object is merely to "spot the winner" among the different treatments, or whether in addition it is desired to find some clues as to why the treatments behave as they do. A good example is the experiment which demonstrates, or so we are informed, that each of the three treatments, whiskey and water, gin and water, and rum and water, taken orally in sufficient quantities, produces some degree of intoxication. By itself the experiment provides no information as to whether the intoxication is due to the water, to the other ingredient, or to the fact that the two are mixed. A more extensive experiment with additional treatments would be necessary to throw some light on this question. Although there are occasions when it is sufficient simply to discover which is the best of the treatments, experience suggests that even in strictly applied research progress is faster if experiments supply fundamental knowledge. Similarly, the criticism that a certain treatment should be omitted because it would not be used in practice is valid if the aim is to find the best of the "practical" treatments, but it is not valid if the "impractical" treatment will supply some needed information about the behavior of other treatments in the experiment.

2. The specification of the treatments may raise difficult questions about the conditions under which the treatments are to be compared. Suppose that the object is to find out the effect of an application of sulphate of ammonia on the yield of some crop. It is well known that

this effect is likely to depend on the amounts of other important plant nutrients that are available to the crop. Consequently, it must be decided whether to add these nutrients so that they will be in abundant supply on all plots, or to test the nitrogen on the soil as it happens to be. The decision must be guided by the objectives of the experiment, and more particularly by the type of population to which inferences are to apply. Sometimes it is advisable to test the nitrogen in both the presence and the absence of the other nutrients. One result of this decision will be that the experiment becomes also a test of the other nutrients, even though this may not have been part of the original plan. Experiments of this type, where one factor is tested over different levels of another factor, are known as *factorial experiments* and have come to play a prominent part in experimentation. They are discussed more fully in chapter 5.

3. In some cases it becomes apparent that the treatments that can be tested in practice are not those that we would like to test. The following example is typical of many fields where human relations are involved. Suppose that it is desired to compare two methods A and B of teaching a foreign language, and that we have defined each method clearly and have decided how to measure the success achieved by each method. However, some of the teachers who are to participate in the test already use a method similar to method A and have strong beliefs that method B is inferior, whereas others already follow method B and have no use for method A. If the teachers are divided at random into the groups that are to teach the two methods, we may expect that some teachers will be assigned to teach a method in which they have no faith, and we may decide that this is not the kind of comparison that we had in mind. On the other hand, if we forsake randomization and allow teachers to use the method that they like, differences attributed to the methods may in fact be due to differences in the ability of the two groups of teachers. Given enough resources, the situation can be more thoroughly explored by having each method taught by those who like it, those who do not, and those who are neutral, so that there are six treatments instead of two. In practice, the resources may not permit this elaboration, and the issues that must be faced are how best to use the resources available, and whether the experiment, as it can be done, is worth doing. Here again, the deciding factor is usually the use to which the results are to be put.

4. Discussion may arise as to the need for a "control," this term being used rather loosely for a treatment in which we are not particularly interested, but which may be needed in order to reveal, by comparison, whether other treatments are effective. Suppose that we wish to com-

pare the effectiveness of three treatments that are qualitatively similar, for example, three nitrogenous fertilizers all of which supply the same amount of nitrogen. The control would be a "no-nitrogen" treatment. Three cases may be distinguished. (i) The effectiveness of this type of treatment has been consistently demonstrated previously, and it remains to discover which of the three qualities is best. There is no need for a control. (ii) The type of treatment is in general effective, but occasionally the conditions of the test are such that it is not. For example, nitrogenous fertilizers may fail to produce responses in fields where the soil fertility is very high. In this case it will be well to add a control, which serves primarily to describe the conditions under which the fertilizers were tested. (iii) It may not be known whether the type of treatment is effective. The control should of course be included, and there may be a case for giving it more replication than the other treatments receive. An experiment of this type is analyzed in chapter 3, where the treatments were 4 types of soil fumigant, and the control (no fumigation) was replicated 16 times as against 4 for the other treatments. The effect is to obtain a more accurate estimate of the *average* response to fumigants, at the expense of some loss in accuracy in the comparisons among the fumigants.

An interesting example of the illumination that is sometimes produced by the inclusion of a control has been reported by Jellinek (1.3). A headache remedy contained 3 ingredients, *a*, *b*, and *c*. In order to test whether ingredients *b* and *c* were necessary, the complete mixture (*abc*) was compared with (*ac*) and (*ab*). There were 199 subjects, each of whom was treated with each drug for a two-week period, the appropriate drug being given whenever the subject complained of a headache. Success was measured by the ratio of the number of headaches relieved to the number tested in the two-week period. The mean success rates were 0.84 for (*abc*), 0.80 for (*ac*), and 0.80 for (*ab*); these figures show very little overall difference in effectiveness.

The experiment also contained a "control"—an inert mixture which looked the same as the others but had no active ingredients. This was tested under the same conditions as the 3 drugs. No less than 120 subjects, or about 60%, reported at least some of their headaches relieved by the control, a result that is itself of some interest. Further, the control enables us to isolate the group of 79 subjects whose headaches were not relieved by an inert mixture. For this group, the mean success rates were 0.88 for (*abc*), 0.67 for (*ac*), and 0.77 for (*ab*), and could be shown to be significantly different. As the author comments: "Banal as it may sound, discrimination among remedies for pain can be made only by subjects who have a pain on which the analgesic action can be tested."

If a control is required, it must be an integral part of the experiment so that results for the control are directly comparable with those for the other treatments. This point tends to be overlooked in experiments with human subjects when it is difficult or troublesome to assemble the desired number of subjects. For example, if a new drug is to be tested in some ward of a hospital, the recovery rate in the ward before the drug was introduced is not a satisfactory control, nor is the recovery rate in a different ward where patients happen to be receiving the standard drug. An observed difference between the effects of the new and the standard drug might be due to differences in the severity of the disease or in the type of patient or in other aspects of the medical care in the two time-periods or the two wards. It is necessary to regard the new and the standard drug as two experimental treatments on an equal footing, and to use randomization, or one of the methods of restricted randomization, in assigning the two drugs to patients.

1.24 Further Details. Other features of the experiment that should be included in the draft of the proposals are the number of replications, the types of experimental material to be used, and the measurements that are to be made. Since these primarily affect the accuracy of the experiment, discussion of them is deferred to chapter 2, which is devoted to the general question of accuracy.

Finally, the draft should describe in some detail the proposed method for drawing conclusions from the results. This is the section that is most frequently omitted, though it is a valuable one. It may include a sketch of the analysis of variance, an indication of the tabular form in which results will be shown, and some account of the tests of significance to be made and the treatment differences that are to be estimated. In this process we verify which treatment comparisons are relevant to each of the stated objectives of the experiment. Attention is drawn to deficiencies in the set of treatments and to treatments that supply no information essential to the purpose of the experiment.

REFERENCES

1.1 Kendall, M. G. *The advanced theory of statistics.* Charles Griffin and Co., London, Vol. II, chapters 19 and 20, 1948.

1.2 Fisher, R. A. *The design of experiments.* Oliver and Boyd, Edinburgh, 4th ed., chapters 2 and 3, 1947.

1.3 Jellinek, E. M. Clinical tests on comparative effectiveness of analgesic drugs. *Biom. Bull.* 2, 87–91, 1946.

CHAPTER 2

Methods for Increasing the Accuracy of Experiments

2.1 Introduction

The results of experiments are affected not only by the action of the treatments, but also by extraneous variations which tend to mask the effects of the treatments. The term *experimental errors* is often applied to these variations, where the word *errors* is not synonymous with "mistakes," but includes all types of extraneous variation. Two main sources of experimental errors may be distinguished. The first is inherent variability in the experimental material to which the treatments are applied. We shall use the term *experimental unit* to denote the group of material to which a treatment is applied in a single trial of the experiment. The unit may be a plot of land, a patient in a hospital, or a lump of dough, or it may be a group of pigs in a pen, or a batch of seed. It is characteristic of such units that they produce different results even when subjected to the same treatment: these differences, whether large or small, contribute to the experimental errors. The second source of variability is lack of uniformity in the physical conduct of the experiment, or in other words, failure to standardize the experimental technique.

Neither the presence of experimental errors nor their causes need concern the investigator, provided that his results are sufficiently accurate to permit definite conclusions to be reached. In many fields of research, however, with the time and labor that can be given to an experiment, the results are so greatly influenced by experimental errors that only treatment differences that are large can be detected, and even these may be subject to considerable uncertainty. As a consequence, the investigation of methods for increasing the accuracy of experiments has played a prominent part in experimental research in recent years.

The methods may be classified broadly into three types. The first is to increase the size of the experiment, either through the provision of more replicates or by the inclusion of additional treatments. The second is to refine the experimental technique. Thirdly, we may handle the experimental material so that the effects of variability are reduced. This may be done by careful selection of the material, by taking additional measurements that provide information about the material, or finally by skillful grouping of the experimental units in such a way that the units to which one treatment is applied are closely comparable with those to which another treatment is applied. These various methods are discussed in more detail in succeeding sections.

At this point it is advisable to discuss briefly the concepts *accuracy* and *precision*. Although these words are regarded as roughly synonymous in most dictionaries, they have sometimes been assigned technical meanings, particularly in the natural sciences and engineering. The difference between the concepts may be illustrated by an experiment in which children are weighed on a machine that has a bias of 1 lb. That is, if the true weight of a child is 46 lb., repeated weighings of the child give readings which vary around 47 lb. The *accuracy* of a measurement of the weight of the child signifies the closeness with which the measurement approaches the true weight, 46 lb. The term *precision*, on the other hand, is concerned merely with the repeatability of the measurements. Thus the precision of a measurement denotes the closeness with which the measurement approaches the average, 47 lb., of a long series of measurements made under similar conditions. It is clear that, if the bias is large, a measurement may be of high precision but of low accuracy.*

Most of the devices mentioned in this chapter, including replication, additional measurements, and skillful grouping of the experimental material, operate on the precision of the experiment. In so far as the method of measurement remains biased, these procedures do not affect this bias. A refinement in technique may reduce it. Other things being equal, an increase in precision is accompanied by an increase in accuracy, though if a large bias is present a substantial increase in precision may result in a trivial increase in accuracy.

For simplicity in presentation, we shall sometimes speak as if no bias is present, that is, as if precision and accuracy are identical. Thus the phrase "true difference between two treatments" will mean the true difference as recorded by the measuring device that was actually used. This will be identical with the natural concept of a "true difference" only if the measurements are not biased in favor of one of the treatments.

* For a discussion of the difficult problem of measuring precision and accuracy see Shewhart (2.8).

2.2 Number of Replications

Whatever the source of the experimental errors, replication of the experiment steadily decreases the error associated with the difference between the average results for two treatments, provided that precautions (such as randomization) have been taken to ensure that one treatment is no more likely to be favored in any replicate than another, so that the errors affecting any treatment tend to cancel out as the number of replications is increased. The rate at which the experimental error is reduced is predictable from statistical theory. The basic quantity used to measure experimental errors is the *error variance per experimental unit,* which is defined as the expected value of the square of the error that affects the observation for a single experimental unit. The square root of this quantity is called the *standard error per unit.* If σ^2 is the error variance per unit and there are r replications, the error variance of the difference between the means for two treatments is $2\sigma^2/r$, and the corresponding standard error is $\sqrt{2\sigma^2}/\sqrt{r}$. This result is valid unless increased replication necessitates the use of less homogeneous experimental material or of a less careful technique.

In succeeding sections some advice is given on the choice of the number of replications for an experiment. To present realistic advice is not easy. Often the size of an experiment is limited by lack of resources or by the conflicting claims of other experiments. However, even if it is realized that an experiment must fall short of the precision desired, it is a good practice to try to estimate the degree of precision that will be attained and to present this information as part of the proposal for the experiment.

2.21 Number of Replications for Tests of Significance. Consider the difference between the effects of a pair of treatments, which might be a standard procedure and some new procedure which it is hoped will prove better than the standard. The precision desired in an experiment may be specified in either of two ways. We may specify the size of the true difference which the experiment is to *detect* by means of a test of significance, or we may specify how closely we wish to *estimate* the true difference, by stating the width of the confidence interval that we would like to have for the true difference.

Although the specification used is to some extent a matter of individual taste, the approach by means of a test of significance is helpful mainly in the initial stages of a line of research. The reasoning in this approach is roughly as follows. If it can be established that the new procedure is superior to the standard procedure by at least some stated amount, say

20% of their mean, then we will have discovered a useful result. On the other hand, if the first experiment shows no significant difference between the two treatments, we will be discouraged from further investigation of the new treatment. Consequently, the first experiment must be large enough to ensure that if the true difference is 20% or more, it is highly probable that we will obtain a significant difference. Note that in this approach we do not insist that the first experiment give a precise estimate of the true difference, but merely that it give a significant result at the chosen level of significance.

The probability of obtaining a significant result depends on the true standard error σ per unit, the number of replications r, and the number of degrees of freedom n that the experiment provides for estimating the error variance. The exact calculation of the probability is rather complicated; tables have been given by Neyman *et al.* (2.1) and others. The argument used in the following example, though logically faulty, leads to an approximation that is good enough for most practical purposes.

Example. The true standard error per unit is 12% of the mean of the experiment, and there are 8 replicates and 28 d.f. for the estimate of error variance. What is the probability of obtaining a significant result when the true difference is 20%?

Let d be the observed difference and δ the true difference between the mean of the new treatment and that of the standard treatment. The true standard error of d is

$$\sqrt{\frac{2\sigma^2}{r}} = \sqrt{\frac{2(12)^2}{8}}; \text{ i.e., } = 6\%$$

In practice we do not know σ, and in its place we use the estimated standard error per unit s, obtained from the analysis of variance of the results. We will assume that s happens to be equal to σ.

The test of significance of d is made by means of a Student t-test, with 28 d.f. To find the value that d must attain in order to be significant, we multiply the standard error of d, 6%, by the significant value of t. Suppose that the test is made at the 5% significance level, and that it is a one-tailed test.* The 10% value of t in the standard tables, for

* A one-tailed test is used when it is known that the new treatment must be at least as good as the standard procedure. The difference d is declared significant only if d is sufficiently large *and positive;* if d is negative we ascribe the result to experimental errors, since we do not entertain the possibility that the new treatment could be inferior. A two-tailed test is made when we do not know which treatment is better; large values of d, whether positive or negative, are declared significant. All standard tables of the significance levels of t are computed for two-tailed tests. For a one-tailed test we read the value of t corresponding to twice the significance probability.

28 d.f. is 1.701, so that if d is to be significant we must have

$$d \geq \sqrt{\frac{2}{r}}\, st_1 = (6)(1.701) = 10.206$$

The problem is to find the probability that this happens.

The quantity

$$\frac{d - \delta}{\sqrt{\frac{2}{r}}\, s} = \frac{d - 20}{6} = t_2$$

follows the t-distribution with 28 d.f. But it is evident that

$$d \geq 10.206 \quad \text{if and only if} \quad t_2 = \frac{d - 20}{6} \geq \frac{10.206 - 20}{6} = -1.632$$

Hence the probability which we seek is the probability that a value of t, with 28 d.f., should exceed -1.632. In the table of t, the probability that a value lies outside the limits ± 1.632 is about .11. The probability wanted is therefore $[1 - (\frac{1}{2})(0.11)]$, or about .945. It is highly likely that a significant difference would be found in this experiment.

The flaw in the argument is the assumption that the value of s can be assumed equal to σ, whereas s is actually a random variable. If the argument is carried through algebraically, it gives the following rule for calculating the approximate probability P of obtaining a significant result.

Step 1. Find t_2 from the relation

$$\delta = \sqrt{\frac{2}{r}}\, \sigma(t_1 + t_2) \tag{2.1}$$

where δ = true difference that it is desired to detect.
σ = true standard error per unit.
r = number of replications.
t_1 = significant value of t in the test of significance (with care to distinguish between one-tailed and two-tailed tests).

Step 2.

$$P = 1 - (\tfrac{1}{2})p_2 \tag{2.2}$$

where p_2 is the probability corresponding to t_2 in the ordinary t-table. The degrees of freedom in t_1 and t_2 are those available for the estimate of error variance.

TABLE 2.1 NUMBER OF REPLICATIONS REQUIRED FOR A GIVEN PROBABILITY OF OBTAINING A SIGNIFICANT RESULT

Upper figure: Test of significance at 5% level, probability 80%
Middle figure: Test of significance at 5% level, probability 90%
Lower figure: Test of significance at 1% level, probability 95%

One-tailed tests

True difference (δ) as percent of the mean	True standard error per unit (σ) as percent of the mean														
	2	3	4	5	6	7	8	9	10	11	12	14	16	18	20
5	3	6	9	13	19	25	33	41	50						
	4	7	12	18	26	35	45								
	7	13	22	33	47										
10	2	2	3	4	6	7	9	11	13	16	19	25	33	41	50
	2	3	4	5	8	9	12	15	18	22	26	35	45		
	3	4	7	9	13	17	22	27	33	40	47				
15	2	2	2	3	3	4	5	6	7	8	9	12	15	19	23
	2	2	3	3	4	5	6	7	9	10	12	16	21	26	31
	2	3	4	5	7	8	10	13	15	18	22	29	37	47	
20	2	2	2	2	2	3	3	4	4	5	6	7	9	11	13
	2	2	2	2	3	3	4	5	5	6	7	9	12	15	18
	2	3	3	4	4	5	7	8	9	11	13	17	22	27	33
25	2	2	2	2	2	2	3	3	3	4	4	5	6	7	9
	2	2	2	2	2	3	3	3	4	5	5	7	8	10	12
	2	2	3	3	3	4	5	6	7	8	9	11	14	18	22
30	2	2	2	2	2	2	2	2	3	3	3	4	5	6	7
	2	2	2	2	2	2	3	3	3	4	4	5	6	7	9
	2	2	2	3	3	3	4	4	5	6	7	8	10	13	15

The rule may be inverted in order to give the number of replications r required for a given probability P of obtaining a significant result. In this form we have

$$r \geq 2\left(\frac{\sigma}{\delta}\right)^2 (t_1 + t_2)^2 \tag{2.3}$$

where δ = true difference that it is desired to detect.
σ = true standard error per unit.
t_1 = significant value of t in the test of significance.
t_2 = value of t in the ordinary table corresponding to $2(1 - P)$.

TABLE 2.1 (*Continued*)

Two-tailed tests

True difference (δ) as percent of the mean	True standard error per unit (σ) as percent of the mean														
	2	3	4	5	6	7	8	9	10	11	12	14	16	18	20
5	4	7	11	17	24	32	41								
	5	9	15	22	31	42									
	7	14	24	38											
10	2	3	4	5	7	9	11	14	17	20	24	32	41		
	2	3	5	7	9	12	15	18	22	27	31	42			
	3	5	7	11	14	19	24	30	37	45					
15	2	2	3	3	4	5	6	7	8	10	11	15	19	24	29
	2	2	3	4	5	6	7	9	11	13	15	19	25	31	39
	3	3	4	6	7	9	12	14	17	21	24	33	42		
20	2	2	2	3	3	3	4	5	5	6	7	9	11	14	17
	2	2	2	3	3	4	5	6	7	8	9	12	15	18	22
	2	3	3	4	5	6	7	9	11	12	14	19	24	30	37
25	2	2	2	2	2	3	3	3	4	4	5	6	7	9	11
	2	2	2	2	3	3	4	4	5	5	6	8	10	12	15
	2	2	3	3	4	5	5	6	7	9	10	13	16	20	24
30	2	2	2	2	2	2	3	3	3	4	4	5	6	7	8
	2	2	2	2	2	3	3	3	4	4	5	6	7	8	11
	2	2	3	3	3	4	4	5	6	6	7	9	12	14	17

Notes. In constructing the table, it was assumed that the number of degrees of freedom for error is $3(r - 1)$; this would apply in a randomized blocks experiment with 4 treatments.

No entries are given where more than 50 replications are required.

The use of this equation is slightly tricky, because the number of degrees of freedom in t_1 and t_2 depends on r. Trial and error may be used until the smallest r satisfying (2.3) is found.

Example. In the previous example, how many replications are required in order to have a chance of 4 in 5 that a significant result will be obtained? Here we have $\delta = 20$, $\sigma = 12$, and we will suppose that the experiment has 4 treatments in randomized blocks, which gives $n = 3(r - 1)$ degrees of freedom for the estimate of error. In the t-table, t_1 is the value for probability .10, while t_2 is the value for probability $(1 - .80)$ or .40.

In order to start the calculation some guess must be made about the value of n. It does not matter greatly if the guess is inaccurate; we will try $n = 30$. For this,

$$t_1 = 1.697; \quad t_2 = .854; \quad \text{so that} \quad (t_1 + t_2) = 2.551$$

Hence

$$r \geq 2(^{12}/_{20})^2(2.551)^2 = 4.7$$

Consequently our second trial value is $r = 5$. Since this provides only 12 instead of 30 d.f., we must repeat the calculation to verify whether 5 replications are sufficient. For 12 d.f. $(t_1 + t_2)$ is 2.655. Thus

$$2\left(\frac{\sigma}{\delta}\right)^2(t_1 + t_2)^2 = 2\left(\frac{12}{20}\right)^2(2.655)^2 = 5.08$$

Since this is greater than 5, we conclude that 5 replications are not quite enough, and that 6 is the smallest number of replications that will satisfy the conditions. We can be sure without further computation that 6 will satisfy the conditions, because the right-hand side of the inequality always decreases when we increase r.

Table 2.1, computed by this method, shows the minimum number of replications for a range of values of δ and σ, and is intended to be used in the planning of experiments. Usually it is most convenient to think of δ and σ expressed in percentages of the mean per observation for the experiment, and the range in the table has been constructed with this in mind. Since the value of r depends only on the ratio σ/δ, it does not matter what units are used for δ and σ, provided that they are the same for both. There are three entries for any given pair of values of δ and σ. These show, in descending order, the replications required for a test of significance at the 5% level and a probability .80 of getting a significant result, for a 5% level and probability .90, and for a 1% level and probability .95. The top half of the table applies to one-tailed tests, the lower half to two-tailed tests.

Construction of the table requires some assumption about the size of the experiment. In the case presented the experiment has four treatments in randomized blocks, so that $n = 3(r - 1)$. The values in the table change little if the experiment happens to provide more than this. If the experiment has only two or three treatments it is advisable to check the value of r by means of the formula.

In the practical use of the table it is necessary to estimate the value of the true standard error per unit. Often this can be done fairly well from the results of previous experiments with the same kind of material. It should be realized that the probabilities to which the table applies will be correct only if the value of σ is estimated correctly. If σ is underestimated, the probability of obtaining a significant result will of course be smaller than the presumed value.

As an illustration of the reading of the table, consider the case in the last example, where $\delta = 20$, $\sigma = 12$, $P = .80$, and the test is a one-tailed test at the 5% level. For this we read the top entry in the top half of the table for $\delta = 20$, $\sigma = 12$, and find that 6 replications are necessary. From the middle figure we see that an increase to 7 replications increases the probability to .90. If the test is to be two-tailed, we read at the corresponding place in the lower half of the table, and find that 7 replications are required for a probability .80 and 9 for a probability .90.

The table brings out the value of any reduction in the standard error per unit. In fact, we cannot have a high probability of detecting a 5% difference with any reasonable number of replications unless the standard error per unit is 4% or under. Differences of 20% or over can be detected in most cases without excessive replication. No entries have been inserted in the table for situations where more than 50 replications would be required, since experiments of this size are uncommon.

2.21a Number of Replications for Data Arranged in Two Classes. Table 2.1a gives the number of replications required in experiments where the experimental units are classified into two classes, e.g., "Alive" or "Dead," "Healthy" or "Diseased," "Improved" or "Not improved." We are interested in comparing the proportions or percentages that fall into one of the classes under two different treatments.

Example 1. A standard drug for treating some ailment has been successful in about 30% of the patients. A new drug is to be compared with the standard drug in an experiment. If the new drug has a true success rate as high as 40%, the physician would like to be fairly certain to obtain a significant result in the experiment. The test of significance of the difference in success rates is to be a two-tailed test at the 5% level, and the physician is content with a probability of .80 that a significant difference will be found. How many patients does he need?

Enter the table for two-tailed tests with

P_1 = smaller percentage of success = 30

δ = size of difference to be detected

= larger percentage of success − smaller percentage of success

= 40 − 30 = 10

TABLE 2.1a NUMBER OF REPLICATIONS (WITH DATA ANALYZED IN TWO CLASSES) REQUIRED FOR A GIVEN PROBABILITY OF OBTAINING A SIGNIFICANT RESULT

Upper figure: Test of significance at 5% level, probability 80%
Middle figure: Test of significance at 5% level, probability 90%
Lower figure: Test of significance at 1% level, probability 95%

One-tailed tests

P_1 = smaller % success	$\delta = P_2 - P_1$ = larger *minus* smaller percentage of success													
	5	10	15	20	25	30	35	40	45	50	55	60	65	70
5	330	105	55	35	25	20	16	13	11	9	8	7	6	6
	460	145	76	48	34	26	21	17	15	13	11	9	8	7
	850	270	140	89	63	47	37	30	25	21	19	17	14	13
10	540	155	76	47	32	23	19	15	13	11	9	8	7	6
	740	210	105	64	44	33	25	21	17	14	12	11	9	8
	1370	390	195	120	81	60	46	37	30	25	21	19	16	14
15	710	200	94	56	38	27	21	17	14	12	10	8	7	6
	990	270	130	77	52	38	29	22	19	16	13	10	10	8
	1820	500	240	145	96	69	52	41	33	27	22	20	17	14
20	860	230	110	63	42	30	22	18	15	12	10	8	7	6
	1190	320	150	88	58	41	31	24	20	16	14	11	10	8
	2190	590	280	160	105	76	57	44	35	28	23	20	17	14
25	980	260	120	69	45	32	24	19	15	12	10	8	7	..
	1360	360	165	96	63	44	33	25	21	16	14	11	9	..
	2510	660	300	175	115	81	60	46	36	29	23	20	16	..
30	1080	280	130	73	47	33	24	19	15	12	10	8	..	..
	1500	390	175	100	65	46	33	25	21	16	13	11	..	..
	2760	720	330	185	120	84	61	47	36	28	22	19	..	..
35	1160	300	135	75	48	33	24	19	15	12	9	..	..	..
	1600	410	185	105	67	46	33	25	20	16	12	..	..	..
	2960	750	340	190	125	85	61	46	35	27	21	..	..	..
40	1210	310	135	76	48	33	24	18	14	11	..	..	..	..
	1670	420	190	105	67	46	33	24	19	14	..	..	..	..
	3080	780	350	195	125	84	60	44	33	25	..	..	..	..
45	1230	310	135	75	47	32	22	17	13	..	..	..	..	..
	1710	430	190	105	65	44	31	22	17	..	..	..	..	..
	3140	790	350	190	120	81	57	41	30	..	..	..	..	..
50	1230	310	135	73	45	30	21	15	..	..	..	..	..	..
	1710	420	185	100	63	41	29	21	..	..	..	..	..	..
	3140	780	340	185	115	76	52	37	..	..	..	..	..	..

TABLE 2.1a (*Continued*)

Two-tailed tests

P_1 = smaller % success	$\delta = P_2 - P_1$ = larger *minus* smaller percentage of success													
	5	10	15	20	25	30	35	40	45	50	55	60	65	**70**
5	420	130	69	44	31	24	20	16	14	12	10	9	9	**7**
	570	175	93	59	42	32	25	21	18	15	13	11	10	**9**
	960	300	155	100	71	54	42	34	28	24	21	19	16	**14**
10	680	195	96	59	41	30	23	19	16	13	11	10	9	**7**
	910	260	130	79	54	40	31	24	21	18	15	13	11	**10**
	1550	440	220	135	92	68	52	41	34	28	23	21	18	**15**
15	910	250	120	71	48	34	26	21	17	14	12	10	9	**8**
	1220	330	160	95	64	46	35	27	22	19	16	13	11	**10**
	2060	560	270	160	110	78	59	47	37	31	25	21	19	**16**
20	1090	290	135	80	53	38	28	22	18	15	13	10	9	**7**
	1460	390	185	105	71	51	38	29	23	20	16	14	11	**10**
	2470	660	310	180	120	86	64	50	40	32	26	21	19	**15**
25	1250	330	150	88	57	40	30	23	19	15	13	10	9	..
	1680	440	200	115	77	54	40	31	24	20	16	13	11	..
	2840	740	340	200	130	92	68	52	41	32	26	21	18	..
30	1380	360	160	93	60	42	31	23	19	15	12	10	..	..
	1840	480	220	125	80	56	41	31	24	20	16	13	..	..
	3120	810	370	210	135	95	69	53	41	32	25	21	..	..
35	1470	380	170	96	61	42	31	23	18	14	11	..	..	..
	1970	500	225	130	82	57	41	31	23	19	15	..	..	..
	3340	850	380	215	140	96	69	52	40	31	23	..	..	..
40	1530	390	175	97	61	42	30	22	17	13	..	..	..	..
	2050	520	230	130	82	56	40	29	22	18	..	..	..	..
	3480	880	390	220	140	95	68	50	37	28	..	..	..	..
45	1560	390	175	96	60	40	28	21	16	..	..	..	..	..
	2100	520	230	130	80	54	38	27	21	..	..	..	..	..
	3550	890	390	215	135	92	64	47	34	..	..	..	..	..
50	1560	390	170	93	57	38	26	19	..	..	..	..	..	..
	2100	520	225	125	77	51	35	24	..	..	..	..	..	..
	3550	880	380	210	130	86	59	41	..	..	..	..	..	..

Read the upper line, which corresponds to a 5% test with .80 probability. This gives

$$r = 360$$

This is the size of *each* sample, so that 720 patients will be needed.

Example 2. The physician reports that he will be able to find at most 200 suitable patients in a reasonable length of time. What discriminating power does an experiment of this size have?

Here we are given $P_1 = 30$, $r = 100$ (in each group), and want to find δ. Looking further along the row in which $P_1 = 30$, we see that when $\delta = 20$, $r = 93$, the reading nearest to $r = 100$. With $\delta = 20$, $P_2 = 30 + 20 = 50$. An experiment with 200 patients has 4 chances in 5 of finding a significant result only if the new treatment gives a success rate close to 50%. The physician must decide whether an experiment with this degree of ability to discriminate is worth doing.

As the examples show, use of the table necessitates an estimate of the percent of success expected for one of the treatments. If this treatment has been employed for some time, the estimate can be made from past experience. We must also insert the size of true difference δ that it is desired to detect. It is sometimes difficult for the experimenter to decide what value of δ is of practical importance to him. Several values can be tried; inspection of the necessary sample sizes may aid in reaching a decision.

In the table, values of P_1 are given only up to 50%. If the success rates to be compared both exceed 50%, the table is entered in terms of failure rates.

Example 3. A standard treatment gives 75% success and it is desired to "detect" a new treatment that has 85% success. For a one-tailed 5% test with probability 90%, we take

$$P_1 = \text{smaller percentage of failure} = 15$$

$$\delta = \text{size of difference} = 85 - 75 = 10$$

$$r = 270 \text{ in each sample (from the middle line of the one-tailed table)}$$

The sample sizes required in table 2.1a often come as an unpleasant shock. To detect a difference of 5% in success rates usually requires a total sample size running into the thousands. Small sample sizes will detect only gross differences.

Table 2.1a assumes that the experimental units have been assigned at random to the two treatments. If the units have been grouped into pairs, such that the two members of a pair are highly comparable (see section 4.21), the required values of r are smaller than those read from the table. The reduction is not great unless the pairing is markedly effective.

The table was computed from an approximation given in (2.1a).

$$r = \frac{(K_1 + K_2)^2}{2(\sin^{-1}\sqrt{P_1} - \sin^{-1}\sqrt{P_2}\,)^2}$$

where K_1 and K_2 are the normal deviates corresponding to the significance levels α (the level of the test of significance) and $2(1 - P)$, respectively, P being the desired probability of obtaining a significant result. Values of r less than 30 were inserted by a computer program for the exact power, developed and kindly provided by M. H. Gail and J. J. Gart. The approximation was fairly consistently too low by 1 unit in this region. Computed values of r up to 20 were therefore all raised by 1.

2.22 Number of Replications for Prescribed Limits of Error. The confidence limits for δ, the true difference between the effects of two treatments, are

$$\delta = d \pm \sqrt{\frac{2}{r}}\, st \tag{2.4}$$

where t is the value in the t-table corresponding to the confidence probability chosen and to the number of degrees of freedom n in the estimate of error. The quantity $\sqrt{2}st/\sqrt{r}$ may be called the *limit of error*, since it measures the maximum distance between d and δ for the confidence probability chosen. Since the limit of error depends on s, it is not known until the experiment is completed. If s is replaced by its average or expected value, we obtain the *expected* limit of error

$$L = \sqrt{\frac{2}{r}}\, f\sigma t \tag{2.5}$$

The factor f enters because the average value of s is slightly less than σ; to a close approximation, f equals $\left(1 - \dfrac{1}{4n}\right)$.

From relation (2.5) table 2.2 was constructed to show the numbers of replications required for various values of L and for confidence probabilities of .80, .90, and .95. As in table 2.1, the experiment was assumed to have $3(r - 1)$ degrees of freedom in the estimate of error; no entries were

TABLE 2.2 NUMBER OF REPLICATIONS REQUIRED FOR GIVEN LIMITS OF ERROR IN THE DIFFERENCE BETWEEN TWO TREATMENTS

Upper figure: Confidence probability .80
Middle figure: Confidence probability .90
Lower figure: Confidence probability .95

Limit of error ($\pm L$) as percent of the mean	True standard error per unit (σ) as percent of the mean														
	2	3	4	5	6	7	8	9	10	11	12	14	16	18	20
3	3	4	7	10	14	19	24	31	37	45					
	4	6	11	16	23	31	40	50							
	5	9	15	23	32	43									
4	2	3	4	6	8	11	14	18	21	26	31	41			
	3	4	6	9	13	18	23	29	35	42	50				
	3	6	9	13	19	25	32	40	50						
5	2	2	3	4	6	7	9	12	14	17	20	27	35	44	
	2	3	5	6	9	12	15	19	23	27	32	44			
	3	4	6	9	12	16	21	26	32	39	46				
6	2	2	3	3	4	5	7	8	10	12	14	19	24	31	38
	2	3	4	5	6	8	11	13	16	19	23	31	40	50	
	2	3	5	7	9	12	15	19	23	27	32	43			
8	2	2	2	2	3	3	4	5	6	7	8	11	14	18	21
	2	2	3	3	4	5	6	8	9	11	13	18	23	29	35
	2	3	3	4	6	7	9	11	13	16	19	25	32	40	49
10	2	2	2	2	2	3	3	4	4	5	6	7	9	12	14
	2	2	2	3	3	4	5	5	6	8	9	12	15	19	23
	2	2	3	3	4	5	6	7	9	11	12	16	21	26	32

inserted where more than 50 replications are necessary, and the values of L and σ are expressed as percentages of the mean for the experiment.

This table may be used in the planning of an experiment where the object is to estimate the difference in the effects of two treatments. Suppose that the true standard error per unit is expected to be about 6%, and that preliminary experiments have indicated that the true difference between the effects of the two treatments is around 25%. In this case we would probably not be satisfied with a value of L any higher than 5%; if L is 5% and if the observed difference is found to be 22%, we are confident that the true difference lies between $(22 \pm 5)\%$, that is, between 17 and 27%. From the entry for $\sigma = 6$, $L = 5$, we find that 6 replications suffice for a confidence probability of 80%, 9 for 90%, and 12 for 95%.

It is important to reiterate that this table does not *guarantee* that the limits of error will be of the desired size, even if the value of σ is forecasted correctly. The actual limits will depend on s and may be either larger or smaller than L. They will be about the desired size *on the average* if σ is guessed correctly. For methods giving more definite guarantees see (2.3) and (2.4.) Furthermore, the table makes it clear that small treatment differences of about 5% cannot be accurately estimated in a single experiment, at least with the values of σ and r that normally prevail in many lines of research. For an accurate estimate of a 5% true difference, we would probably not want L any larger than 1%. This value was not included in the table since it demands great numbers of replications.

With data arranged in two classes, an approximate formula for r, the number of replicates for each treatment, is

$$r = \frac{K^2(P_1Q_1 + P_2Q_2)}{L^2}$$

where K is the normal deviate corresponding to the chosen confidence probability and $Q = 1 - P$.

2.23a Sequential Experimentation. Although this term does not have a precise technical meaning, it is used generally for experiments in which (i) the treatments are applied to the experimental units in some definite time-sequence, and (ii) the process of measurement is rapid, so that the yield or response on any unit is known before the experimenter treats the next unit in the time-sequence. A typical example is the experiment described at the beginning of this book (section 1.11). In this, 10 calculations were done by a single computer on each of two desk computing machines in order to estimate the difference in speed between the two machines. The 20 calculations were performed one after the other in a definite order, and the time taken for any calculation was known before the next calculation was started.

When an experiment is sequential, the experimenter can stop after every observation and examine the accumulated results to date before deciding whether to continue the experiment. In other words, the *analysis* can be done sequentially as well as the experiment. It seems plausible that a sequential analysis should be more efficient than the method given in the preceding sections, in which the number of replications needed is estimated in advance and no examination of the results is made until this number of replications has been completed. For instance, the advance calculation might call for 20 replications, but the superiority of one of the treatments might be fully evident on inspection after six replications.

A general method of sequential analysis was developed by Wald (2.2a). This procedure gives rules that determine, after any number of observations, whether to terminate or continue the experiment. If termination is indicated, the rules indicate the conclusion to be drawn. These rules provide the same guarantees of sensitivity that we were seeking in the calculations of number of replications in sections 2.21 and 2.21a. For instance, with data classified in two classes, the experimenter may choose the quantities

$$P_1 = \text{percent success under the standard treatment,}$$

$$\delta = P_2 - P_1 = \text{degree of superiority of the new treatment which it is desired to detect}$$

A sequential paired experiment can be constructed such that if the new treatment actually *is* superior by δ percent, the probability of obtaining a significant difference in favor of the new treatment will be at least 80% or 90% or any desired value.

The chief practical advantage of sequential analysis is that for the same guarantee of sensitivity fewer replications are required, on the average, than with an experiment of pre-determined size. The extent of the saving in replications varies from case to case, but it appears to average about 30 to 40%. A reduction of this amount is highly acceptable when observations are expensive.

So far as we have been able to discover, sequential experimentation of this type has not been widely used. Several explanations can be put forward. Although the method is some 10 years old, it may not yet be well known to experimenters. The available techniques have thus far been restricted to experiments with two treatments. Experiments in which the observations become known in a strictly sequential fashion may be infrequent.

With data arranged in two classes, the sequential technique is described in (2.2a), (2.3a) and (2.4a). The last two references discuss possible applications to medical research. Tables for sequential analysis with continuous data (the sequential t-test) are given in (2.5a) and (2.6a). In all these cases the data must be paired into separate replications. Rushton (2.10a) gives a sequential t-test for unpaired samples.

Sequential methods have also been produced for certain specialized types of experiments. An example is the "up and down" or "staircase" method of sensitivity testing (2.7a, 2.8a). The objective is to estimate the amount of some stimulus to which 50% of the experimental subjects give a positive response. A series of increasing stimuli, s_0, $s_0 + d$, $s_0 + 2d$, etc., equally spaced on some scale, is selected. The subjects are tested in turn, each subject being exposed to a single stimulus. The stimulus s_i for the ith subject is determined by the following sequential rule. Let the stimulus for the preceding subject be s_{i-1}. Then

$$s_i = \begin{cases} s_{i-1} + d, & \text{if the preceding subject fails to respond} \\ s_{i-1} - d, & \text{if the preceding subject responds} \end{cases}$$

This rule automatically concentrates testing in the neighborhood of the stimulus level giving 50% response, and makes for efficient estimation.

Another example is found in the comparison of the lengths of life of two pieces of equipment under operating conditions. Goodman (2.9a) has pointed out the advantages of an experimental plan in which each piece of equipment that fails is replaced by a piece of the opposite type.

In the broadest sense of the term, all experimentation is sequential. Before deciding what scientific questions he will attempt to answer in an experiment, the competent worker considers the knowledge and speculations gained from his own and others' previous experiments.

2.3 Other Methods for Increasing Accuracy

2.31 The Measurement of Relative Efficiency. We have seen that the error variance of the difference between the means for two treatments is $2\sigma^2/r$, where r is the number of replications. This result provides a simple means for comparing, in concrete terms, the relative precision of two experiments. Suppose that one design gives a true error variance of 1.0 per experimental unit, while a second gives 0.5. The two experiments would give equally precise comparisons among the treatment means if the amount of replication in the first were twice that in the second. From a knowledge of the error variances per unit, comparisons are thus easily made in terms of the relative amounts of replication required to attain the same degree of precision. The inverse ratio of the variances per unit is sometimes called the *relative efficiency* of the two designs; in the example above, the relative efficiency of the second experiment to the first is 2.0 or 200%. Similarly we might speak of the *relative gain in efficiency* as 1.0 or 100% in this case. These terms occur frequently in the literature on experimental designs.

One further factor must be taken into account. For the same numbers of treatments and replicates, the number of degrees of freedom in the estimate of error changes with the design. This number enters into many of the uses to which the experimental results are put. When the number of degrees of freedom becomes smaller, the limits of error for a true difference are increased and the probability of obtaining a significant result is decreased; in other words, the sensitivity of the experiment is decreased.

Thus the numbers of degrees of freedom for error are relevant to the comparison of two designs. A change in design which decreases the error degrees of freedom as well as the error variance may not be advantageous. This leads to the question: by how much must the experimental error variance be reduced so as to balance a given reduction in the error degrees of freedom? The question has been discussed by Neyman *et al.* (2.1), Fisher (2.5), and Walsh (2.6). The answer depends on the use to which the results are put. From one point of view, two experiments may be regarded as equal in sensitivity when the limits of error for a given confidence probability are equal. Alternatively, equal sensitivity may imply an equal probability of detecting as significant a given real difference between the effects of two treatments. Other definitions could be set up.

Table 2.3 illustrates the situation. An experiment with a true error variance per unit equal to 1 (i.e., with an infinite number of degrees of freedom) was chosen as the standard. The table shows, for a range of

TABLE 2.3 EFFECT OF NUMBER OF DEGREES OF FREEDOM IN ERROR ON THE SENSITIVITY OF THE EXPERIMENT. ERROR VARIANCES PER UNIT WHICH GIVE EQUAL SENSITIVITY ACCORDING TO FOUR DEFINITIONS

Definition	Number of error degrees of freedom (n)											
	1	2	3	4	5	6	8	10	15	20	30	∞
1	0.04	0.26	0.45	0.56	0.64	0.70	0.77	0.81	0.87	0.90	0.94	1.00
2	0.27	0.59	0.72	0.79	0.83	0.86	0.90	0.92	0.94	0.96	0.97	1.00
3	0.22	0.47	0.61	0.69	0.75	0.79	0.84	0.86	0.91	0.93	0.95	1.00
4	0.10	0.40	0.57	0.67	0.73	0.78	0.83	0.87	0.91	0.93	0.96	1.00

values of the error degrees of freedom (n), the true error variances per unit which are required to equal the standard in sensitivity according to four definitions of this term. In case 1 the average limits of error for a confidence probability .95 are made equal. In case 2 the same criterion is used for a confidence probability .80. In case 3 all experiments have the same probability .25 of detecting as significant a constant difference between two treatments. Case 4 shows the same comparison for a larger constant difference where the probability is .75. The tests of significance are one-tailed t-tests at the 5% level.

For example the figure 0.70 for $n = 6$ (definition 1) means that with a true error variance per unit of 0.70 and 6 d.f. in the estimate of error, the limits of error for a confidence probability .95 are the same on the average as when σ^2 is 1 and n is infinite. A 30% decrease in error variance is needed to compensate for the lack of degrees of freedom.

The wide divergence between the results for definition 1 and definition 2 is noteworthy since the only change in criterion is that from a confidence probability .95 to one of .80. Definitions 3 and 4 exhibit much closer mutual agreement and give results intermediate between those of definitions 1 and 2. In practical applications the discrepancies among the four methods will be smaller than table 2.3 suggests, because comparable designs for the same experiment seldom differ greatly in the degrees of freedom available for error.

Fisher's approach (2.5) is somewhat different. In effect, he calculates the "amount of information" which the estimated difference d between two treatment means supplies about the true difference δ, the calculation being made from the t-distribution of the quantity

$$\frac{\sqrt{r}(d - \delta)}{\sqrt{2}s}$$

He finds the information to be $(n + 1)/(n + 3)s^2$, whereas if σ were known exactly the information would be $1/\sigma^2$. The estimated variances which give equal amounts of information are shown in table 2.4.

TABLE 2.4 ESTIMATED VARIANCES WHICH PROVIDE EQUAL AMOUNTS OF INFORMATION

Number of error degrees of freedom (n)															
1	2	3	4	5	6	7	8	9	10	11	12	15	20	30	∞
0.500	0.600	0.667	0.714	0.750	0.778	0.800	0.818	0.833	0.846	0.857	0.867	0.889	0.913	0.939	1

This table agrees closely with the results for cases 3 and 4, except when n is below 5. For general purposes it is suggested that this table be used to take account of the differences in degrees of freedom for error in two designs that are being compared. Suppose that a design with $n = 6$ is compared with one in which $n = 12$. The former gives more information only if the error variance s_1^2 is less than $(0.778/0.867)s_2^2$.

Thus the relative efficiency of the *first* design to the *second* is estimated as $0.778s_2^2/0.867s_1^2$, or more generally as

$$\frac{(n_1 + 1)(n_2 + 3)s_2^2}{(n_2 + 1)(n_1 + 3)s_1^2}$$

The adjustment is of importance only if n_1 and n_2 are small.

This expression applies to the simple case in which the two experiments have the same set of treatments. The experiments may, however, have *different* sets of treatments, although each provides an estimate of some specific treatment effect in which we are interested. Alternatively, a different method of estimation of the treatment effect may have been employed in the two experiments. In such cases we may wish to assess the relative accuracy with which this treatment effect is estimated in the two experiments. The expression given above still applies, except that s_1^2 and s_2^2 become the estimated variances of the treatment effect as found in the two experiments, while n_1 and n_2 are as before the numbers of degrees of freedom in s_1^2 and s_2^2 respectively.

In succeeding sections a number of other devices which affect the accuracy of an experiment are presented briefly.

2.32 Selection of Treatments. In certain cases the selection of the treatments has a substantial effect on the precision of an experiment. As Fisher has stressed, striking gains in precision may be achieved by testing different types of treatment in the same experiment, instead of conducting a separate experiment for each type. For instance, one type of treatment or *factor* might be the depth of ploughing in the preparation of a field for a crop, while another might be the addition of a nitrogenous fertilizer. We might conduct one experiment in which deep (*D*) and shallow (*S*) ploughing are compared, and another in which no fertilizing (*O*) is compared with the addition of a specified amount of fertilizer (*N*). In a *factorial* experiment both factors would be tested simultaneously by means of the 4 "treatments"—(*DO*), (*SO*), (*DN*), and (*SN*), where the symbol (*DO*) implies deep ploughing with no addition of nitrogen, etc. The average response to nitrogen is assessed by comparing the last two treatments with the first two. This comparison gives as precise an estimate of the average response to nitrogen as if the whole experiment had been devoted to that factor alone. The same property holds for the estimation of the average difference in effect between deep and shallow ploughing. Thus the experiment, as it were, does double duty. The use of factorial experimentation is discussed more fully in chapter 5.

There are many specialized problems where the choice of the proper amounts of some ingredient is important. As a very simple example, suppose that the response to increasing amounts of the ingredient is known to be linear and that the purpose is to determine the slope of the line. The most accurate experiment contains only two amounts of ingredient. These should be placed at the ends of the range within which the response is linear and experimentation is feasible. With three amounts of ingredient, two at the ends and one in the middle, the variance of the estimated slope, for the same total size of experiment, is $1\frac{1}{2}$ times as large, and if more amounts are used it becomes larger still. A more complex example occurs in the estimation of the amount of a virus in a solution by preparing a series of dilutions of the solution and counting the numbers of lesions produced on leaves which are rubbed with the dilutions (2.7). The accuracy of the estimate is known to depend both on the dilution ratio that is chosen and on the number of dilutions that are used.

2.33 Refinements of Technique. Since *technique* is the responsibility of the experimenter, its importance need not be elaborated. The principal objectives of a good technique are as follows.

i. To secure uniformity in the application of the treatments. In pig-feeding experiments, for instance, a uniform amount of food cannot be supplied to each animal without the provision of individual feeding boxes. In the testing of insect sprays, delicate apparatus is required in order to subject each batch of insects to a desired dose of spray.

ii. To exercise sufficient control over external influences so that every treatment produces its effects under comparable and desired conditions. It is difficult to generalize about the degree of control needed; a balance must be struck between the cost incurred and the gain in precision obtained. The artificial production of diseases for experiments on resistance to infection exemplifies a case where experimentation cannot proceed rapidly without such control over external conditions.

iii. To devise suitable unbiased measures of the effects of the treatments. Often the appropriate measurements are readily apparent; sometimes the development of a satisfactory method of measurement requires years of research, as in the estimation of certain of the vitamins, in soil analysis, and in sociological investigations.

iv. To prevent gross errors, from which no type of experimentation seems to be entirely free. Adequate supervision and checking of the work of assistants and a scrutiny by the experimenter of the data from every experimental unit will go far towards the discovery and rectification of errors.

Faulty technique may increase the real experimental errors in two ways. It may introduce additional fluctuations of a more or less random nature. Such fluctuations, if they are substantial, should reveal themselves in the estimate of error as calculated in the analysis of variance (chapter 3). Where his estimated standard errors are consistently higher than those of other workers with similar material, the experimenter is advised to seek the reason, which may lie in differences in technique. In addition, faulty technique may result in measurements that are consistently biased. The estimate of error does not take account of such biases, since it is derived from comparisons of the measurements with one another; in other words, it estimates precision rather than accuracy. The principal safeguards against such biases are care and skill in the construction and handling of measuring devices, plus the intelligent use of randomization.

It is worth while to consider from time to time whether simplifications can be brought into the technique without undue loss of accuracy. Many experiments involve chemical determinations which vary little from unit to unit within the same treatment. For instance, in sugar-beet experiments, the experimental error of the amount of sugar per acre is due mainly to the error in the weight of roots per plot. The amount contributed by the sugar content percent is nearly negligible.

With compound measurements of this type, the labor devoted to the less variable component can be reduced with little loss of accuracy. If the chemical measurement is expensive, a considerable saving in cost may be made. Tests of significance of the chemical composition need not be sacrificed. For example, if the experiment has six replicates, a chemical analysis might be made for each treatment on bulked material from the first three replicates and another on bulked material from the last three replicates. These data provide a valid test of significance based on two "replicates."

2.34 Selection of Experimental Material. The choice of the experimental unit may be of importance. In the planning of field experiments numerous studies have been made of the variability among crop yields on plots of various sizes and shapes under uniform treatment. From these data the best size and shape are selected. The criterion should be to obtain the maximum accuracy for a given expenditure of time and labor.

Frequently, uniform material is prepared specially for experimental purposes, as in the development of inbred lines by laboratories engaged in animal experimentation. Alternatively, the experiment may be confined to a sample chosen for its homogeneity from a large batch of experimental material. If the results of the experiment are to be applied to unselected material, these types of specialization have potential disadvantages. Responses obtained to the treatments in highly selected material may not be the same as the responses in unselected material.

Selectivity in the material is difficult to avoid where experiments are conducted in the field of economics or sociology. In testing some method of farm management, for instance, the experiment may require the active participation of a group of farmers. The success of such experiments depends greatly on the tact and resourcefulness of the investigator in persuading farmers to cooperate so that the participants are a representative sample of the population about which generalizations are to be drawn.

2.35 Additional Measurements. In the course of an experiment it may be possible to take supplementary measurements which predict at least to some extent the performances of the experimental units. In an experiment to measure the effects of different diets on the weights of children, their weights at the start of the experiment would be a supplementary measurement of this kind, since the increase in weight of a child during the experiment is probably correlated with his initial weight.

By a technique known as the *analysis of covariance,* we may estimate from the data the extent to which the observations were influenced by the variations in these supplementary measurements. The average response to each treatment can then be adjusted so as to remove the experimental error arising from this source. In the school experiment, the *adjusted* responses to treatments represent approximately the responses that would be obtained if all the children had the same initial weight. Thus the effects of variations in the initial weights are largely eliminated from the experimental error, without the necessity of equalizing the initial weights in the planning of the experiment.

Since the analysis is a statistical technique of comparatively recent origin (it was first presented by Fisher about 1932), it may be unfamiliar to many investigators who could use it to obtain a substantial increase in precision with little effort. Its purpose is to remove experimental errors arising from extraneous sources of variation which it is impractical or too costly to control by a more refined technique. Cases are fairly common where the use of covariance has more than doubled the efficiency of the experiment. The extra work involved consists in taking the subsidiary measurements and in applying the adjustments to the treatment means. An example of the computations is given in section 3.8.

2.36 Planned Grouping: Complete Blocks. Finally, we may attempt to minimize the experimental errors through the choice of the experimental plan. The prospects of increased accuracy by this means have been widely explored during the past thirty years. This book is intended as a source of reference to the numerous procedures that have been developed.

The basic idea is simple. Consider an experiment of which a number of separate replications have been conducted. The experimental errors of the results from any replicate can arise only from sources of variation that affect the units within that replicate. Consequently, the error of the difference between two treatment means taken over a number of replicates must also arise solely from variations *within* the individual replications. Variations from one replicate to another do not contribute to the errors. In carrying out an experiment we utilize this simple fact by trying to control sources of variation that affect different units in the same replicate. We need not attempt to reduce differences among the replicates. For example, if the experimental units form a very heterogeneous batch, we attempt to group them so that units in the same replicate are similar; we do not worry if the units in one replicate are not at all similar to those in another. By this device, precise experiments can often be made from what at first sight appears an unpromising batch of material. Similarly, if a uniform experimental technique cannot be maintained throughout the experiment, the important point is to keep the technique uniform within a replication; changes should be made when moving from one replicate to another. This type of design, known in agriculture as *randomized blocks*, is discussed in section 4.2.

The idea is carried a stage further in the latin square (section 4.3). The treatments are arranged diagrammatically so that each appears once in every row and once in every column of a square array. Variations among the groups of experimental units which correspond to the rows and also among those which correspond to the columns are eliminated from the experimental errors. For example, in a comparison of 5 different ways of performing some manual operation, the rows may represent 5 operators, each of whom carries out all the methods in turn. The columns, numbered from left to right, may prescribe the order in which each operator performs the 5 tasks. The plan of a latin square design for this experiment might read as follows.

	Order of procedure				
Operator	(1)	(2)	(3)	(4)	(5)
I	*D*	*A*	*B*	*E*	*C*
II	*A*	*B*	*C*	*D*	*E*
III	*B*	*E*	*A*	*C*	*D*
IV	*C*	*D*	*E*	*B*	*A*
V	*E*	*C*	*D*	*A*	*B*

The letters A, B, C, D, E represent the 5 methods. Thus operator I carries out method D first, method A next, and so on. With this arrangement, the experimental errors are unaffected by differences among the abilities of the men and also by systematic variations introduced by the order in which the methods are performed. If the opportunity presents itself, this type of control may include a third factor by the use of a graeco-latin square (section 4.5).

2.37 Planned Grouping: Incomplete Blocks. In the randomized block design, the precision depends to some extent on the number of treatments. As this number increases, it becomes more difficult to keep the variations between experimental units within each replicate small, and the error variance per unit tends to increase. In experiments with a large number of treatments, numerous attempts have been made to avoid this loss of precision by the formation of groups of experimental units which do not contain all the treatments and therefore can remain small. These groups are called *incomplete blocks.* The groups are so constructed that we can still remove the effect of the differences among groups from the experimental errors, just as such differences are automatically removed in the randomized blocks design. With incomplete blocks, however, this removal usually requires the application of adjustments to the treatment means, so that the statistical analysis of the results becomes more complex.

The best method of constructing the incomplete blocks varies with the nature of the experiment. Sometimes certain comparisons among treatments are of greater interest than others. For instance, in a field experiment containing a number of early-maturing and an equal number of late-maturing varieties of a crop, comparisons between pairs of varieties which mature at about the same date will presumably be more important than comparisons between an early and a late variety. An incomplete block might contain either all the early or all the late varieties, a pair of blocks forming a complete replication. This arrangement, of which several variations are possible, is known as the *split-plot* design (section 7.1).

A similar situation occurs in factorial experiments. Certain comparisons, e.g., the average effects produced by varying one factor, are nearly always of immediate interest. Others, which represent rather complex relationships among the effects of different factors, may be of minor importance, often because previous experiments have shown that these interrelationships, called *interactions,* exert little or no influence. It has been found that, by a deliberate sacrifice of precision in the estimates of these interactions, the size of the group or block can be reduced so as to produce an increase in the precision with which the average effects of the factors are estimated. This principle, known as *confounding,* can also be combined with the use of the latin square (chapters 6 and 8).

If all comparisons between pairs of treatments are potentially of equal importance, a different method is used in forming the blocks. When treatments are arranged in incomplete blocks, two treatments which occur in the same block are more precisely compared than two which are placed in different blocks, at least in so far as the investigator has succeeded in decreasing the variation within blocks. Therefore the goal is to construct a design such that *any pair of treatments* occurs equally often within some block. This type of problem has been investigated previously for its mathematical interest. A solution can be found for any number of treatments and any size of block, but most of the solutions require too many replications for the usual conditions of experimentation. The solutions available to date (called *balanced incomplete block designs* (chapter 11)) are given for all cases in which the number of replications does not exceed ten. Here again the latin square principle can sometimes be used (chapter 13) to eliminate variation amongst two different types of grouping of the experimental units.

For a given number of treatments and a given size of incomplete block, balanced designs allow little choice in the number of replications. Thus, with 64 treatments and blocks of 8 units each, 9 replications are required to balance the design. In order to provide designs for small numbers of replications, a number of additional types have been developed; these are similar to balanced designs except that they lack complete symmetry.

One group, called *lattice* designs (chapter 10), has been used widely in agricultural experiments during the past 15 years. For these designs, the number of treatments must be an exact square, e.g., 25 or 49, while the number of units in the block is the corresponding square root, 5 or 7, respectively. A further group, known as *cubic lattices* (section 10.4), is useful when the number of treatments exceeds 100 yet it is desirable to have a small block. In this case the number of treatments is the cube of the number of units per block, and the number of replications may be any multiple of 3.

2.38 Summary. There are numerous methods available for increasing the accuracy of an experiment. Frequently the same end may be reached in several different ways. Thus, in the experiment which tests the effects of a number of diets on the weights of children, the disturbing effects of variations in their initial weights can be reduced by selecting for the experiment a number of children of approximately the same initial weight, or by the use of the analysis of covariance, or by a planned grouping in which children within the same replicate are of approximately equal initial weight, or simply by having a large number of replications. The method adopted should be that for which the desired standard of accuracy can be attained with the smallest expenditure of time and effort. There is no special merit in either a complicated experimental plan or a highly refined technique if equally accurate results can be secured with less effort in some other way. A good working rule is to use the simplest experimental design that meets the needs of the occasion. This is not to say that the more complex designs will be used only rarely; in fact, they have already demonstrated their utility in fields of research where no other equally practicable method appears to exist for reaching the same results.

2.4 The Grouping of Experimental Units

2.41 Investigation of Methods for Grouping. In order to take full advantage of the opportunities for increased precision by suitable grouping of the units, the investigator must know the best criteria for grouping. Frequently these are suggested from previous observations on the experimental material or on disturbing factors that become evident in the course of an experiment. Where knowledge is less definite, as in a new line of research, there are two ways in which information may be accumulated.

One is to devote an experiment specifically to this problem. In this case it is usually advisable to subject all units to the same treatment, carrying out a so-called *uniformity trial.* The word *uniformity* refers to uniformity of treatment; the experimental material and technique should be representative of those which are used in actual experiments. The amount of material should be sufficient to enable the experimenter to superimpose on the results a hypothetical trial of the size which he customarily performs. It is worth while to record any auxiliary measurements which might predict the performance of the experimental units. From the results of these trials any proposed grouping of the units can be formed and the amount of variation within and among the groups can be calculated. If it is desired to compare different experimental plans, each may be superimposed on the results of the uniformity trial. By the method indicated in section 2.31, the investigator may estimate the relative numbers of replications that are needed to reach the same degree of precision with the various plans.

Secondly, useful information can sometimes be secured without any special trial or technique by examination of the results of *actual experiments.* With an experiment which is suitably designed, it is possible to estimate from an analysis of variance of the results what the error variance would have been if any particular grouping had not been used. This estimate, being derived from the same data, is directly comparable with the error variance obtained by the use of the grouping and enables the investigator to evaluate the success of the grouping. Consequently, for experiments in a new line of research, the experimenter may try any method of grouping that *might* be effective, in the knowledge that its appropriateness will be revealed by the experiments themselves. Instructions for performing such comparisons, which are easily made when the analysis of variance has been completed, are given in the notes which accompany each type of design.

2.42 Criteria for Grouping. Many factors have been used as the basis for grouping. In agricultural experimentation it has been repeatedly demonstrated that plots close together tend to be more similar in their yields than plots farther apart. The almost universal practice is to put neighboring plots in the same group, with the groups approximately square in shape whenever practicable. Contiguity frequently forms a good basis for grouping in greenhouse experiments also, where differences may exist along the bench in temperature, sunlight, air currents, or accessibility for watering. Alternatively, in many plant experiments, and more particularly in animal experiments, some characteristic of the plant or animal is much more useful, the physical location of the experimental units throughout the course of the trial having a relatively minor influence on the results. Age, weight, vigor, sex, and genetic constitution are some of the factors most commonly used. The animal itself may constitute the block in cases where the treatments can be applied in succession to the same animal without producing residual effects which obscure the results.

Grouping is helpful in coping with the effects of time trends in experiments that extend over a number of work periods. It is often found, in both the biological and the industrial laboratory, that the level of yield shifts slightly from day to day and, to a greater extent, from week to week and from month to month. If, for example, five days are required to complete the work of applying the treatments in an experiment that has five replicates, the wise procedure is to complete a single replicate on each day. In this way any changes in the level of yield that occur from day to day will not influence comparisons made among the treatments. The alternative method of applying all replicates of treatment 1 on the first day, treatment 2 on the second day, and so on, is an open invitation to bias, since any day-to-day shifts in level will show up as treatment effects.

A single grouping, as in the randomized block design, can be used to eliminate simultaneously variation from a number of different sources. For example, suppose that an experiment were planned to investigate whether some treatment applied to rats enabled the animals to withstand subjection to a dose of a poisonous gas. The replication would contain two animals, a treated rat and an untreated rat to be included as a control. Given a sufficient stock of experimental animals, the pairs in any replication might be animals of the same litter (and consequently of the same age), of the same sex, and of approximately the same weight and vigor. If it were convenient to test two animals at the same time, the pair could be put into the gas chamber together, thus eliminating also the effects of variations in the strength of the dose or in the time for which it was supplied.

The freedom with which the size of block can be varied depends on the type of research. In field experiments the block can be laid down so as to accommodate any number of treatments, though, as previously stated, the standard error per plot may be expected to increase slowly as the size of block increases. When the use of designs with incomplete blocks is contemplated for field experiments, the principal question is to decide for what number of treatments the reduction in block size becomes worth while. On the other hand, the number of leaves on a plant, or of animals of the same litter, usually varies within rather narrow limits. In such cases, where the most homogeneous grouping fixes the size of the block, incomplete block designs may be of special interest.

The gains in efficiency obtainable by skillful construction of the blocks, when measured in terms of increased replication, are often large; examples can be cited in which the effective replication has been increased from two- to tenfold. Frequently, apart from the preliminary care and thought involved, an accurate design uses the same material and involves the same physical operations as the design which it replaced. In some of the more complex designs the computations required to estimate the treatment averages and to perform tests of significance become rather laborious and involved. The experimenter's reaction to these difficulties will depend on his aptitude for numerical computations. At a first trial considerable time may be spent in mastering some of the methods, but the processes become easier with familiarity.

REFERENCES

2.1 Neyman, J., Iwaszkiewicz, K., and St. Koldziejczyk. Statistical problems in agricultural experimentation. *Jour. Roy. Stat. Soc. Suppl.* 2, 114–136, 1935.

2.2 Fisher, R. A., and Yates, F. *Statistical tables for biological, agricultural and medical research.* Oliver and Boyd, Edinburgh, 4th ed., 1953.

2.3 Harris, M., Horvitz, D. G., and Mood, A. M. On the determination of sample sizes in designing experiments. *Jour. Amer. Stat. Assoc.* 43, 391–402, 1948.

2.4 Stein, C. A two-sample test for a linear hypothesis whose power is independent of the variance. *Ann. Math. Stat.* 16, 243–259, 1945.

2.5 Fisher, R. A. *The design of experiments.* Oliver and Boyd, Edinburgh, 4th ed., §74, 1947.

2.6 Walsh, J. E. On the "information" lost by using a *t*-test when the population variance is known. *Jour. Amer. Stat. Assoc.* 44, 122–125, 1949.

2.7 Price, W. C., and Spencer, F. L. Accuracy of the local lesion method for measuring virus activity. *Amer. Jour. Botany* 30, 720–735, 1943.

2.8 Shewhart, W. A. *Statistical method from the viewpoint of quality control.* U. S. Dept. of Agric. Graduate School, chapter 4, 1939.

2.1a Paulson, E., and Wallis, W. A. Planning and analyzing experiments for comparing two percentages. In *Techniques of statistical analysis.* McGraw-Hill, New York, 1947.

2.2a Wald, A. *Sequential analysis.* John Wiley and Sons, New York, 1947.

2.3a Bross, I. D. J. Sequential medical plans. *Biometrics* 8, 188–205, 1952.

2.4a Armitage, P. Sequential tests in prophylactic and therapeutic trials. *Quart. Jour. Med.* 23, 255–274, 1954.

2.5a Tables to facilitate sequential *t*-tests. *Nat. Bur. Stand. Appl. Math. Series* 7, 1951.

2.6a Davies, O. L. (Ed.) *Design and analysis of industrial experiments.* Hafner Co., New York, 1954.

2.7a Dixon, W. J., and Mood, A. M. A method for obtaining and analyzing sensitivity data. *Jour. Amer. Stat. Assoc.* 43, 109–126, 1948.

2.8a Brownlee, K. A., *et al.* The up-and-down method with small samples. *Jour. Amer. Stat. Assoc.* 48, 262–277, 1953.

2.9a Goodman, L. A. Methods for measuring useful life of equipment under operational conditions. *Jour. Amer. Stat. Assoc.* 48, 503–550, 1953.

2.10a Rushton, S. On a two-sided sequential *t*-test. *Biometrika* 39, 302-308, 1952.

ADDITIONAL READING

Cochran, W. G. Catalogue of uniformity trial data. *Jour. Roy. Stat. Soc. Suppl.* 4, 233–253, 1937.

Smith, H. F. An empirical law describing soil heterogeneity in the yields of agricultural crops. *Jour. Agr. Sci.* 28, 1–23, 1938.

Sukhatme, P. V., and Panse, V. G. Size of experiments for testing sera or vaccines. *Ind. Jour. Vet. Sci. and Animal Husb.* 13, 75–82, 1943.

CHAPTER 3

Notes on the Statistical Analysis of the Results

3.1 Introduction

With regard to the statistical analysis for a design, the procedure adopted in this book is to present the computations, with worked numerical examples for the designs likely to be most frequently used. To supply a complete mathematical justification of the analysis for each type of experiment, although in many respects desirable, would have unduly increased the amount of material. The same general technique governs the analysis of all designs, differences arising only in the adaptation of the technique to the particular structure of a design. In the present chapter this technique—the method of least squares—is described and applied to one of the simplest designs. With perseverance the method may be used to verify the computing instructions for other designs in this book. It should be pointed out that the computing methods that are easiest in practice are not necessarily those that flow from a straightforward application of the theory, so that at first sight some of the instructions may appear different from those given by theory.

Some account will also be given of certain statistical techniques, such as the technique for missing values and the analysis of covariance, that are particularly useful in handling the results of experiments, though they are perhaps too specialized to be included as part of a general introductory course in statistics.

3.2 The General Method of Analysis

3.21 The Mathematical Model. The illustrative data come from an experiment conducted in 1935 by the Rothamsted Experimental Station (3.1). The object of the experiment was to measure the effectiveness of 4 soil fumigants in keeping down the numbers of eelworms in the soil. The fumigants were chlorodinitrobenzene (CN), carbon disulphide jelly (CS), and two proprietary preparations, "Cymag" (CM) and "Seekay" (CK). Each fumigant was tested both in a single and a double dose. These comprize 8 distinct treatments. The control (no fumigant) was the ninth treatment. Normally the replication would have consisted of 9 plots, but in this case 4 control plots were placed in each "replicate," which actually contained 12 plots. The purpose was to supply a fairly accurate standard against which to measure the performance of the fumigants.

The experiment was laid out in 4 separate blocks of 12 plots each, so that there were 4 replications for each dose of each fumigant and 16 replications of the control. Within a block the 12 "treatments" were allotted to individual plots at random. The experiment is of the familiar type known in agriculture as randomized blocks (see chapter 4). The fumigants were ploughed in during spring, after which a crop of oats was sown. After harvest, a sample of 400 grams of soil was taken from each plot and the number of eelworm cysts counted for each sample. The data are those labelled "second count" in table 3.1.

TABLE 3.1 Plan and numbers of eelworm cysts per 400 grams of soil, first count above, second below

	0 269 466	2CK 283 280	1CN 252 398	1CM 212 386	2CM 95 199	2CS 127 166	2CK 80 142	0 134 590	
Block 1 Totals 2587 4383	1CS 138 194	0 100 219	0 197 421	2CM 263 379	1CK 107 236	1CN 89 332	1CM 41 176	0 74 137	Block 2 Totals 964 3075
	2CS 282 372	1CK 230 256	0 216 708	2CN 145 304	0 88 356	0 25 212	2CN 42 308	1CS 62 221	
	1CK 124 268	0 211 505	1CS 194 433	2CK 222 408	2CK 193 292	0 209 352	1CK 109 132	1CM 153 454	
Block 3 Totals 1743 4752	0 102 363	2CN 193 561	2CS 128 311	1CN 42 222	0 29 254	2CN 9 92	2CS 17 28	0 19 106	Block 4 Totals 872 2470
	2CM 162 365	0 191 563	1CM 107 415	0 67 338	1CS 23 80	1CN 19 114	0 44 268	2CM 48 298	

In a formal analysis of the results the first step is to set up an equation for every observation. This equation expresses the observation as the sum of four components: (i) a general average about which the observations are presumed to be fluctuating; (ii) a component representing the effect of the treatment applied; (iii) a component representing certain environmental effects which the design of the experiment enables us to isolate; and (iv) a residual component, representing all other sources that influence the observation, and generally referred to as the "experimental error."

Component (iii) cannot be fully appreciated without some knowledge of the technique employed in the construction of designs. This technique consists in arranging the conduct of the experiment so that, in the subsequent analysis of the results, differences among certain groups of observations can be measured. What is more important, the effects of such differences can be eliminated from the estimates of the treatment effects. If this elimination were not feasible, these differences would contribute to the "experimental error," component (iv), with the consequence that less accurate estimates of the treatment effects would be obtained. In the present example, as will be seen, these differences are differences among the groups of 12 observations that constitute a block. With each design that we describe, the environmental components that are eliminated in this way are pointed out.

We proceed to write down the mathematical model as applied to the example. Instead of numbering the observations from 1 to 48, it is convenient to use a subscript i to denote the treatment applied, j to denote the block in which the observation lies, and k to denote the order within the block. For the plots that receive fumigants, the subscripts i and j are sufficient to define each plot uniquely. The subscript k is needed to distinguish between different control plots within the same block. In this notation the equation for any observation may be written

$$y_{ijk} = \mu + \tau_i + \beta_j + e_{ijk} \tag{3.1}$$

where μ represents the general mean, τ_i the effect of the treatment, β_j that of the block, and e_{ijk} the experimental error.

3.22 Assumptions Made in the Model. There is already implicit in the model one assumption that may not be true in the data obtained from experiments. This assumption is that the treatment effect and the block effect are *additive*. If we take the difference between two observations in the same block, say y_{1jk_1} and y_{2jk_2}, we have, as a deduction from the model,

$$y_{1jk_1} - y_{2jk_2} = \tau_1 - \tau_2 + e_{1jk_1} - e_{2jk_2} \tag{3.2}$$

The block effects have disappeared from the right-hand side; that is, the model implies that the difference between the true effects of two treatments is the same in all blocks. On a block with a high eelworm infestation a successful fumigant is presumed to reduce the eelworm numbers by the same amount as on a block with a low eelworm infestation. This is unlikely to be so, except as a first approximation. Little systematic study of the accuracy of this approximation in practice has been made. From general experience it appears that the approximation works well in a large number of experiments; nevertheless, cases are not infrequent where the approximation is poor.

In addition, several assumptions are made about the residual effects e_{ijk}. These are taken to be independent from observation to observation, and to be distributed with zero mean and the same variance σ^2. Further, for tests of significance and the estimation of confidence limits, the e_{ijk} are assumed to follow the normal or Gaussian frequency distribution. In practice, these assumptions are only approximately fulfilled. The consequences of errors in the assumptions will be indicated later (section 3.9).

Apart from changes in detail, the type of model and assumptions presented for this example apply to all experimental arrangements included in this book, except for two types of design for which a more complex model is required. These are the *split-plot* designs (chapter 7), where two or more different error variances are postulated, and *incomplete block designs* (chapter 9), where additional assumptions are made about the nature of the environmental effects β_j.

3.23 Estimation of the Treatment Effects. When the assumptions in the previous section are valid, there is a well-known result in statistical theory that the best estimates of the unknowns μ, τ_i, and β_j are obtained by the *method of least squares*. This method chooses estimates m, t_i, and b_j, respectively, which minimize the sum of squares of the residuals

$$\sum (y_{ijk} - m - t_i - b_j)^2 \tag{3.3}$$

taken over all observations.

In order to find these minimizing values we differentiate the sum of squares with respect to each unknown in turn, and set the derivative equal to zero. Consider a specific unknown, e.g., t_i. If any square contains t_i, its derivative is

$$-2(y_{ijk} - m - t_i - b_j)$$

If the square does not contain t_i, the derivative is zero. Hence, if the derivative of the sum of squares in (3.3) is set equal to zero, the resulting equation may be written

$$\sum (y_{ijk} - m - t_i - b_j) = 0$$

or

$$\sum (m + t_i + b_j) = \sum y_{ijk} \tag{3.4}$$

where the sum is over all observations whose equations contain t_i, or more generally over all observations whose equations contain the parameter to be estimated.

Equations (3.4) are called the equations of estimation, or the *normal* equations. As we have seen, the equation of estimation for any parameter is constructed by equating the total of the observed values y_{ijk} to the total of the expected values $(m + t_i + b_j)$ over all observations whose equation contains the parameter.

We now write down some normal equations for the example. Since m occurs in all 48 observations, its normal equation is

$$48m + 4(t_1 + t_2 + \cdots + t_8) + 16t_9 + 12(b_1 + b_2 + \cdots + b_4) = G \quad (3.5)$$

where t_9 denotes the control treatment, and G the grand total of all 48 observations. At this point a simplification may be introduced. In the model the quantities t_i are needed only to indicate by how much an individual treatment differs from the average of all treatments in the experiment. Accordingly the model is not changed in any essential feature if we assume that the mean of all the t_i is zero. It may also be shown that any convenient linear function of the t_i may be assumed to be zero. The same remarks apply to the block values b_j. From the form of equation (3.5) it appears that convenient assumptions of this kind are

$$t_1 + t_2 + \cdots + t_8 + 4t_9 = 0 \quad (3.6)$$

$$b_1 + b_2 + b_3 + b_4 = 0 \quad (3.7)$$

With these assumptions equation (3.5) reduces to $48m = G$, so that m becomes the mean value of all observations in the experiment.

By summing over the observations that receive a particular treatment t_i we find that its normal equation is

$$4m + 4t_i + b_1 + b_2 + b_3 + b_4 = T_i \quad (i = 1, 2, \cdots, 8)$$

and for the control,

$$16m + 16t_9 + 4(b_1 + b_2 + b_3 + b_4) = T_9$$

where the T's denote observed treatment totals. From relation (3.7), these equations reduce to

$$t_i = \frac{T_i}{4} - m \quad (i = 1, 2, \cdots, 8); \qquad t_9 = \frac{T_9}{16} - m \quad (3.8)$$

As might be expected, the effect of any treatment is estimated by taking the difference between the mean of the observations which receive that treatment and the general mean.

Let us express these estimates in terms of the original model (3.1). By adding the relevant equations we find

$$T_i = 4\mu + 4\tau_i + (\beta_1 + \beta_2 + \beta_3 + \beta_4) + \sum e_{ijk}$$

$$T_9 = 16\mu + 16\tau_9 + 4(\beta_1 + \beta_2 + \beta_3 + \beta_4) + \sum e_{9jk}$$

Hence, using (3.8),

$$m + t_i = \mu + \tau_i + \tfrac{1}{4}(\beta_1 + \beta_2 + \beta_3 + \beta_4) + \bar{e}_i$$

where $\bar{e}_i$ is the average of the residual errors on plots that receive treatment i. For the difference between a pair of treatments we deduce

$$t_i - t_u = (\tau_i - \tau_u) + (\bar{e}_i - \bar{e}_u) \tag{3.9}$$

This shows that the estimated difference equals the true difference plus an error of estimate $(\bar{e}_i - \bar{e}_u)$. Note that the error of estimate does not contain any of the environmental (block) effects arising from component (iii). This result exemplifies a valuable property of least squares solutions, namely, that the estimate of the difference between the effects of any two treatments is not influenced by the environmental effects, but only by the residual effects e_{ijk}. It is in this sense that we say that environmental effects are *eliminated* from the estimates of the treatment effects.

Estimates of the block or environmental effects are usually of less interest. It may be verified that the least squares estimate of β_j is

$$b_j = \frac{B_j}{12} - m$$

where B_j is the observed block total.

3.24 Tests of Significance and Confidence Limits. When the estimates of the treatment effects have been obtained, the details of the subsequent study of them vary with the type of experiment. Usually, however, some of the following procedures form the basis for the final conclusions: (i) a test of significance of the null hypothesis that the effects τ_i and τ_u of two treatments are identical, or, more generally, of the hypothesis that some linear combination of the τ_i is zero; (ii) the construction of confidence limits for the difference $(\tau_i - \tau_u)$ between the effects of two treatments, or for a linear function of the treatment effects; (iii) a test of significance of the null hypothesis that a group of treatments τ_1, $\tau_2, \cdots, \tau_p$ *all* have identical effects. In this section the theory governing these procedures is presented briefly, without proof. The arithmetical methods are described in the next section.

i. Test of significance of the difference between two treatment effects. From equation (3.9) we have

$$t_i - t_u = (\tau_i - \tau_u) + (\bar{e}_i - \bar{e}_u) \tag{3.9}$$

Since the e's are normally distributed with zero means, it follows by theory that $(t_i - t_u)$ is normally distributed about the true difference $(\tau_i - \tau_u)$. Further, the variance of $(t_i - t_u)$ may be shown to be

$$\sigma^2 \left(\frac{1}{r_i} + \frac{1}{r_u}\right)$$

where r_i and r_u are the numbers of replications of t_i and t_u, respectively. Since it is useful to be able to calculate the variance of any comparison in which we are interested, the rules for doing this are given in section 3.5.

In order to proceed we need an estimate of the experimental error variance σ^2. From theory, the best estimate is known to be

$$s^2 = \sum \frac{(y_{ijk} - m - t_i - b_j)^2}{n_e}$$

where the residual sum of squares is taken over all observations. The divisor n_e, called the *number of degrees of freedom* (d.f.) in the estimated error, is given by the rule

n_e = (Total number of observations) − (number of independent parameters that were estimated)

The word *independent* is introduced because we nearly always include in the original model more parameters than are strictly necessary. In the example we found that we could assume one linear relation (3.6) among the t's and one (3.7) among the b's. Consequently the number of independent parameters is

$$1 \text{ (for } m) + 8 \text{ (for the } t\text{'s)} + 3 \text{ (for the } b\text{'s)} = 12$$

Since there are 48 observations, n_e is 36.

Hence, if the null hypothesis is true, i.e., $\tau_i = \tau_u$, then $(t_i - t_u)$ is normally distributed with mean zero and estimated variance

$$s^2 \left(\frac{1}{r_i} + \frac{1}{r_u}\right)$$

The ratio

$$\frac{t_i - t_u}{s\sqrt{\dfrac{1}{r_i} + \dfrac{1}{r_u}}}$$

is known to follow Student's t-distribution with n_e, or 36 d.f. This is the quantity used for a test of significance of the null hypothesis.

With a more complicated function of the estimated treatment effects, say $L = (w_1t_1 + w_2t_2 + \cdots + w_kt_k)$, where the w's are any set of numbers, the procedure is essentially the same. The estimated variance of L is s^2f, where f is a numerical factor found by the rules given in section 3.5. For a test of the null hypothesis that the true value of L is zero, we use the ratio $L/s\sqrt{f}$, which is distributed as Student's t with n_e degrees of freedom.

ii. The construction of confidence limits. The theory for a test of significance also leads to confidence limits for the unknown true difference $(\tau_i - \tau_u)$. For, when τ_i and τ_u are not necessarily equal, the quantity

$$\frac{(t_i - t_u) - (\tau_i - \tau_u)}{s\sqrt{\dfrac{1}{r_i} + \dfrac{1}{r_u}}}$$

may be shown to follow Student's t-distribution with n_e degrees of freedom. Hence the confidence limits are given by

$$(\tau_i - \tau_u) = (t_i - t_u) \pm sd(n_e, \alpha)\sqrt{\frac{1}{r_i} + \frac{1}{r_u}}$$

where $d(n_e, \alpha)$ is the value of Student's t corresponding to n_e degrees of freedom and the chosen confidence probability α. A similar method is used for the more general function L.

iii. A test of the identity of a group of treatment effects. Quite frequently the first test to be made is that of the hypothesis that all k treatments have produced identical effects, i.e., that all τ_i are equal. In other cases we are more interested in the hypothesis that some subgroup of the treatments, say treatments 1 to p, has produced the same effects. We consider the latter test, since it reduces to the first test if p is put equal to k.

This test is provided by a general theorem that has many applications. First, we find the least squares estimates of all treatment and environmental effects in the usual way, and compute the residual sum of squares

$$S_1^2 = \sum (y_{ijk} - m - t_i - b_j)^2$$

Next we start again and rewrite the mathematical model, inserting the restriction that $\tau_1 = \tau_2 = \cdots = \tau_p$. That is, wherever any of these τ_i appears in the equation for an observation, we replace the τ_i by a common symbol, say τ'. The least squares estimates are computed for this new set of equations, using the same observations. Again we compute the residual sum of squares which may be denoted by

$$S_2^2 = \sum (y_{ijk} - m' - t' - b_j')^2$$

This will always be found to be at least as large as S_1^2. Finally, the theorem states that, if the null hypothesis is true, the quantity

$$\frac{S_2^2 - S_1^2}{p - 1} \div \frac{S_1^2}{n_e}$$

follows Snedecor's F-distribution with $(p - 1)$ and n_e degrees of freedom.

Although the procedure may appear rather complex, the test criterion used is a reasonable one. S_1^2 is the sum of squares of deviations of the observations from the values predicted for them by the original model, so that it measures how closely the original model agrees with the data. S_2^2 plays the same role for the restricted model. Consequently, if S_2^2 is much larger than S_1^2, we are inclined to think that the restricted model does not fit the data nearly so well as the original model, and therefore to reject the null hypothesis that the restricted model is the correct one. However, the value of $(S_2^2 - S_1^2)$ *alone* does not provide a measure of the improvement in the fit with the original model. If, for instance, this difference is 10, then, other things being equal, we should regard the improvement in fit as greater when S_1^2 is 1 than when S_1^2 is 1000. It is this type of consideration that leads to the use of the ratio $(S_2^2 - S_1^2)/S_1^2$ as the essential part of the test criterion.

Although the theorem postulates two separate sets of least squares solutions, nearly all designs are constructed so that only one set of solutions must be found in practice. A number of designs which could have been included in this book were omitted because the least squares analysis seems too cumbersome for frequent use.

3.25 The Analysis of Variance. All the procedures (t-tests, F-tests, and construction of confidence limits) use the residual sum of squares, which will often be called the *error* sum of squares. This quantity could be found by calculating for each observation y_{ijk} the value $(m + t_i + b_j)$ predicted by the least squares solution. The sum of the squares of the differences between observed and predicted values could then be obtained. This method is slow, and the error sum of squares is much more quickly computed by a technique known as the *analysis of variance.*

In the original model, each observation is represented as the sum of four components due respectively to the general mean, the effect of the treatment, the environmental effect, and the residual effect. In the same way the analysis of variance partitions the sum of squares of the observations into four sums of squares, one attributable to the general mean, one to differences between the estimated effects of the treatments, one to the environmental effects which the experiment is capable of measuring, and lastly one which is the residual or error sum of squares. In most cases we compute the original sum of squares and the first three components, obtaining the error sum of squares by subtraction.

The analysis of variance provides much more than a short-cut method of securing the error sum of squares. The sum of squares due to treatments is the quantity $(S_2^2 - S_1^2)$ needed for the F-test of the hypothesis that no differences exist between the effects of the treatments. By a slight extension the analysis also supplies the sum of squares required for testing the equality of the effects of a subgroup of the treatments. The component due to environmental effects enables us to estimate by how much the accuracy of the experiment has been increased by eliminating these effects from the estimates of the treatment means.

The analysis of variance depends on a number of algebraic relations which will be illustrated for the eelworm experiment. From section 3.23 it will be recalled that the least squares normal equation for any unknown, say t_i, was

$$\sum (y_{ijk} - m - t_i - b_j) = 0$$

summed over all observations which received the ith treatment. If we multiply this equation by t_i and add the equations for different treatments together, we obtain the equation

$$\sum t_i(y_{ijk} - m - t_i - b_j) = 0 \tag{3.10}$$

where the sum is now over *all* observations, since every observation is associated with one and only one t_i. Similarly we may establish the relations

$$\sum m(y_{ijk} - m - t_i - b_j) = 0 \tag{3.11}$$

$$\sum b_j(y_{ijk} - m - t_i - b_j) = 0 \tag{3.12}$$

where both sums extend over all observations.

A few more relations of this type are required. If we add the estimates t_i over all plots in the jth block, the sum is

$$(t_1 + t_2 + t_3 + \cdots + t_8) + 4t_9$$

But this is zero because of equation (3.6) in which a linear relation among the t's was introduced. Hence, if we multiply by b_j and add over all blocks, we establish the relation

$$\sum b_j t_i = 0 \tag{3.13}$$

where the sum is again over all observations. By similar arguments we prove the additional relations

$$\sum mb_j = 0; \qquad \sum mt_i = 0 \tag{3.14}$$

These relations lead to the partition of the sum of squares of the observations. Write

$$y_{ijk} = m + t_i + b_j + (y_{ijk} - m - t_i - b_j)$$

Square both sides, and add over all observations. The six relations (3.10) to (3.14) show that all six sums arising from cross-product terms on the right-hand side add to zero. Consequently, we have the following analysis of sums of squares.

$$\sum y_{ijk}^2 = \sum m^2 + \sum t_i^2 + \sum b_j^2 + \sum (y_{ijk} - m - t_i - b_j)^2 \tag{3.15}$$

This equation is the basis of the analysis of variance of the results.

3.26 Application to the Example. To obtain the left-hand side of (3.15) in practice, we simply compute the sum of the squares of all observations. For the eelworm data this will be found to be 5,481,198. The components due to the mean, to treatments, and to blocks are not usually calculated from (3.15) as it stands, because it is quicker and more accurate to obtain them from *totals* rather than from the estimated effects.

Thus, for the mean, the contribution is $48m^2$, since there are 48 observations. If G is the grand total of all observations, 14,680, this contribution may be written $G^2/48$, or $(14{,}680)^2/48$, which amounts to 4,489,633. This term is sometimes called the correction for the mean.

Written in full, the sum of squares for treatments is

$$4(t_1^2 + t_2^2 + \cdots + t_8^2) + 16t_9^2 \tag{3.16}$$

But from (3.8),

$$t_i = \frac{T_i}{4} - m \quad (i = 1, 2, \cdots, 8); \qquad t_9 = \frac{T_9}{16} - m$$

where T_i is the observed treatment total. By substitution, (3.16) becomes

$$\tfrac{1}{4}[(T_1 - 4m)^2 + (T_2 - 4m)^2 + \cdots + (T_8 - 4m)^2] + \tfrac{1}{16}(T_9 - 16m)^2$$

When each parenthesis is expanded, we have

$$\frac{T_1^2 + T_2^2 + \cdots + T_8^2}{4} + \frac{T_9^2}{16} - 2m(T_1 + T_2 + \cdots + T_8 + T_9) + 48m^2$$

Now the sum of the observed treatment totals T_i is the grand total G, or $48m$. Consequently the last two terms above may be amalgamated to give

$$\frac{T_1^2 + T_2^2 + \cdots + T_8^2}{4} + \frac{T_9^2}{16} - 48m^2 \qquad (3.17)$$

Note that the last term is the correction for the mean, already found to be 4,489,633. Expression (3.17) is an example of the more general expression

$$\sum_{i=1}^{k} \frac{T_i^2}{r_i} - C$$

which gives the treatments sum of squares when the ith treatment is replicated r_i times, C being the correction for the mean.

To obtain the numerical value we first calculate the treatment totals as shown in table 3.2.

TABLE 3.2 TREATMENT TOTALS FROM TABLE 3.1

Level of application	CN	CS	CM	CK	Totals
0		5858			5858
1	1066	928	1431	892	4317
2	1265	877	1241	1122	4505

The treatments sum of squares is given by

$$\frac{(1066)^2 + (1265)^2 + \cdots + (1122)^2}{4} + \frac{(5858)^2}{16} - 4{,}489{,}633 = 157{,}448$$

The blocks sum of squares, $\sum b_j^2$, can likewise be expressed in terms of block totals as follows.

$$\frac{B_1^2 + B_2^2 + B_3^2 + B_4^2}{12} - C$$

$$= \frac{(4383)^2 + (3075)^2 + (4752)^2 + (2470)^2}{12} - 4{,}489{,}633 = 289{,}427$$

where the observed block totals have been inserted from table 3.1. The divisor 12 is the number of observations per block.

The complete analysis of variance is shown in table 3.3. It is not customary to display the correction term for the mean, since this is of no particular interest. The item labelled "total" at the foot of the table is the original sum of squares, *minus* the correction for the mean (5,481,198

$-$ 4,489,633 = 991,565). This may be shown to be equal to the sum of squares of deviations of the observations from their mean; i.e.,

$$\sum y_{ijk}^2 - 48m^2 = \sum (y_{ijk} - m)^2$$

TABLE 3.3 ANALYSIS OF VARIANCE OF THE DATA IN TABLE 3.1

Source of variation	Degrees of freedom (d.f.)	Sum of squares (s.s.)	Mean square (m.s.)
Treatments	8	157,448	19,681
Blocks	3	289,427	96,476
Error (by subtraction)	36	544,690	15,130
Total	47	991,565	

The column which has been computed above is, of course, the column labelled "sum of squares." The other two columns remain to be explained. The "degrees of freedom" associated with any component are the number of independent parameters required to describe that component in the model. In the case of treatments, this always equals one less than the number of treatments, and similarly for blocks. For the total s.s., the number is the number of observations (48) less one representing the contribution of the mean.

The degrees of freedom have two principal uses. First, by subtraction, they give the degrees of freedom for error (36). As mentioned in section 3.24, this is the divisor needed for the error s.s. in order to estimate the error variance σ^2. The estimate is 544,690/36, or 15,130, which is called the *error mean square* and is shown in the right-hand column of table 3.3. Second, the degrees of freedom for treatments are used in an F-test of the hypothesis that all k treatments produced the same effects. In section 3.21 the value of F for testing this hypothesis was given as

$$F = \frac{S_2^2 - S_1^2}{k - 1} \div \frac{S_1^2}{n_e}$$

The denominator S_1^2/n_e is the error m.s. In the numerator the quantity $(S_2^2 - S_1^2)$ may be shown to be the treatments s.s., while $(k - 1)$ is the number of degrees of freedom associated with treatments. Thus the numerator is 157,448/8, or 19,681, called the *treatments mean square.* The F-ratio for this experiment is 19,681/15,130, or 1.30, with 8 and 36 d.f.

In the same way the mean square for blocks forms the numerator for the F-test of the hypothesis that all block effects are identical. Although this test is rather seldom of interest, the mean square for blocks has other uses. By formulae which will be given with the individual designs, this mean square leads to an estimate of the error variance that would have been obtained if the experiment had not been grouped into blocks.

With some designs the structure of the analysis of variance is more complicated than in the example above. What happens is that the treatment and environmental contributions to the sum of squares are entangled. If we calculate the treatments s.s., a different result is obtained according to whether we assume that block effects are present or absent, and vice versa for blocks. In such cases the usual procedure is to present the blocks s.s. as calculated when treatment effects are ignored, and the treatments s.s. as calculated when block effects are taken into account. These items are sometimes called "blocks, ignoring treatments" and "treatments, eliminating blocks," respectively. If calculations are made in this way, the two terms may still be added so as to obtain the error s.s. by subtraction; further, the treatments s.s. is still the appropriate one for an F-test of the treatment effects.

This presentation of the analysis of variance is necessarily very inadequate. For further reading, reference should be made to Fisher (3.2, 3.3) and Snedecor (3.4), whose discussions involve little mathematics. For presentation of points of theory, see Kendall's book (3.5) and papers by Fisher (3.6, 3.7) and Yates (3.8). Unfortunately, no single reference contains a really comprehensive account of the subject.

3.3 Accuracy in Computations

3.31 Original Records. If there exists a series of rules which will guarantee that all computations are accurately done, we do not know them. The following notes may help in maintaining a high standard of accuracy.

The first place where errors may occur is in the original records. These should be made in a clearly legible and permanent form. Where scale readings are being taken, the habit of checking each reading immediately after it has been written down is a useful one. As the observations are being recorded, any that appear anomalous should be examined and the reason sought. Gross errors are often eliminated in this way. If on checking such an observation no error or explanation is found, a note to this effect should be made. Such notes are helpful in situations where the later analysis is done by a different person. Most statisticians who analyze other people's experiments have encountered the perplexing problem where an observation is so completely out of line that a gross error is suspected, yet the original records contain no comment about the observation, and the recorder's memory of the circumstances in which it was taken has faded.

Copying from the original records to a form suitable for analysis may also introduce errors. Since cross-checking is rather a soporific task, mistakes tend to be overlooked even when two persons are used for the check. Sometimes foresight can eliminate the need for copying.

3.32 Checks in the Analysis of Variance. The analysis of variance itself consists of two sorts of operations. First, a number of totals are formed. These are nearly all self-checking, for the fact that the treatment totals and the block totals both add to the grand total may be accepted as proof of their correctness. This is not a watertight check, since compensating errors in two treatment totals pass undetected. Also it does not ensure that treatments and blocks are correctly identified, and cases have occurred where treatment totals were mistakenly labelled as block totals, and vice versa. But apart from such rare events the sum check is as satisfactory as any.

The second stage, the computation of the various sums of squares, is not self-checking and must be checked by recomputation. The most accurate method is to have two computers carry out the analysis independently, comparing results only when both have finished the task. The practice of having one computer "check" the results of the other, though more convenient, always seems to produce a certain number of errors. This is to be expected, because, when the first calculations are nearly always correct, as they should be, it is difficult for the checker to avoid the presupposition that they are correct, with resulting mental laziness. Recomputation by the same person, though often unavoidable, is probably less accurate than the two preceding methods.

An expert who wishes to do so can make the analysis self-checking. For instance, the error s.s. can be computed independently of the other sums of squares, sometimes quite easily and sometimes with more difficulty. If this sum of squares agrees with that found by subtraction, all the items in the "sums of squares" column may be regarded as correct, apart possibly from errors in labelling. However, with the computational methods that we present, checking is required except in a few cases specifically noted.

The steps in the preparation of summary tables (e.g., calculation of treatment means from totals) are usually partly but not entirely self-checking. For example, the fact that treatment means average to the general mean forms a check. Care should be taken to recompute those steps that are not self-checked.

With the more complex designs, where treatment means have to be adjusted in order to eliminate the environmental effects, errors have arisen because the computing instructions were misunderstood. For instance, the adjustments made were ten times as large as they should have been. A person thoroughly familiar with the design would have known that the adjustments were ridiculously large, but someone new to the design may have no idea what size of adjustment it is reasonable to expect. This danger with complex designs should not be ignored. It can be minimized by careful study of the method of analysis for any design that it is proposed to use.

3.33 Number of Figures to be Retained. Since the primary purpose of the analysis of variance is to obtain an accurate value for the error m.s., the number of figures which it is worth retaining during the calculations depends on the size of this figure. This is not known in advance, but often a rough idea of the coefficient of variation (ratio of the standard error to the mean of the experiment) is available from previous experience. For the *original data* from which the analysis of variance is computed, a crude rule which errs on the safe side is as follows. Record the original data to 4 significant figures if the coefficient of variation is between 0.4 and 4%, to 3 if it is between 4 and 40%, and to 2 if it exceeds 40%. Let us apply this rule to the eelworm data. Since experiments on eelworms are not very common, there is not much basis on which to predict the coefficient of variation. Suppose that we guess that it is unlikely to be below 20% (it actually turned out to be 40%). Then, for safety, data should be recorded to 3 significant figures. Since the average eelworm number per plot is around 300, this means that data are recorded to the nearest eelworm, which is as accurately as they could be recorded.

A more precise rule, requiring a little calculation, is that the rounding interval should not exceed one-quarter of the standard error per observation. To apply this rule to the eelworm data, starting with the same premise that the coefficient of variation is 20%, we note that the mean per plot in the experiment is about 300. Thus the predicted value of the standard error per observation is about 60, so that the rounding interval should not exceed 15. It will be quite satisfactory to round each observation to the nearest 10 eelworms. The discrepancy from the result given by the previous rule illustrates the fact that the first rule tends to be too conservative.

The advisability of deliberately rounding records already taken should be left to the judgment of the experimenter. With modern calculating machines it may be more expeditious to carry out the computations with a few unnecessary figures than to take time to perform and check the rounding. The rules are most likely to be useful where computing devices are poor or unavailable.

In the analysis of variance itself, it is generally as well to carry the full number of figures obtained from the uncorrected sum of squares; e.g., if the original data contain one decimal place, the sum of squares and hence the analysis of variance will contain two decimal places. This number will usually be excessive, but no more time is likely to be wasted in writing down the unnecessary digits than in estimating how many digits should be retained.

In the final presentation of results, on the other hand, superfluous digits should be strictly avoided. They make the conclusions more difficult to grasp, impede rapid mental comparisons, and give some readers an erroneous impression of the accuracy of the results. A treatment mean should be rounded to one-tenth of its estimated standard error; that is, if the estimated standard error of the treatment means is 2.56, the means could be rounded to 0.256 and should be rounded to 0.1, i.e., to one decimal place, since this is the nearest rounding interval that is convenient. In some experiments (e.g., those of the split-plot type) different treatment comparisons have different standard errors. Here the rounding interval should be decided from those comparisons that have the smallest error.

3.34 Identification of Data. The practice should be followed of labelling all data, from original records to summary tables, so that another person at a later time can tell what the data mean. Much data, expensive to collect and potentially valuable for some research, have had to be discarded or destroyed because the original collector cannot be reached, and new investigators are unable to discover what measurements were taken and under what circumstances. Sometimes the units in which the observations or summary tables are presented cannot be found, or the crop on which a field experiment was carried out, or the nature of the experimental treatments. Frequently, data serve their most fruitful purpose when some later investigator gathers from many places all the material bearing on some question, and it should be a habit to facilitate rather than hinder such research.

3.4 Subdivision of the Sum of Squares for Treatments

3.41 Reasons for Subdivision. Often an experiment is planned to provide the answers to a number of different questions, not necessarily connected with one another. The eelworm experiment is a partial though not an ideal example. One way in which the questions asked might be framed is as follows. (i) Is soil fumigation effective? Here we are interested in the average reduction in eelworm numbers taken over all fumigants. (ii) Are there differences in the effectiveness of different fumigants? For this question, comparisons of the reductions due to different fumigants are relevant. (iii) What is the relative effectiveness of single and double doses? Since the simplest reasonable hypothesis is that the reduction in eelworm numbers to the double dose is twice that to the single dose, we might narrow this question by asking: Is the average response to fumigation proportional to the amount of dressing, or in other words is the average response curve linear? (iv) Finally, if there is some indication of curvature in the responses, we might ask whether the amount of curvature is the same for all fumigants.

With experiments of this type the F-test of the complete treatments m.s. is not particularly helpful because it is directed, as it were, at a mixture of several diverse questions. By an extension of the analysis of variance, we can subdivide the treatments s.s. into a number of components that are more relevant to the individual questions. Moreover, an F-test can be made on the mean square for each component. The rules for subdivision are given in succeeding sections. Effective use of the device requires long practice applied to a considerable number of experiments.

3.42 Subdivision into Single Components.

The simplest case, where all treatments have the same number r of replications, is considered first. The calculations are best made from the treatment totals T_i. Examples of the sort of quantity that we are interested in studying are

$$T_1 - T_2; \qquad \frac{T_1 + T_2}{2} - T_3; \qquad T_1 + T_2 - (T_3 + T_4)$$

Quantities of this type are called *linear functions* of the T's. Note that the sum of the coefficients of the T's is always zero, as it must be if the quantity is to represent a comparison among the T's. These ideas may be formalized as follows.

Definition. Any linear function

$$z_w = l_{w1}T_1 + l_{w2}T_2 + \cdots + l_{wk}T_k$$

is called a *comparison* among the T's if

$$l_{w1} + l_{w2} + \cdots + l_{wk} = 0$$

Rule 1. If z_w is any comparison among the T's, the quantity

$$\frac{z_w^2}{D_w}, \text{ where } D_w = r(l_{w1}^2 + l_{w2}^2 + \cdots + l_{wk}^2)$$

is a component of the sum of squares for treatments and represents 1 d.f.

This rule gives the divisor D_w to be used so that the comparison z_w may be included as a part of the treatments s.s. in the analysis of variance. Incidentally, the rule implies that z_w^2/D_w cannot exceed the treatments s.s. If it is found to do so, this is a sure sign of a mistake in computation.

Definition. Two comparisons z_1 and z_2 are said to be *orthogonal* if

$$l_{11}l_{21} + l_{12}l_{22} + \cdots + l_{1k}l_{2k} = 0$$

This "sum of products" relation is very important. By theory, it ensures that z_1 and z_2 are distributed independently of each other. Its relevance to the partition of the treatments s.s. appears in rule 2.

Rule 2. If z_1 and z_2 are orthogonal, then z_2^2/D_2 is a component of

$$\text{Treatments s.s.} - \frac{z_1^2}{D_1}$$

This means that, if we divide the treatments s.s. into the contribution from a comparison z_1 and the remainder, and now wish to subdivide the remainder, we must choose comparisons that are orthogonal to z_1. Similarly, after removing the contribution of z_2, the next comparison z_3 must be orthogonal to both z_1 and z_2, and so on. This leads to the next rule.

Rule 3. If the comparisons $z_1, z_2, \cdots, z_{k-1}$ are mutually orthogonal, i.e., every pair is orthogonal, then

$$\text{Treatments s.s.} = \frac{z_1^2}{D_1} + \frac{z_2^2}{D_2} + \cdots + \frac{z_{k-1}^2}{D_{k-1}}$$

This algebraic identity partitions the treatments s.s. [which has (k — 1) degrees of freedom] into $(k - 1)$ components, each representing a single degree of freedom.

Starting with a specified z_1, we can always find $z_2, z_3, \cdots, z_{k-1}$ so as to construct a complete orthogonal set. In fact, it may be proved that there is considerable freedom of choice in this process. When all but the final z_{k-1} have been selected, there is only one possible choice for z_{k-1}, but at any previous stage many z's can be found that are orthogonal to all preceding z's. The experimenter must choose the z's that are most relevant for purposes of interpretation.

Table 3.4 shows the single components most frequently used for interpretative purposes when there are three or four treatments. It is assumed that all treatments have the same number of replicates. The divisors D are those required for inserting the square of z in the analysis

TABLE 3.4 SAMPLE SETS OF SINGLE COMPONENTS

Three treatments

i. Equally spaced increments of one ingredient

	z_1	z_2
T_1	−1	1
T_2	0	−2
T_3	1	1
Component	Lin.	Quad.
divisor D	$2r$*	$6r$

ii. Two qualities of an ingredient and a control

	z_1	z_2
T_1 (0)	−2	0
T_2 (a_1)	1	−1
T_3 (a_2)	1	1
Component	Effect of a	Quality diff.
divisor D	$6r$	$2r$

Four treatments

i. Equally spaced increments of one ingredient

	z_1	z_2	z_3
T_1	−3	+1	−1
T_2	−1	−1	+3
T_3	+1	−1	−3
T_4	+3	+1	+1
Component	Lin.	Quad.	Cubic
D	$20r$	$4r$	$20r$

ii. Two comparable types of ingredient, a and b, and two qualities of each type

	z_1	z_2	z_3
T_1 (a_1)	+1	+1	0
T_2 (a_2)	+1	−1	0
T_3 (b_1)	−1	0	+1
T_4 (b_2)	−1	0	−1
Component	a vs. b	Quality diff. within a	Quality diff. within b
D	$4r$	$2r$	$2r$

iii. Different levels of each of two different ingredients a and b

	z_1	z_2	z_3
T_1 (a_1b_1)	−1	−1	+1
T_2 (a_2b_1)	+1	−1	−1
T_3 (a_1b_2)	−1	+1	−1
T_4 (a_2b_2)	+1	+1	+1
Component	Average response to a	Average response to b	Interaction
D	$4r$	$4r$	$4r$

r* = number of replicates, assumed the same for all treatments. The coefficients given above are not valid if r varies from treatment to treatment.

of variance. The reader should verify that these divisors satisfy rule 1, and that each set of z's forms a complete orthogonal set.

With three treatments the two sets presented are essentially the same, though the meaning attached to each z is different. Case i applies in an experiment where we have, say, zero, single, and double dressings of a fumigant, and wish to isolate the linear and quadratic components of the response curve. The response found by fitting a straight line to the three levels is represented by z_1, while z_2 measures the deviation from a linear response. Case ii arises when we have comparable dressings of two different fumigants, and wish to examine the average effect of fumigation (z_1) and the difference between the fumigants (z_2).

With four treatments, case i shows zero, single, double, and triple applications of the same ingredient; here the response curve can be divided into its linear, quadratic, and cubic components. Alternatively, in case ii, we might be testing the wearing qualities of two types of "100% wool" suits and two types of suits made of wool-rayon mixtures. Case iii is used, for instance, when we examine the effects of two different amounts of sugar and two different amounts of vanilla in the preparation of a cake. Experiments of this type are discussed more fully in chapter 5.

When different treatments have differing numbers of replicates, the rules are changed slightly. If the ith treatment has r_i replicates, we have shown that the treatments s.s. is

$$\frac{T_1^2}{r_1} + \frac{T_2^2}{r_2} + \cdots + \frac{T_k^2}{r_k} - \frac{(T_1 + T_2 + \cdots + T_k)^2}{(\sum r_i)}$$

The changes to be noted are:

i. A linear function

$$z_w = l_{w1}T_1 + l_{w2}T_2 + \cdots + l_{wk}T_k$$

is a comparison among the treatment totals T_i if

$$r_1 l_{w1} + r_2 l_{w2} + \cdots + r_k l_{wk} = 0$$

ii. The divisor required for z_w^2 is

$$D_w = r_1 l_{w1}^2 + r_2 l_{w2}^2 + \cdots + r_k l_{wk}^2$$

iii. Two comparisons z_1 and z_2 are orthogonal if

$$r_1 l_{11} l_{21} + r_2 l_{12} l_{22} + \cdots + r_k l_{1k} l_{2k} = 0$$

3.43 Incomplete Subdivisions. The subdivision need not be complete in the sense that it is composed entirely of single components. In the eelworm experiment we suggested that the treatments s.s., with 8 components, might be divided into the following parts: (i) a single component representing the average response to fumigation; (ii) a single component representing the deviation of the average response from linearity; (iii) a part representing differences in the responses to the different fumigants. Since there are 4 fumigants, this part comprizes 3 components; (iv) a part representing differences in the curvature of the response curves for different fumigants: this also has 3 components. Some rules for partial subdivision will be given and applied to this example. The most general rule is as follows.

Rule 4. A set of g quantities Q_i are independent components of the treatments s.s., with n_i degrees of freedom respectively, if

$$Q_i = \frac{{z_{i1}}^2}{D_{i1}} + \frac{{z_{i2}}^2}{D_{i2}} + \cdots + \frac{{z_{in_i}}^2}{D_{in_i}} \qquad (i = 1, 2, \cdots g)$$

where all z_{ij} are mutually orthogonal comparisons, and the D_{ij} are the appropriate divisors. Further, the remainder

$$\text{Treatments s.s.} - \sum Q_i$$

is an independent component of the treatments s.s. with

$$[(k - 1) - \sum n_i] \text{ degrees of freedom}$$

This rule, which is a deduction from previous rules, is not very helpful to the beginner. With experience it often becomes easy, without carrying out the details, to see whether a set of Q_i can be expressed in this way. Some more specialized rules follow.

Rule 4a. If T_1, T_2, $\cdots$, T_p are the totals for any set of treatments, then

$$Q = \frac{{T_1}^2}{r_1} + \frac{{T_2}^2}{r_2} + \cdots + \frac{{T_p}^2}{r_p} - \frac{(T_1 + T_2 + \cdots + T_p)^2}{\sum r_i}$$

is a component of the treatments s.s., with $(p - 1)$ degrees of freedom.

This rule states that the sum of squares of deviations among any subgroup of the treatments, with proper divisors, is part of the treatments s.s.

Rule 4b. Sometimes the treatments (or part of them) can be divided into a number of groups. The number of treatments in a group need not be constant, and different treatments may have different amounts of replication. Let T_{ij} be the total for the jth treatment in the ith group, and let

$$S_i = T_{i1} + T_{i2} + \cdots + T_{ip_i};$$

$$R_i = r_{i1} + r_{i2} + \cdots + r_{ip_i} \qquad (i = 1, 2, \cdots, g)$$

Then let

$$Q_i = \frac{T_{i1}^2}{r_{i1}} + \frac{T_{i2}^2}{r_{i2}} + \cdots + \frac{T_{ip_i}^2}{r_{ip_i}} - \frac{S_i^2}{R_i} \qquad (i = 1, 2, \cdots, g)$$

$$Q_{g+1} = \frac{S_1^2}{R_1} + \cdots + \frac{S_g^2}{R_g} - \frac{(S_1 + \cdots + S_g)^2}{R_1 + \cdots + R_g}$$

In this case Q_i represents the sum of squares of deviations among the treatments in the ith group, with $(p_i - 1)$ degrees of freedom, while Q_{g+1} is the sum of squares of deviations among the group totals with $(g - 1)$ degrees of freedom. These Q's are independent components of the treatments s.s.

Rule 4c. This applies when all treatments have the same number r of replicates, and every group contains p treatments. Let

$$z_i = l_1 T_{i1} + l_2 T_{i2} + \cdots + l_p T_{ip}$$

where the set of l's remains the same in all groups. Then

$$Q = \sum \frac{(z_i - \bar{z})^2}{r \sum l_i^2}$$

is a component of the treatments s.s., with $(g - 1)$ degrees of freedom. This rule is useful when we wish to compare linear functions of the treatments within each group. Note that the divisor in Q is the same as the divisor that would be used for a single z_i.

Applications. In the eelworm experiment, the *average* effects of fumigation are obtained from a comparison of the total for the control with the totals for the single and double levels of fumigation. Consequently, we might regard the 9 treatments as divided into 3 groups: the control, the single levels, and the double levels. By rule 4*b*, this gives the following partition of the treatments s.s. (the treatment totals are found in table 3.2, p. 56).

Between fumigants (single level):
(3 d.f.)

$$\frac{(1066)^2 + (928)^2 + (1431)^2 + (892)^2}{4} - \frac{(4317)^2}{16} = 45{,}461$$

Between fumigants (double level):
(3 d.f.)

$$\frac{(1265)^2 + (877)^2 + (1241)^2 + (1122)^2}{4} - \frac{(4505)^2}{16} = 23{,}641$$

Between levels:
(2 d.f.)

$$\frac{(5858)^2 + (4317)^2 + (4505)^2}{16} - \frac{(14{,}680)^2}{48} = 88{,}347$$

The three sums of squares add to 157,449, in agreement with the treatments s.s., 157,448, as given in table 3.3, p. 57.

The next step is to divide the sum of squares between levels into a component representing the average response to fumigation and one representing the deviation from a linear response. By least squares theory, the former is given by a comparison of the double level with the zero level. That is, we may take

$$z_1 = S_2 - S_0 = 4505 - 5858 = -1353$$

where the subscript refers to the level. The deviation from linearity is measured by the orthogonal comparison

$$z_2 = (S_2 - S_1) - (S_1 - S_0) = S_2 + S_0 - 2S_1$$
$$= 4505 + 5858 - 2(4317) = 1729$$

Since each S is a total over 16 plots, the divisors are 32 and 96 respectively. Hence the contributions to the sum of squares are:

$$\text{Average linear response:} \quad \frac{(1353)^2}{32} = 57{,}207$$

$$\text{Average curvature:} \quad \frac{(1729)^2}{96} = 31{,}140$$

The subdivision found thus far might be presented as shown below.

	d.f.	s.s.	m.s.
Between treatments	8	157,449	19,681
Average linear response	1	57,207	57,207
Average curvature	1	31,140	31,140
Between single levels of fumigants	3	45,461	15,154
Between double levels of fumigants	3	23,641	7,880

This is not the subdivision envisaged at the beginning of this section. Instead, it was proposed to divide the final 6 d.f. into 3 representing differences in the responses to individual fumigants and 3 representing differences in the curvatures of the individual response curves. This separation is more difficult.

Consider a comparison of (CN) and (CS). Estimates of the difference between these fumigants are available both at the single (1CN) and the double (2CN) level. If the effects of both fumigants are proportional to the amounts of dressing, the true difference at the double level will be twice that at the single level. On this assumption, two independent estimates of the difference between (CN) and (CS), for a single dressing, are

$$[(1CN) - (1CS)] \quad \text{and} \quad \frac{[(2CN) - (2CS)]}{2}$$

The second estimate is considerably more accurate than the first, since its variance is only ¼ as large. Statistical theory shows that such estimates are combined by weighting each inversely as its variance.

Hence, on the assumption of linearity, the most accurate estimate of the difference per unit dressing is

$$\frac{[(1CN) - (1CS)] + (4)(\frac{1}{2})[(2CN) - (2CS)]}{1 + 4}$$

$$= \frac{[(1CN) - (1CS)] + 2[(2CN) - (2CS)]}{5} \quad (3.18)$$

This shows that differences in the linear responses to the 4 fumigants are measured by comparisons of the quantities (1CN) + 2(2CN), etc. These quantities are given below.

(Single level) + 2(double level)

(CN)	(CS)	(CM)	(CK)	Total
3596	2682	3913	3136	13,327

It is for quantities of this kind that rule 4*c* is useful. This rule shows that the sum of squares of deviations of these quantities, when divided by $r\sum l_i^2$, or (4) $[(1)^2 + (2)^2]$, i.e., 20, is a component of the treatments s.s. This gives 43,408, representing differences in linear responses to the fumigants.

We now consider the final three components. The curvature of an individual response curve is measured by a comparison of the type

$$[(2CN) - (1CN)] - [(1CN) - (0)] = [(2CN) + (0) - 2(1CN)]$$

where (0) represents the total number of eelworms over 4 plots for the control treatment. When the difference between two fumigants is taken, the (0) term disappears and we obtain the comparison

$$[(2CN) - 2(1CN)] - [(2CS) - 2(1CS)] \quad (3.19)$$

It is easy to verify that this is orthogonal with comparisons like (3.18). Consequently differences in curvature are compared by means of the quantities shown below.

2(Single level) − (double level)

(CN)	(CS)	(CM)	(CK)	Total
867	979	1621	662	4129

Rule 4*c* again applies. The divisor is 20, and the sum of squares contributes 25,693. The final separation is given in table 3.5. It is evident

from F-tests that none of the mean squares is significant at the 5% level, and only that for the average linear response approaches near to the 5% level. These results must not be taken as final conclusions for this experiment. As will be seen in section 3.8, additional data were recorded in this experiment which permit a more accurate analysis.

TABLE 3.5 SUBDIVISION OF THE TREATMENTS S.S. FOR THE EELWORM DATA

Source of variation	d.f.	s.s.	m.s.
Average linear response	1	57,207	57,207
Average curvature	1	31,140	31,140
Differences in linear response	3	43,408	14,469
Differences in curvature	3	25,693	8,564
Error	36	544,690	15,130

A more extensive discussion of this general topic, with a number of examples, is given in Snedecor's book (3.4), chapter 15. It is worth repeating that the amount and type of subdivision that should be done depend on the experiment, and to some extent on individual taste. As will be seen in the next section, any single component can be tested by means of a t-test derived from the treatment means, and some workers prefer to make the test in this way.

3.5 Calculation of Standard Errors for Comparisons among Treatment Means

3.51 Rules. We first give a rule (rule 5) that is more general than we need, because it is sometimes useful for other purposes.

Rule 5. The standard error of any linear function

$$z = l_1 y_1 + l_2 y_2 + \cdots + l_N y_N$$

of the individual observations is

$$\sigma_z = \sigma \sqrt{l_1^2 + l_2^2 + \cdots + l_N^2}$$

The *estimated* standard error is obtained by substituting s for σ, where s is the square root of the error m.s. in the analysis of variance. This rule supplies the standard error of any kind of linear function, provided that it has been expressed in terms of the *individual observations*. For linear functions of the *treatment means*, one of two special cases of the rule is used.

Rule 5a. If all treatments have the same number r of replications, the standard error of a linear function

$$z = l_1 \bar{y}_1 + l_2 \bar{y}_2 + \cdots + l_k \bar{y}_k$$

of the treatment means $\bar{y}_i$ is

$$\sigma_z = \frac{\sigma}{\sqrt{r}} \sqrt{l_1^2 + l_2^2 + \cdots + l_k^2}$$

Rule 5b. If the ith treatment has r_i replications. the standard error of z is

$$\sigma_z = \sigma \sqrt{\frac{l_1^2}{r_1} + \frac{l_2^2}{r_2} + \cdots + \frac{l_k^2}{r_k}}$$

As in the general case, we substitute s for σ in order to obtain the estimated standard error. Finally, a rule that may save labor in complex cases is as follows.

Rule 6. If the linear functions $z_1, z_2, \cdots, z_p$ are mutually orthogonal, the standard error of

$$z = l_1 z_1 + l_2 z_2 + \cdots + l_p z_p$$

is

$$\sigma_z = \sqrt{l_1^2 \sigma_1^2 + l_2^2 \sigma_2^2 + \cdots + l_p^2 \sigma_p^2}$$

where σ_i^2 is the variance of z_i.

This rule enables the work to be done in two stages. Sometimes, in a large experiment, it is relatively easy to express the function z in terms of a number of familiar orthogonal functions, but rather tedious to write z in terms of original observations. The variances of the z_i may be found by rule 5 and substituted in rule 6. Note that different z_i may involve the same set of observations, provided that the z_i are orthogonal.

3.52 Examples. These rules will now be illustrated by application to the eelworm experiment. The treatment means are shown in table 3.6, the figures in parentheses denoting the order in which the treatments are numbered.

TABLE 3.6 MEAN NUMBERS OF EELWORMS PER PLOT

	Fumigant				
Level	(CN)	(CS)	(CM)	(CK)	Means
0		366(9)			366
1	266(5)	232(6)	358(7)	223(8)	270
2	316(1)	219(2)	310(3)	280(4)	281

From table 3.5 the standard error per plot is $\sqrt{(15{,}130)}$, or $s = 123$ with 36 d.f.

Example 1. Standard error of the average linear response. This is measured by the average response to the double dressing, i.e.,

$$z = \tfrac{1}{4}(\bar{y}_1 + \bar{y}_2 + \bar{y}_3 + \bar{y}_4) - \bar{y}_9 = -85$$

Since the first 4 means are based on 4 replicates, while y_9 has 16 replicates, we apply rule 5*b*, which gives

$$s_z = (123)\sqrt{\left(\frac{1}{16}\right)\left(\frac{1}{4}\right)+\left(\frac{1}{16}\right)\left(\frac{1}{4}\right)+\left(\frac{1}{16}\right)\left(\frac{1}{4}\right)+\left(\frac{1}{16}\right)\left(\frac{1}{4}\right)+(1)\left(\frac{1}{16}\right)}$$

$$= \frac{123}{\sqrt{8}} = 43.5$$

The value of Student's t is 85/43.5, or 1.95.

As mentioned previously, the t-test of any comparison, as made above, is identical with the F-test of the corresponding component in the analysis of variance. The basic relationship is $F = t^2$; that is, if the 5% values of t for n degrees of freedom are read from the table and squared, their squares are the 5% F values for 1 and n degrees of freedom, and similarly for any other significance level. This result may be verified in the present instance. The value of t^2 is 3.80, with 36 d.f. From the analysis of variance in table 3.5, the F value for the average linear response is 57,207/15,130 or 3.78, with 1 and 36 d.f., the two values agreeing apart from rounding errors. The reader may check that the same agreement holds for the average curvature.

Example 2. Response to the single dressing of (*CN*). The response is measured by

$$z = \bar{y}_5 - \bar{y}_9$$

Once again rule 5*b* is appropriate.

$$s_z = 123\sqrt{\frac{1}{4} + \frac{1}{16}} = \frac{123\sqrt{5}}{4} = 68.8$$

Example 3. Difference between the linear responses to (*CN*) *and* (*CS*). In section 3.43 it was shown that this difference is

$$z = \frac{[\bar{y}_5 - \bar{y}_6 + 2\bar{y}_1 - 2\bar{y}_2]}{5}$$

Since all means have 4 replicates, rule 5*a* may be used.

$$s_z = \frac{123}{(2)(5)}\sqrt{1 + 1 + 4 + 4} = (12.3)(\sqrt{10}) = 38.9$$

Example 4. Although the experiment is not complex enough to exhibit a profitable application of rule 6, this example shows that the rule agrees with the other rules. Suppose that at the double level we had compared (CN) with (CS), and also the two chemicals with the two proprietary mixtures. The comparisons are

$$z_1 = \bar{y}_1 - \bar{y}_2; \qquad z_2 = \frac{\bar{y}_1 + \bar{y}_2}{2} - \frac{\bar{y}_3 + \bar{y}_4}{2}$$

By rule 5*a* their standard errors are

$$\sigma_{z_1} = \frac{\sigma}{2}\sqrt{2}; \qquad \sigma_{z_2} = \frac{\sigma}{2}$$

and it will be noted that the two functions are orthogonal.

If now we wished to compare the double dressing of (CN) with the mean of the 3 other double dressings, the comparison would be

$$z = \bar{y}_1 - \frac{\bar{y}_2 + \bar{y}_3 + \bar{y}_4}{3}$$

By a direct use of rule 5*a*, we have

$$\sigma_z = \frac{\sigma}{2}\sqrt{1 + \frac{1}{9} + \frac{1}{9} + \frac{1}{9}} = \frac{\sigma}{\sqrt{3}}$$

But alternatively we may write

$$z = \frac{2z_1}{3} + \frac{2z_2}{3}$$

and apply rule 6 to give

$$\sigma_z = \sqrt{\frac{4}{9}\sigma_{z_1}^2 + \frac{4}{9}\sigma_{z_2}^2} = \sigma\sqrt{\left(\frac{4}{9}\right)\left(\frac{1}{2}\right) + \left(\frac{4}{9}\right)\left(\frac{1}{4}\right)} = \frac{\sigma}{\sqrt{3}}$$

in agreement with the more direct method.

3.53 Testing Effects Suggested by the Data. In order that F- and t-tests be valid, the tests to be made in an experiment should be chosen before the results have been inspected. The reason for this is not hard to see. If tests are selected *after* inspection of the data, there is a natural tendency to select comparisons that appear to give large differences. Now large apparent differences may arise either because there are large real effects, or because of a fortuitous combination of the experimental errors. Consequently, in so far as differences are selected just because they seem to be large, it is likely that an undue proportion of the cases selected will be those where the errors have combined to make the differences large. The extreme case most commonly cited is that of the experimenter who always tests, by an ordinary t-test, the difference between the highest and the lowest treatment means. If the number of treatments is large, this difference will be substantial even when the treatments produce no real differences in effect. It may be shown that with 3 treatments the observed value of t will exceed the 5% level in the table about 13% of the time. With 6 treatments the figure is 40%, with 10 treatments 60%, and with 20 treatments 90%. When the experimenter thinks that he is making a t-test at the 5% level, he is actually testing at the 13% level, or the 40% level, and so on. To summarize, the selection of those differences that look large and therefore "interesting" invalidates the ordinary tests of significance. The effect is to obtain too many significant results, or to raise the significance level of the test from the presumed 5% to some higher level, usually unknown.

Sometimes the difficulty can be resolved by using for the test of significance a special table which takes account of the selection that has been exercised. For the problem discussed above, there is a table of the significance levels of the ratio of the range of a group of treatment means to an independent estimate of the standard error of a treatment mean. This ratio is called the *studentized range* (3.1a). The table furnishes a fully valid test of significance of the difference between the highest and the lowest treatment means. More generally, the table gives a conservative test of the difference between any pair of treatment means that are picked out because their difference looks interesting or large.

Occasionally, one treatment mean is found to differ by an unexpectedly large amount from all the other treatment means. In this event we may want to test the difference between the aberrant mean and the average of all the other treatment means, with due allowance for the fact that the largest difference of this type has been selected. For this, the table of the extreme studentized deviate from the sample mean can be used (3.2a). In our notation this is a table of the significance levels of the absolute value of

$$(\bar{y}_1 - \bar{\bar{y}})/s'$$

where $\bar{y}_1$ is the outlying mean, $\bar{\bar{y}}$ is the mean of *all* the treatments, and s' is the estimated standard error of a treatment mean. Another useful table (3.3a) applies when a group of treatment means are all being compared with a standard treatment.

The most general test of this type (3.3, 3.4a) applies to any linear comparison among the treatment totals (or means) that has been picked out after inspection of the results. If z is the comparison, in the notation of section 3.42, the standard error of z is of the form $s\sqrt{D}$, where s is the estimated standard deviation per unit, based on n d.f. The significance level of the ratio $z/s\sqrt{D}$ can be taken as $\sqrt{(t-1)F}$, where t is the number of treatments and F has $(t-1)$ and n d.f.

To illustrate, suppose that the treatment *totals* over 4 replicates are as follows:

T_1	T_2	T_3	T_4	T_5
16.2	17.1	18.4	24.9	25.3

The experimenter is interested in the unexpected disparity between the pair T_4, T_5 and the rest of the treatments. He takes

$$z = 3(T_4 + T_5) - 2(T_1 + T_2 + T_3) = 47.2$$

By the rule in section 3.42,

$$D = 4(3^2 + 3^2 + 2^2 + 2^2 + 2^2) = 120$$

If $s = 1.23$, with 12 d.f.,

$$\frac{z}{s\sqrt{D}} = \frac{47.2}{(1.23)\sqrt{120}} = 3.50$$

For 4 and 12 d.f., the 5% level of F is 3.26. The 5% significance level of this ratio is therefore $\sqrt{(4)(3.26)} = 3.61$. The comparison in question falls just short of significance.

This test requires high ratios in order to attain significance, because it allows for the possibility that we may have picked out the comparison z for which $z/s\sqrt{D}$ has the highest attainable value. If the comparison is the difference between a pair of treatments, or between one treatment and the remainder, the tests described earlier are more sensitive.

3.54a Multiple Comparisons. In an experiment with more than two treatments, the investigator frequently makes several 5% t-tests in the process of summarizing the results. If there are no differences between the true treatment means, the probability of obtaining an apparently significant result in one of these tests is 0.05. It has long been recognized, however, that if several t-tests have been performed, the probability that at least one of these is apparently significant is greater than 0.05. If the t-tests are independent, this probability is 0.23 for 5 tests, 0.40 for 10 tests, and 0.64 for 20 tests. These results raise the question whether we ought to try to control the frequency with which *any* wrong statement is made in the summary of an experiment, rather than the frequency with which a wrong statement is made in an individual t-test.

The same issue appears when several confidence intervals are computed. The probability that at least one of the intervals does not enclose the true treatment difference is greater than the stipulated 0.05.

Recent research (3.5a, 3.6a, and 3.7a) has clarified this and related questions. One problem to which the research has contributed is that of arranging a set of treatment means in order of performance. Since significant differences rarely exist between every pair of treatments, the most that can usually be done is to create a partial separation, such as, with four treatments,

$$\underline{A \quad C} \quad B \quad D$$
$$A \quad \underline{C \quad B} \quad D$$

The line joining A and C denotes that these two treatments have not been shown to be significantly different; A is, however, significantly different from B and D. Similarly, C differs significantly from D but not from B, and B differs significantly from D.

In the past, this separation has often been made by first performing an F-test on all the treatment means. If F is not significant, no t-tests are made, the four means being regarded as indistinguishable. If F is significant, the ordinary t-test for the difference between two means is applied to every pair of means. It saves time to compute the *least significant difference*, $\sqrt{2}st_{.05}/\sqrt{r}$. Any two means whose difference exceeds this value are declared significantly different.

Duncan (3.5a) has examined the type of protection which this method gives the experimenter against erroneously finding significant results. If all four true means are in fact identical, the F-test assures 95 chances in 100 that the means will be declared indistinguishable. If a specified pair of true means are identical, the combined F- and t-tests give a protection level of at least 95 chances in 100 that the means will be declared indistinguishable. If, however, three of the true means are identical, the protection level for the trio depends on the position of the fourth mean. When the fourth mean is far removed from the three identical means, the level is about 87%. As the fourth mean moves nearer to the other three, the level lies between 87% and 95%.

In the newer methods, the protection levels for sets of two, three, four, etc., means are kept at or above preselected values. The Newman-Keuls method (3.5a) keeps all levels at 95%. Duncan's multiple range test (3.5a) makes the level 95% for sets of two means. For sets of three means, the level is 90.25%, i.e. $(0.95)^2$; for sets of four means it is $(0.95)^3$, or 85.7%, and so on. This decrease in the level of protection makes the multiple range test more sensitive than the Newman-Keuls test in detecting real differences, at the expense of lowered protection. As Duncan suggests, his test appears to be a reasonable working compromise for the experimenter.

3.6 Subdivision of the Sum of Squares for Error

3.61 Reasons for Subdivision. The sum of squares for error can be partitioned into components in the same way as that for treatments. Although such subdivisions are not so frequently required as with treatments, they have a number of uses. Sometimes there is reason to believe that the error s.s. is not homogeneous; that is, the residual errors e do not all have the same variance σ^2 as postulated in the mathematical model. In this case, as will be shown in section 3.63, subdivisions of the error may be necessary in order to obtain valid t-tests. Occasionally a gross error in some observation may be suspected, and it is helpful to calculate the contribution of this observation to the error s.s. In addition, a subdivision may help in understanding the nature of the error m.s.

3.62 Rules for Subdivision. These follow the same general pattern as with treatments, though they are a trifle more complicated. It is necessary to go back to the normal equations (3.4) from which the treatment and block effects were estimated. These equations took the form

$$\sum (y_{ijk} - m - t_i - b_j) = 0 \tag{3.4}$$

In other words, the residuals that provide the estimates of error must add to zero over any treatment or over any block. Consequently, if the linear function

$$z = \sum l_{ijk} y_{ijk}$$

is to be a component of the error, the coefficients l_{ijk} must sum to zero over all observations that receive any specified treatment and also over all that are in any specified block. These are the tests by which we tell whether any proposed z is part of the error.

For any z which satisfies these tests, the contribution to the error s.s. is z^2/D, where by rule 5

$$D = \sum {l_{ijk}}^2$$

We now examine some components of the error s.s. in the eelworm experiment. Obviously, the difference z between the eelworm numbers on any two control plots in the same block satisfies the conditions. Similarly any comparison among the 4 control observations in a block is a component of the error. It follows that the sum of squares of deviations of these 4 observations from their mean contributes 3 d.f. to the error s.s. The 4 blocks together contribute 12 d.f. of this type of error. The sum of squares will be found to be 307,312. This component is a measure of the amount of variation among observations that receive the same treatment and lie in the same block, and is the type of component that we should expect to constitute the error. However, this type of component is encountered only when some treatments have been replicated *within* the block.

Consider now the difference between two other treatments, say $(1CN)$ and $(1CS)$. Let

$$z_1 = (1CN)_1 - (1CS)_1$$

where the subscript 1 denotes that the two observations come from block 1. This function is not a component of the error, for, while the coefficients add to zero over every block, they do not do so over treatments $(1CN)$ and $(1CS)$. But if we take

$$z_1 - z_2 = (1CN)_1 - (1CS)_1 - (1CN)_2 + (1CS)_2$$

the conditions are satisfied. This expression shows how the difference between two treatments changes from one block to another. It is a measure of the effect of the blocks on the treatment difference, and is usually called an interaction of treatments with blocks. Such interactions are the typical components of error.

If z_i is the difference between $(1CN)$ and $(1CS)$ in the ith block, any comparison among the z_i satisfies the rules required for a component of the error. Hence the sum of squares of deviations of the z_i, when divided by 2, contributes 3 d.f. to the error s.s. A similar result holds if we take *any* comparison among the treatments and calculate it separately for each block. Since the 9 treatments provide 8 independent comparisons, we see that the 24 d.f. for the interactions of treatments with blocks may be divided into 8 sets of 3 d.f. The sum of squares for these 24 d.f. can of course be found by subtraction as shown in table 3.7.

TABLE 3.7 SUBDIVISION OF THE SUM OF SQUARES FOR ERROR

	d.f.	s.s.	m.s.
Error	36	544,690	15,130
Among controls	12	307,312	25,609
Treatments × blocks	24	237,378	9,891

The F-ratio of the two components of error is 2.59, with 12 and 24 d.f. In the ordinary table this corresponds to a significance level of about 2½%. It is probably more correct to make a two-tailed test by doubling this probability. Either test indicates that the controls are more variable than the other treatments. This is not very surprising, since in data of this type the variance may tend to increase as the mean increases. In any event the result throws doubt on the suitability of the original mathematical model; it would seem better to ascribe different error variances to the control and the fumigants.

The control also contributes to the treatments × blocks component of the error. The 8 comparisons among the 9 treatments can be divided into 1 between the control and all fumigants, and 7 which are comparisons among the fumigants themselves. The former will contribute 3 d.f. to the treatments × blocks interaction; the latter, 21 d.f. If the control is more variable than the fumigants, we would expect the mean square for the 3 d.f. to be larger than that for the 21 d.f., though it should be remembered that a mean square with only 3 d.f. is poorly determined. In the next section these two components will be computed separately.

3.63 Calculation of a Separate Error for a Treatment Comparison. As the eelworm example illustrates, certain treatments may be erratic in their effects, while others in the same experiment show more stable results. When this occurs, comparisons involving the erratic treatments have a higher experimental error variance than those among the more stable treatments. Since the error m.s. in the analysis of variance is some weighted average of these different variances, its use for individual t-tests may not be valid. In such cases it is helpful to be able to calculate a separate error for any specific treatment comparison. The procedure will be illustrated for the comparison between the control and the fumigants.

The first step is to compute the comparison separately for each block.

	Totals in block				
	1	2	3	4	Total
Control	1814	1295	1769	980	5858
Fumigants	2569	1780	2983	1490	8822
2 (Control)—(Fumigants)	1059	810	555	470	2894

Since there are 4 control plots and 8 fumigated plots in a block, the control total is multiplied by 2 in forming the comparison. The contribution to the error s.s. is

$$\frac{(1059)^2 + (810)^2 + (555)^2 + (470)^2}{24} - \frac{(2894)^2}{96} = 8862$$

The divisor 24 is found by the usual rule. The mean square is 2954, with 3 d.f. By subtraction, the mean square for the interaction of fumigants with blocks (21 d.f.) is found to be 10,882. Contrary to anticipation, the component involving the control has the smaller mean square, though the difference does not approach significance.

As often happens when the simple model is inadequate, the best approach for a more accurate analysis is questionable. It is possible to compute a separate error, by the method just given, for each t-test that is made. This is the procedure least open to criticism, since it makes no assumption that errors are constant from one treatment to another. Since, however, each t-test would be based on only 3 d.f., these tests are insensitive, so that this method should not be used without good reason. In the present case it is probably justifiable to postulate only two error variances. The first, appropriate to the controls, is estimated by the mean square 25,609, with 12 d.f.; the second, for the fumigants, by the mean square 10,882, with 21 d.f. By this approach the variance of the mean response to the double dressing, for example, is estimated as

$$\frac{10{,}882}{16} + \frac{25{,}609}{16} = 2281$$

since each mean is taken over 16 plots.

Separation of the error s.s. into *single* components is seldom required except in special studies or as an exercise. One method is to set up 8 orthogonal comparisons among the treatments. Calculate each comparison separately in each block, and let z_{ij} be the value obtained for the ith comparison ($i = 1, 2, \cdots, 8$) in the jth block. Then find

$$w_{i1} = z_{i1} - z_{i2}; \qquad w_{i2} = z_{i1} + z_{i2} - 2z_{i3};$$

$$w_{i3} = z_{i1} + z_{i2} + z_{i3} - 3z_{i4}$$

Each w is a component of the error, and the whole 24 w's are mutually orthogonal. The squares of the w's, each with its proper divisor, give a separation of the error s.s. into single components.

3.7 Missing Data

3.71 Method of Handling Missing Data in the Analysis. From time to time certain observations are missing, through failure to record, gross errors in recording, or accidents. The omissions naturally affect the method of analysis. With each of the common designs we give computational instructions for analyzing data that contain gaps. The object of this section is to indicate the theoretical basis for these methods.

When certain observations are absent, the correct procedure is to write down a mathematical model *for all observations that are present.* The least squares normal equations are then constructed in the usual way. These take exactly the same general form as when all observations are present; i.e.,

$$\sum (y_{ijk} - m - t_i - b_j) = 0 \tag{3.20}$$

over all observations whose equations contain any specified parameter that is to be estimated. Since, however, the terms in the equation corresponding to missing observations are absent, the system of equations loses some of the symmetry that it possesses when all observations are present, and the solutions are more difficult. The same general procedure supplies F- and t-tests of hypotheses about the nature of the treatment effects as described in section 3.24, though again the details become more complicated.

In the analysis of variance, two changes may be noted. Owing to the missing observations, the treatments and blocks s.s. become entangled, so that the treatments s.s. must be computed after allowing for block effects, as mentioned at the end of section 3.26. Secondly, if a observations are absent, the total number of degrees of freedom is reduced by a. Unless one or more complete treatments or blocks is missing, the number of parameters required to describe these effects will be the same as before. Consequently, the missing degrees of freedom all come from the error s.s., which now represents $(n_e - a)$ degrees of freedom. In short, missing data may be handled by applying the standard least squares procedure to all observations that are *not* missing. For future reference, this method will be called the "correct least squares procedure."

To the experimenter it may be a difficult business to carry out the construction and solution of a set of unfamiliar normal equations, even though he is quite competent to analyze a set of complete data. For this reason Yates (3.9), following a suggestion by Fisher, considered inserting values for the missing observations so as to obtain a set of complete data. Suppose that only a single observation is missing, and that a value x is substituted for this observation. If the analysis of variance is calculated in the usual manner for complete data, the error s.s. is found to be of the form

$$Ax^2 - 2Bx + C$$

where A, B, and C are numbers that are determined by the type of design and the values for the other observations (A is always positive).

In order to find a numerical value for x, Yates proposed to use the value that minimizes the error s.s. This is $x = B/A$. If this value is inserted in place of the missing observation, and if the data are analyzed as if no observations were absent, Yates showed that several important properties hold. (i) The estimates of treatment and block effects are exactly the same as those obtained by the correct least squares procedure. (ii) The error s.s. is exactly the same as given by the correct procedure. (iii) To obtain the correct partition of the degrees of freedom, we subtract 1 from the total s.s. and 1 from the error s.s.

Yates also showed that the method of insertion fails to agree with the correct least squares procedure in two respects. The treatments s.s., as obtained in the analysis of variance of the "complete" data, is always slightly larger than the correct treatments s.s. for an F-test of the treatments. Unless an appreciable fraction of the total observations is missing, this overestimation is unlikely to be large; further, the exact F-test can be obtained by means of some additional calculations. The second defect of the method of insertion is that it may not give proper t-tests. That this will happen is clear, because in the analysis of "complete" data r replications are ascribed to the treatment that contains the missing observation, whereas there are only $(r - 1)$ replications. To allow for this disturbance we give special rules which provide t-tests that are approximately correct.

Thus the method adopted with a missing value is first to "estimate" this value by means of the formula B/A, which will be presented for the common designs. This estimate is used in place of the missing value, and the rest of the analysis is conducted as if the data were complete (except for the changes in degrees of freedom). Special methods are available for exact F-tests and for t-tests. If several observations are absent, a repeated application of the formula enables values to be substituted for each missing observation.

This method is essentially an ingenious computational device whose purpose is to enable the easy computations that apply to complete data to be used even when data are incomplete. Substitution of estimates for the missing data does not in any way recover the information that is lost through loss of data, as some experimenters have suggested, usually facetiously; it merely attempts to reproduce the results obtained by an application of the least squares method to the data that are present. The only complete solution of the "missing data" problem is not to have them.

3.8 The Analysis of Covariance

3.81 Purpose of the Technique. As indicated in section 3.2, experiments can be planned so that certain types of environmental effect are eliminated from the estimates of the treatment effects, with the result that these estimates are made more accurate. In the eelworm experiment, where the plots were grouped into blocks of 12 plots each, any differences from block to block in the severity of eelworm infestation were eliminated in this way. On the other hand, differences in infestation *from plot to plot within the same block* are not controlled by the design and do contribute to the experimental errors, since treatments were assigned at random to plots within each block. Accordingly, before the fumigants were applied, samples were taken in order to estimate the natural infestation on each plot.

The analysis of covariance shows how to use these supplementary data to reduce the experimental errors by eliminating the effects of variations in the initial infestation within a block. The technique is potentially very useful. It often happens that some source of variation which cannot be controlled by the design can be measured by taking additional observations. Whenever this is so, the analysis of covariance can be utilized, often to great advantage. One caution is that, since the additional observations are to measure *environmental* effects, they must not be influenced by the treatments. The situation where such measurements are influenced by the treatments is discussed in section 3.88.

3.82 Initial Steps in the Analysis. The first step is to construct a new mathematical model. If y_{ijk} refers to the final eelworm count and x_{ijk} to the initial, the relation is

$$y_{ijk} = \mu + \tau_i + \beta_j + \gamma(x_{ijk} - \bar{x}) + e_{ijk} \tag{3.21}$$

The only change is the introduction of a new term to describe the effect of the initial eelworm number. We have assumed that the effect is linear, i.e., it is a constant multiple γ of the amount by which the initial eelworm number x_{ijk} on the plot differs from the average initial number $\bar{x}$ for the whole experiment.

As before, the unknowns are estimated by least squares. In this case we minimize

$$\sum [y_{ijk} - m - t_i - b_j - c(x_{ijk} - \bar{x})]^2 \tag{3.22}$$

The normal equation for m is again of the form

$$\sum [y_{ijk} - m - t_i - b_j - c(x_{ijk} - \bar{x})] = 0 \tag{3.23}$$

over all observations. If the same linear restrictions as before are applied to the t_i and the b_j, equation (3.23) implies that m is the mean of the y_{ijk}. The equation for t_i is the same as (3.23) except that the sum is over those observations that receive t_i. The equation may be rearranged as

$$r_i m + r_i t_i = T_{iy} - c(T_{ix} - r_i \bar{x}) \tag{3.24}$$

where r_i is the number of replications, and T denotes a treatment total. This gives

$$m + t_i = \frac{T_{iy}}{r_i} - c\left(\frac{T_{ix}}{r_i} - \bar{x}\right) \tag{3.25}$$

The important feature of this equation is that in order to obtain t_i, the observed treatment mean of the y's (T_{iy}/r_i) is *adjusted*, the adjustment depending on the treatment mean of the x's. It is this adjustment that removes the effect of the initial infestation.

The equation for c, the adjustment factor, will not be developed in full. It turns out that c is given by the ratio of the error sum of products of y and x to the error s.s. of x.

3.83 Computations. In practice we start with a joint analysis of the sums of squares and products of y and x. In order to illustrate certain features of the covariance technique, the treatments s.s. will be subdivided into "linear" and "curvature" components, as was done for y in table 3.5 (p. 70). The original values for x are given above the y values in table 3.1 (p. 46). The treatment totals and the quantities

TABLE 3.8 TREATMENT TOTALS FOR NUMBERS OF CYSTS *

	Before fumigation (x)					After fumigation (y)				
Level	CN	CS	CM	CK	Total	CN	CS	CM	CK	Total
0		1975			1975		5858			5858
1	402	417	513	570	1902	1066	928	1431	892	4317
2	389	554	568	778	2289	1265	877	1241	1122	4505
2(2) + (1): L	1180	1525	1649	2126	6480	3596	2682	3913	3136	13327
2(1) − (2): Q	415	280	458	362	1515	867	979	1621	662	4129

* Numbers of cysts per plot were the numbers found in 400 grams of soil. Thus the totals in the table represent 1600 grams of soil, except for the no-fumigant totals, which represent 6400 grams.

needed for isolating the linear response and the curvature are reproduced in table 3.8. The analysis of the sum of squares of x follows by the same methods as given previously for y.

To analyze the sum of products, we carry out the same operations as for a sum of squares, except that at every stage a square is replaced by the corresponding product. A few examples will suffice.

$$\text{Total: } (269)(466) + (283)(280) + \cdots + (48)(298) - \frac{(6166)(14{,}680)}{48} = 355{,}929$$

$$\text{Treatments: } \frac{(1975)(5858)}{16} + \frac{(402)(1066)}{4} + \cdots + \frac{(778)(1122)}{4} - \frac{(6166)(14{,}680)}{48} = -9222$$

$$\text{Average linear response: } \frac{(2289 - 1975)(4505 - 5858)}{32} = -13{,}276$$

The complete analysis of covariance is shown in table 3.9.

TABLE 3.9 Sums of squares and products (x = before, y = after, fumigation)

	d.f.	(x^2)	(yx)	(y^2)
Blocks	3	159,618	175,873	289,427
Treatments	8	29,142	−9,222	157,448
Average linear response	1	3,081	−13,276	57,207
Average curvature	1	2,204	8,285	31,140
Differences in linear	3	22,975	−6,837	43,408
Differences in curvature	3	882	2,606	25,693
Error	36	121,408	189,278	544,690
Total	47	310,168	355,929	991,565

If we denote the sums of squares and products for error by E_{xx}, E_{yy}, and E_{yx}, respectively, the adjustment factor c, or regression coefficient of y on x, is given by

$$c = \frac{E_{yx}}{E_{xx}} = \frac{189{,}278}{121{,}408} = 1.559024$$

It may seem surprising that the coefficient is greater than 1. This is probably explained by the great seasonal increase in the eelworm numbers, obvious from table 3.8.

The residual error s.s. may now be found. The original error s.s. is 544,690, with 36 d.f. To remove the effect of the regression on the initial eelworm numbers, we subtract

$$\frac{E_{yx}^2}{E_{xx}} = \frac{(189{,}278)^2}{121{,}408} = 295{,}089$$

$$E_{yy} - \frac{E_{yx}^2}{E_{xx}} = 544{,}690 - 295{,}089 = 249{,}601$$

Thus the residual s.s. is 249,601, with 35 d.f., since 1 d.f. must be subtracted for the additional parameter γ. The residual m.s. s_{yx}^2 is 249,601/35, or 7131. This is less than half the original mean square of 15,130, and indicates that the use of covariance has approximately doubled the accuracy of the experiment.

One feature of the covariance method deserves comment at this point. In the model the effect of x was assumed to be linear, but no assumption was made about the *strength* of the effect. That is left to be determined by the data. If the eelworm numbers on each plot had remained unchanged throughout the season, except for the influence of the treatments, and if the eelworm numbers had been accurately measured, the residual s.s. would have been zero. The residual s.s. as found presumably represents a contribution due to seasonal variations in infestation and one due to the fact that the eelworm numbers were not estimated accurately, being obtained only from small samples of soil. If the variable x had had *no* linear effect on y, the residual m.s. would have been the same as the original mean square, apart from sampling fluctuations.

3.84 The Adjusted Treatment Means. A little time is saved if the adjusted treatment totals are found first. From equation (3.24) the adjustment to the ith treatment total of y is

$$-c(T_{ix} - r_i\bar{x})$$

where r_i is the number of replicates, and $\bar{x}$ the general mean of x (128.46). Thus for the control the adjusted total is

$$5858 - 1.5590(1975 - 16 \times 128.46) = 5858 + 125 = 5983$$

and for $(1CN)$

$$1066 - 1.5590(402 - 4 \times 128.46) = 1066 + 174 = 1240$$

The *means* are shown in table 3.10.

TABLE 3.10 Adjusted treatment means (eelworms per 400 gm. soil)

Level of dressing	Fumigant			
	CN	*CS*	*CM*	*CK*
0		374		
1	310	270	358	201
2	365	204	289	178

3.85 *t*-tests. These tests are slightly complicated by the fact that account must be taken of the sampling error of the adjustment factor c. The difference between two adjusted means may be written

$$\bar{y}_1 - \bar{y}_2 - c(\bar{x}_1 - \bar{x}_2)$$

From regression theory the variance of this quantity is given by

$$\sigma_{y\cdot x}^2\left(\frac{1}{r_1}+\frac{1}{r_2}\right)+(\bar{x}_1-\bar{x}_2)^2\sigma_c^2=\sigma_{y\cdot x}^2\left[\frac{1}{r_1}+\frac{1}{r_2}+\frac{(\bar{x}_1-\bar{x}_2)^2}{E_{xx}}\right] \quad (3.26)$$

where $\sigma_{y.x}^2$ is the residual error variance, and E_{xx} the error s.s. for x. As an estimate of $\sigma_{y.x}^2$ we use the residual error m.s., in this case 7131 with 35 d.f.

Example. A t-test of the reduction to the double dressing of CS. The reduction is (374 − 204), or 170 eelworms per sample. The estimated variance is

$$(7131)\left[\frac{1}{16}+\frac{1}{4}+\frac{(123.44-138.50)^2}{121{,}408}\right]=(7131)(0.31437)=2242 \quad (3.27)$$

the means of the x's being found by division from table 3.8. Hence

$$t=\frac{170}{\sqrt{2242}}=\frac{170}{47.35}=3.59$$

with 35 d.f., which is highly significant.

For a more complicated comparison of the adjusted means,

$$\sum l_i[\bar{y}_i-c(\bar{x}_i-\bar{x})]$$

the estimated variance is

$$s_{y\cdot x}^2\left[\sum\frac{l_i^2}{r_i}+\frac{(\sum l_i\bar{x}_i)^2}{E_{xx}}\right] \quad (3.28)$$

(The term in $\bar{x}$ vanishes because for any comparison the sum of the l's must be zero.)

One annoying feature of these tests is that the x values enter into the variance, so that every comparison necessitates a separate computation of the variance. As a time-saving approximation, Finney (3.10) has proposed that an average value for the contribution of the term in the x's may be used. This amounts to using

$$s_{y\cdot x}^2\left[1+\frac{t_{xx}}{E_{xx}}\right]=(7131)\left[1+\frac{3643}{121{,}408}\right]=(7131)(1.030)=7345$$

as the *effective* residual error m.s., where t_{xx} is the treatments *mean square* for x. The term in brackets represents the average contribution from the x's, or in other words the contribution from sampling errors in the adjustment factor c.

Thus, for the variance of the difference between the adjusted means of the control and $2CS$, we would use

$$(7345)\left(\frac{1}{r_1}+\frac{1}{r_2}\right) = (7345)\left(\frac{1}{16}+\frac{1}{4}\right) = 2295$$

instead of the value 2242 given by the more exact expression (3.27). Similarly, instead of (3.28), we use

$$(7345)\sum\left(\frac{l_i^2}{r_i}\right)$$

This approximation is usually good enough if the number of error degrees of freedom exceeds 20, since in such cases the contribution from errors in c is small. The more exact test may be used if n_e is small.

3.86 *F*-tests. The fact that the same adjustment factor appears in every treatment mean also influences F-tests. The procedure is shown in table 3.11.

TABLE 3.11 F-TEST WITH THE ANALYSIS OF COVARIANCE

					Residuals		
	d.f.	(x^2)	(yx)	(y^2)	d.f.	s.s.	m.s.
Treatments	8	29,142	−9,222	157,448	8	237,192	29,649
Error	36	121,408	189,278	544,690	35	249,601	7,131
T + E	44	150,550	180,056	702,138	43	486,793	

The figures to the left of the vertical line are from the previous analysis (table 3.9), and in practice would not be recopied. Form a new line in the analysis by addition of the items for treatments and error. From this value for (y^2), i.e., 702,138, subtract the contribution due to a regression on x,

$$\frac{(180{,}056)^2}{150{,}550} = 215{,}345$$

The remainder, 486,793, is entered in the column headed "residuals s.s." and carries 43 d.f., 1 being subtracted for the regression. The same process is completed for the error line; actually, this was already done in computing the residual error s.s. The residual s.s. for treatments is found by subtracting that for error from that for treatments + error

$$237{,}192 = 486{,}793 - 249{,}601$$

It always has the same number of degrees of freedom as the original treatments s.s. The F-test of the adjusted treatment means is given by the ratio of the residual m.s.

$$F = \frac{29{,}649}{7131} = 4.16$$

with 8 and 35 d.f.

If it is desired to test some component of the treatments s.s., the same calculation is made with the component in place of the treatments s.s. throughout. If several components are to be tested, this becomes rather tedious. A useful approximation is to construct from the original analysis of covariance in table 3.9 an analysis of $(y - cx)^2$. This is most easily done by multiplying each term in (x^2) by c^2, or 2.43048, each term in (yx) by $-2c$, or -3.11805, and adding the two products to (y^2). The results are shown in table 3.12.

TABLE 3.12 ANALYSIS OF VARIANCE OF $(y - cx)$

	d.f.	s.s.	m.s.	F	F'
Treatments	8	257,032	32,129		
Average linear response	1	106,090	106,090	14.88	14.51
Average curvature	1	10,664	10,664	1.50	1.47
Differences in linear	3	120,566	40,189	5.64	5.05
Differences in curvature	3	19,711	6,570	0.92	0.92
Error	35	249,592	7,131		

It will be noted that this calculation gives the correct residual error m.s. However, sums of squares for components of the treatments are always larger than those given by the more roundabout correct procedure. Thus the F values shown on the right are all too large. The overestimation is seldom great, and the F values serve for a preliminary inspection. Those that are just beyond the significance level may be recomputed by the correct procedure if it is thought worth while. For comparison, the F' values shown above are those obtained by the correct procedure (it is a useful exercise to check them). With either F or F' there is a significant average linear response and significant differences among fumigants in their linear responses, while the curvature terms do not approach significance. In the analysis made without covariance (table 3.5, p. 70) no component was significant.

3.87 The Increase in Accuracy Due to Covariance. From table 3.9 we see that with no covariance the error s.s. for y is 544,690, with 36 d.f., giving an error m.s. of 15,130. A comparable figure when covariance is used is the effective residual m.s., 7345. This is preferable to the residual m.s. itself (7131) because it makes allowance for the sampling error of the adjustment factor c. The accuracy obtained with covariance relative to that without covariance is estimated by 15,130/7345, or 2.06. The use of covariance appears to have had about the same effect as doubling the number of replicates. In making this comparison, we ignored the effect of the reduction in error d.f. from 36 to 35, because it is negligible.

As in ordinary regressions, the x variable may be transformed to another scale if this is likely to produce a more linear relation with y. The use of log x, for instance, is common in biological work. Two or more different x variables may be used. The calculations for this case are described by Snedecor (3.4, chapter 13).

3.88 The Case Where the *x* Variable is Influenced by the Treatments. In a covariance analysis the treatment mean $\bar{y}_i$ is adjusted by the amount $-c(\bar{x}_i - \bar{x})$. The effect of the adjustment is to change each $\bar{y}_i$ to the value that it would be expected to have if all treatments had the same x mean. It is in this way that the technique removes the effect of variations in the $\bar{x}_i$. If, however, the differences among the $\bar{x}_i$ are in part produced by the treatments, the adjustment removes part of the treatment effect and its interpretation is changed. Consequently, in the standard use of covariance, it is important to be sure that the treatments did not affect the x values. Sometimes this is obvious, as in the eelworm experiment, because the x values were recorded before the treatments were applied. A more doubtful case is that of the number of plants per plot in a field experiment, counted after the application of the treatments, which may or may not influence plant numbers. An F-test of the x values is helpful in such cases.

Where the treatments do affect x, a covariance analysis may add information about the way in which the treatments produced their effects. In the eelworm experiment, the yields of the oats which were grown on the plots were obtained. Since eelworms attack oats, it would be interesting to know whether the effects of the treatments on the oats were simply a reflection of their effects on the eelworms. This is examined by a covariance analysis in which the oats yields are the y values, and the eelworm numbers after harvest are the x values. If the F-test of the adjusted treatment means still shows significance, the conclusion is that not all the treatment effect on the oats can be attributed to the reduction in eelworm numbers. This happened in the present instance, because some treatments supplied nitrogen to the crop, and therefore acted in part as fertilizers as well as fumigants. As Bartlett (3.11) has pointed out, the interpretation of this use of covariance requires care, since a hidden extrapolation may be involved.

When the x variables are influenced by treatments, the short-cut method for t-tests by use of the effective residual error and the approximate F-tests by means of an analysis of $(y - cx)$ should be avoided, since they may be seriously in error.

3.9 Effects of Errors in the Assumptions Underlying the Analysis of Variance

The assumptions made in the analysis of variance are that treatment and environmental effects are additive, and that the experimental errors are independently distributed in the normal distribution, with a common variance. In practice we can never be sure that these assumptions all hold, and often there is good reason to suspect that some are false. The consequences of failures in the assumptions and the remedial steps to be taken have been summarized by Eisenhart (3.12), Cochran (3.13), and Bartlett (3.14). Only a few comments will be given here.

As a rule, the failure of an assumption will affect both the significance levels and the sensitivity of F- and t-tests. When the experimenter thinks that he is testing at the 5% level, he may actually be testing at the 8% level. Usually, though not invariably, the true significance probability is larger than the apparent one; that is, too many significant results are obtained. Also, there is usually a loss of sensitivity, in the sense that a more powerful test than the analysis of variance F-test could be constructed if the correct mathematical model were known. There is a corresponding loss of accuracy in the estimates obtained for the treatment effects, since these, too, could be made more accurate if the correct model were known.

Although generalization is hazardous, experience suggests that in the majority of experiments, at least in the field of biology, these disturbances are not sufficiently great to invalidate the technique. They do imply, however, that significance levels and confidence limits must be considered approximate rather than exact. For the same reason, the inflexible use of say the 5% significance level to divide the effects into those that are regarded as "real" and those that are not is hardly justifiable.

The most serious disturbances appear to arise when the experimental error variance is not constant over all observations. Sometimes this happens, as mentioned previously, because certain treatments are erratic in their effects. In such cases, the appropriate error variance for comparing one pair of treatments might be four times as large as that for another pair, and the use of the same estimated variance for both comparisons would lead to t-tests that were completely erroneous. Where this type of disturbance is suspected, the remedy is to divide the error s.s. into components each of which is homogeneous, as indicated in section 3.63.

The same problem may arise because the experimental errors follow a distribution that is decidedly skew. In such distributions the error variance for a treatment tends to be a function of the mean produced by the treatment. If the nature of the functional relationship is known, a transformation can be found that will place the data on a scale on which the error variance is more nearly constant. This transformation is then made on the observations before starting the analysis. The principal transformations that have been found useful are discussed by Bartlett (3.14); they include logs, square roots, and (for data expressed in fractions) inverse sines.

Such transformations may also be useful in cases where treatment and environmental effects are not additive. If, for instance, a treatment increases all observations by 20%, irrespective of the initial level, a change to logs will introduce additivity. When transformations are made for this purpose, it should be realized that they will also affect the distribution of the experimental errors. Fortunately, it often happens that such transformations also bring the distribution of errors closer to normality.

Finally, the assumption that the errors are independent from observation to observation may be obviously untenable. It is well known that crop yields on neighboring plots tend to be positively correlated, and in laboratory experiments observations made by the same person at about the same time tend to exhibit the same type of correlation. These correlations might completely vitiate tests of significance. The remedy in this case is the proper use of randomization, which, as it were, introduces independence in the assignment of treatments to the experimental units or in the assignment of the order in which observations are made, so that the errors may effectively be regarded as independent. For further discussion of this question, see Yates (3.15), Fisher (3.3), Bartlett (3.16), and Welch (3.17).

REFERENCES

3.1 *Rothamsted Experimental Station Report*, Harpenden, Herts, England, p. 176, 1935.

3.2 Fisher, R. A. *Statistical methods for research workers.* Oliver and Boyd, Edinburgh, 10th ed., chapters VII and VIII, 1946.

3.3 Fisher, R. A. *The design of experiments.* Oliver and Boyd, Edinburgh, 4th ed., 1947.

3.4 Snedecor, G. W. *Statistical methods.* Iowa State College Press, Ames, Iowa, 4th ed., chapters 10 and 11, 1946.

3.5 Kendall, M. G. *The advanced theory of statistics.* Charles Griffin and Co., London, Vol. 2, chapters 23 and 24, 1946.

3.6 Fisher, R. A. On a distribution yielding the error functions of several well-known statistics. *Proc. Inter. Math. Cong.*, 805–813, 1924.

3.7 Fisher, R. A. Applications of Student's distribution. *Metron* V, 90–104, 1926.

3.8 Yates, F. Orthogonal functions and tests of significance in the analysis of variance. *Jour. Roy. Stat. Soc. Suppl.* 5, 177, 1938.

3.9 Yates, F. The analysis of replicated experiments when the field results are incomplete. *Emp. Jour. Exp. Agr.* 1, 129–142, 1933.

3.10 Finney, D. J. Standard errors of yields adjusted for regression on an independent measurement. *Biom. Bull.* 2, 53–55, 1946.

3.11 Bartlett, M. S. A note on the analysis of covariance. *Jour. Agr. Sci.* 26, 488, 1936.

3.12 Eisenhart, C. The assumptions underlying the analysis of variance. *Biometrics* 3, 1–21, 1947.

3.13 Cochran, W. G. Some consequences when the assumptions for the analysis of variance are not satisfied. *Biometrics* 3, 22–38, 1947.

3.14 Bartlett, M. S. The use of transformations. *Biometrics* 3, 39–52, 1947.

3.15 Yates, F. The formation of latin squares for use in field experiments. *Emp. Jour. Exp. Agr.* 1, 235–244, 1933.

3.16 Bartlett, M. S. The effect of non-normality on the t-test. *Proc. Camb. Phil. Soc.* 31, 223, 1935.

3.17 Welch, B. L. On the z-test in randomized blocks and latin squares. *Biometrika* 29, 21–52, 1937.

3.1a *Biometrika tables for statisticians.* Vol. I. Table 29. Cambridge University Press, 1954.

3.2a *Ibid.*, table 26.

3.3a Dunnett, C. W. A multiple comparison procedure for comparing several treatments with a control. *Jour. Amer. Stat. Assoc.* 50, 1096–1121, 1955.

3.4a Scheffé, H. A method for judging all contrasts in the analysis of variance. *Biometrika* 40, 87–104, 1953.

3.5a Duncan, D. B. Multiple range and multiple F tests. *Biometrics* 11, 1–42, 1955.

3.6a Federer, W. T. *Experimental design.* Macmillan Co., New York, 1955.

3.7a Hartley, H. O. Some recent developments in analysis of variance. *Comm. on Pure and App. Math.* 8, 47–72, 1955.

ADDITIONAL READING

Anderson, R. L., and Bancroft, T. A. *Statistical theory in research.* McGraw-Hill, New York, 1952.

Anderson, R. L., and Houseman, E. E. Tables of orthogonal polynomial values extended to $N = 104$. *Iowa Agr. Exp. Sta. Res. Bull.* 297, 1942.

Anscombe, F. J. The validity of comparative experiments. *Jour. Roy. Stat. Soc. A*, 111, 182–211, 1948.

Beall, G. The transformation of data from entomological field experiments so that the analysis of variance becomes applicable. *Biometrika* 32, 243–262, 1942.

BOSE, S. S., and MAHALANOBIS, P. C. On estimating individual yields in the case of mixed-up yields of two or more plots in field experiments. *Sankhya* 4, 103–120, 1938.

BOX, G. E. P. Effects of inequality of variance and of correlation between errors in the two-way classification. *Ann. Math. Stat.* 25, 484–498, 1954.

COCHRAN, W. G. Analysis of variance for percentages based on unequal numbers. *Jour. Amer. Stat. Assoc.* 38, 287–301, 1943.

CORNISH, E. A. Analysis of covariance in quasi-factorial designs. *Ann. Eugen.* 10, 269–279, 1940.

DE LURY, D. B. The analysis of covariance. *Biometrics* 4, 153–170, 1948.

FEDERER, W. T., and SCHLOTTFELDT, C. S. The use of covariance to control gradients in experiments. *Biometrics* 10, 282–290, 1954.

KEMPTHORNE, O. *The design and analysis of experiments.* John Wiley and Sons, New York, 1952.

NAIR, K. R. The application of the technique of analysis of covariance to field experiments with several missing or mixed-up plots. *Sankhya* 4, 581–588, 1940.

PEARCE, S. C. Randomized blocks with interchanged and substituted plots. *Jour. Roy. Stat. Soc. B*, 10, 252–256, 1948.

PEARSON, E. S. The analysis of variance in cases of non-normal variation. *Biometrika* 23, 114–133, 1931.

PITMAN, E. J. G. Significance tests which may be applied to samples from any populations. III. The analysis of variance test. *Biometrika* 29, 322–335, 1939.

TUKEY, J. W. One degree of freedom for non-additivity. *Biometrics* 5, 232–242, 1949.

WALKER, H. M. Degrees of freedom. *Jour. Ed. Psych.* 31, 253–269, 1940.

WARD, G. C., and DICK, I. D. Non-additivity in randomized block designs and balanced incomplete block designs. *N. Z. Jour. Sci. Tech.* 33, 1952.

CHAPTER 4

Completely Randomized, Randomized Block, and Latin Square Designs

4.1 Completely Randomized Designs

4.11 Description. The simplest type of layout is that in which treatments are allotted to the units entirely by chance. More specifically, if a treatment is to be applied to four units, for example, the randomization gives every group of four units in the experimental material an equal probability of receiving the treatment. In addition the units should be processed in random order at all subsequent stages of the experiment where this order is likely to affect the results.

This design has several conveniences:

1. Complete flexibility is allowed. Any number of treatments and of replicates may be used. The number of replications can be varied at will from treatment to treatment (though such variation is not recommended without good reason). All the available experimental material can be utilized—an advantage in small preliminary experiments where the supply of material is scarce.

2. The statistical analysis is easy even if the numbers of replicates are not the same for all treatments or if the experimental errors differ from treatment to treatment.

3. The method of analysis remains simple when the results from some units or from whole treatments are missing or are rejected. Moreover, the relative loss of information due to missing data is smaller than with any other design.

The principal objection to a completely randomized design is on the grounds of accuracy. Since the randomization is not restricted in any way to ensure that the units which receive one treatment are similar to those which receive another treatment, the whole of the variation among the units enters into the experimental error. For this reason the error can often be reduced by the use of a different design, unless the units are highly homogeneous or the experimenter has no information by which to arrange or handle the units in more homogeneous groups.

Complete randomization seems the obvious procedure for many laboratory experiments, e.g., in physics, chemistry, or cookery, where a quantity of material, after thorough mixing, is divided into small samples or batches to which the treatments are applied. On the other hand, these designs are seldom used in field experiments, the method of randomized blocks having been found consistently more accurate.

One fact compensates to some extent for the higher experimental errors as compared with other designs. For a given number of treatments and a given number of experimental units, complete randomization provides the maximum number of degrees of freedom for the estimation of error. As pointed out in section 2.31, the sensitivity of the experiment increases as the number of error degrees of freedom is increased. Consider, for example, an experiment with 2 treatments in 4 replicates. The degrees of freedom for error are 6 under complete randomization as against 3 if the method of pairing (section 4.2) is used. From section 2.31 it appears that the paired experiment must produce about a 14% reduction in the error variance in order to offset the additional unreliability in the estimate of error. This point is worth bearing in mind with small experiments.

To summarize, complete randomization may be appropriate (i) where the experimental material is homogeneous, (ii) where an appreciable fraction of the units is likely to be destroyed or to fail to respond, and (iii) in small experiments where the increased accuracy from alternative designs does not outweigh the loss of error degrees of freedom.

4.12 Randomization. If the number of units does not exceed 16, tables 15.6 and 15.7 may be used. Suppose that there are 3 treatments, of which two have 4 replicates while the third has 8 replicates. Numbers 1–16 are assigned to the units in any convenient order. A random permutation is drawn from table 15.7, say

9, 13, 8, 5, 12, 1, 14, 16, 6, 7, 3, 4, 10, 11, 15, 2

The first treatment is applied to units 9, 13, 8, 5; the second to units 12, 1, 14, 16; and the third to the remainder.

With more than 16 units in the experiment, numbered discs, beans, or a book of random numbers may be used, as described in chapter 15.

4.13 Statistical Analysis. Table 4.1 shows some of the results of an experiment on the effects of applications of sulphur in reducing scab disease of potatoes. The object in applying sulphur is to increase the acidity of the soil, since scab does not thrive in very acid soil. In addition to untreated plots which serve as a control, 3 amounts of dressing were compared—300, 600, and 1200 lb. per acre. Both a fall and a spring application of each amount was tested, so that in all there were 7 distinct treatments. The sulphur was spread by hand on the surface of the soil, and then disced in to a depth of about 4 inches. The quantity to be

TABLE 4.1 FIELD PLAN AND SCAB INDICES FOR A COMPLETELY RANDOMIZED EXPERIMENT ON POTATOES

F3 9	0 12	S6 18	F12 10	S6 24	S12 17	S3 30	F6 16
0 10	S3 7	F12 4	F6 10	S3 21	0 24	0 29	S6 12
F3 9	S12 7	F6 18	0 30	F6 18	S12 16	F3 16	F12 4
S3 9	0 18	S12 17	S6 19	0 32	F12 5	0 26	F3 4

Notation: F = fall, S = spring application, 0 = control. The numbers 3, 6, 12 are the amounts of sulphur in 100 lb. per acre.

Results grouped by treatments

	0	F3	S3	F6	S6	F12	S12	
	12 30	9	30	16	18	10	17	
	10 18	9	7	10	24	4	7	
	24 32	16	21	18	12	4	16	
	29 26	4	9	18	19	5	17	
Totals	181	38	67	62	73	23	57	$G = 501$
Means	22.6	9.5	16.8	15.5	18.2	5.8	14.2	

Analysis of variance

Source of variation	d.f.	s.s.	m.s.	F
Treatments	$(t - 1) = 6$	972.3	162.0	3.61 *
Error	$(N - t) = 25$	1122.9	44.9	
Total	$(N - 1) = 31$	2095.2		

* Denotes significance at the 5% level.

analyzed is the "scab index." This is, roughly speaking, the percentage of the surface area of the potato that is infected with scab. It is obtained by examining 100 potatoes at random from each plot, grading each potato on a scale from 0 to 100% infected, and taking the average.

The design was a completely randomized one, with 4 replications of each sulphur dressing and 8 replications of the control, which received extra replication in order to obtain a fairly good estimate of the natural infestation. Incidentally, a randomized blocks layout might have been superior.

The computations are very simple. Let y denote an observation, T_i a treatment total, G the grand total, r_i the number of replications of the ith treatment, $N = \sum r_i$ the total number of observations, and t the number of treatments.

Step 1. Find the treatment totals and the grand total.

Step 2. The sums of squares are computed as follows.

$$\text{Correction factor: } C = \frac{G^2}{N} = \frac{(501)^2}{32} = 7843.8$$

$$\text{Total: } \sum y^2 - C = (9)^2 + (12)^2 + \cdots + (4)^2 - 7843.8 = 2095.2$$

$$\text{Treatments: } \sum \frac{T_i^2}{r_i} - C = \frac{(181)^2}{8} + \frac{(38)^2 + (67)^2 + \cdots + (57)^2}{4} - 7843.8 = 972.3$$

$$\text{Error: (Total s.s.)} - \text{(treatments s.s.)} = 2095.2 - 972.3 = 1122.9$$

The analysis of variance is given at the foot of table 4.1. The F-ratio for treatments is significant at the 5% level. From the treatment means it appears that all dressings had some beneficial effect, and that the fall application was more effective than the spring one. There is little or no evidence that the higher dressings were more effective than the lowest dressing. If the summary is conducted from this point of view, we should isolate and test two individual components of the treatments s.s.: (i) a component measuring the average effect of all dressings, (ii) a component comparing the fall and spring applications. The following computations are required.

Average effect of sulphur. The total over all dressings is 320, representing 24 plots. Since the control total, 181, represents 8 plots, the comparison is

$$3(181) - 320 = 223$$

The contribution to the sum of squares in the analysis of variance is $(223)^2/96$, or 518.0. By the rules in section 3.42, the divisor, 96, is $[(9)(8) + 24]$.

Fall versus spring application. The comparison is

$$(38 + 62 + 23 - 67 - 73 - 57) = -74$$

It is obviously orthogonal to the previous comparison. The contribution to the sums of squares is $(74)^2/24$, or 228.2. These calculations lead to the analysis of variance in table 4.2.

The remaining four components of the treatments s.s. must represent comparisons among the levels of sulphur. The average reduction in scab due to sulphur is significant at the 1% level, while the superiority of the fall application is also significant. Differences among the levels show no sign of significance.

TABLE 4.2 SUBDIVISION OF THE TREATMENTS SUM OF SQUARES IN THE EXPERIMENT ON POTATO SCAB

Source of variation	d.f.	s.s.	m.s.	F
Treatments	6	972.3	162.0	3.61 *
Control vs. sulphur	1	518.0	518.0	11.54 **
Fall vs. spring application	1	228.2	228.2	5.08 *
Comparisons among levels	4	226.1	56.5	1.26
Error	25	1122.9	44.9	

* Denotes significance at the 5% level.
** Denotes significance at the 1% level.

The conclusions might be phrased as follows. "The application of sulphur produced a significant decrease in the scab index, the averages being 22.6 for the untreated plots, 16.4 for plots with the spring application, and 10.2 for plots with the fall application. The fall application proved significantly better than the spring application. There was no indication that the higher levels of dressing were more effective than the lowest level." It should be remarked that a continuation of this experiment and other experiments caused these conclusions to be modified, the higher levels having shown to better effect.

A more extensive discussion of the analysis of experiments of this type is given by Snedecor (4.1). Note that, if certain observations are missing or have to be rejected, the method of computation remains exactly the same except for the slight complication that different treatments will usually have different numbers of replications.

The practice of calculating a separate standard error for each treatment is still found in some types of experimentation. Unless the treatments have different error variances, the use of a pooled error, as given by the analysis of variance, is recommended since more sensitive tests of significance are obtained. In cases where the experimental error appears to vary considerably from treatment to treatment, a test of homogeneity of the error variance, due to Bartlett (4.2), may be made. This test is also described in Snedecor (4.1, section 10.13).

It is worth emphasizing that, even if the treatments are randomly assigned to the units, the above methods of analysis do not apply when restrictions on the randomization are introduced at a later stage in the experiment. Consider a chemical experiment in which samples of liquid from the same bottle are treated in four different ways. The samples, being mutually indistinguishable, are assigned at random to the treatments. It is, however, practicable to treat only a few samples during each work period, and the investigator decides to complete a single replication at each session. The result of this additional control is that the differences among replicates are eliminated from the experimental errors and must equally be eliminated from the estimated errors. The analysis should follow the method described in section 4.2. This point will arise frequently in experiments where many physical operations must be carried out.

4.14 Standard Errors. The estimated standard error of the difference between two treatment means is

$$s_d = \sqrt{s^2\left(\frac{1}{r_1} + \frac{1}{r_2}\right)}$$

This formula applies when a pooled error is used, s^2 being the pooled error m.s. per unit, and r_1 and r_2 the numbers of replicates for the two treatments. If $r_1 = r_2 = r$, the formula reduces to $\sqrt{\frac{2s^2}{r}}$. The degrees of freedom for t-tests are those in s^2.

Example. The average effect of the sulphur has already been tested by an F-test. We will perform the same test by means of a t-test applied to the treatment means. Since the control has 8 replications, and the mean of all sulphur dressings has 24 replications,

$$s_d = \sqrt{(44.9)(\tfrac{1}{8} + \tfrac{1}{24})} = 2.736$$

The mean scab index for the control is 22.62, and that for the dressings is 13.33. Hence

$$t = \frac{22.62 - 13.33}{2.736} = 3.395$$

with 25 d.f. The value of t^2 is 11.53, in agreement with the F found in the analysis of variance.

If the true error variances per unit are considered to be different for the two treatments, the appropriate formula is

$$s_d = \sqrt{\frac{s_1^2}{r_1} + \frac{s_2^2}{r_2}}$$

where s_1^2, s_2^2 are the respective error m.s. per unit for the two treatments

In this case the ratio of the treatment difference to s_d does not follow Student's t-distribution except in special instances. The development of correct significance levels for this case presents a problem that has stimulated much discussion in recent years. Various methods that have been suggested agree closely in practical application except when the numbers of degrees of freedom in s_1^2 and s_2^2 are small. Fisher and Yates (4.3, tables V1 and V2) give tables of the significance levels of d/s_d derived from Fisher's theory of fiducial probability. (Note that their s_1^2 corresponds to our s_1^2/r_1, etc.) Where these tables are not readily accessible, the following approximation is suggested. It probably errs slightly on the conservative side, in the sense that the value of t required for significance may be slightly too high.

Let n_1, n_2 be the numbers of degrees of freedom in s_1^2, s_2^2, respectively. From the ordinary t-table record the significance levels t_1, t_2 corresponding to n_1 and n_2 degrees of freedom, respectively. The approximate significance level for the ratio d/s_d is

$$t' = \frac{w_1 t_1 + w_2 t_2}{w_1 + w_2}$$

where

$$w_1 = \frac{s_1^2}{r_1}; \qquad w_2 = \frac{s_2^2}{r_2}$$

Since t' always lies between the ordinary t values for n_1 and n_2 degrees of freedom, this calculation is needed only for those occasional cases where d is close to the borderline of significance. Further, when $n_1 = n_2 = n$, t' is the ordinary t value for n degrees of freedom.

Example. This experiment is a case where one might suspect that the error variance would not be homogeneous. If a treatment is very successful in reducing scab, the scab indices for that treatment will all be close to zero; hence their variance must be small. With a treatment such as the control the scab index has a greater possible range of variation. Consequently, we might expect the error variance for a treatment to depend on the mean produced by the treatment. With only 3 d.f. for the estimated error variance per treatment, we cannot hope to detect small differences in the true error variances. Perhaps the most important comparison is that between the variance for the control and the average variance for the sulphur dressings. The error m.s. will be found to be 70.0 for the control (with 7 d.f.) and 35.2 for the dressings (with 6×3 or 18 d.f.). The F-ratio, 1.99, falls just short of the 10% level of significance.

Consequently, the conclusions will be on a sounder basis if we ascribe different true error variances to the control and the dressings. For a

t-test of the average effect of sulphur we now take

$$s_d = \sqrt{\frac{70.0}{8} + \frac{35.2}{24}} = \sqrt{8.75 + 1.47} = 3.197$$

The value of t is 9.29/3.197, or 2.906. To test this at the 1% level by the approximate test above we have

$$n_1 = 7; \quad n_2 = 18; \quad t_1 = 3.499; \quad t_2 = 2.878; \quad w_1 = 8.75; \quad w_2 = 1.47$$

Hence the value required for significance is

$$t' = \frac{(8.75)(3.499) + (1.47)(2.878)}{10.22} = 3.410$$

The average effect of sulphur now falls short of the 1% significance level. It remains significant at the 5% level. For comparisons among the different sulphur dressings we would use ordinary t-tests, where s^2 is taken as 35.2 with 18 d.f. The effect will be to enhance the significance of comparisons among the dressings, though the conclusions quoted previously are not altered.

As mentioned in section 3.9, it is usually better to handle this type of experiment, where the error variance is thought to be a function of the mean, by converting the original data to a scale on which the error variances will be more nearly constant. On the supposition that the error variance may be proportional to the mean, the square roots of the scab indices might have been analyzed. As a matter of interest this analysis is shown in table 4.3; square roots were recorded to only one decimal place.

TABLE 4.3 ANALYSIS OF VARIANCE OF THE SQUARE ROOTS OF THE SCAB INDICES

Source of variation	d.f.	s.s.	m.s.	F
Treatments	6	18.22	3.04	4.19
Control vs. sulphur	1	8.17	8.17	11.27
Fall vs. spring application	1	4.59	4.59	6.33
Comparisons among levels	4	5.46	1.36	1.88
Error	25	18.12	0.725	
Error for controls	7	6.27	0.896	
Error for dressings	18	11.85	0.658	
Total	31	36.34		

The error m.s. for the controls is now closer to that for the dressings, the F-ratio being 1.36 against 1.99 in the original analysis. Further, the F-ratios for most components of the treatments s.s. are higher than in the original analysis, suggesting an increase in the sensitivity of the analysis.

4.15a Statistical Analysis with Data Arranged in Two Classes. In many completely randomized experiments, the data are not measured on a continuous scale, but are merely classified into two classes, e.g., "success" or "failure," "dead" or "alive." In this event the statistical analysis is carried out by the standard χ^2 tests for a $2 \times t$ contingency table. Table 4.1a shows the results of an experiment at the Eversley Childs Sanatorium, Cebu, Philippine Islands, in which 6 drugs for the treatment of leprosy were compared (4.1a). At the beginning of the experiment and at the end of 48 weeks of treatment each patient was examined by a physician who did not know which drug the patient was receiving. The data for each treatment are the numbers of patients classified as "improved" or "not improved." Initially the experiment was arranged in randomized blocks (section 4.21), but the grouping into blocks appeared to be without effect.

TABLE 4.1a NUMBERS OF PATIENTS "IMPROVED" AND "NOT IMPROVED"

	Treatment						
	A	*B*	*C*	*D*	*E*	*F*	Total
Imp. (a_i)	16	13	15	8	15	2	69 (Σa_i)
Not	38	34	37	45	34	50	238
Total	54	47	52	53	49	52	307
Proportion improved (p_i)	0.2963	0.2766	0.2885	0.1509	0.3061	0.0385	0.22476 ($\bar{p}$)

The proportion p_i of patients improved under each treatment is shown at the foot of the table. These are the figures by which the clinical effectiveness of a treatment is judged.

The χ^2 test for this 2×6 contingency table is the analogue of the F-test with continuous data: it tests the null hypothesis that there are no differences in the true proportions improved under the different treatments. The value of χ^2 is often computed from the observed (O) and the expected (E) numbers in each cell, using the formula

$$\chi^2 = \sum (O - E)^2/E$$

Since the proportions improved p_i must be calculated in order to look at the treatment effects, the equivalent formula of Brandt and Snedecor is more expeditious (4.1).

$$\chi^2 = \frac{\sum a_i p_i - \bar{p} \sum (a_i)}{\bar{p}\bar{q}}$$

where a_i = number improved under the ith treatment.

$\bar{p} = \frac{69}{307}$ = overall proportion improved.

$\bar{q} = 1 - \bar{p}$.

For these data,

$$\chi^2 = \frac{(16)(0.2963) + (13)(0.2766) + \cdots + (2)(0.0385) - (69)(0.22476)}{(0.22476)(0.77524)}$$

$$= 17.40 \text{ (5 d.f.)}$$

This is significant at the 5% level. With this method the p_i must be carried to extra decimal places to avoid loss of accuracy in χ^2.

Note that if all treatments have the same number r of replicates, the Brandt-Snedecor formula simplifies to

$$\chi^2 = \sum (a_i - \bar{a})^2/r\bar{p}\bar{q}$$

The next step is to compare the individual treatments. This is done by breaking down the contingency table into smaller tables. Treatments A and B were two types of sulfone drug; in treatments C and E the active ingredient was dihydrostreptomycin sulphate; treatment D was the combination of streptomycin plus sulfone; and treatment F was a control that did not contain any active agent against the leprosy bacillus. A rational series of comparisons is as follows:

Active drugs vs. Control	$(A \cdots E)$ vs. F
Sulfones vs. Strep. vs. Combined	(A, B) vs. (C, E) vs. D
Between Sulfones	A vs. B
Between Streps.	C vs. E

The contingency tables required for this breakdown appear in table 4.2a. The value of χ^2 is computed *separately* for each table and is shown under the table. The correction for continuity should be used in all 2×2 tables, but not in larger tables. Note that these χ^2 values will not add up to the total χ^2 of 17.40, even though all 5 d.f. have been accounted for.

The results are that the combined drugs $A \cdots E$ show a significantly higher percentage improvement than the control ($\chi_c^2 = 11.21$, 1 d.f.). No differences in effectiveness among the drugs can be detected. The latter result is fairly obvious from inspection of the original table 4.1a, and the complete breakdown is given here primarily to exhibit the method.

TABLE 4.2a BREAKDOWN OF CONTINGENCY TABLE FOR DETAILED COMPARISONS

Active drugs vs. Control

	$A \cdots E$	F	Total
Imp.	67	2	69
Not	188	50	238
Total	255	52	307

$\chi_c^2 = 11.21$ (1 d.f.)

Sulfones vs. Streps. vs. Combined

	A, B	C, E	D	Total
Imp.	29	30	8	67
Not	72	71	45	188
Total	101	101	53	255

$\chi^2 = 4.34$ (2 d.f.)

Between Sulfones

	A	B	Total
Imp.	16	13	29
Not	38	34	72
Total	54	47	101

$\chi_c^2 = 0.00$ (1 d.f.)

Between Streps.

	C	E	Total
Imp.	15	15	30
Not	37	34	71
Total	52	49	101

$\chi_c^2 = 0.00$ (1 d.f.)

If the treatments represent increasing amounts of a substance, the primary interest usually lies in the curve of relation between the p_i and the corresponding amounts x_i. When this relation is likely to be approximately linear, we may wish to subdivide χ^2 into a single d.f. representing the linear regression and a remainder representing deviations from the regression. The same device is useful when the treatments are ordered (e.g. mild, moderate, and intense background noise under which some task has to be performed) if it is reasonable to assign scores x_i to the levels of treatment.

For the ith treatment, let $p_i = a_i/n_i$. To compute the single d.f. χ^2 for regression, find the quantities

$$N = \Sigma a_i x_i - \frac{(\Sigma a_i)(\Sigma n_i x_i)}{(\Sigma n_i)}$$

$$D = \Sigma n_i x_i^2 - \frac{(\Sigma n_i x_i)^2}{(\Sigma n_i)}$$

These are, respectively, the numerator and denominator of the regression coefficient b of p_i on x_i. Then the χ^2 for linear regression is

$$\chi_1^2 = \frac{N^2}{\bar{p}\bar{q}D}$$

with 1 d.f., where $\bar{p} = (\Sigma a_i)/(\Sigma n_i)$. If the true relation is close to linear, χ_1^2 may reveal a significant relationship when the total χ^2 does not. The remainder, $\chi^2 - \chi_1^2$, with $(t - 2)$ d.f., serves as a test of deviations from linearity, where t is the number of treatments.

4.2 Single Grouping: Randomized Blocks

4.21 Description. The essence of this design is that the experimental material is divided into groups, each of which constitutes a single trial or replication. At all stages of the experiment the object is to keep the experimental errors within each group as small as is practicable. Thus, when the units are assigned to the successive groups, all units which go in the same group should be closely comparable. Similarly, during the course of the experiment, a uniform technique should be employed for all units in the same group. Any changes in technique or in other conditions that may affect the results should be made between groups.

This division into replications need be recognized only at those stages in the conduct of the experiment where the division may help to reduce experimental errors. In agricultural field experiments the division is made at the start when the plots are marked out in the field. Since neighboring plots are known to be more alike in fertility than plots some distance removed, each replicate consists of a compact group of plots and is made approximately square in shape if feasible. Cultivations designed to keep the land clean of weeds will usually be carried out without regard to the replications, because it is not believed that results are affected by the order in which plots are cultivated. Similarly, the plots will generally be harvested in whatever order is most convenient. If, however, harvesting must be spread over a number of days, it is well to harvest the plots replication by replication, in case rainfall or other factors should produce changes in the weight of the crop from day to day. In other types of experimentation no real distinction might be made between replicates until relatively late in the experiment.

The principal advantages of randomized blocks are as follows.

1. By means of the grouping, more accurate results are usually obtained than with completely randomized designs.

2. Any number of treatments and any number of replicates may be included. With the design as described above, each treatment will have the same number of replicates. If extra replication is desired for some treatments, each of these may be applied to two units within every group. This device provides twice the standard number of replicates for the treatments in question, at the expense of some increase in the size of the group, which now contains more than a single replication. Similarly a treatment may be applied three or four times in a group.

3. The statistical analysis is straightforward. Mishaps which necessitate the omission of a complete group or of the entire data from one or more treatments do not introduce any complication in the analysis. When data from some individual units are lacking, the "missing-plot"

technique developed by Yates enables the available results to be fully utilized. Some extra computational labor is, however, involved, and if the gaps are numerous the design is less convenient in this respect than complete randomization.

4. If the experimental error variance is larger for some treatments than for others, an unbiased error for testing any specific combination of the treatment means can still be obtained.

No design is more frequently used than randomized blocks. Certainly if a satisfactory degree of precision is reached, there is little need to search for alternative designs.

It is worth noting that the replication means provide unbiased comparisons of the differences among replicates. Occasionally, these differences measure some property of the experimental material that is of interest. The variance-ratio test of the replications against the error m.s. requires some care in its interpretation. In a greenhouse experiment, for instance, a different type of soil might be used in each replication. Significant differences among replications might be due either to differences in soil type or to differences in the positions of the replications within the greenhouse.

4.22 Randomization. When the units have been grouped, the treatments are assigned at random to the units within each group. A new randomization is made for every group. Unless the number of treatments exceeds 16, tables 15.6 and 15.7 are convenient for this operation.

4.23 Statistical Analysis. The example comes from an experiment carried out by the North Carolina Agricultural Experiment Station at Rocky Mount, N. C., in 1944. The experiment tested the effects of 5 levels of application of potash, supplying respectively 36, 54, 72, 108, and 144 lb. K_2O per acre, on the yield and properties of cotton. The measure chosen for analysis is the Pressley strength index. This is found by measuring the breaking strength of a bundle of fibers of a given cross-sectional area. A single sample of cotton was taken from each plot, and 4 determinations were made on each sample. The figures in table 4.4 are the means of these 4 samples.* The experiment was arranged in 3 randomized blocks of 5 plots each.

To make the computing instructions more general we suppose that there are t treatments and r replicates, and that y denotes a typical observation.

* Since the machine which measures the index is calibrated in arbitrary units, no dimensions are ascribed to the data in table 4.4. The index can be converted approximately into pounds per square inch by means of a regression formula.

Step 1. Find the treatment totals (T_i), the replicate totals (R_j), and the grand total (G).

TABLE 4.4 STRENGTH INDEX OF COTTON IN A RANDOMIZED BLOCKS EXPERIMENT

Treatments	Replications				
Pounds K_2O per acre	1	2	3	Totals	Coded dressing
36	7.62	8.00	7.93	23.55	13
54	8.14	8.15	7.87	24.16	8
72	7.76	7.73	7.74	23.23	3
108	7.17	7.57	7.80	22.54	−7
144	7.46	7.68	7.21	22.35	−17
Totals	38.15	39.13	38.55	115.83	

Analysis of variance

Source of variation	d.f.	s.s.	m.s.	F
Replications	$(r-1) = 2$	0.0971		
Treatments	$(t-1) = 4$	0.7324	0.1831	4.19 *
Error	$(r-1)(t-1) = 8$	0.3495	0.0437	
Total	$rt-1 = 14$	1.1790		

Subdivision of the treatments s.s.

	d.f.	s.s.	m.s.	F
Linear response	1	0.5663	0.5663	12.96**
Deviations	3	0.1661	0.0554	1.27

Step 2. The sums of squares are obtained as follows.

Correction factor: $C = \frac{G^2}{tr} = \frac{(115.83)^2}{15} = 894.4393$

Total: $\sum y^2 - C = (7.62)^2 + (8.14)^2 + \cdots + (7.21)^2 - 894.4393 = 1.1790$

Replications: $\sum \frac{R_j^2}{t} - C = \frac{(38.15)^2 + (39.13)^2 + (38.55)^2}{5} - 894.4393 = 0.0971$

Treatments:

$$\sum \frac{T_i^2}{r} - C = \frac{(23.55)^2 + (24.16)^2 + \cdots + (22.35)^2}{3} - 894.4393 = 0.7324$$

Error: (total s.s.) − (replications s.s.) − (treatments s.s.)

$$= 1.1790 - 0.0971 - 0.7324 = 0.3495$$

The F-ratio for treatments (table 4.4) is 0.1831/0.0437, or 4.19, which is significant at the 5% level with 4 and 8 d.f. The treatment totals suggest that the strength decreases with increasing applications of potash, though there is a hint of a maximum in strength for the 54-lb.

application. Accordingly it is worth while to examine the shape of the response curve. We will first fit a linear regression on the amount of dressing. Note that successive increments of dressing are not equal, the last two increments being twice as large as the first two. This is a fairly common practice in cases where the effectiveness of an increase in dressing is expected to be smaller at the higher levels of application.

Although the regression can be calculated by the standard formula, a simple coding of the dressings lightens the work. If we place the lowest dressing (36 lb.) at zero and take 18 lb. as 1 unit on the scale, the dressings may be coded as 0, 1, 2, 4, and 6, respectively. A further device is to subtract a common amount from these values so that their mean is zero. Since the mean of the coded dressings is $1\frac{3}{5}$, this is the amount that must be subtracted. To avoid fractions, we multiply all values by 5 and subtract 13 from each product. The resulting coded dressings are shown in table 4.4; the signs have been changed so that the regression coefficient will be positive.

The regression coefficient is now obtained as

$$b = \frac{(13)(23.55) + (8)(24.16) + (3)(23.23) - (7)(22.54) - (17)(22.35)}{(3)[(13)^2 + (8)^2 + (3)^2 + (7)^2 + (17)^2]}$$

$$= \frac{31.39}{1740} = 0.0180$$

The factor (3) in the denominator is inserted to convert the treatment totals to means. The regression coefficient represents the average decrease in strength for a unit increase on the coded scale. Since 5 units on the coded scale correspond to 18 lb. K_2O, the decrease in strength for each additional 18 lb. K_2O is estimated as 0.090.

The contribution of the regression to the sum of squares in the analysis of variance is (by rule 1, section 3.42)

$$\frac{(31.39)^2}{1740} = 0.5663$$

as shown in the analysis of variance. The contribution from the linear regression is significant at the 1% level. Since the mean square for deviations from the regression, 0.0554, is only slightly above the error m.s., there seems no point in investigating a quadratic regression, though the reader may care to verify that its contribution is small.

The conclusion from the data is that increased dressings of potash produce a weaker fiber, the strength index declining by 0.090 for each 18 lb.-increment in K_2O. As an exercise we will obtain confidence limits for this rate of decline. We first require the estimated standard error for the 0.090 figure. This can be found by expressing the figure as a linear function of the treatment *means* $\bar{y}_i$ and applying rule 5*a* of section 3.51. We have

$$0.090 = \tfrac{5}{580}[13\bar{y}_1 + 8\bar{y}_2 + 3\bar{y}_3 - 7\bar{y}_4 - 17\bar{y}_5]$$

so that the estimated standard error is

$$\frac{s}{\sqrt{r}}\sqrt{\sum l_i^2} = \frac{\sqrt{0.0437}}{\sqrt{3}}\,\frac{5}{580}\sqrt{(13)^2 + (8)^2 + \cdots + (17)^2}$$

$$= 5\,\frac{\sqrt{0.0437}}{\sqrt{1740}} = 0.0251$$

Alternatively, this value could be derived by noting that the square root of the F-value for the linear regression, i.e., $\sqrt{12.96}$, or 3.6, is the ratio of 0.090 to its estimated standard error. Thus the latter must be 0.09/3.6, or 0.0250. Finally, for the 80% confidence limits, we have $(0.090 \pm 0.025 \times 1.397)$, or (0.090 ± 0.035), where 1.397 is the value of t for a probability 0.20 and 8 d.f.

4.24 Standard Errors. The estimated standard error of the difference between two treatment *means* is

$$s_d = \sqrt{\frac{2s^2}{r}}$$

If some treatments receive extra replication, the general formula is

$$s_d = \sqrt{s^2\left(\frac{1}{r_1} + \frac{1}{r_2}\right)}$$

In experiments where the error appears to be heterogeneous, a separate error may be obtained for any pair or for any group of treatments. For this a new randomized blocks analysis is carried out on those treatments which belong to the group under consideration.

4.25 Missing Data. The method of analysis when part of the data is missing is described by Yates (4.4). With a single missing unit, the first step is to calculate a value for the unit by means of the formula

$$y = \frac{rB + tT - G}{(r-1)(t-1)} \tag{4.1}$$

where B is the total of the remaining units in the block where the missing unit appears, T is the total of the yields of this treatment in the other blocks, and G is the grand total; r and t are the numbers of replicates and treatments respectively. The analysis of variance is then carried out as usual except that 1 d.f. is subtracted from the total s.s. and from the error s.s.

The standard error of the difference between the mean of the treatment with a missing value and the mean of any other treatment is

$$\sqrt{s^2\left[\frac{2}{r} + \frac{t}{r(r-1)(t-1)}\right]} \tag{4.2}$$

When there are several missing values, for units $a, b, c, d, \cdots$, we first guess values by inspection for all units except a. Formula (4.1) is then used to find an approximation for a. With this approximation and the values previously assumed for $c, d, \cdots$, we again use formula (4.1) to insert an approximation for b. After a complete cycle of these operations, a second approximation is found for a and so on until the new approximations are not materially different from those found previously. The analysis of variance is then completed; for each missing unit, 1 d.f. is subtracted from the total and error s.s.

Suppose that in the previous example two observations had been missing and that the data had appeared as follows.

	Replications			
Pounds K_2O	1	2	3	Totals
36	a	8.00	7.93	15.93
54	8.14	8.15	7.87	24.16
72	7.76	b	7.74	15.50
108	7.17	7.57	7.80	22.54
144	7.46	7.68	7.21	22.35
Totals	30.53	31.40	38.55	100.48

Since differences between replications are not pronounced, we might take as a trial value for a the mean of 8.00 and 7.93, or 7.96. For estimating b_1, the first trial value of b, by formula (4.1) we now have

$$B = 31.40; \qquad T = 15.50; \qquad G = 100.48 + 7.96 = 108.44$$

so that

$$b_1 = \frac{(3)(31.40) + (5)(15.50) - 108.44}{8} = 7.91$$

For estimating a_2, the second trial value of a, we now have

$$B = 30.53; \qquad T = 15.93; \qquad G = 100.48 + 7.91 = 108.39$$

so that

$$a_2 = \frac{(3)(30.53) + (5)(15.93) - 108.39}{8} = 7.86$$

Taking a_2 as 7.86, we find that the next trial value of b is 7.92. This is so close to the previous value that we may stop. Thus $a = 7.86$, $b = 7.92$.

The analysis of variance is now computed with these values inserted. There will be 12 d.f. in the total s.s., and 6 d.f. in the error s.s. The error m.s. is 0.0491.

To obtain a standard error for the comparison of two treatments, A and B, we assign an "effective" number of replicates to each treatment. Any replicate of treatment A is counted as 1 when both A and B are present in the replicate, as $(t - 2)/(t - 1)$ when A is present and B is not, where t is the total number of treatments in the experiment, and as 0 when A is missing. The same rule is applied in scoring B. Suppose that in the example we are comparing the means of the 36- and 72-lb. dressings. Since $t = 5$, the effective replication for each mean is scored as $1\frac{3}{4}$. The standard error of their difference is taken as $\sqrt{s^2(\frac{4}{7} + \frac{4}{7})}$, or $\sqrt{1.143s^2}$. If we are comparing the mean of the 36-lb. dressing with that of the 54-lb. dressing, which has no missing values, the score for the first mean is 2 and that for the second $2\frac{3}{4}$, so that the standard error of their difference is $\sqrt{s^2(\frac{1}{2} + \frac{4}{11})}$, or $\sqrt{0.944s^2}$.

This useful approximation, due to Taylor (4.17a), is a refinement of an earlier rule given by Yates. The exact formulae are laborious to compute. In the example the correct values for the standard errors of the two differences considered above are $1.069s$ and $0.937s$, as compared with the approximate values of $1.068s$ and $0.972s$, respectively.

As a result of the disturbance introduced by the missing units, the treatments m.s. is slightly too large; however, the variance-ratio test is unlikely to be much in error unless a substantial proportion of the units are missing. Yates (4.4) gives the method for obtaining an exact test.

4.26 Estimation of Efficiency. If E_b and E_e are the block and error m.s. and n_b, n_t, and n_e the block, treatment, and error degrees of freedom,

$$E_{c.r.} = \frac{n_b E_b + (n_t + n_e)E_e}{n_b + n_t + n_e} \tag{4.3}$$

is an estimate of the error variance of a completely randomized design with the same experimental material. Comparison of $E_{c.r.}$ and E_e, taking account of the change in numbers of error degrees of freedom (section 2.31), provides an estimate of the increase in accuracy which results from the grouping into replicates.

The result in (4.3) may be proved in various ways. Since there is no complete discussion of results of this type in the literature, one method of proof will be sketched; the details, which are a matter of algebraic manipulation, will not be given completely. This proof uses the properties of randomization. It is not the easiest proof, but requires very few assumptions.

Let the experiment have t treatments and r replications, and let e_i be the experimental error of the observation on the ith unit, and τ_j the effect of the jth treatment. No assumption is made about the nature or distribution of the errors. For this reason it is not necessary to introduce any specific symbol for the effect of the replication, since any type of effect can be represented by appropriate choice of the e's. Tho treatment effect and the error are assumed to be additive; that is, if the randomization happens to put the jth treatment on the ith unit, their joint effect is $(\tau_j + e_i)$. Throughout the randomization the e_i are regarded as a set of fixed numbers, each associated with a specific unit.

Without loss of generality, we may assume that the total of all observations is zero, and that the totals of the e's and the τ's are both zero. Let

$$S = \sum_{i=1}^{rt} e_i^2; \qquad T = \sum_{j=1}^{t} \tau_j^2$$

Further, for all possible randomizations of the randomized blocks design the replications will of course remain the same; consequently, for any given batch of data the replication totals remain unchanged. Let R denote the sum of squares of these totals, divided by t.

The main part of the proof consists in working out the average values of the mean squares in the analysis of variance, taken over all possible randomizations of each type of design. The results come out as shown in table 4.5.

TABLE 4.5 AVERAGE VALUES OF MEAN SQUARES TAKEN OVER THE RANDOMIZATION SETS

Completely randomized

	d.f.	m.s.
Treatments	$(t-1)$	$\dfrac{rT}{(t-1)} + \dfrac{S}{(rt-1)}$
Error	$t(r-1)$	$\dfrac{S}{(rt-1)}$

Randomized blocks

	d.f.	m.s.
Replications	$(r-1)$	$\dfrac{R}{(r-1)}$
Treatments	$(t-1)$	$\dfrac{rT}{(t-1)} + \dfrac{S}{r(t-1)} - \dfrac{R}{r(t-1)}$
Error	$(r-1)(t-1)$	$\dfrac{S}{r(t-1)} - \dfrac{R}{r(t-1)}$

This analysis leads at once to the result. Since

$$n_b = (r - 1); \qquad n_t + n_e = r(t - 1);$$

it is seen from table 4.5 that for randomized blocks the average value of $n_b E_b$ is R, while that of $(n_t + n_e)E_e$ is $(S - R)$. It follows that the average value of

$$\frac{n_b E_b + (n_t + n_e)E_e}{n_b + n_t + n_e} = \frac{R + S - R}{(rt - 1)} = \frac{S}{(rt - 1)}$$

But this is the average of the error m.s. for a completely randomized design on the same data.

As an illustration of the details, consider the treatments m.s. for randomized blocks, which is probably the hardest term. The square of the first treatment total is

$$T_1^2 = (r\tau_1 + e_1 + e_2 + \cdots + e_r)^2$$

When we average over all possible randomizations, there is no contribution from terms in τe_i, since the treatment appears equally often on all units, and the total of the e's is zero. The average value of the contribution from the r terms in e^2 is the sum of the squares of *all* the e's, divided by t, or S/t, since the treatment appears on any specific unit in a fraction $1/t$ of all randomizations.

The contribution from terms in $e_i e_j$ may be written

$$e_1(e_2 + e_3 + \cdots + e_r) + e_2(e_1 + e_3 + \cdots + e_r) + \cdots + e_r(e_1 + e_2 + \cdots + e_{r-1}) \qquad (4.4)$$

The randomization is restricted so that every e comes from a different replication. If the subscript denotes the replicate, the mean value of $e_i e_j$ is $R_i R_j / t^2$, where R_i is the replication total of the e's. Consequently, the mean value of (4.4) is

$$\frac{1}{t^2}[R_1(R_2 + R_3 + \cdots + R_r) + R_2(R_1 + R_3 + \cdots + R_r) + \cdots + R_r(R_1 + R_2 + \cdots + R_{r-1})] = -\frac{1}{t^2}(R_1^2 + R_2^2 + \cdots + R_r^2) = -\frac{R}{t}$$

since the total of the R_i is zero, and since R will be recognized as the sum of squares of the R_i, divided by t. Hence

$$E(T_1^2) = r^2\tau_1^2 + \frac{S}{t} - \frac{R}{t}$$

If this expression is summed for all t treatments and divided by $r(t - 1)$ to give the treatments m.s., the result in table 4.5 follows.

4.27a Statistical Analysis with Data Arranged in Two Classes. The method of analysis when the results consist of data classified into two classes will be presented first for an experiment with two treatments. Consider an experiment in which the objective was to compare the abilities of two culture media to detect the presence of *Salmonella* organisms in specimens of feces. Samples from each of 222 specimens were grown on each medium. For each sample a record was made as to whether presence (+) or absence (−) of the organisms was found. The experiment was thus a randomized blocks experiment with 222 replications, the pair of samples from each specimen constituting a block or replication. Real differences between blocks would be expected, because many specimens may contain no *Salmonella*, in which case both media should give −, while other specimens may have many organisms present, in which case both media are likely to give + even if one of them is much poorer than the other.

At first sight it seems natural to present the data in the usual 2 × 2 table for a test of the difference between two proportions, as in table 4.3a.

TABLE 4.3a NUMBERS OF (+) AND (−) FOR TWO GROWTH MEDIA

	Medium		
	A	*B*	
+	12	7	19
−	210	215	425
Total	222	222	444

The figures to be compared are the proportions of positives, 12/222 and 7/222, for the two media. However, because of the arrangement of the experiment in randomized blocks, these two proportions are positively correlated, so that the ordinary χ^2 test for the difference between two proportions is invalid (except that this test may be an adequate approximation when block effects are small).

The correct test cannot be made from table 4.3a, since this table does not show the results in individual blocks. Presentation of results for the 222 blocks can be considerably condensed, because in any block only four combinations of results for the two media are possible: ++; +−; −+; or −−. It is sufficient to record the numbers of blocks which contain each of these four possibilities. The data obtained for this experiment are shown in table 4.4a (i). The right half of the table (ii) gives an equivalent presentation that is often used. Note that in the right half of the table the proportions 12/222, 7/222 now lie in the margins.

TABLE 4.4a CONDENSED PRESENTATION OF RESULTS FOR INDIVIDUAL BLOCKS

(i)

Combination A	Combination B	No. of blocks
+	+	7
+	−	5
−	+	0
−	−	210
		222

(ii)

B \ A	+	−	
+	7	0	7
−	5	210	215
	12	210	222

To make the test, ignore all blocks showing ++ or −−. These blocks give no clue as to which treatment is superior and, as we have indicated, their results may be due to the nature of the blocks rather than to the performances of the treatments.

This leaves 5 blocks. If A and B are equally effective, the number of blocks in which A is + and B is − should equal that in which A is − and B is +, apart from sampling errors. Consequently we test the observed partition, 5:0, against an expectation of 2½ in each cell by the usual χ^2 test for a single binomial. If a is the number of (+−) blocks and b the number of (−+) blocks, the value of χ^2, corrected for continuity, simplifies to

$$\chi_c^2 = \frac{(a - b - 1)^2}{a + b} = \frac{(4.0)^2}{5} = 3.2 \quad (1 \text{ d.f.})$$

In the numerator, the absolute value of $(a - b)$ is reduced by 1 before squaring. The probability is 0.074.

The exact significance probability can be calculated from the binomial series $(\frac{1}{2} + \frac{1}{2})^n$, as follows. The probability that A should receive all 5 + signs is $1/2^5$ or 1/32. For a two-tailed test, we double this value, giving a probability of 0.062. The difference in the proportions of + results for the two media is not quite significant. This exact test, known as the *sign* test, has numerous applications in other situations (4.3a).

With more than 2 treatments, the data are arranged as on the left of table 4.5a, which shows results of a comparison of 4 growth media for diphtheria bacilli from throat washings of 69 subjects. Four blocks (subjects) gave + on all 4 media; two gave + on A, B, D and − on C, and so on. For a χ^2 test of the hypothesis that there are no differences among treatments, compute the following quantities:

$$T_i = \text{total number of + in } i\text{th treatment}$$

$$B_j = \text{total number of + in } j\text{th block}$$

TABLE 4.5a PRESENTATION OF RESULTS WITH 4 TREATMENTS

	Medium						
	A	*B*	*C*	*D*	No. of blocks	Freq. dist. of no. of + in blocks	
	+	+	+	+	4	B_j	f_j
	+	+	−	+	2	4	4
	−	+	+	+	3	3	5
	−	+	−	+	1	2	1
	−	−	−	−	59	0	59
Total +	6	10	7	10	69		

A frequency distribution of the B_j (made from the right side of table 4.5a) is shown on the right of this table. From this, compute

$$\sum f_j B_j = 33 = \sum T_i \qquad \sum f_j B_j^2 = 113$$

Then, for t treatments,

$$\chi^2 = \frac{t(t-1)\sum (T_i - \bar{T})^2}{t\sum (f_j B_j) - \sum (f_j B_j^2)} = \frac{(4)(3)(12.75)}{(4)(33) - (113)} = 8.05$$

with $(t - 1)$ or 3 d.f. The significance probability is about 0.045. For more detailed comparisons among a subgroup of the treatments, make a separate table containing the results for this subgroup alone, and apply the same test. If there are only two treatments, this χ^2 test reduces to the one given earlier in this section (without the correction for continuity) (4.4a).

4.3 Double Grouping: Latin Squares

4.31 Description. In the latin square the treatments are grouped into replicates in two different ways. Examples of latin squares are shown for different numbers of treatments in plan 4.1 (p. 145). Note that every row and every column of any square is a complete replication. The effect of the double grouping is to eliminate from the errors all differences among rows and equally all differences among columns. Thus the latin square provides more opportunity than randomized blocks for the reduction of errors by skillful planning.

The experimental material should be arranged and the experiment conducted so that the differences among rows and columns represent major sources of variation. Some examples of the uses of latin squares in various fields of research may indicate the utility of the design. In field experiments the plots are usually laid out in a square formation, so that soil fertility and other variations in two directions are controlled. Variations along and across the greenhouse bench may be similarly handled in greenhouse experiments. Occasionally the latin square is advantageous even when the plots form a continuous line. In this case the rows may be compact blocks of land while the columns specify the order within each block. If the yield gradient is suspected to be in the same direction all along the line, blocks and order in blocks together remove the effects of the gradient more thoroughly than a single control. An example of this type is shown below.

TABLE 4.6 A LATIN SQUARE WITH THE PLOTS IN ONE CONTINUOUS LINE

Seven treatments *A*, *B*, *C*, *D*, *E*, *F*, *G*
(Experimental limitations force the plot units to go at right angles to gradient.)

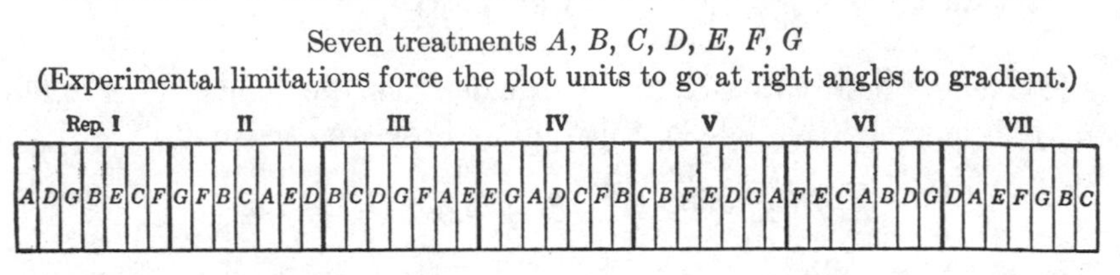

Main and Tippett (4.5) describe how 4 × 4 latin squares were employed in experiments on the weaving of cotton cloth. The purpose of a series of experiments was to investigate the effect of the sizing treatment applied to the warp. The criterion was the number of breaks in the warp during weaving. Four warps, each with a different sizing treatment, were woven simultaneously on 4 different looms, which could be supervised by a single weaver. Then each warp was moved to a different loom of the set so that after 4 periods every sizing treatment had been tested on all 4 looms.

If *A*, *B*, *C*, *D* represent the 4 warps, the latin square used was as follows.

	Looms			
Periods	1	2	3	4
I	*A*	*D*	*B*	*C*
II	*D*	*C*	*A*	*B*
III	*C*	*B*	*D*	*A*
IV	*B*	*A*	*C*	*D*

This arrangement eliminates constant differences among the looms, which were found to be large, and also differences among the 4 periods of weaving. No "period" differences were apparent—a result which might be anticipated, the authors suggest, because humidity and warp tension were controlled throughout the experiment.

Frequently, particularly in industry, an experiment requires a series of operations each of which may introduce variability into the final results. In such cases the latin square may be useful in a preliminary investigation of the sources of variation. For example, in the preparation of an explosive mixture used in primers, variation may occur either in the mixing of the ingredients of the explosive or in the process of charging. One experiment of this type involved 4 mixing-blending teams and 4 charging operators. On each day, the product of each team was sent to a different charging operator, the arrangement being changed daily according to the following 4 × 4 latin square (letters W, X, Y, and Z represent the mixing-blending teams).

	Charging operators			
	1	2	3	4
Monday	W	Z	X	Y
Tuesday	X	W	Y	Z
Wednesday	Y	X	Z	W
Thursday	Z	Y	W	X

The latin square analysis of variance enables us to isolate consistent differences amongst the teams and consistent differences amongst the chargers, as well as day-to-day variations.

As another example of this type, McGehee and Gardner (4.2a) employed a 5 × 5 latin square to measure the effect of factory music on the production of women workers in the operation known as "setting" in rug manufacturing. Four distinct music programs (A, B, C, D) were compared with no music (E). A single replication occupied the 5 working days of the week, a different program being tested on each day. The latin square was used to rotate the programs from week to week, so that over a 5-week period each program appeared once on any specific working day of the week, as shown below.

Week	Mon.	Tues.	Wed.	Thurs.	Fri.
1	A	B	C	D	E
2	B	C	D	E	A
3	C	A	E	B	D
4	D	E	A	C	B
5	E	D	B	A	C

Consistent differences in production rates between weeks and between days of the week are eliminated from the errors of the comparisons among programs. The example illustrates how a valid experiment can be conducted under ordinary factory conditions on a question on which, as the authors point out, little data are available although many pronouncements about the effects of music have been made.

As a more complex example, a 12 × 12 latin square was used by Chen, Bliss, and Robbins (4.6) to estimate the toxicities of 12 poisons when applied to cats. Each drug was injected into the femoral vein of the cat at 1 c.c. per minute until the cat died, the amount required being recorded. Since an observer could administer only two drugs at one time, three observers participated, each treating two cats in the morning and two in the afternoon. The rows of the latin square eliminated systematic differences between observers and between the morning and afternoon injections. The columns represented days. All three major variables—day, time of day, and observer—would have inflated the experimental error if left uncontrolled. The plan is shown below (letters represent drugs).

Time	Observer	Day											
		1	2	3	4	5	6	7	8	9	10	11	12
10:30 A.M.	I	*I* *K*	*J* *G*	*B* *J*	*L* *H*	*H* *I*	*G* *B*	*F* *L*	*K* *C*	*D* *E*	*E* *F*	*A* *D*	*C* *A*
	II	*B* *E*	*L* *D*	*G* *F*	*C* *G*	*D* *J*	*J* *K*	*K* *A*	*E* *L*	*H* *C*	*A* *I*	*F* *B*	*I* *H*
	III	*C* *F*	*K* *H*	*A* *K*	*B* *E*	*F* *G*	*L* *C*	*I* *D*	*D* *B*	*G* *A*	*H* *L*	*J* *I*	*E* *J*
2:30 P.M.	I	*J* *D*	*C* *F*	*E* *I*	*K* *A*	*A* *L*	*I* *E*	*H* *C*	*F* *G*	*B* *J*	*G* *B*	*L* *H*	*D* *K*
	II	*A* *H*	*B* *E*	*C* *L*	*D* *J*	*E* *C*	*F* *A*	*G* *B*	*H* *I*	*I* *K*	*J* *D*	*K* *G*	*L* *F*
	III	*G* *L*	*I* *A*	*D* *H*	*F* *I*	*K* *B*	*H* *D*	*J* *E*	*A* *J*	*L* *F*	*C* *K*	*E* *C*	*B* *G*

4.32 Number of Replications. The chief restriction on the utility of the latin square is that the number of replicates equals the number of treatments; if the latter is considerable, the number of replications required becomes impractical. Squares larger than 12 × 12 are seldom used, while the most common range is from the 5 × 5 to the 8 × 8 square. Latin squares also suffer to some extent from the same disadvantage as randomized blocks in that the experimental error per unit is likely to increase with the size of square.

The small squares provide only a few degrees of freedom for the estimation of error—none with the 2 × 2, two with the 3 × 3, and six with the 4 × 4. This fact precludes the use of single 2 × 2 squares, while the 3 × 3 and 4 × 4 squares must produce a substantial reduction in error over randomized blocks or complete randomization to counterbalance the loss of degrees of freedom. More than one square may, however, be included in the same experiment. Three 3 × 3 squares (9 replicates) furnish 10 error d.f., while two 4 × 4 squares give 15 d.f., provided that in each case the "squares × treatments" interaction can be pooled with error. A considerable number of 2 × 2 squares would be required, since the error degrees of freedom are 1 less than the number of squares.

4.33 Randomization. A complete representation of the squares from 4 × 4 to 6 × 6 and sample squares up to the 12 × 12 is given by Fisher and Yates (4.3). For squares up to the 6 × 6, the randomization procedure given in this reference selects (with a minimum of labor) a square at random from all latin squares of a given size. For a discussion of the theoretical basis of the randomization, see Yates (4.7).

Examples of latin squares are shown in plan 4.1. The method of randomization for plan 4.1 is as follows.

3 × 3. Arrange the columns at random and the last 2 rows at random.

4 × 4. Select at random one of the 4 squares. Arrange at random all columns and the last 3 rows. It is equally good, though not strictly necessary, to randomize *all* rows and columns.

5 × 5 and higher squares. Arrange all rows, columns, and treatments independently at random.

For 3 × 3 and 4 × 4 squares, this procedure selects one square at random from all possible squares. For 5 × 5 and larger squares, some types of squares have no chance of being selected if plan 4.1 is used. Unless latin squares are used very frequently, however, the randomization sets are sufficiently large for experimental plans.

4.34 Statistical Analysis. Several experiments have demonstrated that people find it difficult to select, by personal judgment, unbiased samples even from relatively small populations that can be thoroughly inspected before selection. In this experiment, each population consisted of a small area of wheat, containing about 80 shoots, the shoots being slightly over 2 feet high. There were 12 samplers, all experienced in studying the growth of wheat. Each sampler inspected each area and measured the heights of 8 shoots whose heights were to give a representative sample of the shoot heights in the area. The quantity that will be analyzed is the difference between the mean height of the 8 selected shoots and the true mean height in the corresponding area, in other words the sampler's error.

The samplers were divided into two groups of six, of which we will consider only the first group. The samplers represent the experimental treatments. There were 6 areas, each sampled by all 6 samplers in the group. These areas, which serve as the replications, may be taken as the columns of the 6 × 6 latin square. The rows of the square prescribed the order in which each man sampled the 6 areas (see table 4.7). This

TABLE 4.7 SAMPLER'S ERRORS IN SHOOT HEIGHTS (CENTIMETERS) 6 × 6 LATIN SQUARE

Order	Areas						Totals
	1	2	3	4	5	6	
I	F+ 3.5	B+ 4.2	A+6.7	D+6.6	C+4.1	E+3.8	+28.9
II	B+ 8.9	F+ 1.9	D+5.8	A+4.5	E+2.4	C+5.8	+29.3
III	C+ 9.6	E+ 3.7	F−2.7	B+3.7	D+6.0	A+7.0	+27.3
IV	D+10.5	C+10.2	B+4.6	E+3.7	A+5.1	F+3.8	+37.9
V	E+ 3.1	A+ 7.2	C+4.0	F−3.3	B+3.5	D+5.0	+19.5
VI	A+ 5.9	D+ 7.6	E−0.7	C+3.0	F+4.0	B+8.6	+28.4
Totals	+41.5	+34.8	+17.7	+18.2	+25.1	+34.0	+171.3 = G

Totals for samplers

A	B	C	D	E	F
+36.4	+33.5	+36.7	+41.5	+16.0	+7.2

feature served two purposes. It helped to prevent the men from getting in each other's way, since no two men worked on the same area at the same time, and it permitted an examination of the effects of the order of sampling on the errors.

Since only 3 of the 36 errors in table 4.7 are negative, there is evidence of a rather consistent tendency towards overestimation of the shoot heights. For simplicity, the observations will be referred to as measures of bias, though of course they contain a component due to chance fluctuations. There is also an indication, on casual inspection of the table, that the bias varies from one sampler to another. This point will be tested in the analysis of variance. The computations are as follows. First find the totals for each row, column, and latin letter and the grand total, G, shown in table 4.7. The following sums of squares are then calculated.

$$\text{Correction factor: } C = \frac{G^2}{r^2} = \frac{(171.3)^2}{36} = 815.10$$

$$\text{Total: } (3.5)^2 + (4.2)^2 + \cdots + (8.6)^2 - C = 1144.73 - 815.10 = 329.63$$

$$\text{Rows: } \tfrac{1}{6}[(28.9)^2 + \cdots + (28.4)^2] - C = 843.70 - 815.10 = 28.60$$

$$\text{Columns: } \tfrac{1}{6}[(41.5)^2 + \cdots + (34.0)^2] - C = 893.97 - 815.10 = 78.87$$

$$\text{Treatments: } \tfrac{1}{6}[(36.4)^2 + \cdots + (7.2)^2] - C = 970.70 - 815.10 = 155.60$$

The divisor 6 is replaced by r for an $r \times r$ square.

These sums of squares are entered in the analysis of variance. The error s.s. is found by subtraction.

TABLE 4.8 Analysis of variance for a 6 × 6 latin square

Source of variation	d.f.		s.s.	m.s.	F
Rows (order of sampling)	$(r - 1)$	= 5	28.60	5.720	1.72
Columns (areas)	$(r - 1)$	= 5	78.87	15.774	4.74 **
Treatments (samplers)	$(r - 1)$	= 5	155.60	31.120	9.35 **
Error	$(r - 1)(r - 2)$	= 20	66.56	3.328	
Total	$(r^2 - 1)$	= 35	329.63		

To obtain the mean squares, each sum of squares is divided by the corresponding number of degrees of freedom. The F value for samplers is 31.120/3.328, or 9.35, well beyond the 1% level, which is 4.10 for 5 and 20 d.f. This shows that the extent of the overestimation varies from one sampler to another. The means for the samplers are given below.

A	B	C	D	E	F
+6.07	+5.58	+6.12	+6.92	+2.67	+1.20

The estimated standard error of each mean is $\sqrt{s^2/r}$, where s^2 is the error m.s., or in this case $\sqrt{(3.328)/6} = 0.745$. For testing the difference between a pair of means, the standard error is $\sqrt{2}$ (0.745) = 1.053. Since the 5% t value for 20 d.f. is 2.086, the difference between two means must be at least (2.086)(1.053), or 2.20, in order to attain significance at this level. It appears that the samplers fall into two sets—A, B, C, D and E, F, the biases being significantly smaller for E and F than for the others.

Some further information about the nature of the biases can be obtained from an examination of the row and column totals. From table

4.8 the F value for rows is seen to be 1.72, which is below the 5% value, 2.71. No clear effect of the order in which the areas were sampled has therefore been established; further, the row totals (table 4.7) do not suggest any consistent trend.

On the other hand, the F value for the columns m.s., 4.74, is significant at the 1% level, so that the bias varies from area to area. Previous observations had indicated that the bias might depend on the mean shoot height of the area. The column totals are ranked below in order of increasing true mean shoot height.

Area	True mean shoot height (centimeters)	Column total (centimeters)
1	59.0	+41.5
2	66.2	+34.8
6	72.3	+34.0
4	74.5	+18.2
5	76.0	+25.1
3	76.4	+17.7

Apparently on areas where the shoots are not so high the positive bias is greater than on areas with relatively high shoots The result is not surprising, since the sampler had to stoop in order to do his work. The linear regression of the column totals on the true mean heights contributes 59.88 to the sum of squares for columns, 78.87. The mean square for deviations from this regression (4 d.f.) is only 4.75, which is not significantly greater than the error m.s. The inference is that the large value of the mean square for columns is due mainly to the negative correlation between the bias and the mean height. For further discussion of this experiment, see (4.8).

The analysis of two 4×4 squares is discussed in (4.9). A numerical example with five 3×3 squares will be found in (4.10). It should be noted that with several latin squares the interaction of treatments with squares may be tested. If there is no reason to expect that this interaction is real, the corresponding sum of squares may be combined with error.

4.35 Standard Errors. The formula $\sqrt{\frac{2s^2}{r}}$ gives the estimated standard error of the difference between two treatment means, where s^2 is the mean square per unit and r the number of replicates. In cases where heterogeneity of errors is suspected, the error cannot be subdivided with the same ease as in randomized blocks. It is possible to remove from the error s.s. the contribution of any one of the *treatments*, Yates (4.11). For each yield y_i of the treatment, calculate the quantity

$$d_i = R_i + C_i - ry_i$$

where R_i and C_i are the corresponding row and column totals, and r is the number of treatments. The sum of squares of deviations of the quantities d_i, divided by $r(r - 2)$, is the contribution of this treatment to the error s.s. and has $(r - 1)$ degrees of freedom. For illustration, suppose that we wish to calculate the contribution of sampler F to the error in section 4.34. From table 4.7 the first d value is

$$d_1 = (+28.9) + (+41.5) - 6(+3.5) = +49.4$$

and the remaining values will be found to be +52.7, +61.2, +57.5, +29.5, +49.1, with a total +299.4. The sum of squares of deviations from the mean is 610.34. Divided by 24, this gives 25.43, with 5 d.f., for the contribution of sampler F to the error s.s. Consequently the error s.s. (table 4.8) may be divided as shown below.

	d.f.	s.s.	m.s.
Total error	20	66.56	3.328
Contribution of F	5	25.43	5.086
Remainder	15	41.13	2.742

The error m.s. for the suspected treatment may be compared with the mean square for the remaining $(r - 1)(r - 3)$ degrees of freedom. to test whether the treatment shows abnormally high variation. In this example the variance ratio is 1.85, with 5 and 15 d.f. If it is concluded that the treatment is unusually variable, the mean square with $(r - 1)(r - 3)$ degrees of freedom is used as error for the other treatments. It should be stressed that the ordinary variance-ratio tables can be used in this test only if the decision to make the test was taken before examining the individual results. If the test is made merely because one treatment looks anomalous on inspection, different significance levels that take account of this fact are needed, as discussed in section 3.53.

4.36 Missing Values. Except for changes in the formulae, the procedure is similar to that for randomized blocks (section 4.25). A single missing value is substituted by means of the formula, reference (4.4)

$$y = \frac{r(R + C + T) - 2G}{(r - 1)(r - 2)} \tag{4.5}$$

where R, C, T, and G are respectively the totals of the row, column, and treatment which contain the missing value and the grand total. The estimated standard error of the difference between the corresponding treatment mean and the mean of a treatment with no missing values is

$$\sqrt{s^2\left(\frac{2}{r} + \frac{1}{(r-1)(r-2)}\right)} \tag{4.6}$$

When several values are missing, repeated application is made of formula (4.5) as described in reference (4.4). When values have been inserted for all missing data, the usual analysis of variance is calculated. One degree of freedom is subtracted from the error d.f. for each missing value.

The exact formula for the standard error of a treatment mean is complex. Yates (4.4) proposed a useful approximate rule, which shows the number of replicates to be assigned to any treatment mean for a comparison with another treatment mean. Each observation on one treatment is given one replication if the other treatment is present in both the corresponding row and column. The replication is ⅔ when the other treatment is missing in the row *or* column and is ⅓ when the other treatment is missing in both the row and column. When the treatment itself is missing, the replication is 0. Although complicated at first sight, the rule is not difficult to apply and is reasonably accurate. Consider for instance the following 6 × 6 square with 3 units missing (two *B*'s and one *E*) as indicated by the parentheses ( ).

B	*E*	*C*	*F*	*D*	*A*
F	*D*	(*B*)	*E*	*A*	*C*
C	*A*	*E*	*D*	*B*	*F*
(*E*)	*C*	*A*	(*B*)	*F*	*D*
A	*F*	*D*	*C*	*E*	*B*
D	*B*	*F*	*A*	*C*	*E*

Taking the yields of *B* and *E* in the order of the rows, we ascribe the following numbers of replicates.

$$B: \ \tfrac{2}{3} + 0 + 1 + 0 + 1 + 1 = 3\tfrac{2}{3} = \tfrac{11}{3}$$

$$E: \ 1 + \tfrac{1}{3} + \tfrac{2}{3} + 0 + 1 + 1 = 4$$

Hence the standard error of the difference between the treatment means is $\sqrt{s^2(\frac{3}{11} + \frac{1}{4})}$. For the difference between *A* and *B*, the reader may verify that the standard error is $\sqrt{s^2(\frac{3}{14} + \frac{1}{4})}$. Notice that the effective replication for a treatment may change from one comparison to another.

The analysis required when a single row, column, or treatment is missing is described by Yates (4.11). When more than one row, column, or treatment is missing, consult Yates and Hale (4.12). These methods are also presented by DeLury (4.13).

4.37 Estimation of Efficiency. The effectiveness of either the row or the column grouping may be tested from the results of latin squares. The expression

$$E' = \frac{n_r E_r + (n_t + n_e)E_e}{n_r + n_t + n_e}$$

is an estimate of the error m.s. which would have been obtained if the row grouping had not been used, i.e., if the design had been randomized blocks with the columns as blocks. In the formula, E_r and E_e are the mean squares for rows and error, respectively, and n_r, n_t, n_e are the numbers of degrees of freedom for rows, treatments, and error. The increased number of degrees of freedom with randomized blocks is taken into account by the method of section 2.31. The effectiveness of the column grouping is tested similarly.

A 5 × 5 experiment on potatoes will serve as an illustration. The mean squares for the yields of potatoes were as follows.

	d.f.	m.s.
Rows	4	62.6
Columns	4	104.7
Treatments	4	272.0
Error	12	26.2

If the experiment had been conducted in randomized blocks with the columns as blocks, the estimated error m.s. would be

$$\frac{4(62.6) + 16(26.2)}{20} = 33.5$$

The degrees of freedom available for error would increase from 12 to 16. To take account of the advantage of the additional degrees of freedom (section 2.31), we reduce 33.5 by the multiplier (13)(19)/(17)(15), or 0.969. This gives 32.5 as a comparable mean square for randomized blocks. The row grouping increased the information by an estimated 25%. Of course a considerable number of these comparisons would be needed for drawing general conclusions.

4.4 Cross-over Designs

4.41 Description. In dairy husbandry and biological assay a design has been used which closely resembles the latin square but may have some advantages when the number of treatments is small. In the simplest case where there are two treatments A and B, the units are first grouped into pairs as if a randomized block design were to be used. Suppose that from previous knowledge one unit in each pair is expected to give a higher response than the other and that the difference in favor of the superior unit is expected to be about the same in all pairs. It will clearly be advisable to ensure that each treatment is applied to the "better" member in half the replicates and to the "poorer" member in the other half. The pairs or replicates, which must be even in number, are divided *at random* into two equal sets, the first set to receive treatment A on the superior member of each pair, the second to receive B.

The experiment discussed in chapter 1 provides an example. The data (shown in table 4.10) are the times required to compute the sums of squares of 27 observations on each of two machines whose speeds it was desired to compare. Ten sums of squares were calculated, making 10 replications of the experiment. The cross-over design was used because it was thought that the second computation of a sum of squares might be faster than the first, so that for a fair comparison each machine should be used first in five of the ten replications. The randomization assigned machine A to be used first in replications 1, 3, 6, 8, and 9.

The degrees of freedom in the analysis of variance are as follows.

	d.f.
Columns (pairs or replicates)	9
Rows	1
Treatments	1
Error	8
Total	19

Note that only 1 d.f. is assigned to rows. That is, the arrangement removes from error only the *average* difference between the 2 rows. In so far as the difference is not constant but varies from one replication to another, this variation enters into the experimental error.

In some experiments it is known that this difference *will* vary, and it is possible to estimate in advance (at least roughly) whether the difference will be large or small for a given replicate. Suppose that the difference is expected to decrease steadily from replicate 1 to replicate 10. With this knowledge a more accurate design is obtained by the use of five 2×2 latin squares, as shown below.

TABLE 4.9 Five 2×2 latin squares for 2 treatments in 10 replicates

	Square									
Rows	I		II		III		IV		V	
	1	2	3	4	5	6	7	8	9	10
Better	*A*	*B*	*A*	*B*	*B*	*A*	*A*	*B*	*A*	*B*
Poorer	*B*	*A*	*B*	*A*	*A*	*B*	*B*	*A*	*B*	*A*

In this arrangement we remove from error 5 d.f. for rows, one in each square. Thus not only the average difference is removed, but also the variation in the difference from square to square. The combined analysis of variance of the 5 squares is set out as follows.

	d.f.
Squares	4
Columns within squares	5
Rows within squares	5
Treatments	1
Error	4
Total	19

The 9 d.f. for squares and columns within squares are exactly the same as the 9 d.f. for columns in the cross-over design. Owing to the more complete elimination of the row effects, there are only 4 d.f. for error instead of 8 with the cross-over plan.

To summarize, the cross-over design is particularly appropriate when the difference between the rows is substantially the same in all replicates, for in this case the whole of the real difference between rows is concentrated in the single degree of freedom and the error variance is no larger than that of the latin squares. Even if the difference between rows is known to be variable, the cross-over design may be preferable in small experiments where few degrees of freedom are available for error.

In dairy husbandry the cross-over may be used to compare the effects of two feeding rations on the amount and quality of milk produced by the cow. Since cows vary greatly in their milk production, each ration is tested on every cow by feeding it during either the first or the second half of the period of lactation, so that each cow gives a separate replicate. The milk yield of a cow declines sharply from the first to the second half of its period, so that the first half is always "better," in the sense above. Whether a cross-over is superior to a set of latin squares depends on circumstances. The rate of decline is not constant from cow to cow; it is greater in general for high-yielding than for low-yielding cows. Thus if previous production records are available, the cows may be divided into pairs on the basis of yielding ability, each pair being made a separate 2×2 latin square. This plan is likely to give a smaller error than the cross-over, though sometimes not sufficiently smaller to counterbalance the loss of degrees of freedom.

The design can be used with any number of treatments, subject to the restriction that the number of replicates must be a multiple of the number of treatments. With three treatments, for example, a plan can be drawn up from the 3 cycles *ABC, BCA, CAB,* where the order of the letters denotes the row to which each treatment is applied. Each cycle is allotted at random to one-third of the replicates. For higher numbers of treatments a design is constructed in the same way from the columns of any latin square. When the number of treatments exceeds four, however, the degrees of freedom for error are sufficiently large so that a set of latin squares is usually preferable.

4.42 Statistical Analysis. All sums of squares are calculated in the usual way. For the example (table 4.10), the computations are as follows.

$$\text{Correction factor: } \frac{(352)^2}{20} = 6195.2$$

$$\text{Total: } (30)^2 + (14)^2 + (21)^2 + \cdots + (23)^2 - 6195.2 = 910.8$$

$$\text{Columns: } \tfrac{1}{2}[(44)^2 + (42)^2 + \cdots + (47)^2] - 6195.2 = 357.8$$

$$\text{Rows: } \frac{(194 - 158)^2}{20} = 64.8$$

$$\text{Treatments: } \frac{(216 - 136)^2}{20} = 320.0$$

TABLE 4.10 CROSS-OVER EXPERIMENT FOR COMPARING THE SPEEDS OF TWO COMPUTING MACHINES *A* AND *B*

Time (seconds minus 2 minutes) taken to calculate a sum of squares

Rows	Columns (replications)										Totals
	1	2	3	4	5	6	7	8	9	10	
First (poorer)	*A*30	*B*21	*A*22	*B*13	*B*13	*A*29	*B* 7	*A*12	*A*23	*B*24	194
Second (better)	*B*14	*A*21	*B* 5	*A*22	*A*18	*B*17	*A*16	*B*14	*B* 8	*A*23	158
Totals	44	42	27	35	31	46	23	26	31	47	352

Treatment totals: *A* 216; *B* 136

Analysis of variance

Source of variation	d.f.	s.s.	m.s.
Columns (replications)	9	357.8	39.8
Rows (first versus second)	1	64.8	64.8
Treatments	1	320.0	320.0 **
Error	8	168.2	21.0
Total	19	910.8	

The error s.s. is found by subtraction. The F-ratio for treatments is significant at the 1% level.

We will calculate the confidence limits for the true difference in speed which were quoted in section 1.12. The average observed difference is 8.0 seconds in favor of machine B. The estimated standard error of this difference is $\sqrt{2(21.0)/10}$, or 2.049, based on 8 d.f. Consequently the confidence limits for the true difference are

$$8.0 \pm 2.049t'$$

where t' is the value of Student's t corresponding to $(1 - P)$, where P is the confidence probability chosen. For $P = .95$, $t' = 2.306$, and the limits are (8.0 ± 4.7), or 3.3 and 12.7. For $P = .80$, $t' = 1.397$, and, for $P = .99$, $t' = 3.355$, giving confidence limits of (5.1, 10.9) and (1.1, 14.9), respectively, as quoted.

Worked examples for the case where there are two treatments are also given by Brandt (4.14) and Fieller (4.15); the latter also illustrates the analysis of covariance for this design and the procedure when one column is missing.

In general, if there are t treatments and r replicates, the degrees of freedom subdivide as follows.

	d.f.
Columns	$(r - 1)$
Rows	$(t - 1)$
Treatments	$(t - 1)$
Error	$(t - 1)(r - 2)$
Total	$(tr - 1)$

4.43 Standard Errors. The usual formula $\sqrt{\dfrac{2s^2}{r}}$ holds for the standard error of the difference between two treatment means.

The substitution formula for a missing value is

$$y = \frac{rC + t(R + T) - 2G}{(t - 1)(r - 2)} \tag{4.7}$$

where the capital letters refer respectively to the totals of the column, row, and treatment in which the missing value appears and to the grand total. The difference between the mean of the affected treatment and the mean of a treatment with no missing value has a standard error

$$\sqrt{s^2\left[\frac{2}{r} + \frac{t}{r(t - 1)(r - 2)}\right]} \tag{4.8}$$

4.5 Triple Grouping: Graeco-latin Squares

4.51 Description. In this arrangement the treatments are grouped into replicates in three different ways with the consequence that the effects of three different sources of variation are equalized for all treatments.

TABLE 4.11 A 5 × 5 GRAECO-LATIN SQUARE

Litters	Pen (weight groups)				
	a	*b*	*c*	*d*	*e*
I	A_1	B_3	C_5	D_2	E_4
II	B_2	C_4	D_1	E_3	A_5
III	C_3	D_5	E_2	A_4	B_1
IV	D_4	E_1	A_3	B_5	C_2
V	E_5	A_2	B_4	C_1	D_3

The additional grouping (usually represented by greek letters) is denoted here by subscripts. Each treatment ($A, \cdots, E$) appears once in each row and column and once with each subscript ($1 \cdots 5$).

One example is an arrangement proposed by Dunlop (4.16) for testing 5 feeding treatments on pigs. The arrangement requires 5 pigs from each of 5 litters (I, II, III, IV, and V), the effects of litter differences being equalized by making litters correspond to the rows of the square. The animals are housed in 5 pens (*a*, *b*, *c*, *d*, and *e*) which constitute the columns. The columns are used also to control differences in weight; i.e., the heaviest animal in each litter is assigned to the first pen, the next heaviest to the second pen, and so on; consequently the columns eliminate simultaneously the pen differences and the principal variations in weight within litters. Thus far the design is an ordinary latin square.

The subscripts signify the positions of the five feeding crates within each pen. This extra control, though possibly unnecessary, was suggested for several reasons. The first and fifth crates in each pen were of different construction from the other crates; moreover the pigs occupying the end crates have less "company" when feeding.

The graeco-latin square has not been used often, probably because the units can seldom be balanced conveniently in all three groupings. In the example discussed above, the practical difficulty was to secure five litters of which each contained five or more pigs. An interesting industrial application is described by Tippett (4.17).

The designs have been constructed for all numbers of treatments from 3 to 12, except 6 and 10; examples are shown in plan 4.2 (p. 146).

4.52 Randomization. Arrange the rows and columns independently at random. Assign the latin letters and the subscripts at random to their respective classifications.

4.53 Statistical Analysis. The sums of squares for rows, columns, treatments, and subscripts are all obtained in the usual way. For an $r \times r$ square the error has $(r-1)(r-3)$ degrees of freedom. This number is rather inadequate when r is less than 6.

4.6a Designs for Estimating Residual Effects When Treatments Are Applied in Sequence

4.61a Adjustment for Residual Effects. One of the most common uses of cross-over and latin square designs is in experiments in which the different treatments are applied in sequence to the same subject, animal, machine, or plot. The rows of the square represent the successive periods of application, while the columns represent the subjects, animals, machines, or plots. The standard methods of analysis which we have described involve the implicit assumption that there is no residual or carry-over effect of any treatment into the succeeding period. Where there seems to be some risk of residual effects, a common practice is to separate any two periods of treatment by an interval of time long enough for the residual effects to have died out.

To allow a long enough "rest period" is sometimes not feasible, however, or is undesirable on other grounds. In agricultural field experiments we should not want the rest period to be longer than the interval between the harvesting of one year's crop and the sowing of the next crop. In some dairy cow feeding experiments the whole experiment must be completed in one lactation, so that the total time available for treatment periods plus rest periods is limited. And in general, the shorter the rest period, the sooner the experiment is finished.

Another solution is to change the method of analysis of the results so that the treatment means can be adjusted for residual effects. To see what this implies, suppose that the experiment consists of the two 3×3 latin squares shown in table 4.6a.

Besides producing *direct* effects t_a, t_b, and t_c during the period in which they are applied, the treatments also produce *residual* effects r_a, r_b, and r_c, respectively, in the period immediately following the one in which they are applied. Thus for the third period in sequence I, the predicted

TABLE 4.6a A DESIGN FOR THE ESTIMATION OF RESIDUAL EFFECTS

Period	Sequence I	II	III	IV	V	VI
1	*A*	*B*	*C*	*A*	*B*	*C*
2	*B*	*C*	*A*	*C*	*A*	*B*
3	*C*	*A*	*B*	*B*	*C*	*A*

total treatment effect is $(t_c + r_b)$, since treatment C is given in the third period and treatment B in the period immediately preceding. For the second period in this sequence the total treatment effect is likewise $(t_b + r_a)$. In the first period the treatment effect is simply t_a, since it is presumed that all the experimental material was treated alike prior to the start of the experiment. If the quantities t_a, t_b, and t_c are estimated from this model, they provide measures of the direct effects that are free from the disturbance due to residual effects.

The statistical analysis is made easier if any treatment is preceded equally often in the design by each of the other treatments. The design in table 4.6a has this property, every treatment being preceded twice by each of the other treatments. Designs with this property have been called *balanced* with respect to residual effects, although the balance is incomplete because a treatment is never preceded by itself. Williams (4.5a) has shown that if the number of treatments is even, this type of balance can be achieved by the suitable choice of a single latin square. When the number of treatments is odd, two latin squares are required for balance. Balanced designs for 4, 5, and 6 treatments are given in table 4.7a.

TABLE 4.7a DESIGNS BALANCED FOR RESIDUAL EFFECTS

Four treatments

Period				
1	*A*	*B*	*C*	*D*
2	*B*	*C*	*D*	*A*
3	*D*	*A*	*B*	*C*
4	*C*	*D*	*A*	*B*

Six treatments

Period						
1	*A*	*B*	*C*	*D*	*E*	*F*
2	*C*	*D*	*E*	*F*	*A*	*B*
3	*B*	*C*	*D*	*E*	*F*	*A*
4	*E*	*F*	*A*	*B*	*C*	*D*
5	*F*	*A*	*B*	*C*	*D*	*E*
6	*D*	*E*	*F*	*A*	*B*	*C*

Five treatments (use both squares)

Period										
1	*A*	*B*	*C*	*D*	*E*	*A*	*B*	*C*	*D*	*E*
2	*B*	*C*	*D*	*E*	*A*	*C*	*D*	*E*	*A*	*B*
3	*D*	*E*	*A*	*B*	*C*	*B*	*C*	*D*	*E*	*A*
4	*E*	*A*	*B*	*C*	*D*	*E*	*A*	*B*	*C*	*D*
5	*C*	*D*	*E*	*A*	*B*	*D*	*E*	*A*	*B*	*C*

Since the 4 × 4 square provides only 3 d.f. for the experimental error, at least two squares are recommended with this design. With 3 or 5 treatments, the number of squares should be a multiple of 2. Treatments are assigned to letters at random, and the columns (sequences) are randomized. If the experiment has more than one square, the experimenter must decide whether to remove the period effects separately in each square, as is best if period effects are likely to differ from square to square, or to remove only the overall period effects, as in a cross-over design. In the former case the squares are kept separate and columns are randomized within each square: in the latter, all columns are randomized.

4.62a Statistical Analysis When Residual Effects Are Present.

The analysis follows methods developed by Williams (4.5a). The numerical

TABLE 4.8a PLAN AND MILK YIELDS PER PERIOD (IN 10 LB. − 100)

	Sequence (cow)							
Period	I	II	III	Totals	IV	V	VI	Totals
1	*A* 38	*B* 109	*C* 124	271	*A* 86	*B* 75	*C* 101	262
2	*B* 25	*C* 86	*A* 72	183	*C* 76	*A* 35	*B* 63	174
3	*C* 15	*A* 39	*B* 27	81	*B* 46	*C* 34	*A* 1	81
Totals	78	234	223	535	208	144	165	517

Computation of direct and residual effects

	T	R	F	$\hat{T} = 24\hat{t}$	Direct effect means	$\hat{R} = 24\hat{r}$	$\hat{r}$	$\hat{R}'$
A	271	162	399	−383	42.5	−193	−8.0	+46
B	345	137	431	−56	56.1	−100	−4.2	−83
C	436	220	222	+439	76.7	+293	+12.2	+37
	1052	519	1052	0		0	0.0	0

Analysis of variance

Source of variation	d.f.		s.s.	m.s.
Sequences (cows)	$mn - 1$	5	5,781.1	
Periods within squares	$m(n - 1)$	4	11,489.1	
{Direct effects (unadj.)	$n - 1$	2	2,276.8	
{Residual effects (adj.)	$n - 1$	2	616.2	308.1
{Residual effects (unadj.)	$n - 1$	2	38.4	
{Direct effects (adj.)	$n - 1$	2	2,854.6	1,427.3
Error	$(n - 1)(mn - m - 2)$	4	199.2	49.8
Total	$mn^2 - 1$	17	20,362.4	

example in table 4.8a comes from an experiment on the feeding of dairy cows (4.6a), the treatments being as follows:

$$A = \text{Roughage}; \qquad B = \text{Limited grain}; \qquad C = \text{Full grain}$$

The data are the milk yields per period (6 weeks). The original experiment consisted of six 3×3 latin squares, of which two have been chosen to illustrate the computations.

Compute the marginal totals shown in table 4.8a. Additional symbols and formulae needed are as follows:

$$n = \text{number of treatments} = 3$$

$$m = \text{number of squares} = 2$$

For each treatment, compute

T = treatment total

R = total of the yields in periods immediately following the application of this treatment. For treatment A,

$$R = 25 + 27 + 76 + 34 = 162$$

F = total of the sequences (columns) in which this treatment is the final one. For treatment A,

$$F = 234 + 165 = 399$$

If $P_1 = 533$ is the total of all yields in the first period and $G = 1052$ is the grand total, we also require the two quantities

$$P_1 - nG = P_1 - 3G = 533 - 3(1052) = -2623$$

$$nP_1 - (n + 2)G = 3P_1 - 5G = 3(533) - 5(1052) = -3661$$

For each treatment, the values of T, R, and F and the subsequent computations are set out under the plan in table 4.8a. For the estimated direct effect $\hat{t}$ of a treatment, the general formula is

$$mn(n^2 - n - 2)\hat{t} = (n^2 - n - 1)T + nR + F + (P_1 - nG)$$

$$24\hat{t} = 5T + 3R + F - 2623$$

The values of $24\hat{t}$ should add exactly to zero over all treatments. The figures are then divided by 24, and the general mean $1052/18 = 58.44$ is added to each quotient to give the direct effect means, which can be regarded as the treatment means per period, adjusted so as to remove residual effects from the previous period.

The estimated residual effect $\hat{r}$ of a treatment is given by the equation

$$mn(n^2 - n - 2)\hat{r} = nT + n^2R + nF + nP_1 - (n + 2)G$$

$$24\hat{r} = 3T + 9R + 3F - 3661$$

These figures also sum to zero over the treatments. The $\hat{r}$ values are obtained on division by 24. The general mean can be added to each of the $\hat{r}$ values if desired.

It may help to clarify the process if we show how the direct effect means can be obtained from the unadjusted treatment means. For treatment A, the unadjusted mean is 271/6, or 45.2. In terms of the model, this quantity is

$$45.2 = \frac{6 \text{ (direct effect mean)} + 2\hat{r}_b + 2\hat{r}_c}{6}$$

since treatment A is twice preceded by B and twice by C. Hence

$$\text{Direct effect mean} = 45.2 - \tfrac{1}{3}(\hat{r}_b + \hat{r}_c) = 45.2 - \tfrac{1}{3}(-4.2 + 12.2)$$

$$= 42.5$$

as already found.

In the analysis of variance (table 4.8a), the total sum of squares and the sums of squares for sequences (cows) and periods are found in the usual way.

Correction factor: $C = \dfrac{G^2}{mn^2} = \dfrac{(1052)^2}{18} = 61{,}483.6$

Total: $(38)^2 + (109)^2 + \cdots + (1)^2 - C = 20{,}362.4$

Sequences: $\frac{1}{3}[(78)^2 + (234)^2 + \cdots + (165)^2] - C = 5{,}781.1$

Periods: $\frac{1}{3}[(271)^2 + (183)^2 + \cdots + (81)^2] - \frac{1}{9}[(535)^2 + (517)^2]$

$$= 11{,}489.1$$

The sum of squares for Periods is computed above as "BetweenPeriods within squares," since the experiment was planned as a set of latin squares and it was intended to remove period effects separately in each square.

Owing to the fact that the direct and residual effects are to some degree entangled, their sums of squares will not add up correctly unless they are calculated appropriately. The total sum of squares for treatment effects may be computed either as

Direct effects (unadjusted) + Residual effects (adjusted)

or as

Direct effects (adjusted) + Residual effects (unadjusted)

It is best to obtain all four sums of squares, since this supplies a check. The quantities to be squared will be found in the "Computation" section under the plan in table 4.8a.

Direct (unadjusted):

$$\sum \frac{T^2}{mn} - C = \frac{1}{6}[(271)^2 + (345)^2 + (436)^2] - C = 2{,}276.8$$

Residual (adjusted):

$$\frac{\sum \hat{R}^2}{mn^3(n^2 - n - 2)} = \frac{1}{216}[(193)^2 + (100)^2 + (293)^2] = 616.2$$

Direct (adjusted):

$$\frac{\sum \hat{T}^2}{mn(n^2 - n - 1)(n^2 - n - 2)} = \frac{1}{120}[(383)^2 + (56)^2 + (439)^2]$$
$$= 2{,}854.6$$

Residual (unadjusted):

For this we need the auxiliary quantities

$$\hat{R}' = \hat{R} + G - nT = \hat{R} + G - 3T$$

shown on the extreme right of the "Computation" section. The sum of squares is

$$\frac{\sum \hat{R}'^2}{mn^3(n^2 - n - 1)} = \frac{1}{270}[(46)^2 + (83)^2 + (37)^2] = 38.4$$

As a check, note that

$$2{,}276.8 + 616.2 = 2{,}854.6 + 38.4$$

The Error sum of squares is found by subtraction. If only the overall sum of squares for Periods is removed, the d.f. for Error become $(n - 1)(mn - 3)$.

In making F-tests of the direct and residual effects, the adjusted mean squares must be used. In this example the direct effects give a significant F-value, but residual effects do not attain significance, although they did so in the larger experiment from which the data are taken.

4.63a Standard Errors. The standard error of the difference between two direct effect means is

$$\sqrt{\frac{2s^2}{r} \cdot \frac{(n^2 - n - 1)}{(n^2 - n - 2)}} = \sqrt{\frac{2(49.8)}{6} \cdot \frac{(5)}{(4)}} = 4.56$$

where $r = mn$ is the number of replicates of each treatment. For the difference between two residual effect means, we have

$$\sqrt{\frac{2s^2}{r} \cdot \frac{n^2}{(n^2 - n - 2)}} = \sqrt{\frac{2(49.8)}{6} \cdot \frac{(9)}{(4)}} = 6.11$$

4.64a Some Further Developments. Like the latin squares, these designs are restricted as to the number of replications and the number of periods. Patterson (4.10a) has investigated the incomplete block designs that can be balanced with respect to residual effects: these designs will be useful when the number of treatments exceeds the number of periods.

In the statistical analysis given above, adjustment is made only for residual effects from the period immediately preceding. It is assumed that no residual effects persist from treatments applied two periods back. Designs can be constructed that are balanced both for treatments immediately preceding and for treatments applied two periods back (Williams, 4.5a, 4.8a). For n treatments, $(n - 1)$ squares must be used. A completely orthogonal set of latin squares supplies the necessary design. This is a set of $(n - 1)$ squares such that when any two squares are superimposed, each letter of one square occurs once with every letter of the other square. A doubly balanced design also exists for $n = 6$, for which there is no completely orthogonal set of squares.

When the period means exhibit a marked trend, as in the case of dairy cow milk yields within a lactation, Patterson (4.7a) has shown that a more complex mathematical model may be required in order to obtain unbiased estimates of error from the analysis. If the trend in the period means is analyzed into its linear, quadratic, and higher-degree components, the variance from cow to cow of the linear component may be substantially larger than that of the other components. When this happens, the standard error obtained for direct effect means from the analysis given above is subject to a bias, which can be removed by applying the more realistic model. From an examination of data from dairy cow experiments, Lucas (4.9a) concluded that the bias did not appear to be of practical importance in 3×3 experiments, but might be so in 4×4 experiments.

4.65a Designs for More Thorough Investigation of Residual Effects. From the standard error formulae in section 4.63a, it is seen that the standard error of a difference is greater for residual than for direct effects in the ratio $n/\sqrt{(n^2 - n - 1)}$, so that these designs give less precise estimates of residual than of direct effects. This is not undesirable if the experimenter uses the residual effects solely in order to obtain unbiased estimates of the direct effects. In some lines of work, however, estimates of the sizes of the residual effects are of as much interest as estimates of the sizes of the direct effects. For this purpose, another drawback of the designs given above is that there is a positive correlation between the estimated residual and direct effects of a treatment. This implies that the tests of residual and direct effects are not quite independent.

Designs which give independent estimates of direct and residual effects, of approximately equal precision, are obtained by adding an extra period to the designs above (4.13*a*). In the new final period, the treatments that were applied in the previous final period are repeated. Table 4.9a illustrates the design for three treatments, obtained by extending the two 3 × 3 latin squares.

TABLE 4.9a AN ALTERNATIVE DESIGN FOR ESTIMATING RESIDUAL AND DIRECT EFFECTS

	Sequence					
Period	I	II	III	IV	V	VI
1	*A*	*B*	*C*	*A*	*B*	*C*
2	*B*	*C*	*A*	*C*	*A*	*B*
3	*C*	*A*	*B*	*B*	*C*	*A*
4	*C*	*A*	*B*	*B*	*C*	*A*

Any treatment is now preceded equally often by every treatment, including itself. This makes the estimates of direct and residual effects orthogonal. In the notation of section 4.62a, table 4.10a gives the

TABLE 4.10a FORMULAE FOR DIRECT AND RESIDUAL EFFECTS

	Direct effects	Residual effects
Estimated mean	$\hat{t} = \dfrac{(n+1)T - F - G}{mn(n+2)}$	$\hat{r} = \dfrac{nR + P_1 - G}{mn^2}$
s.s. in analysis of variance	$\dfrac{\Sigma[(n+1)T - F - G]^2}{mn(n+1)(n+2)}$	$\dfrac{\Sigma(nR + P_1 - G)}{mn^3}$
s.e. of diff. between 2 means	$\sqrt{\dfrac{2s^2(n+1)}{mn(n+2)}}$	$\sqrt{\dfrac{2s^2}{mn}}$

formulae needed in the analysis. Although direct effects are more precisely estimated than residual effects, the difference in precision is much reduced as compared with previous designs. Note that since direct and residual effects are orthogonal, their sums of squares appear only once in the analysis of variance.

Numerous variations in the design are possible. If the magnitude and duration of the residual effects are of primary interest, we might use a design in which after the first period each treatment is followed for several periods by a standard treatment S or by no treatment. With three treatments, the sequences would run $ASSS$, $BSSS$, $CSSS$. Sometimes interest lies in the *cumulative* effect of repeated application of a treatment: for this, sequences AAA, BBB, CCC are used. The experiment may contain some cumulative sequences and some balanced sequences, giving estimates of direct, residual, and cumulative effects. These designs have been used to some extent in experiments on perennial crops and on crop rotations in agriculture (4.11a, 4.12a).

A special problem arises in dairy cow feeding experiments. In practice, the farmer would probably give a successful feed to the cows during the whole of their lactations. Consequently, the type of experiment that simulates practical conditions is one in which each feed is given during the whole of the lactation. Such experiments tend to have high standard errors, however, because the variation between cows enters into the error. Consequently the latin square design is sometimes used as a substitute that may give more precise results because variation among cows does not contribute to the error. For this purpose, however, the quantity that we should try to estimate from the latin square is not the direct or residual effect, but the *cumulative* effect that would result if a treatment were applied over all periods. Under certain assumptions the average cumulative effect per period is approximated by the sum of the direct and the residual effects.

4.66a Switchback or Reversal Designs. During the period of lactation the milk yield of a cow rises at first but thereafter exhibits a decline. The rate of decline varies markedly from cow to cow. With latin square designs, these variations in rate of decline contribute to the experimental error. For instance, if treatment C is given in the final period to a group of cows that have small declines, while A is given last to cows that have steep declines, the experiment favors C over A.

The switchback design copes with this source of error. To compare two treatments, three periods are used. Half of the animals receive the sequence $A_1B_2A_3$, the other half the sequence $B_1A_2B_3$, where the suffixes denote the periods. For an animal in the first group, the difference per period between A and B is estimated as

$$\frac{A_1 + A_3}{2} - B_2$$

The important point about this estimate is that it is orthogonal to the linear component of the period effects. If the decline is linear, the rate of decline, whether small or large, does not contribute to the error of the comparison of A with B. For an animal in the second group, the treatment difference is estimated as $\{A_2 - (B_1 + B_3)/2\}$, and the same property holds.

With four periods, the double reversal design uses the two sequences $A_1B_2A_3B_4$ and $B_1A_2B_3A_4$. For a cow in the first group, the estimate

$$\frac{A_1 + 3A_3 - 3B_2 - B_4}{4}$$

is orthogonal to both the linear and the quadratic components of the period trend. The treatment comparison is unaffected by that part of the trend which can be represented by a second-degree curve in time.

The analysis of these designs has been described, with examples, by Brandt (4.14) and later by Ciminera and Wolfe (4.14a). Extensions of the designs to more than two treatments are given by Taylor and Armstrong (4.15a) and Lucas (4.16a). Since the designs have a specialized objective, it is advisable to compare their precision with that of the latin square before adopting them as the best tool.

REFERENCES

4.1 Snedecor, G. W. *Statistical methods.* Iowa State College Press, Ames, Iowa, 5th ed., chapter 10, 1956.

4.2 Bartlett, M. S. Some examples of statistical methods of research in agriculture and applied biology. *Jour. Roy. Stat. Soc. Suppl.* 4, 158–159, 1937.

4.3 Fisher, R. A., and Yates, F. *Statistical tables for biological, agricultural and medical research.* Oliver and Boyd, Edinburgh, 3rd ed., 1948.

4.4 Yates, F. The analysis of replicated experiments when the field results are incomplete. *Emp. Jour. Exp. Agr.* 1, 129–142, 1933.

4.5 Main, V. R., and Tippett, L. H. C. The design of weaving experiments. *Shirley Inst. Mem.* 18, 109–120, 1941.

4.6 Chen, K. K., Bliss, C. I., and Robbins, E. B. The digitalis-like principles of *calotropis* compared with other cardiac substances. *Jour. Pharm. and Exp. Ther.* 74, 223–234, 1942.

4.7 Yates, F. The formation of latin squares for use in field experiments. *Emp. Jour. Exp. Agr.* 1, 235–244, 1933.

4.8 Cochran, W. G., and Watson, D. J. An experiment on observer's bias in the selection of shoot-heights. *Emp. Jour. Exp. Agr.* 4, 69–76, 1936.

4.9 Wishart, J. *Field trials: their lay-out and statistical analysis.* Imp. Bur. Plant Breed. and Genetics, Cambridge.

4.10 Paterson, D. D. *Statistical techniques in agricultural research.* McGraw-Hill, New York, 172, 1939.

4.11 Yates, F. Incomplete latin squares. *Jour. Agr. Sci.* 26, 301–315, 1936.

4.12 Yates, F., and Hale, R. W. The analysis of latin squares when two or more rows, columns, or treatments are missing. *Jour. Roy. Stat. Soc. Suppl.* 6, 67–79, 1939.

4.13 DeLury, D. B. The analysis of latin squares when some observations are missing. *Jour. Amer. Stat. Assoc.* 41, 370–389, 1946.

4.14 Brandt, A. E. Tests of significance in reversal or switchback trials. *Iowa Agr. Exp. Sta. Res. Bull.* 234, 1938.

4.15 Fieller, E. C. The biological standardization of insulin. *Jour. Roy. Stat. Soc. Suppl.* 7, 1–64, 1940.

4.16 Dunlop, G. Methods of experimentation in animal nutrition. *Jour. Agr. Sci.* 23, 580–614, 1933.

4.17 Tippett, L. H. C. *Applications of statistical methods to the control of quality in industrial production.* Manchester Stat. Soc., 1934.

4.1a Doull, J. Clinical evaluation studies in lepromatous leprosy, first series. *Int. Jour. Leprosy* 22, 377–402, 1954.

4.2a McGehee, W., and Gardner, J. E. Music in a complex industrial job. *Personnel Psychology* 2, 405–417, 1949.

4.3a Dixon, W. J., and Mood, A. M. The statistical sign test. *Jour. Amer. Stat. Assoc.* 41, 557–566, 1946.

4.4a Cochran, W. G. The comparison of percentages in matched samples. *Biometrika* 37, 257–266, 1950.

4.5a Williams, E. J. Experimental designs balanced for the estimation of residual effects of treatments. *Australian Jour. Sci. Res. A*, 2, 149–168, 1949.

4.6a Cochran, W. G., Autrey, K. M., and Cannon, C. Y. A double change-over design for dairy cattle feeding experiments. *Jour. Dairy Sci.* 24, 937–951, 1941.

4.7a Patterson, H. D. The analysis of change-over trials. *Jour. Agr. Sci.* 40, 375–380, 1950.

4.8a Williams, E. J. Experimental designs balanced for pairs of residual effects. *Australian Jour. Sci. Res. A*, 3, 351–363, 1950.

4.9a Lucas, H. L. Bias in estimation of error in change-over trials with dairy cattle. *Jour. Agr. Sci.* 41, 146–148, 1951.

4.10a Patterson, H. D. The construction of balanced designs for experiments involving sequences of treatments. *Biometrika* 39, 32–48, 1952.

4.11a Cochran, W. G. Long-term agricultural experiments. *Jour. Roy. Stat. Soc. Suppl.* 6, 104–148, 1939.

4.12a Yates, F. The design of rotation experiments. *Commonwealth Bur. Soil Sci. Tech. Comm.* No. 46, 1949.

4.13a Lucas, H. L. Extra-period latin square change-over designs. Not yet published.

4.14a Ciminera, J. L., and Wolfe, E. K. An example of the use of extended cross-over designs in the comparison of *NPH* insulin mixtures. *Biometrics* 9, 431–446, 1953.

4.15a Taylor, W. B., and Armstrong, P. J. The efficiency of some experimental designs used in dairy husbandry experiments. *Jour. Agr. Sci.* 43, 407–412, 1953.

4.16a Lucas, H. L. Switch-back trials for more than two treatments. *Jour. Dairy Sci.* 39, 146–154, 1956.

4.17a Taylor, J. Errors of treatment comparisons when observations are missing. *Nature* 162, 262–263, 1948.

ADDITIONAL READING

GRANT, D. A. The latin square principle in the design and analysis of psychological experiments. *Psych. Bull.* 45, 427–442, 1948.

KEMPTHORNE, O., and BARCLAY, W. D. The partition of error in randomized blocks. *Jour. Amer. Stat. Assoc.* 48, 610–614, 1953.

PLANS

Plan 4.1 Selected latin squares

3 × 3

A B C
B C A
C A B

4 × 4

1

A B C D
B A D C
C D B A
D C A B

2

A B C D
B C D A
C D A B
D A B C

3

A B C D
B D A C
C A D B
D C B A

4

A B C D
B A D C
C D A B
D C B A

5 × 5

A B C D E
B A E C D
C D A E B
D E B A C
E C D B A

6 × 6

A B C D E F
B F D C A E
C D E F B A
D A F E C B
E C A B F D
F E B A D C

7 × 7

A B C D E F G
B C D E F G A
C D E F G A B
D E F G A B C
E F G A B C D
F G A B C D E
G A B C D E F

8 × 8

A B C D E F G H
B C D E F G H A
C D E F G H A B
D E F G H A B C
E F G H A B C D
F G H A B C D E
G H A B C D E F
H A B C D E F G

9 × 9

A B C D E F G H I
B C D E F G H I A
C D E F G H I A B
D E F G H I A B C
E F G H I A B C D
F G H I A B C D E
G H I A B C D E F
H I A B C D E F G
I A B C D E F G H

10 × 10

A B C D E F G H I J
B C D E F G H I J A
C D E F G H I J A B
D E F G H I J A B C
E F G H I J A B C D
F G H I J A B C D E
G H I J A B C D E F
H I J A B C D E F G
I J A B C D E F G H
J A B C D E F G H I

11 × 11

A B C D E F G H I J K
B C D E F G H I J K A
C D E F G H I J K A B
D E F G H I J K A B C
E F G H I J K A B C D
F G H I J K A B C D E
G H I J K A B C D E F
H I J K A B C D E F G
I J K A B C D E F G H
J K A B C D E F G H I
K A B C D E F G H I J

Plan 4.1 (*Continued*)

Selected latin squares

12 × 12

A	B	C	D	E	F	G	H	I	J	K	L
B	C	D	E	F	G	H	I	J	K	L	A
C	D	E	F	G	H	I	J	K	L	A	B
D	E	F	G	H	I	J	K	L	A	B	C
E	F	G	H	I	J	K	L	A	B	C	D
F	G	H	I	J	K	L	A	B	C	D	E
G	H	I	J	K	L	A	B	C	D	E	F
H	I	J	K	L	A	B	C	D	E	F	G
I	J	K	L	A	B	C	D	E	F	G	H
J	K	L	A	B	C	D	E	F	G	H	I
K	L	A	B	C	D	E	F	G	H	I	J
L	A	B	C	D	E	F	G	H	I	J	K

Plan 4.2

Graeco-latin squares

3 × 3

A_1	B_3	C_2
B_2	C_1	A_3
C_3	A_2	B_1

4 × 4

A_1	B_3	C_4	D_2
B_2	A_4	D_3	C_1
C_3	D_1	A_2	B_4
D_4	C_2	B_1	A_3

5 × 5

A_1	B_3	C_5	D_2	E_4
B_2	C_4	D_1	E_3	A_5
C_3	D_5	E_2	A_4	B_1
D_4	E_1	A_3	B_5	C_2
E_5	A_2	B_4	C_1	D_3

7 × 7

A_1	B_5	C_2	D_6	E_3	F_7	G_4
B_2	C_6	D_3	E_7	F_4	G_1	A_5
C_3	D_7	E_4	F_1	G_5	A_2	B_6
D_4	E_1	F_5	G_2	A_6	B_3	C_7
E_5	F_2	G_6	A_3	B_7	C_4	D_1
F_6	G_3	A_7	B_4	C_1	D_5	E_2
G_7	A_4	B_1	C_5	D_2	E_6	F_3

8 × 8

A_1	B_5	C_2	D_3	E_7	F_4	G_8	H_6
B_2	A_8	G_1	F_7	H_3	D_6	C_5	E_4
C_3	G_4	A_7	E_1	D_2	H_5	B_6	F_8
D_4	F_3	E_6	A_5	C_8	B_1	H_7	G_2
E_5	H_1	D_8	C_4	A_6	G_3	F_2	B_7
F_6	D_7	H_4	B_8	G_5	A_2	E_3	C_1
G_7	C_6	B_3	H_2	F_1	E_8	A_4	D_5
H_8	E_2	F_5	G_6	B_4	C_7	D_1	A_3

9 × 9

A_1	B_3	C_2	D_7	E_9	F_8	G_4	H_6	I_5
B_2	C_1	A_3	E_8	F_7	D_9	H_5	I_4	G_6
C_3	A_2	B_1	F_9	D_8	E_7	I_6	G_5	H_4
D_4	E_6	F_5	G_1	H_3	I_2	A_7	B_9	C_8
E_5	F_4	D_6	H_2	I_1	G_3	B_8	C_7	A_9
F_6	D_5	E_4	I_3	G_2	H_1	C_9	A_8	B_7
G_7	H_9	I_8	A_4	B_6	C_5	D_1	E_3	F_2
H_8	I_7	G_9	B_5	C_4	A_6	E_2	F_1	D_3
I_9	G_8	H_7	C_6	A_5	B_4	F_3	D_2	E_1

Plan 4.2 (Continued) Graeco-latin squares

11 × 11

A_1	B_7	C_2	D_8	E_3	F_9	G_4	H_{10}	I_5	J_{11}	K_6
B_2	C_8	D_3	E_9	F_4	G_{10}	H_5	I_{11}	J_6	K_1	A_7
C_3	D_9	E_4	F_{10}	G_5	H_{11}	I_6	J_1	K_7	A_2	B_8
D_4	E_{10}	F_5	G_{11}	H_6	I_1	J_7	K_2	A_8	B_3	C_9
E_5	F_{11}	G_6	H_1	I_7	J_2	K_8	A_3	B_9	C_4	D_{10}
F_6	G_1	H_7	I_2	J_8	K_3	A_9	B_4	C_{10}	D_5	E_{11}
G_7	H_2	I_8	J_3	K_9	A_4	B_{10}	C_5	D_{11}	E_6	F_1
H_8	I_3	J_9	K_4	A_{10}	B_5	C_{11}	D_6	E_1	F_7	G_2
I_9	J_4	K_{10}	A_5	B_{11}	C_6	D_1	E_7	F_2	G_8	H_3
J_{10}	K_5	A_{11}	B_6	C_1	D_7	E_2	F_8	G_3	H_9	I_4
K_{11}	A_6	B_1	C_7	D_2	E_8	F_3	G_9	H_4	I_{10}	J_5

12 × 12

A_1	B_{12}	C_6	D_7	I_5	J_4	K_{10}	L_{11}	E_9	F_8	G_2	H_3
B_2	A_{11}	D_5	C_8	J_6	I_3	L_9	K_{12}	F_{10}	E_7	H_1	G_4
C_3	D_{10}	A_8	B_5	K_7	L_2	I_{12}	J_9	G_{11}	H_6	E_4	F_1
D_4	C_9	B_7	A_6	L_8	K_1	J_{11}	I_{10}	H_{12}	G_5	F_3	E_2
E_5	F_4	G_{10}	H_{11}	A_9	B_8	C_2	D_3	I_1	J_{12}	K_6	L_7
F_6	E_3	H_9	G_{12}	B_{10}	A_7	D_1	C_4	J_2	I_{11}	L_5	K_8
G_7	H_2	E_{12}	F_9	C_{11}	D_6	A_4	B_1	K_3	L_{10}	I_8	J_5
H_8	G_1	F_{11}	E_{10}	D_{12}	C_5	B_3	A_2	L_4	K_9	J_7	I_6
I_9	J_8	K_2	L_3	E_1	F_{12}	G_6	H_7	A_5	B_4	C_{10}	D_{11}
J_{10}	I_7	L_1	K_4	F_2	E_{11}	H_5	G_8	B_6	A_3	D_9	C_{12}
K_{11}	L_6	I_4	J_1	G_3	H_{10}	E_8	F_5	C_7	D_2	A_{12}	B_9
L_{12}	K_5	J_3	I_2	H_4	G_9	F_7	E_6	D_8	C_1	B_{11}	A_{10}

CHAPTER 5

Factorial Experiments

5.1 Description

5.11 A 2^2 Factorial Experiment. In a factorial experiment the effects of a number of different factors are investigated simultaneously. The treatments consist of all combinations that can be formed from the different factors. To illustrate the simplest case, consider an experiment on sugar beet with 2 factors. These were nitrogen, none (n_0) versus 3 cwt. sulphate of ammonia per acre (n_1), and depth of winter ploughing (7 in. versus 11 in.). Ploughing took place in late January, the nitrogen was applied in late April, and the seed was sown early in May. Since both factors occur at 2 levels or variations, the experiment is described as a 2 × 2 factorial experiment. The 4 treatment combinations are shown below, with the mean yields of sugar per acre (cwt.) underneath.

Treatment combinations and yields of sugar (cwt. per acre)

1	2	3	4
(n_0, 7 in.)	(n_1, 7 in.)	(n_0, 11 in.)	(n_1, 11 in.)
40.9	47.8	42.4	50.2

The yields may be placed in the following 2 × 2 table.

Depth	Nitrogen n_0	Nitrogen n_1	Mean	Response to n_1
7 in.	40.9	47.8	44.4	+6.9
11 in.	42.4	50.2	46.3	+7.8
Mean	41.6	49.0	45.3	
11 in. minus 7 in.	+1.5	+2.4		

The results might be summarized as follows. Considering the effect of nitrogen, we might report that the application of nitrogen increased yields by 6.9 cwt. with shallow ploughing and by 7.8 cwt. with deep ploughing. These figures are called the *simple* effects of nitrogen. They represent the type of information that would be wanted, for instance, in giving advice to a farmer who always used shallow ploughing but was doubtful whether to apply nitrogen. For the simple effects of depth of ploughing, we might report that 11 in. ploughing was superior to 7 in. by 1.5 cwt. in the absence of nitrogen and by 2.4 cwt. when nitrogen was applied.

There is another way of looking at the results. It sometimes happens that the effects of the factors are *independent*. By this we mean that the response to nitrogen is the same whether ploughing is shallow or deep, and that the difference between the effects of deep and shallow ploughing is the same whether nitrogen is present or not. In this event the two simple effects of nitrogen, 6.9 cwt. and 7.8 cwt., are estimates of the same quantity and differ only by experimental errors. On this supposition we would naturally average the two figures in order to estimate the response to nitrogen. The average, 7.4 cwt., is called the *main* effect of nitrogen. It can be derived alternatively as the difference between the two column means in the table, 41.6 and 49.0. Similarly the main effect of depth of ploughing (11 in. minus 7 in.) is the average of 1.5 cwt. and 2.4 cwt., or 1.9 cwt. Note that a main effect, being an average of the simple effects, is more precisely estimated than the latter. In this experiment the standard error of a main effect is $1/\sqrt{2}$ times that of a simple effect.

Consequently, if we are sure that the factors operate independently, the summary that was given above in terms of simple effects may be replaced by another that is both more concise and more accurate. This might read as follows. "The application of nitrogen increased the yield of sugar by 7.4 cwt., while 11 in. ploughing increased the yield by 1.9 cwt. as compared with 7 in. ploughing." It is worth repeating that when the factors are independent the figure 7.4 cwt. is the best estimate not only of the *average* response to nitrogen, but also of the response on plots ploughed to 7 in. and of that on plots ploughed to 11 in. In other words, the whole of the information in the experiment is contained in the main effects.

The question arises: How do we know whether the factors are independent? Frequently the answer is suggested by knowledge of the processes by which the factors produce their effects. In the present case an agronomist might reason that deep ploughing should enable the plant to develop a more vigorous root system. With this the plant should be able to utilize more effectively any added nutrient such as nitrogen. Thus he might predict that the response to nitrogen would be greater with deep than with shallow ploughing, though he probably would not expect it to be much greater. In short, he would predict that the two factors would not be quite independent in their effects.

In addition to the information that may be available from such reasoning, a factorial experiment itself provides a test of the assumption of independence. For, if the depth of ploughing does affect the response to nitrogen, the difference between 7.8 cwt. (the response to nitrogen with deep ploughing) and 6.9 cwt. (the response with shallow ploughing) is an estimate of this effect. The difference, 0.9 cwt., can be tested in the usual way by a t-test. If it proves significant, the assumption of independence is rejected by the data. The difference (sometimes divided by a numerical factor) is called the *interaction* between nitrogen and depth of ploughing.

Interchanging the roles of the two factors, we may also consider whether the superiority of deep over shallow ploughing is affected by the presence of nitrogen. To measure the interaction in this case, we subtract 1.5 cwt. (superiority of deep ploughing when no nitrogen is added) from 2.4 cwt. (superiority when nitrogen is added). The difference is again 0.9 cwt. It is easy to see that this equality always holds with a 2×2 experiment. Each difference is equal to the sum of the observations in one diagonal of the 2×2 table, minus the corresponding sum for the other diagonal. Such interactions are called *two-factor*, or *first-order*, interactions.

5.12 Advantages of Factorial Experimentation When Factors Are Independent.

The advantages of factorial experimentation naturally depend on the purpose of the experiment. We suppose for the present that the purpose is to investigate the effects of each factor over some preassigned range that is covered by the levels of that factor which are used in the experiment. In other words the object is to obtain a broad picture of the effects of the factors rather than to find, say, the combination of the levels of the factors that give a maximum response. One procedure is to conduct separate experiments each of which deals only with a single factor. Another is to include all factors simultaneously by means of a factorial experiment.

If all factors are independent in their effects, the factorial approach will result in a considerable saving of the time and material devoted to the experiments. The saving follows from two facts. First, as we have seen, when factors are independent all the simple effects of a factor are equal to its main effect, so that main effects are the only quantities needed to describe fully the consequences of variations in the factor. Secondly, in a factorial experiment, each main effect is estimated with the same precision as if the whole experiment had been devoted to that factor alone. Thus, in the preceding example, half the plots receive nitrogen and half do not. Consequently, the main effect of nitrogen is estimated just as precisely as it would be in a simple experiment of the same size devoted to nitrogen alone. The same result holds for the effect of depth of ploughing. The two single-factor experiments would require *twice* the total number of plots in order to equal the precision obtained by the factorial experiment. If there were n factors, all at two levels and all independent, the single-factor approach would necessitate n times as much experimental material as a factorial arrangement of equal precision. The gain from factorial arrangements in this case is very substantial.

Practical considerations may diminish this gain. The experimenter frequently lacks the resources to conduct a large experiment with many treatments, and must proceed with only one or two factors at a time. Further, it has been pointed out previously that, as the number of treatment combinations in an experiment is enlarged, the standard error per unit increases. This standard error is therefore likely to be higher for a large factorial experiment than for a comparable single-factor experiment. This increase in standard errors can usually be kept small by the device known as confounding, described in chapter 6.

5.13 Factorial Experimentation When Factors Are Not Independent. We assume that the purpose is still to investigate each factor over the range represented by its levels. When factors are not independent, the simple effects of a factor vary according to the particular combination of the other factors with which these are produced. In this case the single-factor approach is likely to provide only a number of disconnected pieces of information that cannot easily be put together. In order to conduct an experiment on a single factor A, some decision must be made about the levels of other factors B, C, D, say, that are to be used in the experiment (e.g., whether all plots should be ploughed 7 in., 9 in., or 11 in. deep in an experiment on nitrogen). The experiment reveals the effects of A for this particular combination of B, C, and D, but no information is provided for predicting the effects of A with any other combination of B, C, and D. With a factorial approach, on the other hand, the effects of A are examined for every combination of B, C, and D that is included in the experiment. Thus a great deal of information is accumulated both about the effects of the factors and about their interrelationships.

In this connection, Fisher (5.1) has pointed out that it is sometimes advisable to introduce into an experiment an extra factor that is not itself of interest, in order that the experiment may form the basis for sounder recommendations about the other factors. In agricultural experimentation in Britain, farmyard manure has served as a subsidiary factor of this kind. Any recommendations made to farmers about other factors will be put to the test in some fields where the farmer has applied manure and in others where he has not, so that it is well to investigate the other factors in both the presence and the absence of manure.

5.14 Summary Comments. To summarize, the following are some instances where factorial experimentation may be suitable:

1. In exploratory work where the object is to determine quickly the effects of each of a number of factors over a specified range.

2. In investigations of the interactions among the effects of several factors. From their nature, interactions cannot be studied without testing *some* of the combinations formed from the different factors. Frequently, information is best obtained by testing all combinations.

3. In experiments designed to lead to recommendations that must apply over a wide range of conditions. Subsidiary factors may be brought into an experiment so as to test the principal factors under a variety of conditions similar to those that will be encountered in the population to which recommendations are to apply.

On the other hand, if considerable information has accumulated, or if the object of the investigation is specialized, it may be more profitable to conduct intensive work on a single factor or on a few combinations of factors. For instance, some investigations are directed towards finding the combination of the levels of the factors that will produce a maximum response. Friedman and Savage (5.2) present arguments to show that a well-planned series of single-factor experiments will often reach the maximum more quickly than a single large factorial experiment. An alternative strategy for this problem, due to Box and Wilson (5.1a), does use experiments in which all the factors appear, but for studying the nature of the response in the neighborhood of the maximum these authors developed special designs that differ from the standard factorials discussed here. Chapter 8A contains a fuller account of this problem.

When the factors to be investigated are numerous, the chief disadvantage of a single factorial experiment lies in its size and complexity. The magnitude of the task can be reduced by having only a single replication (section 6.24). It is even possible (chapter 6A) to obtain most of the desired information by testing only a fraction (e.g. one-half or one-quarter) of the total number of treatment combinations, although this course is not without its risks. The difficulties arising from large experiments should not be considered a criticism of the factorial method, whose efficiency is greatest when the factors are numerous. The basis of the difficulty is that a major research program is being undertaken.

Experimenters sometimes find the results of factorial experiments difficult to interpret, because they appear to present a bewildering variety of treatment comparisons. It is true that the competent summary of a large factorial experiment demands an orderly procedure and often takes considerable time. If the factors are for the most part independent, the method of analysis by means of main effects and interactions (to be illustrated later) will reduce the data to manageable proportions. If the numerous factors interact in a puzzling manner, prolonged study of the results and further experimentation may be needed before the facts are mastered. The trouble in this case is that the phenomena are complex, not that the experimentation is faulty.

5.2 Calculation of Main Effects and Interactions

5.21 Notation for the 2^n Series. The object of section 5.2 is to explain how main effects and interactions are calculated, and how they are represented in the analysis of variance of the results. We begin with the 2^n series, where each factor occurs at only two levels.

For this system the notation used is similar to that of Yates (5.3). Letters A, B, C, $\cdots$ denote the factors. The letters a, b, c, $\cdots$ denote *one* of the two levels at which the corresponding factor occurs; for purposes of clarity this level will be called the second level. The first level is signified by absence of the corresponding letter. Thus the treatment combination *bd*, in a 2^4 factorial experiment, means the treatment combination which contains the first levels of factors A and C, and the second levels of factors B and D. The treatment combination which consists of the first level of *all* factors is denoted by the symbol (1).

The symbol (ab) will denote the *mean* of all observations which receive the treatment combination *ab*. The letters A, B, and AB, when they refer to numbers, will represent, respectively, the main effects of A and B and the A by B interaction.

5.22 The 2^2 Factorial Experiment. As already shown in section 5.11, we have

$$A = \tfrac{1}{2}[(ab) - (b) + (a) - (1)]$$

$$B = \tfrac{1}{2}[(ab) + (b) - (a) - (1)]$$

Yates (5.3) introduces the same multiplier $\frac{1}{2}$ for the interaction, which he defines as

$$AB = \tfrac{1}{2}[(ab) - (b) - (a) + (1)]$$

These quantities and the general mean M are shown in terms of the means for the treatment combinations in table 5.1.

TABLE 5.1 MAIN EFFECTS AND INTERACTIONS EXPRESSED IN TERMS OF INDIVIDUAL TREATMENT MEANS: 2^2 FACTORIAL

Factorial effect	Treatment combination				Divisor
	(1)	(a)	(b)	(ab)	
M	+	+	+	+	4
A	−	+	−	+	2
B	−	−	+	+	2
AB	+	−	−	+	2

The rows of the table express the factorial effects in terms of the original means. If the equations represented by the table are solved for (ab), (b), etc., in terms of M, A, etc., it will be found that the columns of the table enable us to express the original means in terms of the factorial effects. For example, from the column for (ab),

$$(ab) = M + \tfrac{1}{2}[A + B + AB]$$

The only point to remember is that the factor $\frac{1}{2}$ occurs with all terms except M. From these results, simple effects may be calculated from factorial effects. Thus

$$(a) - (1) = \text{simple effect of } A \text{ when } B \text{ is at the } \textit{first} \text{ level} = A - AB$$

$$(ab) - (b) = \text{simple effect of } A \text{ when } B \text{ is at the } \textit{second} \text{ level} = A + AB$$

These relations are useful when an experiment has been summarized in terms of the factorial effects, and it is later desired to estimate some of the simple effects. From the above example it may be noted that the quantity AB measures the error that is committed in estimating the simple effects of A if the two factors are erroneously assumed to be independent.

The functions A, B, and AB satisfy the conditions for an orthogonal set of functions (section 3.42). Consequently the squares of the factorial effects, when suitably multiplied, divide the treatments s.s. into three single components, each with 1 d.f. In practice these components will usually be calculated from the treatment totals $[ab]$, $[b]$, etc., rather than from means. If we define factorial effect *totals* as illustrated below,

$$[A] = [ab] - [b] + [a] - [1]$$

then the sum of squares for A in the analysis of variance is $[A]^2/4r$, where r is the number of replicates. Corresponding formulae hold for the other factorial effects.

5.23 The 2^3 Factorial Experiment. In this case there are *eight* treatment combinations: (1), a, b, c, ab, ac, bc, and abc. The simple effect of A is determined for each of *four* combinations of the other factors: (1), b, c, and bc. As before, the main effect of A is defined to be the average of these four simple effects.

$$A = \tfrac{1}{4}[(abc) - (bc) + (ab) - (b) + (ac) - (c) + (a) - (1)]$$

Similar definitions hold for B and C.

The interaction of A with B is now measured separately at each of the two levels of C.

$$AB \text{ (}C\text{ at second level)} = \tfrac{1}{2}[(abc) - (bc) - (ac) + (c)]$$

$$AB \text{ (}C\text{ at first level)} = \tfrac{1}{2}[(ab) - (b) - (a) + (1)]$$

As would be expected, the quantity AB is taken as the average of these two effects. Thus

$$AB = \tfrac{1}{4}[(abc) - (bc) - (ac) + (c) + (ab) - (b) - (a) + (1)]$$

There are two other first-order interactions, AC and BC, which are defined similarly.

In addition, we encounter a new interaction. Separate estimates were given above for AB at each of the two levels of C. The difference between these two estimates measures the effect of C on the AB interaction. This difference, with the conventional factor $\frac{1}{4}$, is

$$\tfrac{1}{4}[(abc) - (bc) - (ac) + (c) - (ab) + (b) + (a) - (1)]$$

and may be called the interaction of AB with C. If the algebra is carried out, it will be found that the same expression measures the interaction of AC with B, and that of BC with A. Hence the quantity is called the ABC interaction. It is a *three-factor*, or *second-order*, interaction.

Three-factor interactions are more difficult to understand than two-factor interactions. Fortunately, in practice three-factor interactions are often small relative to main effects and two-factor interactions; and quite frequently they can be neglected for the purposes to which the results are to be put. Occasionally, cases arise where they are important. It might happen, for instance, that factor A does not exert any influence unless factors B and C are present in the combination (bc). In this event the interaction ABC is as large as the main effect of A or the interactions AB and AC. The same effect may occur in less extreme cases, where the combination bc is particularly favorable to the response to factor A. If three-factor interactions are found to be substantial, a careful scrutiny of the simple effects is usually helpful in interpretation.

The expressions for the factorial effects in terms of the treatment means are summarized in table 5.2. The rows of the table give the

TABLE 5.2 MAIN EFFECTS AND INTERACTIONS EXPRESSED IN TERMS OF INDIVIDUAL TREATMENT MEANS: 2^3 FACTORIAL

Factorial effect	Treatment combination								Divisor
	(1)	(a)	(b)	(ab)	(c)	(ac)	(bc)	(abc)	
M	+	+	+	+	+	+	+	+	8
A	−	+	−	+	−	+	−	+	4
B	−	−	+	+	−	−	+	+	4
C	−	−	−	−	+	+	+	+	4
AB	+	−	−	+	+	−	−	+	4
AC	+	−	+	−	−	+	−	+	4
BC	+	+	−	−	−	−	+	+	4
ABC	−	+	+	−	+	−	−	+	4

factorial effects in terms of the treatment means, while the columns give the treatment means in terms of the factorial effects. For example,

$$(a) = M + \tfrac{1}{2}[A - B - C - AB - AC + BC + ABC]$$

$$(a) - (1) = A - AB - AC + ABC$$

$$(abc) - (1) = A + B + C + ABC$$

As before, the factor $\frac{1}{2}$ appears with all terms except M in the expression for a treatment mean. The difference between two treatment means, of which two examples are given above, is easily found by noting the signs in the two columns in question.

As the reader may verify, the 7 factorial effects are mutually orthogonal, and each is orthogonal to M. If r is the number of replicates, each factorial effect has the same variance, $\dfrac{8\sigma^2}{4^2 r}$, or $\dfrac{\sigma^2}{2r}$. The contribution of any effect to the sum of squares for treatments is $[\;\;]^2/8r$, where $[\;\;]$ denotes the factorial effect *total*; e.g.,

$$[A] = -[1] + [a] - [b] + [ab] - [c] + [ac] - [bc] + [abc]$$

5.24 The 2^n Factorial Experiment. A few formulae will be given which apply to any number of factors, all at two levels. With n factors, there are n main effects, $n(n-1)/2$ two-factor interactions, $n(n-1)(n-2)/6$ three-factor interactions, and so on. The successive numbers are the coefficients in the expansion of $(1+1)^n$, omitting the first coefficient, unity.

Two equivalent methods for writing out any factorial effect in terms of the original treatment means are available. Each is simple to remember, though the actual writing may take some time in a large experiment. They will be illustrated by finding the $BCDE$ interaction in a 2^6 experiment, with factors A, B, C, D, E, and F.

Rule 1. Evens versus odds. In every factorial effect, half the treatment combinations receive a + sign and half a − sign. Those which receive one sign are those which contain an *even* number of the letters that appear in the factorial effect. In the $BCDE$ interaction 4 letters, b, c, d, and e, appear. There are 8 combinations that contain an even number of letters: (1), bc, bd, be, cd, ce, de, and $bdce$. Each of these can be combined with any one of 4 combinations of the remaining letters a and f, namely, (1), a, f, and af. The 32 terms appear below. Since the terms contain $abcdef$, they receive by convention a + sign.

(1)	*a*	*f*	*af*
bc	*abc*	*bcf*	*abcf*
bd	*abd*	*bdf*	*abdf*
be	*abe*	*bef*	*abef*
cd	*acd*	*cdf*	*acdf*
ce	*ace*	*cef*	*acef*
de	*ade*	*def*	*adef*
bcde	*abcde*	*bcdef*	*abcdef*

The 32 combinations with a − sign are those which have an *odd* number of the letters b, c, d, or e. In detail, these are

b	*ab*	*bf*	*abf*
c	*ac*	*cf*	*acf*
d	*ad*	*df*	*adf*
e	*ae*	*ef*	*aef*
bcd	*abcd*	*bcdf*	*abcdf*
bce	*abce*	*bcef*	*abcef*
bde	*abde*	*bdef*	*abdef*
cde	*acde*	*cdef*	*acdef*

When calculated from the treatment *means*, this difference between the sums for the two groups of 32 combinations is divided by 2^{n-1} to give the $BCDE$ effect.

Rule 2. Algebraic. In this rule the $BCDE$ interaction is expressed formally as

$$BCDE = \frac{1}{2^{n-1}}(a+1)(b-1)(c-1)(d-1)(e-1)(f+1)$$

Note that − signs appear with the factors that enter into the interaction and + signs with those that do not. If this expression is expanded algebraically, it gives the interaction as a linear function of the treatment means.

As in the 2^2 and 2^3 cases, all factorial effects are orthogonal to one another and to the mean. The contribution of any effect to the sum of squares in the analysis of variance is $[\ \ ]^2/2^n r$, where $[\ \ ]$ denotes the effect total.

5.24a Yates' Method of Computing Factorial Effect Totals. Yates (5.3) gives an automatic method of computing all the factorial effect totals without writing down the algebraic expressions for the factorial effects. If all or most of the factorial effects are wanted for study, Yates' method is the most expeditious. If only a minority of the factorial effects are of interest, e.g., the main effects and two-factor interactions in a 2^6 factorial, a separate calculation of these effects may save time.

TABLE 5.1a APPLICATION OF YATES' METHOD TO A 2^4 FACTORIAL m = MANURE, n = NITROGEN, p = PHOSPHORUS, k = POTASSIUM

Yields per plot (total over 6 harvests, km. per 3-meter row)

	Rep. 1	Rep. 2	Rep. 3	Rep. 4	Total
(1)	32	43	27	19	121
m	47	41	48	45	181
n	26	36	24	18	104
mn	61	76	56	64	257
p	29	39	27	28	123
mp	51	34	40	48	173
np	36	31	32	30	129
mnp	76	65	70	63	274
k	35	42	56	35	168
mk	63	41	60	53	217
nk	80	68	75	67	290
mnk	100	68	87	66	321
pk	40	44	53	36	173
mpk	64	39	75	72	250
npk	105	99	74	73	351
$mnpk$	90	82	89	101	362
Total	935	848	893	818	3494

TABLE 5.1a (*Continued*)

Yates' method

	Treatment totals	(1)	(2)	(3)	Effect total (4)		Effect mean
(1)	121	302	663	1362	3494	G	54.6
m	181	361	699	2132	576	M	18.0 **
n	104	296	996	408	682	N	21.3 **
mn	257	403	1136	168	104	MN	3.2
p	123	385	213	166	176	P	5.5 *
mp	173	611	195	516	−10	MP	−0.3
np	129	423	80	188	112	NP	3.5
mnp	274	713	88	−84	−46	MNP	−1.4
Total		3494	4070	4856	5088		
k	168	60	59	36	770	K	24.1 **
mk	217	153	107	140	−240	MK	−7.5 **
nk	290	50	226	−18	350	NK	10.9 **
mnk	321	145	290	8	−272	MNK	−8.5 **
pk	173	49	93	48	104	PK	3.2
mpk	250	31	95	64	26	MPK	0.8
npk	351	77	−18	2	16	NPK	0.5
$mnpk$	362	11	−66	−48	−50	$MNPK$	−1.6
Odds	1459	1642	2312	2192	5704		
Evens	2035	2428	2544	2896	88		
Total	3494	4070	4856	5088	5792		

** Double asterisk denotes significance at the 1% level.
* Single asterisk denotes significance at the 5% level.

The upper half of table 5.1a shows the yields of a 2^4 experiment, in four randomized blocks, on the effects of fertilizers on grass. In the lower half, Yates' method is applied to the treatment totals. The steps are as follows:

1. Write down the treatment combinations in the systematic order shown. The letters m, n, p, k are introduced in turn. The introduction of any letter is followed by its combinations with all previous treatment combinations. Thus p is followed by mp, np, and mnp in turn. Place the corresponding treatment totals in the next column.

2. In the top half of column (1), add the yields in pairs.

$$302 = 121 + 181; \quad 361 = 104 + 257; \quad \text{etc.}$$

In the lower half of column (1), subtract the *first* member of each pair from the *second*.

$$60 = 181 - 121; \quad 153 = 257 - 104; \quad \text{etc.}$$

3. Apply the same process to column (1) obtaining column (2); then to column (2) obtaining column (3); and finally to column (3), obtaining

column (4), which contains the factorial effect totals. The first number in column (4) is the grand total G. The succeeding factorial effects come out in the order in which the treatment combinations were written down. With n factors, the process is continued until we reach column (n). To obtain the factorial effect means, divide by $2^{n-1}r$, or in this case 32. G is divided by $2^n r$, or 64, to give the mean yield of the whole experiment.

Checks. The sum of squares of the factorial effect totals, divided by $2^n r$, should equal the treatments sum of squares, computed in the usual way. This check verifies the whole process except for errors in the signs of the factorial effect totals and errors in labeling the factors.

A series of intermediate checks is recommended for the larger factorials. In the "Treatment totals" column, add separately the *odd* numbered and the *even* numbered totals, i.e.

$$\text{Odds:} \quad 121 + 104 + 123 + \cdots + 351 = 1459$$

$$\text{Evens:} \quad 181 + 257 + 173 + \cdots + 362 = 2035$$

as shown at the foot of the column in table 5.1a.

Do the same in column (1), and also add the top half of this column. The latter total, 3494, should agree with the total 3494 of the previous column. A second check on column (1) is provided by the relation

$$4070 - 3494 = 2035 - 1459$$

These two relations constitute a satisfactory check on the values in column (1). In the same way, column (2) is checked from column (1) and so on.

TABLE 5.2a ANALYSIS OF VARIANCE

	d.f.	s.s.	m.s.
Replications	3	493.3	
Treatments	15	26,791.9	1,786.1
Error	45	4,074.2	90.5
Total	63	31,359.4	

In the analysis of variance (table 5.2a), it is unnecessary to divide the treatments sum of squares into single d.f., since tests of significance can be applied directly to the factorial effect totals or means. If s^2 is the error variance per unit, the standard errors for factorial effect totals and means are as follows:

$$\text{Effect totals:} \quad \text{s.e.} = \sqrt{2^n r s^2} = \sqrt{(64)(90.5)} = 76.1$$

$$\text{Effect means:} \quad \text{s.e.} = \sqrt{s^2/2^{n-2}r} = \sqrt{(90.5)/16} = 2.38$$

For 45 d.f., the 5% and 1% t-values are 2.014 and 2.690, respectively. A factorial effect mean must exceed (2.38)(2.014) = 4.79 for 5% significance, and similarly 6.40 for 1% significance. All main effects, and the interactions MK, NK, MNK are significant. The relationships among the effects of M, N, and K are interesting to examine from the three-way table of means for these factors.

The alternative method of computing factorial effect totals, recommended when it is not desired to isolate the high-order interactions, is illustrated later in section 6.14. The first step is to write down the + and − signs for the main effects, M, N, P, and K. These are the only expressions that need be remembered. The expression for MN is found by multiplying corresponding signs in M and N. Similarly MNP is given by the product of corresponding signs in MN and P, and so on.

5.25 Factors at More than Two Levels: a 4 × 2 Factorial. Various notations are used. For example, the 3 levels of a factor A may be denoted by a_0, a_1, a_2, or by a_1, a_2, a_3, or simply by the numbers 0, 1, 2. For illustration, we give below the treatment totals in a 4 × 2 experiment on sugar cane. The treatments were 4 levels of dressing of potash, k_0, k_1, k_2, and k_3, in arithmetic progression, and 2 levels of phosphate, p_0 and p_1. There were 5 replicates in randomized blocks.

TABLE 5.3 Total yields of sugar cane (tons per acre)

	k_0	k_1	k_2	k_3	Total
p_0	180	248	277	285	990
p_1	251	307	342	346	1246
Totals	431	555	619	631	2236

The main effect of phosphate (P), which occurs at only 2 levels, is of course derived from the comparison of the 2 marginal totals, 990 and 1246. For the main effect of potash (K) there are 4 marginal totals which may be compared. It will be recalled (section 3.42) that 3 independent comparisons may be made amongst 4 totals, and that an infinite number of such sets of 3 may be chosen. Thus the main effect of K comprises 3 independent comparisons.

With the 2^n system, we were able to define every main effect as a specific linear combination of the treatment means. In the present case we could select a particular set of 3 independent comparisons, each of which would be a specific linear combination of the treatment means. These could be defined as the "components" of the main effect of K. However, the particular set that is most useful for the interpretation of the results will change from experiment to experiment, so that a formal definition of this type would be of limited utility. The experimenter should use whichever set appears most relevant.

Now consider the interaction between P and K. From the differences between the two rows in table 5.3, the effect of P is estimated separately at each level of K, the estimates being

	k_0	k_1	k_2	k_3
$(p_1 - p_0)$	71	59	65	61

Any comparison among these 4 figures is a measure of the effect of K on the response to P, and therefore is a component of the interaction between P and K. Consequently the interaction between P and K consists of 3 independent comparisons, and is said to have 3 d.f. As with the main effect of K, the particular set of 3 components that will be of interest varies with the type of experiment.

We may also wish to consider the interaction as the effect of P on the response to K. As we have seen, the response to K has 3 components. Since there are 4 increasing levels of potash dressing, we might choose as components the linear, quadratic, and cubic components of the response curve. Apart from a divisor, the linear component is

$$-3(180) - 1(248) + 1(277) + 3(285) = 344$$

at the lower level of P and

$$-3(251) - 1(307) + 1(342) + 3(346) = 320$$

at the higher level of P. The difference between these 2 figures, -24, estimates the effect of P on the linear response to K, and is a part of the interaction between K and P. The other two components are the effects of P on the quadratic and cubic responses to K.

In a 2^n system, the interaction between P and K is identical with the interaction between K and P. In the more general case the corresponding result is that any component of the interaction between K and P can be derived from the components of the interaction between P and K. For example, the effect of P on the linear response to K can be written

$$-24 = -3(71) - 1(59) + 1(65) + 3(61)$$

so that it is a linear function of the responses to P at the 4 levels of K. In this more general sense, the two interactions are still equivalent.

We will consider the analysis of variance of the 8 treatment totals in some detail. With 5 replicates, the total s.s., on a single-plot basis, is

$$\frac{(180)^2 + (251)^2 + (248)^2 + \cdots + (346)^2}{5} - \frac{(2236)^2}{40} = 4165.2$$

The 7 d.f. subdivide into 1 for the main effect of P, 3 for that of K, and 3 for the PK interaction. In practice the first two terms are calculated directly and the interaction obtained by subtraction. The computations are

$$P: \frac{(1246 - 990)^2}{40} = 1638.4$$

$$K: \frac{(431)^2 + (555)^2 + (619)^2 + (631)^2}{10} - \frac{(2236)^2}{40} = 2518.4$$

TABLE 5.4 PRELIMINARY ANALYSIS OF VARIANCE OF DATA IN TABLE 5.3

	d.f.	s.s.	m.s.
P	1	1638.4	1638.4
K	3	2518.4	839.5
PK	3	8.4	2.8
Total	7	4165.2	

As is common in agricultural experiments, the interaction m.s. is small compared with those for the main effects (in fact it is below the mean square for error).

As an exercise we will divide the K main effect and its interaction with P into linear, quadratic, and cubic components. The first step is to calculate the totals for these effects separately at each level of P. The results are shown below, the multipliers for the 4 levels of K being shown in parentheses. These multipliers are obtained from a table of orthogonal polynomials (5.9).

	K_l	K_q	K_c
	(−3, −1, +1, +3)	(+1, −1, −1, +1)	(−1, +3, −3, +1)
p_0	+344	− 60	+18
p_1	+320	− 52	−10
Sum $(p_1 + p_0)$	+664	−112	+ 8
Diff. $(p_1 - p_0)$	− 24	+ 8	−28

From the *sum* line, we obtain the 3 components of the K main effect,

$$K_l = \tfrac{1}{200}(664)^2 = 2204.5, \quad K_q = \tfrac{1}{40}(112)^2 = 313.6$$

$$K_c = \tfrac{1}{200}(8)^2 = 0.3$$

Similarly, from the *difference* line, the components of the PK interaction are

$$PK_l = \tfrac{1}{200}(24)^2 = 2.9, \quad PK_q = \tfrac{1}{40}(8)^2 = 1.6$$

$$PK_c = \tfrac{1}{200}(28)^2 = 3.9$$

The divisors are found from the usual rule for linear functions (section 3.42). The three K components add to 2518.4, the total s.s. for K, while the PK components add to 8.4, the sum of squares for PK. The detailed analysis of variance is shown in table 5.5.

TABLE 5.5 More detailed analysis of variance of the data in table 5.3

	d.f.	s.s. or m.s.
P	1	1638.4
K_l	1	2204.5
K_q	1	313.6
K_c	1	0.3
PK_l	1	2.9
PK_q	1	1.6
PK_c	1	3.9

Since every component has 1 d.f., the mean squares are the same as the sums of squares. The error m.s. is 16.35, with 28 d.f. The quadratic component of K is significant, indicating a falling off in the response at the higher levels of application. Neither the cubic component nor any of the interactions approaches significance.

5.26 A 3^2 Factorial Experiment. As a second example, the treatment totals are shown in table 5.6 for an experiment with 3 levels of nitrogen fertilizer and 3 of phosphate fertilizer. The data are the numbers of lettuce plants that emerged from the ground and are totals over 12 plots

TABLE 5.6 Numbers of lettuce plants emerging *

	n_0	n_1	n_2	Totals
p_0	449 ($A\alpha$)	413 ($C\beta$)	326 ($B\gamma$)	1188
p_1	409 ($B\beta$)	358 ($A\gamma$)	291 ($C\alpha$)	1058
p_2	341 ($C\gamma$)	278 ($B\alpha$)	312 ($A\beta$)	931
Totals	1199	1049	929	3177

* The use of the latin and greek letters will be explained later.

each. Both nitrogen and phosphate appear to have had a deleterious effect on emergence (the subscript 2 denotes the largest application). The main effects of N and P both comprize two independent comparisons, and thus have 2 d.f. each. Since the amounts of N and P were in arithmetic progression, it would probably again be appropriate, as in the previous example, to choose the linear and quadratic components of the regression on amount of dressing as the individual components of the main effects.

The N_l component can be estimated separately at each level of P. This means that there are 2 d.f. which measure the effect of P on the linear response to N. These are a part of the NP interaction. Another 2 d.f. are supplied by the effect of P on the N_q component, so that the NP interaction contains 4 d.f.

From inspection of the margins of table 5.6, it is evident that the main effects of both N and P are approximately linear. Consequently, the most interesting single degree of freedom from the interactions is likely to be the interaction of N_l with P_l. In many agricultural experiments this type of interaction is the only one that approaches significance. It is worth while calculating this interaction separately in the analysis of variance. The other three components, N_lP_q, N_qP_l, and N_qP_q, will also be obtained. The initial computations appear in table 5.7.

TABLE 5.7 CALCULATION OF LINEAR AND QUADRATIC EFFECTS FOR THE ANALYSIS OF VARIANCE

	N_l (−1, 0, 1)	N_q (1, −2, 1)		P_l (−1, 0, 1)	P_q (1, −2, 1)
p_0	−123	− 51	n_0	−108	−28
p_1	−118	− 16	n_1	−135	−25
p_2	− 29	+ 97	n_2	− 14	+56
Sum	$-270(N_l)$	$+30(N_q)$	Sum	$-257(P_l)$	$+3(P_q)$
P_l	$+94(N_lP_l)$	$+148(N_qP_l)$	N_l	$+94(P_lN_l)$	$+84(P_qN_l)$
P_q	$+84(N_lP_q)$	$+78(N_qP_q)$	N_q	$+148(P_lN_q)$	$+78(P_qN_q)$

The left side of the table shows the N_l and N_q effects for each level of P, while the right side shows the P_l and P_q effects for each level of N. For instance, from table 5.6,

$$N_lp_0 = 326 - 449 = -123; \qquad P_qn_0 = 449 - 2(409) + 341 = -28$$

The column sums give the individual components of the N and P main effects. Now consider the difference between the third and the first row. For the N_l column, this difference (+94) gives the linear effect of P on N: i.e., the N_lP_l interaction. From the P_l column, the difference gives the linear effect of N on P_l, or the P_lN_l interaction, which is exactly the same as the N_lP_l interaction. The other two columns provide the N_qP_l and the N_lP_q effects.

Finally, the sum of the first and third rows, minus twice the second row, leads to the components of interaction that contain a quadratic term. It will be seen that all four components can be obtained from either the left or the right half of the table, so that in practice only one half is required. By computing both halves we verify the symmetry of the components with respect to N and P.

The squares of these quantities, with appropriate divisors, will give an analysis of variance of the 8 d.f. amongst the treatment totals into 8 single components. Since an individual entry in table 5.6 is the total of 12 plots, the divisors may be verified to be as shown below.

	N_l or P_l	N_q or P_q	N_lP_l	N_lP_q	N_qP_l	N_qP_q
Divisor	72	216	48	144	144	432

For instance, the divisor for N_qP_q may be worked out as follows. The three N_q figures in table 5.7 (i.e., -51, -16, $+97$) each have divisor $12(1^2 + 2^2 + 1^2)$, or 72. Since the N_qP_q total is a linear function of these three figures, with coefficients 1, -2, and 1, this total has divisor $72(1^2 + 2^2 + 1^2)$, or 432. The analysis of variance is shown in table 5.8.

TABLE 5.8 SUBDIVISION OF THE TREATMENT S.S.

	d.f.	s.s. or m.s.
N_l	1	1012.50
N_q	1	4.17
P_l	1	917.35
P_q	1	0.04
N_lP_l	1	184.08
N_lP_q	1	49.00
N_qP_l	1	152.11
N_qP_q	1	14.08

The error m.s. in this experiment is about 59. The linear effects of both fertilizers are significant, with no indication of curvature. The N_lP_l effect is significant at the 10% level, but not at the 5% level.

In table 5.6, latin and greek letters were superimposed so as to form a 3 × 3 graeco-latin square. This square leads to another method of calculating the sum of squares for the interactions (4 d.f.). Although the method is not likely to be of use for purposes of interpretation, it has formed the basis of some ingenious devices in the construction of designs. In the square, the column totals represent the main effects of N and the row totals those of P. Since the latin letter totals are orthogonal to rows and columns, it seems reasonable to suppose that they must represent two of the 4 components of the NP interaction. Similarly, the greek letter totals represent the remaining 2 components. These totals are shown below.

A	B	C	Total	α	β	γ	Total
1119	1013	1045	3177	1018	1134	1025	3177

Each figure is now a total of 36 plots. The sum of squares of deviations of the latin letter totals, divided by 36, is 164.22, and the corresponding figure for the greek letters is 235.06. These add to 399.28, which is the same as the total s.s. for the interactions in table 5.8, apart from rounding differences.

5.27 General Method of Analysis. Suppose that there are three factors, A, B, C, which occur at α, β, and γ levels respectively. The main effects have $(\alpha - 1)$, $(\beta - 1)$, and $(\gamma - 1)$ components or degrees of freedom respectively. Each component of the main effect of A can be estimated separately at each of the β levels of B. Thus each component of A contributes $(\beta - 1)$ degrees of freedom to the AB interaction. This means that the AB interaction contains a total of $(\alpha - 1)(\beta - 1)$ components or degrees of freedom. Similarly the AC interaction has $(\alpha - 1)(\gamma - 1)$ degrees of freedom, and the BC interaction has $(\beta - 1)(\gamma - 1)$ degrees of freedom.

To compute the sums of squares for the main effects and first-order interactions, we form two-way tables for *each pair of factors.* Consider the A by B two-way table. The total s.s. among cells has $(\alpha\beta - 1)$ degrees of freedom. From the marginal totals in the table we compute the sum of squares for the main effect of A, with $(\alpha - 1)$ degrees of freedom, and that for the main effect of B, with $(\beta - 1)$ degrees of freedom. By subtraction, the sum of squares for the AB interaction, with $(\alpha - 1)(\beta - 1)$ degrees of freedom, is obtained.

There remains the three-factor, or ABC, interaction. Now each of the $(\alpha - 1)(\beta - 1)$ components in the AB interaction is estimated separately at each level of C. It will therefore contribute $(\gamma - 1)$ degrees of freedom to the ABC interaction, so that the latter contains in all $(\alpha - 1)(\beta - 1)(\gamma - 1)$ degrees of freedom. The sum of squares for ABC is also obtained most easily by subtraction. From the total s.s. amongst treatments, with $(\alpha\beta\gamma - 1)$ degrees of freedom, subtract the sums of squares for A, B, C, AB, AC, and BC. The remainder will be the sum of squares for ABC.

It is hoped that the reader will find no difficulty in extending these methods to the case where there are more than three factors. In general, the sums of squares for main effects are calculated directly, and those for interactions are calculated by subtraction. To compute an $ABCD$ interaction, for instance, we require a four-way table for the four factors represented. From the sum of squares for this table we subtract the sum of squares for all main effects and two- and three-factor interactions among the factors in question. Calculations must be checked by recomputation. Of course, if the experimenter subdivides any interaction into single components as in the previous section, a check is provided by this process.

In the general case it remains true that any component of a factorial effect is orthogonal to any component of any *other* factorial effect. Thus any component of the BC interaction is orthogonal to any component of the A main effect, or of the BCD interaction, etc. Two components of the *same* factorial effect may or may not be independent. For instance, if A occurs at three levels, the comparison $(a_2 - a_0)$ is independent of the comparison $(a_2 - 2a_1 + a_0)$, but is not independent of $(a_1 - a_0)$.

The following selected references contain a discussion of factorial experiments with some worked examples.

5.3 YATES, F. The design and analysis of factorial experiments. *Imp. Bur. Soil Sci. Tech. Comm.* 35, 1937. This gives the most comprehensive account that is available, with numerous worked examples.

5.4 YATES, F. Complex experiments. *Jour. Roy. Stat. Soc. Suppl.* 2, 181–247, 1935. An earlier reference, with examples of 2^2, 2^3, 4×3, and 3^3 factorials.

5.5 DAVIES, O. L. (Ed.). *Design and analysis of industrial experiments.* Hafner Co., New York, 1954. An excellent account of the industrial applications of factorial designs.

5.6 LINDQUIST, E. F. *Design and analysis of experiments in psychology and education.* Houghton Mifflin Co., Boston, 1953. Contains a detailed discussion of the use and interpretation of factorial experiments in psychology and education.

5.7 KEMPTHORNE, O. *The design and analysis of experiments.* John Wiley & Sons, New York, 1952. Covers about the same ground as this book, with more emphasis on the theory underlying the analysis and on methods of constructing designs.

5.28 Interpretation of the Analysis: First Example. The separation of the treatment comparisons into main effects and interactions is a convenient and powerful method of analysis in cases where interactions are small relative to main effects. When interactions are large, this analysis must be supplemented by a detailed examination of the nature of the interactions. It may, in fact, be found that an analysis into main effects and interactions is not suited to the data at hand. There is sometimes a tendency to apply the factorial method of analysis mechanically without considering whether it is suitable or not, and also a tendency to rely too much on the initial analysis of variance alone when writing a summary of the results. Below two examples are presented where the initial analysis of variance is not very informative, and where the results can be summarized better in terms of simple rather than factorial effects.

The first experiment was conducted by the Wailuku Sugar Company. Three varieties of sugar cane were compared, in combination with three levels of nitrogen (150, 210, and 270 lb. N per acre respectively). The crop was the second harvesting, or the first ratoon crop. In table 5.9 only the relevant part of the analysis of variance is shown: i.e., the sum of squares for the main effects of V and N and for the VN interaction, plus the error s.s. The data are in tons of cane per acre. The conclusions

TABLE 5.9 ANALYSIS OF VARIANCE OF A 3 × 3 SUGAR CANE EXPERIMENT

	d.f.	s.s.	m.s.
V	2	319.38	159.69 *
N	2	56.54	28.27
VN	4	559.79	139.95 *
Error	24	1053.84	43.91

from the analysis of variance are that the main effects of V and the VN interaction are both significant, but there is no sign of a main effect due to N. This statement tells little about the results of the experiment.

The treatment totals (over 4 replications) are shown below with their standard errors. Since the standard error per plot is $\sqrt{43.91} = 6.626$, the standard error for a treatment total is 13.3 as shown.

TABLE 5.10 TREATMENT TOTALS (TONS) IN A 3 × 3 SUGAR CANE EXPERIMENT (±13.3)

	n_0	n_1	n_2	Total	s.e.
v_1	266.1	275.9	303.8	845.8	
v_2	245.8	250.2	281.7	777.7	±23.0
v_3	274.4	258.1	231.6	704.1	
Total	786.3	784.2	817.1	2387.6	
s.e.		±23.0			

Instead of having no effect, nitrogen has apparently given a steady increase in yields with the first two varieties, but a steady decrease with the third variety. Further, the significant main effects of varieties apply only to the average varietal yields over all 3 dressings of N, and not to yields with a particular rate of dressing. On the average v_1 gives a substantially higher yield than v_3, but at the lowest level of N, v_3 is slightly above v_1.

The subsequent analysis may be made either by means of t-tests applied to table 5.10 or by means of a further subdivision of the analysis of variance. It is of interest to examine the response to N for each variety separately. To test the linear responses by means of a t-test, we require the standard error for $(n_2 - n_0)$ as computed for each variety. This standard error is 13.3 × 1.414, or 18.8, and since the 5% t-value for 24 d.f. is 2.064, the quantities $(n_2 - n_0)$ must attain the value 18.8 × 2.064, or 38.8, in order to be significant at the 5% level. The actual values are

$$(n_2 - n_0)v_1 = +37.7; \qquad (n_2 - n_0)v_2 = +35.9;$$
$$(n_2 - n_0)v_3 = -42.8$$

Thus neither of the increases with v_1 and v_2 quite reaches the significance level, though both are close to it. The decrease with v_3 is significant.

The test of linearity of the response curves is made by means of the quantity $(n_2 + n_0 - 2n_1)$, calculated for each variety. The standard error of this quantity is $13.3 \times \sqrt{6}$, or 32.6. The reader may verify from table 5.10 that none of the curvature terms even exceeds its standard error.

It is not quite so clear what tests are appropriate for appraising the varietal differences. Since, however, the difference between v_1 and v_2 is very consistent at all levels of N, a t-test of the total difference is suggested. This difference is 68.1, and since the value required for significance is $23.0 \times 1.414 \times 2.064$, or 67.1, the superiority of v_1 over v_2 is just significant. Further, it is evident on inspection that v_3 does not differ significantly from the other varieties at either of the two lower levels of N. At the highest level, v_3 is significantly below both v_1 and v_2.

When interactions are large, much care is required in the preparation of a statement that summarizes the results, and it is not easy to reach a form that is free from criticism. The following is a suggestion.

"The increase in yield of cane to the highest dressing of N (270 lb. per acre) over the lowest dressing (150 lb. per acre) was 9.4 tons per acre with v_1 and 9.0 tons per acre with v_2. Both increases just failed to be significant at the 5% level. With v_3, on the other hand, the highest dressing of N decreased the yield significantly by 10.7 tons per acre as compared with the lowest dressing. For all three varieties the effects of N appeared to be proportional to the amount applied, within the range investigated.

"Variety 1 gave a higher yield than variety 2 for all levels of N, the average difference, 5.7 tons, being just significant at the 5% level. The yields for variety 3 did not differ significantly from those of the other varieties at the two lower levels of N. At the highest level of N, the yield for variety 3 was significantly lower than that for the other varieties." *

5.29 Interpretation of the Analysis: Second Example. This example is more complex, mainly because the factorial (a 3×3) is of an unusual type. The data come from a long-term experiment on meadow hay,

* The explanation for the harmful effect of N with v_3 is that this variety ripened earlier than the others. By the time the experiment was harvested, much cane had fallen to the ground through overripeness on the plots with v_3 that received N. Had the variety been harvested at the optimum time, results would have been different.

conducted at Lady Manner's School, Bakewell, England, with the co-operation of Rothamsted Experimental Station. The yields are for the 1937 season. The two factors are shown schematically below.

First factor (3 levels)		Second factor (3 levels)	
No manure Mixed artificials 8 tons compost	Applied in 1936	No manure Mixed artificials 8 tons compost	Applied in 1937

The nature of the experiment may become clear from a discussion of some of the individual treatment combinations. In any replicate there are nine plots, of which three received mixed artificial fertilizers in 1937. In the previous year, 1936, one of these three plots received no manure, one received mixed artificials, and one received compost. Thus these three plots enable us to compare the *residual* effects of the 1936 applications of artificials and compost, on plots which received artificials in the current year, 1937. Similarly, from other treatment combinations, we can compare the residual effects of artificials and compost on plots which were unmanured in the current year and on plots which received compost in the current year.

Further, in any replicate there are three plots which were unmanured in 1936. Of these, one had no manure, one had artificials, and one had compost in 1937. Consequently we may also assess the *direct* effects of artificials and compost applied in 1937 on plots which were unmanured in the previous year, and likewise on plots which received artificials or compost in the previous year. Both the artificials and compost contain the three common plant nutrients, nitrogen, phosphorus, and potash.

The system of treatments is an ingenious one, designed to measure direct and residual effects at the same time. The experiment was started in 1932, but for simplicity we will ignore any effects of treatments applied prior to 1936.

The treatment means and the analysis of variance are shown in tables 5.11 and 5.12. The analysis of variance was calculated from the treatment means, rather than from single plots, and the error m.s. has been adjusted so as to apply to a treatment mean. The direct effects are highly significant, the interactions are significant at the 5% level, but residual effects are not significant. In considering the nature of the effects leading to these results, it is again convenient to think in terms of simple effects rather than of main effects and interactions. It is worth noting that the standard error of a single entry in the two-way table is $\sqrt{6.656}$, or 2.58, while that for the difference between two entries is 3.65.

With regard to *residual* effects, table 5.11 shows that compost produced a large and highly significant increase, 13.4 cwt. ± 3.65, on the plots that received no manure in the current year. On plots that received manure in 1937, either artificials or compost, there is no suggestion of any residual effect of compost. None of the residual effects of artificials approaches the level required for significance. These results are

TABLE 5.11 TREATMENT MEAN YIELDS (HAY, CWT. PER ACRE) (±2.58)

Direct effects of 1937 treatments	Residual effects of 1936 treatments: None	Artificials	Compost	Mean
None	53.6	56.8	67.0	59.1
Artificials	80.8	82.3	80.5	81.2
Compost	74.3	69.1	70.0	71.1
Mean	69.6	69.4	72.5	70.5

TABLE 5.12 ANALYSIS OF VARIANCE OF TREATMENT MEANS

	d.f.	s.s.	m.s.
Direct effects	2	732.28	366.14 **
Residual effects	2	18.24	9.12
Interaction	4	97.01	24.25 *
Error	24	159.74	6.656

in accord with general fertilizer experience, since a compost is more likely to give residual effects than an inorganic fertilizer, and since residual effects would be expected to show up most clearly on plots which have no manures during the current year.

So far as *direct* effects are concerned, artificials were superior to compost whatever the residual treatment.

1936 manuring	1937 artificials − 1937 compost (cwt.)
No manure	+6.5
Artificials	+13.2
Compost	+10.5

The differences among these three figures are within the limits of experimental error. To test this point, we find the sum of squares of deviations of these differences, which comes to 22.73. This is divided by 2 to make it comparable with the analysis of variance, giving 11.36, which represents two of the 4 d.f. for interactions. The mean square, 5.68, is slightly below the error m.s. Thus the superiority of artificials appears to be consistent.

The increases for artificials and compost over no manure are much reduced on plots that received compost in the previous year.

1936 manuring	(1937 artificials − 1937 none)	(1937 compost − 1937 none)
Compost	+13.5	+3.0
Artificials and none (averaged)	+26.4	+16.5

The striking reduction on plots with 1936 compost is simply a reflection of the nature of the residual effect of compost as previously examined.

The conclusions from the experiment might be summarized as follows.

"Artificials applied in 1937 increased yields by 13.5 cwt. on plots that received compost in the previous year, and by 26.4 cwt. on plots that did not receive compost in the previous year. The corresponding increases due to 1937 compost were 3.0 cwt. and 16.5 cwt., respectively. The superiority of 1937 artificials over 1937 compost, which averaged 10.1 cwt., appeared to be independent of the type of manuring during the previous year.

"As regards residual effects, compost applied in 1936 increased yields by 13.4 cwt. on plots that were unmanured in 1937, but gave no apparent increase on plots manured in 1937. There were no significant residual effects of 1936 artificials."

It seems evident that the significant interaction m.s. in the analysis of variance must arise mainly from the fact that compost had a residual effect only when no manure was applied currently. As an exercise it may be instructive to isolate the part of the interaction s.s. that is due to this effect. Since artificials appeared to have no residual effect, we will combine the 1936 artificials with the 1936 unmanured plots. Consequently, the residual effects of compost are obtained from the comparison

$$2(\text{1936 compost}) - (\text{1936 artificials}) - (\text{1936 none})$$

This quantity may be calculated separately for each of the 1937 treatments. Its values are

$$(\text{1937 none}) = +23.6; \qquad (\text{1937 artificials}) = -2.1;$$

$$(\text{1937 compost}) = -3.4$$

The contrast between the residual effect of compost on plots without 1937 manures and that on plots with 1937 manures may be estimated from the comparison

$$2(+23.6) - (-2.1) - (-3.4) = +52.7$$

The square of this quantity, with a suitable divisor, is a single component of the interaction s.s. Since the divisor for the quantities 23.6, etc., is 6, the divisor needed is 6 × 6, or 36. Hence the sum of squares for this component is $(52.7)^2/36$, or 77.15. The remaining 3 d.f. for interactions have a sum of squares equal to 19.86. The mean square, 6.62, is no larger than the error m.s. This verifies the suggestion that the significance of the interaction m.s. can be attributed to the type of residual effect of compost.

Some readers may have difficulty in satisfying themselves that the component isolated above really is a part of the interaction s.s. It may be helpful to express the basic quantity, 52.7, as a linear function of the original mean yields in table 5.11. The multipliers of the means will be found to be as shown below.

1937 treatments	1936 treatments		
	None	Artificials	Compost
None	−2	−2	4
Artificials	1	1	−2
Compost	1	1	−2

The sums of the coefficients are zero over every row and column so that the expression is orthogonal to both sets of main effects. Hence it must be a part of the interaction. The coefficients also enable us to verify the divisor, 36, which is equal to the sum of the squares of the coefficients.

Since the interactions can be attributed to the behavior of a single group of plots (those with compost in 1936 and no manuring in 1937), it might be suspected that this treatment had been allotted by chance a favorable set of plots. The residual increase, 13.4 cwt., does seem rather large in relation to the direct effect of compost. An examination of previous results does not lend much weight to this suspicion. Over the 4 preceding years, the average residual response to compost, with no current manure, was 11.1 cwt. In 2 years, 1935 and 1933, the plots involved in this comparison were the same as those in 1937. For these years the average residual effect was 6.2 cwt. In the other 2 years, when a different set of compost plots is involved, the average residual effect was 16.0 cwt.

The two preceding examples are intended to illustrate the fact that the most informative subdivision of the treatment comparisons depends on the type of experiment. An analysis copied from a model that appears similar in form may be inappropriate or even meaningless. The experimenter should first decide which comparisons are necessary for the interpretation of the results. The subsequent analysis, either by t-tests or by subdivision of the analysis of variance, should be directed towards these comparisons.

When all the factors represent continuous variables like temperature, time, or the amount of an added substance, it may be natural to think of the response as a mathematical function of the levels of the factors. The type of analysis needed to fit polynomial functions of the first and second degree is outlined in chapter 8A.

5.3 Designs for Factorial Experiments

5.31 Factorials in Complete Block Designs. Most types of experimental plan are suitable for factorial experiments. In particular, if the total number of treatment combinations is not large, the designs described in chapter 4 are frequently used. The relative advantages of complete randomization, randomized blocks, and latin squares are the same with factorial as with non-factorial sets of treatments. For illustration, we show arrangements for a 4 × 2 factorial (i) in 8 randomized blocks and (ii) in an 8 × 8 latin square. In the former case, only the first replicate is given.

TABLE 5.13 Factorial experiment arranged in (i) randomized blocks and (ii) a latin square

8 replications

8 treatment combinations
A at 4 levels (1,2,3,4)
B at 2 levels (1,2)

i. In randomized blocks

Rep. I
12
31
21
32
42
11
22
41

Analysis of variance

Replications		7
Treatment combinations		7
A	3	
B	1	
AB	3	
Error		49
Total		63

ii. In a latin square

42	11	22	12	31	41	32	21
21	32	31	41	22	12	11	42
12	41	42	11	32	21	22	31
31	22	21	42	11	32	41	12
32	12	11	31	21	22	42	41
41	42	32	22	12	31	21	11
11	21	12	32	41	42	31	22
22	31	41	21	42	11	12	32

Analysis of variance

Rows		7
Columns		7
Treatment combinations		7
A	3	
B	1	
AB	3	
Error		42
Total		63

TABLE 5.14 WEIGHTS OF DENERVATED (y) AND CORRESPONDING NORMAL (x) MUSCLE (unit = 0.01 gram)

Length of treatment (minutes)	Type of current	One (a_1)		Three (a_3)		Six (a_6)	
		y	x	y	x	y	x
Rep. I							
1 (b_1)	Galvanic	72	152	74	131	69	131
	Faradic	61	130	61	129	65	126
	60 cycle	62	141	65	112	70	111
	25 cycle	85	147	76	125	61	130
2 (b_2)	Galvanic	67	136	52	110	62	122
	Faradic	60	111	55	180	59	122
	60 cycle	64	126	65	190	64	98
	25 cycle	67	123	72	117	60	92
3 (b_3)	Galvanic	57	120	66	132	72	129
	Faradic	72	165	43	95	43	97
	60 cycle	63	112	66	130	72	180
	25 cycle	56	125	75	130	92	162
5 (b_5)	Galvanic	57	121	56	160	78	135
	Faradic	60	87	63	115	58	118
	60 cycle	61	93	79	126	68	160
	25 cycle	73	108	86	140	71	120
Rep. II							
1 (b_1)	Galvanic	46	97	74	131	58	81
	Faradic	60	126	64	124	52	102
	60 cycle	71	129	64	117	71	108
	25 cycle	53	108	65	108	66	108
2 (b_2)	Galvanic	44	83	58	117	54	97
	Faradic	57	104	55	112	51	100
	60 cycle	62	114	61	100	79	115
	25 cycle	60	105	78	112	82	102
3 (b_3)	Galvanic	53	101	50	103	61	115
	Faradic	56	120	57	110	56	105
	60 cycle	56	101	56	109	71	105
	25 cycle	56	97	58	87	69	107
5 (b_5)	Galvanic	46	107	55	108	64	115
	Faradic	56	109	55	104	57	103
	60 cycle	64	114	66	101	62	99
	25 cycle	59	102	58	98	88	135

5.32 Numerical Example: a $4 \times 4 \times 3$ Factorial in Randomized Blocks. A number of experiments have indicated that electrical stimulation may be helpful in preventing the wasting away of muscles that are denervated. A factorial experiment on rats was conducted by Solandt, DeLury, and Hunter (5.8) in order to learn something about the best type of current and the most effective method of treatment. The factors and their levels are shown below.

A	*B*	*C*
Number of treatment periods daily	Length of treatment (minutes)	Type of current
1	1	Galvanic
3	2	Faradic
6	3	60 cycle alternating
	5	25 cycle alternating

Treatments were started on the third day after denervation and continued for 11 consecutive days. There are 48 different combinations of methods of treatment, each of which was applied to a different rat. Two replications were conducted, using 96 rats in all.

The muscles denervated were the gastrocnemius-soleus group on one side of the animal, denervation being accomplished by the removal of a small part of the sciatic nerve. The measure used for judging the effects of the treatments was the weight of the denervated muscle at the end of the experiment. Since this depends on the size of the animal, the weight of the corresponding muscle on the other side of the body was included as a covariance variate. The data are shown in table 5.14.

For a covariance analysis we require analyses of variance for both the denervated muscle (y) and the normal muscle (x), and the analysis of the product (yx). From the discussion in section 5.27 and preceding sections, the reader should have little difficulty in completing the analysis. The replicate totals, treatment-combination totals, and the grand total are first computed. From the treatment totals, the three two-way tables are constructed, as given in table 5.15. Since each set of main effect totals is obtained twice, this part is self-checking (though only one set of main effect totals need be recorded). To illustrate the computations made on the $A \times B$ table, we have for the analysis of y

Total s.s. for $A \times B$ table

$$= \tfrac{1}{8}[(510)^2 + (543)^2 + \cdots + (546)^2] - \tfrac{1}{96}[(6069)^2] = \quad 1039$$

$$A = \tfrac{1}{32}[(1936)^2 + (2028)^2 + (2105)^2] - \tfrac{1}{96}[(6069)^2] = \quad 447$$

$$B = \tfrac{1}{24}[(1565)^2 + (1488)^2 + (1476)^2 + (1540)^2] - \tfrac{1}{96}[(6069)^2] = \quad 223$$

By subtraction, the AB sum of squares is found to be 369. For the sum of products, we replace each y^2 by the corresponding yx product. Thus

$$\text{Total s.p.} = \tfrac{1}{8}[(510)(1030) + (543)(977) + \cdots + (546)(985)] - \tfrac{1}{96}(6069)(11{,}307) = 1151$$

Finally, the ABC sum of squares is found by calculating the total s.s. among all 48 treatment combinations, and subtracting the sum of squares for A, B, C, AB, AC, and BC.

TABLE 5.15 Two-way tables in a $4 \times 4 \times 3$ factorial

		y				x			
		a_1	a_3	a_6	Total	a_1	a_3	a_6	Total
B	b_1	510	543	512	1,565	1,030	977	897	2,904
Length of	b_2	481	496	511	1,488	902	1,038	848	2,788
treatment	b_3	469	471	536	1,476	941	896	1,000	2,837
	b_5	476	518	546	1,540	841	952	985	2,778
C	G	442	485	518	1,445	917	992	925	2,834
Type of	F	482	453	441	1,376	952	969	873	2,794
current	60	503	522	557	1,582	930	985	976	2,891
	25	509	568	589	1,666	915	917	956	2,788
A totals		1,936	2,028	2,105	6,069	3,714	3,863	3,730	11,307

		y				x			
		b_1	b_2	b_3	b_5	b_1	b_2	b_3	b_5
C	G	393	337	359	356	723	665	700	746
Type of	F	363	337	327	349	737	729	692	636
current	60	403	395	384	400	718	743	737	693
	25	406	419	406	435	726	651	708	703

The analyses of y^2, yx, and x^2 are given in table 5.16. The regression coefficient of weight of denervated on weight of normal muscle is 3977/16,013, or 0.248361. The next step is to form the analysis of $(y - bx)^2$. This is done by adding to any y^2 value $(-2b)$ times the corresponding yx value, plus (b^2) times the corresponding x^2 value. The mean squares for $(y - bx)$ are the quantities used to test the significance of treatment effects on the weights of denervated muscles, adjusted for their regression on the weights of the normal muscles. As shown in section 3.86, the treatment m.s. is slightly inflated because of the sampling error of the regression coefficient. The inflation is sufficiently small, however, that only effects which appear to be on the borderline of significance need be tested by the exact method (section 3.86), which is more laborious to calculate.

In the analysis of variance (table 5.16) only two factorial effects are significant: those for the main effects of C (type of current used) and for the main effects of A (number of treatment periods daily). The AC interaction m.s. is somewhat higher than the error m.s., but does not approach the 5% significance level. These results suggest that in preparing

TABLE 5.16 Analyses of Variance

		y^2	yx	x^2	$(y - bx)^2$	
	d.f.	s.s.	s.p.	s.s.	s.s.	m.s.
Replications	1	605	2,503	10,354	0	....
A (number of treatments)	2	447	64	418	441	220.5 *
B (length of period)	3	223	137	415	181	60.4
C (type of current)	3	2,145	104	281	2,111	703.7 **
AB	6	369	950	5,202	218	36.3
AC	6	645	476	1,014	471	78.5
BC	9	299	11	2,015	418	46.4
ABC	18	1,050	666	5,198	1,040	57.8
Error	47 [1]	3,199	3,977	16,013	2,211	48.1
Total	95 [2]	8,982	8,888	40,910	7,091	.....

[1] 46 d.f. for $(y - bx)^2$.
[2] 94 d.f. for $(y - bx)^2$.

Note. In reference (5.8), the sums of squares for AC and ABC are in error.

summary tables for further examination and for presentation of the results, we probably need only the AC two-way table. The averages for B (length of an individual treatment) should also be considered in case they indicate a trend effect that was not marked enough to attain significance. These data appear in table 5.17.

TABLE 5.17 Weights of Denervated Muscle, Adjusted for Regression on Weight of Normal Muscle (unit = 0.01 gram)

B means	
b_1	64.4
b_2	62.4
b_3	61.4
b_5	64.7
s.e.	±1.43

AC two-way table (s.e. ±2.47)

		Number of treatments			
		a_1	a_3	a_6	Means
Type of current	G	56.0	59.1	65.2	60.1
	F	60.0	55.8	57.3	57.7
	60	63.2	63.9	68.6	65.2
	25	64.5	71.8	73.2	69.8
Means		60.9	62.6	66.1	63.2

s.e. ±1.43 (type of current means)

s.e. ±1.24 (number of treatments means)

The means have been adjusted in the usual way for the regression on the weight of the normal muscle. Because of this adjustment, every mean has a slightly different standard error. However, as pointed out in section 3.85, it is sufficiently accurate to use an average standard error. This is computed from the *effective* error m.s., which is 48.1 × 1.019, or 49.0, as described in section 3.85. Thus the standard error for an entry in the two-way table is $\sqrt{49.0/8}$, or 2.47.

The 25 cycle alternating current gave a significantly higher mean weight than any other type of current. The 60 cycle alternating current was superior to both galvanic and faradic currents, the last two not being significantly different. The weights increased as the number of treatment periods daily increased. On inspection the increase appears approximately linearly related to the number of treatment periods.

To test this supposition, we may fit a regression of the adjusted A totals on the number of treatment periods (the adjusted A means do not carry enough decimals to check with the analysis of variance). The relevant data are shown below.

	Adjusted A totals	z = number of periods
a_1	1950	1
a_3	2005	3
a_6	2115	6

$$\sum (a - \bar{a})(z - \bar{z}) = 421.667$$
$$\sum (z - \bar{z})^2 = 12.667$$

By the usual formula, the contribution of the regression to the sum of squares for A is $(421.667)^2/(12.667)(32)$, or 439. The last divisor, 32, is required because the A totals are totals over 32 rats. The sum of squares for regression accounts for practically all the sum of squares for A (441), so that there is no indication of deviation from linearity. The regression coefficient for a single rat is $421.667/(12.667)(32)$ or 1.0 unit. The conclusion is that, within the range of periods tested, each additional treatment period per day increased the weight by 1 unit.

The B means in table 5.17 do not indicate any consistent effect of length of an individual treatment period (1 to 5 minutes). The major part of the AC interaction seems to come from the anomalous behavior of the faradic current, which produced a drop in weight from a_1 to a_3 and a_6. Since the interaction as a whole was not significant, it does not seem worth while to examine the statistical significance of this effect.

5.33 Other Designs. The total number of treatment combinations increases rapidly as the number of factors or the number of levels of a factor is increased. In this event the latin square necessitates an amount of replication that is usually impracticable, while it becomes difficult to assemble homogeneous replications for a randomized blocks arrangement. Consequently, the experimental error per unit tends to increase. In order to avoid this increase in error, a number of designs have been produced. The basic principle, which is common to all these designs, is the use of a "block" that is smaller than a complete replication. That is, the arrangements are such that the differences among these "incomplete" blocks are eliminated from error, just as the differences among replicates are eliminated from error in a randomized blocks design. Unfortunately, as will be explained later, this reduction in block size can be accomplished only by the deliberate sacrifice of accuracy on certain treatment comparisons. The designs fall into three main groups according to the particular treatment comparisons that are sacrificed in this way.

In the *first group*, the treatment comparisons that are sacrificed are the high-order interactions. These designs are appropriate in lines of research where experience has shown that high-order interactions are nearly always negligible. The designs available in randomized incomplete blocks are described in chapter 6 and those which can be placed in latin squares in chapter 8.

In the *second group*, known in agriculture as split-plot experiments, a factor, or a group of factors and their interactions, are sacrificed. These designs, which are very widely used, are discussed in chapter 7. Finally, in certain cases it is possible to construct designs so that the loss in accuracy is spread evenly over all factors and their interactions. Within this group there are several types: balanced lattices (chapter 10), balanced incomplete blocks (chapter 11) and balanced lattice squares (chapter 12). In cases where the effects of the factors and the sizes of the interactions are rather unpredictable, these designs may be suitable.

REFERENCES

5.1 FISHER, R. A. *The design of experiments.* Oliver and Boyd, Edinburgh, 4th ed., 1947.

5.2 FRIEDMAN, M., and SAVAGE, L. J. Planning experiments seeking maxima. In *Techniques of statistical analysis.* McGraw-Hill, New York, 1947.

5.3 YATES, F. The design and analysis of factorial experiments. *Imp. Bur. Soil Sci. Tech. Comm.* 35, 1937.

5.4 YATES, F. Complex experiments. *Jour. Roy. Stat. Soc. Suppl.* 2, 181–247, 1935.

5.5 DAVIES, O. L. (Ed.). *Design and analysis of industrial experiments.* Hafner Co., New York, 1954.

5.6 LINDQUIST, E. F. *Design and analysis of experiments in psychology and education.* Houghton Mifflin Co., Boston, 1953.

5.7 KEMPTHORNE, O. *The design and analysis of experiments.* John Wiley and Sons, New York, 1952.

5.8 SOLANDT, D. Y., DELURY, D. B., and HUNTER, J. Effect of electrical stimulation on atrophy of denervated skeletal muscle. *Arch. Neur. and Psych.* 49, 802–807, 1943.

5.9 FISHER, R. A., and YATES, F. *Statistical tables for biological, agricultural and medical research.* Oliver and Boyd, Edinburgh, 3rd ed., 1948.

5.1a BOX, G. E. P., and WILSON, K. B. On the experimental attainment of optimum conditions. *Jour. Roy. Stat. Soc. B*, 13, 1–45, 1951.

ADDITIONAL READING

BAINBRIDGE, J. R. Factorial experiments in pilot plant studies. *Ind. Eng. Chem.* 43, 1300–1306, 1951.

BLISS, C. I. 2 × 2 factorial experiments in incomplete groups for use in biological assays. *Biometrics* 3, 69–88, 1947.

FINNEY, D. J. Main effects and interactions. *Jour. Amer. Stat. Assoc.* 43, 566–571, 1948.

WILLIAMS, E. J. The interpretation of interactions in factorial experiments. *Biometrika* 39, 65–81, 1952.

YOUDEN, W. J. Multiple factor experiments in analytical chemistry. *Analyt. Chem.* 20, 1136–1140, 1948.

CHAPTER 6

Confounding

6.1 The Principle of Confounding

6.11 The 2^3 Factorial Experiment with Complete Confounding. This chapter deals with designs where the size of block is reduced by the sacrifice of accuracy on certain high-order interactions. The method will be illustrated first for a 2^3 experiment with three factors, A, B, and C, each at two levels. Since there are only 8 treatment combinations, the replication is not particularly large and in practice this experiment would most frequently be arranged in ordinary randomized blocks or perhaps in an 8×8 latin square. The example is chosen because of its simplicity.

The interaction of highest order is the ABC interaction. It will be recalled (section 5.23) that this interaction is estimated from the comparison

$$(abc) + (a) + (b) + (c) - (ab) - (ac) - (bc) - (1)$$

Suppose that each replicate in the experiment is divided into 2 blocks of 4 units each, such that one block contains abc, a, b, and c, while the other contains ab, ac, bc, and (1). With 3 replicates the plan (before randomization) would be as follows.

TABLE 6.1 2^3 Experiment in blocks of 4 units, with ABC confounded

	Rep. I		Rep. II		Rep. III	
Block	1	2	3	4	5	6
	abc	ab	abc	ab	abc	ab
	a	ac	a	ac	a	ac
	b	bc	b	bc	b	bc
	c	(1)	c	(1)	c	(1)

There are two important properties of this plan. The total of blocks 1, 3, and 5, minus the total of blocks 2, 4, and 6, is the ABC interaction total. Thus the ABC interaction is one of the components of the comparisons amongst blocks. It is said to be *completely confounded* with blocks. Secondly, the other six factorial effects, A, B, C, AB, AC, and BC, will all be found to be orthogonal with the block totals. For instance, the AB interaction may be written (apart from the divisor)

$$[(abc) + (c) - (a) - (b)] + [(ab) + (1) - (ac) - (bc)]$$

Of the 4 units in any block, two carry a (+) sign in this expression and two carry a (−) sign. Consequently, if we increase all the observations in a selected block by any amount, say 50, the estimate of AB remains unchanged, and similarly for the other 5 factorial effects. This means that these 6 factorial effects are not influenced by differences amongst blocks. Various phrases are used to describe this property: the effects may be said to be unconfounded with blocks, or free from block effects, or to be composed entirely of *within-block* comparisons.

Thus differences amongst blocks of 4 units are eliminated from the experimental errors of the main effects and two-factor interactions, whereas with randomized blocks only differences amongst blocks of 8 units are eliminated. The reduction in effective block size is attained by making ABC the same as one of the block comparisons. There is no within-block information available about ABC.

The degrees of freedom in the analysis of variance separate as follows.

	d.f.
Blocks	5
A, B, C, AB, AC, BC	6
Error	12
Total	23

All sums of squares are calculated in the usual way, so that there is no complexity in the computations.

The composition of the error s.s. is perhaps worth noting. Consider the 3 blocks (1, 3, and 5 in table 6.1) that contain the treatments *abc*, *a*, *b*, and *c*. These may be regarded as a randomized blocks experiment with 4 treatments and 3 blocks. Consequently the interaction of treatments with blocks contains six components. Similarly in blocks 2, 4, and 6 the interaction of the other set of treatments with blocks provides six components. These two sets of six components constitute the error for the complete experiment. In other words, the error term in a confounded factorial is made up of interactions between treatments and incomplete blocks. This remains true in the more complex designs that appear later in this chapter, even though in these cases the composition of the error is less easy to detect.

ABC does not appear explicitly in the analysis above. Actually, the experiment, if properly randomized, provides an estimate of error and a test of significance for *ABC*. In practice it is seldom worth while making this test, which is usually very insensitive. The test may, however, throw additional light on the nature of confounding. Let us ignore all other factorial effects and consider table 6.1 as the plan of an experiment to determine *ABC*, i.e., to determine the difference between the group of treatments *abc, a, b, c* and the group *ab, ac, bc,* (1). From this point of view the block becomes the "experimental unit" and the experiment is of the ordinary randomized blocks type, having 2 "treatments" and 3 replicates. The total s.s. among the 6 "units" is, of course, the blocks s.s. in the previous analysis. Thus the blocks s.s. may be subdivided into:

	d.f.
Replicates	2
ABC	1
Error for *ABC*	2
Total	5

If confounding has been effective, the error for *ABC*, being composed of comparisons among blocks, will be larger than the error (with 12 d.f.) which applies to the rest of the experiment. Moreover, the error for *ABC* is estimated from only 2 d.f., so that the test of *ABC* is a poor one. The test might be worth making with a considerable number of experiments of the same type, where it is desired to examine the average *ABC* effect over the whole group.

6.12 The 2^3 Factorial Experiment with Partial Confounding.

Any of the seven factorial effects may be confounded with blocks in this way. The rest will then be free from block effects. These facts enable us to spread the confounding in an experiment among several factorial effects. A plan of this type is shown in table 6.2.

TABLE 6.2 2^3 EXPERIMENT IN BLOCKS OF 4 UNITS, WITH *ABC, AC, BC* PARTIALLY CONFOUNDED

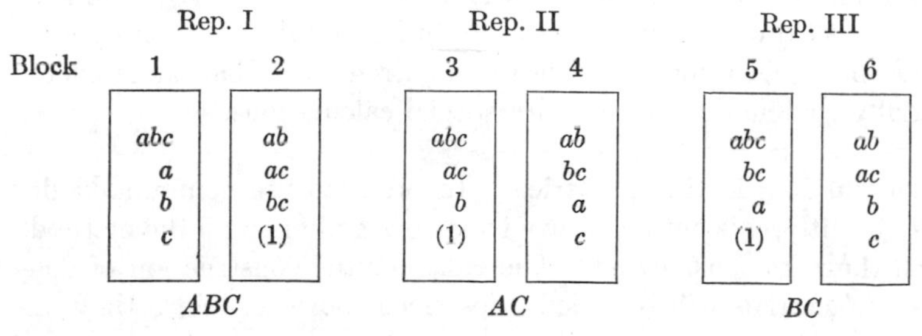

	Rep. I		Rep. II		Rep. III	
Block	1	2	3	4	5	6
	abc	*ab*	*abc*	*ab*	*abc*	*ab*
	a	*ac*	*ac*	*bc*	*bc*	*ac*
	b	*bc*	*b*	*a*	*a*	*b*
	c	(1)	(1)	*c*	(1)	*c*
	ABC		*AC*		*BC*	

The difference between the totals of blocks 1 and 2 represents the *ABC* interaction. In replication II, however, the block composition has been changed so that the difference between the block totals is the *AC* interaction. In replication III the difference is the *BC* interaction.

With this plan, *A*, *B*, *C*, and *AB* are entirely free from block effects. *ABC* is completely confounded with blocks in replicate I, but in the other two replicates the estimate of *ABC* is orthogonal with blocks. Thus a "within-block" estimate of *ABC* may be obtained from replicates II and III. Similarly, a "within-block" estimate of *AC* is available from replicates I and III, and one of *BC* from I and II. These three effects are *partially* confounded with blocks, since an estimate that is free from block effects can still be made for each effect. Each "within-block" estimate is derived from 2 out of the 3 replicates in the experiment. The ratio ⅔ serves as a measure of the extent of the confounding. Yates (6.1) calls this ratio the *relative information* on the confounded interaction. The ratio gives the amount of information available on the partially confounded effect, relative to that available on an unconfounded effect.

In the analysis of variance, all factorial effects appear.

	d.f.
Blocks	5
A	1
B	1
C	1
AB	1
AC	1′
BC	1′
ABC	1′
Error	11
Total	23

The sums of squares for blocks and for the unconfounded effects *A*, *B*, *C*, and *AB* are found in the usual way. The sum of squares for *AC* is calculated only from replicates I and III, the divisor for $[AC]^2$ being 16 instead of 24. Corresponding rules apply to *BC* and *ABC*. In the analysis of variance, the primes are inserted as a reminder that these effects are partially confounded and require special calculations.

6.13 Confounding in the 2^n Series. In order to use confounded designs intelligently, it is not necessary to understand in detail the methods by which they are constructed. The rules for the construction of confounded 2^n factorials will be briefly described, however, since they are fairly simple and may help to explain the choice of plans presented at the end of this chapter.

If a replicate is to be divided into 2 blocks, *any* factorial effect can be confounded with blocks. In most cases we would confound the highest-order interaction in all replicates. Suppose that each incomplete block is to be further divided into 2, so that there will be 4 blocks (I $\cdots$ IV) in a replicate. Having chosen one factorial effect to represent the comparison (I + II − III − IV), we may easily verify that *any other* factorial effect may be made to represent the comparison (I − II + III − IV). That is, to arrange a 2^n factorial in blocks of 2^{n-2} units, we may confound any two factorial effects that we choose.

The remaining block comparison (I − II − III + IV) represents a third factorial effect, which is also confounded with blocks. Barnard (6.2) has shown that this effect is uniquely determined by the two effects that were chosen. It is always their "generalized interaction," and is found by combining all the letters that appear in the two chosen factorial effects, and cancelling all letters that enter twice. Thus, if a 2^5 factorial is to be confounded in blocks of 2^3, or 8, units, we might be inclined to choose *ABCDE* and *BCDE* as the first two factorial effects to be confounded. The generalized interaction of these two effects is *ABBCCDDEE* or, after cancellation, is *A*. Consequently, if these two effects are chosen for confounding, the main effect of *A* is also automatically confounded. These results may be summarized as follows. "If a replicate of a 2^n factorial is arranged in blocks of 2^{n-2} units, three factorial effects are confounded with blocks. Of these, two may be chosen at will: the third is their generalized interaction."

To proceed, a plan with 8 blocks in the replicate can be obtained from one with 4 blocks in the replicate by dividing each block into two parts. In making this division we may confound any factorial effect, except one of the three that is already confounded. Thus, in the 2^5 example, where *ABCDE*, *BCDE*, and *A* were confounded in the division into 4 blocks, we might choose say *ACE* for the division into 8 blocks. But with 8 blocks in the replicate, there are seven components of the comparisons among blocks, and each of these must represent one factorial effect. Hence, in addition to the four factorial effects that are known to be confounded, three others are confounded. These three are the generalized interactions of *ACE* with *ABCDE*, *BCDE* and *A*, or *BD*, *ABD*, and *CE*, respectively.

The general rule should be clear from this case. Suppose that a replicate in a 2^n factorial is to be divided into blocks of 2^{n-k} units, so that there will be 2^k blocks in the replicate. Then we may select any k factorial effects to be confounded, subject only to the restriction that none of these must be a generalized interaction of any group of the others. A further $(2^k - k - 1)$ effects are automatically confounded. These are all the effects which can be expressed as generalized interactions of the group of effects selected for confounding.

In the construction of designs for practical use, it follows that the effects to be confounded must not be selected without examination of the other effects that are automatically confounded. If confounding is to be restricted to high-order interactions, we cannot choose *ABCDE* and *BCDE* in a 2^5 factorial, since *A* would then be confounded. In order to avoid confounding any main effects or first-order interactions in this case, a little trial will show that the only possible choice is two second-order interactions and one third-order interaction: for instance *ABC*, *ADE*, and their generalized interaction *BCDE*. The plans at the end of this chapter contain what appear to be the best choices for general use.

When the interactions to be confounded have been selected, there remains the problem of writing out the plan showing the treatment combinations that appear in each block. This may be done by the procedure used in the present discussion. The replicate is first divided into two by confounding one effect, then into four by confounding a second, and so on. The rules which express each factorial effect in terms of the original treatment combinations (section 5.24) decide the composition of the blocks at each stage. For experiments with a large number of factors, arranged in blocks of small size, Fisher (6.3) has given an alternative method that is more expeditious [see also Finney (6.4)]. In this method a group of 2^{n-k} letter combinations is constructed, such that the generalized interaction of this group with any treatment combination gives the members of the block that contains this treatment combination.

Fisher (6.3) has examined the general problem of confounding when the object is to keep all main effects and first-order interactions completely clear of block effects. He has shown that experiments up to the 2^7 factorial (128 treatment combinations) may be placed in blocks of 8 units, while blocks of 16 units can accommodate experiments up to the 2^{15} factorial (32,768 treatment combinations).

The preceding discussion applied to a single replication. With more than one replicate we may repeat the first replicate, in which case the confounding is complete for the interactions involved. Alternatively, the effects that are confounded may be changed from one replicate to the next, so that some intra-block information is available on all effects.

6.14 Example of a 2^4 Factorial Confounded in Blocks of 8 Units. Table 6.3 contains the plan and yields of a 2^4 field experiment on beans

TABLE 6.3 PLAN AND YIELDS (BEANS IN POUNDS) OF A 2^4 FACTORIAL EXPERIMENT

	Rep. I					Rep. II			
	p	*k*	*d*	*npk*		*npk*	*d*	*p*	*dnk*
Block *a*	45	55	53	36	Block *a*	43	42	39	34
375	*dnk*	*dnp*	*dpk*	*n*	351	*n*	*dnp*	*k*	*dpk*
	41	48	55	42		47	52	50	44
	dp	*nk*	*dk*	*pk*		*nk*	*dp*	(1)	*np*
Block *b*	50	44	43	51	Block *b*	43	52	57	39
381	*dnpk*	(1)	*dn*	*np*	395	*pk*	*dk*	*dnpk*	*dn*
	44	58	41	50		56	52	54	42

conducted by the Rothamsted Experimental Station in 1936. The factors were

Dung (D): none, 10 tons per acre.
Nitrochalk (N): none, 0.4 cwt. N per acre.
Superphosphate (P): none, 0.6 cwt. P_2O_5 per acre.
Muriate of potash (K): none, 1.0 cwt. K_2O per acre.

With 16 treatment combinations in blocks of 8 plots, only one factorial effect is confounded in each replication. From the plan it will be seen that $DNPK$, the highest-order interaction, was confounded in each of the two replicates. The computations proceed as follows.

Step 1. Calculate the totals for each treatment combination, the block and replicate totals, and the grand total. The total s.s. and the sums of squares for replicates and for blocks within replicates should present no difficulty.

Step 2. This deals with the computation of the factorial effect totals. Write down each factorial effect total as a linear function of the treatment combination totals, as shown in table 6.4. The treatment combinations and their total yields are placed in a systematic order in the first two columns, and the remaining columns give the factorial effects. Although the writing of the table requires some time, the procedure is quite simple. When the columns for D and N have been written down, the column for DN is obtained as the *product* of corresponding signs in the D and N columns. Similarly, later in the table DNP may be obtained as the product of the signs of D and NP, or of N and DP, and so on. Thus it is necessary to know only the expressions for the main effects.

The factorial effect totals are shown at the bottom of each column.

TABLE 6.4 Calculation of factorial effect totals in a 2^4 experiment

Treatment combination	Total yield	Factorial effect														
		D	*N*	*DN*	*P*	*DP*	*NP*	*DNP*	*K*	*DK*	*NK*	*DNK*	*PK*	*DPK*	*NPK*	*DNPK*
(1)	115	−	−	+	−	+	+	−	−	+	+	−	+	−	−	+
d	95	+	−	−	−	−	+	+	−	−	+	+	+	+	−	−
n	89	−	+	−	−	+	−	+	−	+	−	+	+	−	+	−
dn	83	+	+	+	−	−	−	−	−	−	−	−	+	+	+	+
p	84	−	−	+	+	−	−	+	−	+	+	−	−	+	+	−
dp	102	+	−	−	+	+	−	−	−	−	+	+	−	−	+	+
np	89	−	+	−	+	−	+	−	−	+	−	+	−	+	−	+
dnp	100	+	+	+	+	+	+	+	−	−	−	−	−	−	−	−
k	105	−	−	+	−	+	+	−	+	−	−	+	−	+	+	−
dk	95	+	−	−	−	−	+	+	+	+	−	−	−	−	+	+
nk	87	−	+	−	−	+	−	+	+	−	+	−	−	+	−	+
dnk	75	+	+	+	−	−	−	−	+	+	+	+	−	−	−	−
pk	107	−	−	+	+	−	−	+	+	−	−	+	+	−	−	+
dpk	99	+	−	−	+	+	−	−	+	+	−	−	+	+	−	−
npk	79	−	+	−	+	−	+	−	+	−	+	−	+	−	+	−
dnpk	98	+	+	+	+	+	+	+	+	+	+	+	+	+	+	+
	1502	−8	−102	32	14	88	50	8	−12	−14	−32	18	28	−22	−32	50

The total for *DNPK* is not required, since owing to the confounding *DNPK* will not appear explicitly in the analysis of variance. The *DNPK* column, however, allows a check to be made, as follows. If the 15 totals are added, it may be verified by inspection of table 6.4 that their sum, +66, should equal 16 times the (*dnpk*) treatment total, minus the grand total, i.e., 16(98) − 1502 = +66. Alternatively, Yates' automatic method of obtaining the factorial effect totals (section 5.24a) can be used in this step.

Step 3. The contribution of each factorial effect to the treatments s.s. in the analysis of variance is now obtained. Since there are 32 plots, the square of each effect is divided by 32. The analysis of variance appears in table 6.5. Note that *DNPK* is omitted, since it already appears

TABLE 6.5 ANALYSIS OF VARIANCE

	d.f.	s.s.	m.s.
Replications	1	3.1	
Blocks in reps.	2	123.2	61.6
D	1	2.0	
N	1	325.1 **	
P	1	6.1	
K	1	4.5	
DN	1	32.0	
DP	1	242.0 **	
NP	1	78.1	
DK	1	6.1	
NK	1	32.0	
PK	1	24.5	
DNP	1	2.0	
DNK	1	10.1	
DPK	1	15.1	
NPK	1	32.0	
Error	14	340.0	24.29
Total	31	1277.9	

in the blocks s.s. As a check, the total of the 14 treatment s.s., 811.6, plus that for *DNPK*, 78.1, should equal the treatments s.s. as found in the ordinary way from the totals for the 16 treatment combinations. The error s.s. is found by subtraction.

Only two effects are significant: the main effect of nitrochalk, which produced a depression of the yield, and the *DP* interaction. Since neither *D* nor *P* produced significant main effects, it is advisable not to lay much stress on the *DP* interaction in the absence of other confirmatory evidence.

Step 4. The best presentation of the results will depend on their nature and on the audience for whom they are intended. Table 6.6 gives

TABLE 6.6 Differential responses (pounds per plot)

Factor	Mean response	Response with							
		Dung (*D*)		Nitrochalk (*N*)		Super. (*P*)		Mur. pot. (*K*)	
		Abs.	Pres.	Abs.	Pres.	Abs.	Pres.	Abs.	Pres.
Dung (*D*)	−0.5	......	......	−2.5	+1.5	−6.0	+5.0	+0.4	−1.4
Nitrochalk (*N*)	−6.4	−8.4	−4.4	......	......	−9.5	−3.2	−4.4	−8.4
Super. (*P*)	+0.9	−4.6	+6.4	−2.2	+4.0	......	......	−0.9	+2.6
Mur. pot. (*K*)	−0.8	+0.1	−1.6	+1.2	−2.8	−2.5	+1.0	......	......

s.e.: ±2.46 for differential response; ±1.74 for mean responses.

a concise presentation of the main effects and two-factor interactions in a form that has been used frequently at Rothamsted Experimental Station. The response to each factor is shown separately for each level of every other factor. Thus the row labelled "Dung" contains the mean response and the differential responses to dung. The figure −2.5 (response to dung with nitrochalk absent) is the average response to dung over all plots that did not receive nitrochalk. The table enables a quick appraisal to be made of the nature of the main effects and two-factor interactions. In the present case, of course, the table is scarcely needed because of the dearth of effects.

The data in the table are easily calculated from the factorial effect totals in table 6.4. For instance,

$$\text{Mean response to dung} = \tfrac{1}{16}[D] = -\tfrac{8}{16} = -0.5$$

$$\text{Mean response to dung (nitrochalk absent)} = \tfrac{1}{16}\{[D] - [DN]\} = -\tfrac{40}{16} = -2.5$$

$$\text{Mean response to dung (nitrochalk present)} = \tfrac{1}{16}\{[D] + [DN]\} = +\tfrac{24}{16} = +1.5$$

These relations may be verified from the signs in the columns in table 6.4.

The standard error per plot is $\sqrt{24.29}$, or 4.93. Since each differential response in table 6.6 is the difference between the means of two groups of 8 plots, its standard error is (4.93)/2, or 2.46. The standard error for a *mean* response is $\dfrac{1}{\sqrt{2}}$ times this value, or 1.74.

6.15 Confounding of a 3^2 Factorial. Let the symbol ij denote the treatment combination that has the ith level of A and the jth level of B ($i, j = 0, 1, 2$). As before, we wish to keep the main effects clear of block effects, that is, to confound only the AB interaction. Since there are 9 treatment combinations in a replicate and since 3 is the only factor of 9, the size of the incomplete block must be 3 units.

We have seen (section 5.26) that the main effects of A and B both contain two components, while the AB interaction has four components. Further, the main effects and the interaction can be divided into single components in various ways, and the division that is most appropriate for interpretation of the results will change from experiment to experiment. This suggests that the particular components of the interaction which we should desire to confound will vary from case to case. Actually, the possibilities for confounding are much more restricted with the 3^n than with the 2^n system: in fact, only one set of components of AB lends itself readily to confounding.

The confounding is based on the properties of the 3×3 graeco-latin square. Suppose that the 9 treatment combinations are set out as in table 6.7, on which a 3×3 square is superimposed. Following the nota-

TABLE 6.7 Use of a 3×3 graeco-latin square to obtain the AB interaction

	b_0	b_1	b_2
a_0	$(00)I_1J_1$	$(01)I_3J_2$	$(02)I_2J_3$
a_1	$(10)I_2J_2$	$(11)I_1J_3$	$(12)I_3J_1$
a_2	$(20)I_3J_3$	$(21)I_2J_1$	$(22)I_1J_2$

tion used by Yates (6.1), we denote the latin letters in the square by I_1,* I_2, and I_3, respectively, and the greek letters by J_1, J_2, and J_3, respectively.

It is clear that comparisons among the row totals of the square give the two components of the main effect A, while the comparisons among the column totals give the main effect of B. Now consider the I totals:

$$I_1 = (00) + (11) + (22); \qquad I_2 = (10) + (21) + (02);$$
$$I_3 = (20) + (01) + (12)$$

By the basic property of the latin square, comparisons among these totals are orthogonal to both rows and columns: that is, to the main effects of A and B. Hence, the comparisons among the I totals must

* In this edition the I and J values agree with those used by Yates.

represent two of the four components of the AB interaction. The same argument shows that the two remaining components of AB are derived from the comparisons among the J totals, where

$$J_1 = (00) + (21) + (12); \qquad J_2 = (10) + (01) + (22);$$
$$J_3 = (20) + (11) + (02)$$

A numerical verification of this fact was made at the end of section 5.26, where the sum of squares for AB was calculated from the latin and greek letter totals which correspond to the I and J totals.

The application of this result to confounding is shown in table 6.8.

TABLE 6.8 3 × 3 EXPERIMENT WITH AB PARTIALLY CONFOUNDED

Plan (i)

Incomplete blocks

(i)	(ii)	(iii)
(00) (11) (22)	(10) (21) (02)	(20) (01) (12)
I_1	I_2	I_3
I components confounded		

Plan (ii)

Incomplete blocks

(i)	(ii)	(iii)
(00) (21) (12)	(10) (01) (22)	(20) (11) (02)
J_1	J_2	J_3
J components confounded		

There are two possibilities. In plan (i) the I components of AB are completely confounded with incomplete blocks, since the block totals have been made the same as the I totals. The main effects and the J components of AB are unconfounded. In plan (ii), the J components of AB are completely confounded. It should be noted that neither the I nor the J components are easy to interpret in terms of the results of an actual experiment. They are selected for confounding because they are convenient for this purpose.

If the number of replicates in an experiment is even, we may use plan (i) in half the replicates and plan (ii) in the other half. With this arrangement the I components may be estimated (clear of block effects) from those replicates in which plan (ii) is used, and vice versa for the J components. The relative accuracy is ½ for each set of components. It may be shown that the relative accuracy is also ½ for *any* component of AB in which we may be interested. The confounding is said to be *balanced* with respect to AB. Balanced confounding is preferable, for there is no reason to confound I more heavily than J.

This design has limited practical utility. It may be used where the factors are known to operate independently, so that AB will be negligible. If information about AB is wanted, a randomized blocks design is advisable unless there will be a substantial reduction in error variance from the use of incomplete blocks. The same principle of construction, however, leads to a very useful design for the 3^3 factorial.

6.16 Confounding of a 3^3 Factorial. The same notation will be used. Thus, (021) denotes the treatment combination $a_0b_2c_1$. Since there are 27 treatment combinations, the possible sizes of incomplete block are 3 and 9 units. The plan for blocks of 3 units necessitates confounding of the two-factor interactions and is not given here. With 9 units per block, only ABC need be confounded. In the case of the 3^2 factorial we were able to divide the 9 treatment combinations into groups of three (the I and J groups), such that comparisons among the group totals gave the components of AB. We shall show that the 27 treatment combinations in a 3^3 factorial can be divided into groups of *nine* such that comparisons among the group totals give the components of ABC. Each set of three groups will contribute two components of ABC. Since ABC has in all eight components or degrees of freedom, there will be four such sets.

The AB interaction can be calculated separately for each level of C. As before, we will obtain the AB interaction from its I and J components. Table 6.9 shows the I components for each level of C.

TABLE 6.9 THE I COMPONENTS OF AB SHOWN FOR EACH LEVEL OF C

	C_0	C_1	C_2
I_1	(000) + (110) + (220)	(001) + (111) + (221)	(002) + (112) + (222)
I_2	(100) + (210) + (020)	(101) + (211) + (021)	(102) + (212) + (022)
I_3	(200) + (010) + (120)	(201) + (011) + (121)	(202) + (012) + (122)

We may regard table 6.9 as a 3×3 table, in which each entry is the total of three treatment combinations. The row totals of the table give the I components of AB, while the column totals give the main effect of C. Further, we may take I and J totals from this table just as in table 6.7. These totals will be called $I - I_1$, $I - J_1$, etc., since they come from the I components of AB. Thus

$I - I_1$: (000) + (110) + (220) + (101) + (211) + (021) + (202) + (012) + (122)

$I - I_2$: (100) + (210) + (020) + (201) + (011) + (121) + (002) + (112) + (222)

$I - I_3$: (200) + (010) + (120) + (001) + (111) + (221) + (102) + (212) + (022)

By the same argument as for the 3×3 design, the comparisons among these totals of 9 treatment combinations must represent two of the components of the interaction of AB with C; that is, of the ABC interaction.

Consequently if we put all treatment combinations in $I - I_1$ into the first block, those in $I - I_2$ into the second, and those in $I - I_3$ into the third, we have a plan which completely confounds two of the eight components of ABC, and leaves all other factorial effects unconfounded. This arrangement appears as replication I in plan 6.7 for the 3^3 factorial at the end of this chapter.

The second set of three groups of 9 treatment combinations is obtained by taking the J totals from table 6.9.

$I - J_1$: (000) + (110) + (220) + (201) + (011) + (121) + (102) + (212) + (022)

$I - J_2$: (100) + (210) + (020) + (001) + (111) + (221) + (202) + (012) + (122)

$I - J_3$: (200) + (010) + (120) + (101) + (211) + (021) + (002) + (112) + (222)

This grouping constitutes replication II in plan 6.7. The remaining sets are obtained by forming a 3×3 table similar to table 6.9 for the J components of AB at each level of C. The reader may verify that the $J - I_1$, $J - I_2$, and $J - I_3$ groups are as shown in replicate III, and that the $J - J_1$, $J - J_2$, and $J - J_3$ groups are as shown in replicate IV of plan 6.7.

6.17 Example of a 3^3 Factorial Confounded in Blocks of 9 Units.

This experiment, conducted by the Seed Laboratory, Iowa State College, tests the effects of 3 levels of nitrogen, 3 of phosphorus, and 3 of potash on the germination of lettuce seedlings. The seed was thoroughly mixed, and divided into 108 samples of about 60 seeds each. Each sample was planted in a copper box, 6 in. square and 1½ in. deep, in a mixture of soil and sand. The boxes were placed in a germinator, at a temperature of about 32°C. At the end of 5 to 7 days the seedlings were classified as normal, abnormal, hard, or dead. The data in table 6.10 show the numbers of normal lettuce plants.

There were 4 replications, each placed on a different shelf in the germinator. On a shelf the boxes were placed in 3 columns of 9 boxes each, each column being an incomplete block. It will be noted that all 3 fertilizers had a deleterious effect on emergence.

TABLE 6.10 NUMBER OF LETTUCE PLANTS EMERGING IN A 3^3 EXPERIMENT

Rep. 1

Blocks	A npk	A No.	B npk	B No.	C npk	C No.	
	012	11	201	42	111	46	
	122	11	121	20	102	48	
	−220	13	210	24	221	58	
	−202	12	011	38	001	53	
	101	11	112	39	010	54	
	021	30	100	61	212	37	
	−000	41	+002	40	−022	41	
	110	21	+222	46	120	25	
	211	21	+020	44	+200	32	
		171		354		394	919

Rep. 2

A npk	A No.	B npk	B No.	C npk	C No.	
121	39	100	32	010	26	
−220	34	012	37	120	16	
102	31	210	37	+200	6	
201	34	221	33	021	13	
−000	40	−202	27	211	12	
−022	31	+020	21	+222	12	
212	26	111	12	+002	16	
011	35	001	30	112	19	
110	38	122	22	101	14	
	308		251		134	693

Rep. 3

Blocks	A npk	A No.	B npk	B No.	C npk	C No.	
	101	26	+020	19	−220	19	
	210	17	211	25	122	19	
	221	21	121	18	100	42	
	−000	38	110	27	201	11	
	112	22	+222	20	010	36	
	011	19	001	40	111	37	
	120	18	012	30	+002	29	
	−202	22	102	28	021	27	
	−022	14	+200	25	212	13	
		197		232		233	662

Rep. 4

A npk	A No.	B npk	B No.	C npk	C No.	
112	37	+020	47	+222	15	
121	31	+002	33	201	42	
010	44	221	15	012	41	
−022	25	122	29	210	24	
100	39	011	38	−000	52	
211	28	212	27	111	28	
−202	35	+200	38	120	30	
001	37	101	31	102	50	
−220	26	110	32	021	29	
	302		290		311	903

The use of the + and − signs is discussed later in this section.

Grand total 3177

TABLE 6.11 Additional summary totals

Treatment totals

	n_0			n_1			n_2		
	p_0	p_1	p_2	p_0	p_1	p_2	p_0	p_1	p_2
k_0	171(000)	160(010)	131(020)	174(100)	118(110)	89(120)	101(200)	102(210)	92(220)
k_1	160(001)	130(011)	99(021)	82(101)	123(111)	108(121)	129(201)	86(211)	127(221)
k_2	118(002)	119(012)	111(022)	157(102)	117(112)	81(122)	96(202)	103(212)	93(222)

Two-way tables

	n_0	n_1	n_2	
p_0	449	413	326	1188
p_1	409	358	291	1058
p_2	341	278	312	931
k_0	462	381	295	1138
k_1	389	313	342	1044
k_2	348	355	292	995
	1199	1049	929	3177

	p_0	p_1	p_2	
k_0	446	380	312	1138
k_1	371	339	334	1044
k_2	371	339	285	995
	1188	1058	931	3177

NPK components

	1*A*	1*B*	1*C*	Total
From treat. totals	944	1102	1131	3177
From rep. 1	171	354	394	919
Difference	773	748	737	2258
	2*A*	2*B*	2*C*	Total
From treat. totals	1119	1113	945	3177
From rep. 2	308	251	134	693
Difference	811	862	811	2484
	3*A*	3*B*	3*C*	Total
From treat. totals	1025	1073	1079	3177
From rep. 3	197	232	23[illegible]	662
Difference	828	841	846	2515
	4*A*	4*B*	4*C*	Total
From treat. totals	1104	991	1082	3177
From rep. 4	302	290	311	903
Difference	802	701	771	2274

The computations proceed as follows.

Step 1. Form the block and replicate totals and the grand total (shown in table 6.10) and also the totals for each treatment combination and the three two-way tables (shown in table 6.11).

Step 2. These data enable us to calculate the total s.s. and the sum of squares for replications, blocks within replications, and for the N, P, K, NP, NK, and PK factorial effects. All are obtained in the usual way and are entered in the preliminary analysis of variance (table 6.12).

TABLE 6.12 ANALYSIS OF VARIANCE FOR A $3 \times 3 \times 3$ FACTORIAL

	d.f.	s.s.	m.s.
Replications	3	2,041.88	
Blocks within replications	8	5,008.15	626.02
N	2	1,016.67	508.34 **
P	2	917.39	458.70 **
K	2	293.39	146.70
NP	4	399.27	99.82
NK	4	589.61	147.40
PK	4	212.89	53.22
NPK: confounded in replications			
1	2′	25.21	12.60
2	2′	64.22	32.11
3	2′	6.39	3.20
4	2′	198.30	99.15
Error	70	4,146.88	59.24
Total	107	14,920.25	

Subdivision of part of the treatments s.s.

		d.f.	s.s.	m.s.
N:	L	1	1,012.50	1,012.50 **
	Q	1	4.17	4.17
P:	L	1	917.35	917.35 **
	Q	1	0.04	0.04
K:	L	1	284.01	284.01 *
	Q	1	9.37	9.37
NP:	$L \times L$	1	184.08	184.08
	Rest	3	215.19	71.73
NK:	$L \times L$	1	256.69	256.69 *
	Rest	3	332.92	110.97
NPK:	$L \times L \times L$	1′	59.12	59.12

Notice that we do not calculate the total s.s. for treatments (with 26 d.f.), since this contains part of the blocks s.s.

Step 3. There remains the calculation of the contribution from the *NPK* interactions. Before doing this, it may be remarked that sometimes, from the nature of the factors or from previous experience, there is good reason to believe that the three-factor interactions will be negligible. In this case the experimenter may decide to pool the sum of squares for *NPK* with the error s.s. without troubling to compute the sum of squares for *NPK*. The pooled error would have 78 d.f. and would be obtained by subtraction of the sum of squares calculated in step 2 from the total s.s. The danger from this procedure is that if three-factor interactions are present the experimenter may not detect them, and moreover his estimate of error will be inflated. For many types of field experimentation in agriculture the procedure seems reasonably safe. When there is doubt it is better to isolate *NPK*.

Consider the two components of *NPK* that are confounded in replication 1. The contribution of these components to the sum of squares for *NPK* must be calculated from the remaining 3 replicates, in which they are unconfounded with blocks. The totals needed are shown in table 6.11 under the heading "*NPK* components." From the treatment totals, compute the total (944) of the 9 treatment combinations that appear in block 1*A*. Thus

$$\text{Block } \mathit{1A} = \mathit{012} + \mathit{122} + \mathit{220} + \mathit{202} + \mathit{101} + \mathit{021} + \mathit{000} + \mathit{110} + \mathit{211}$$

$$944 = 119 + 81 + 92 + 96 + 82 + 99 + 171 + 118 + 86$$

In the same way we obtain 1102 for the total over all treatments that appear in block 1*B* and 1131 for block 1*C*. Underneath these figures we place the respective totals for blocks 1*A*, 1*B*, and 1*C*. By subtraction we obtain the totals 773, 748, and 737. These are the totals of the groups of treatment combinations taken over replications 2, 3, and 4. The sum of squares of deviations of these quantities from their mean is divided by 27, since each figure contains 27 observations. The result, 25.21, is the contribution of the two components of *NPK* to the sum of squares for *NPK*. The remaining six components are found similarly from replications 2, 3, and 4. Of course, all eight components could be computed in one step as

$$\tfrac{1}{27}[(773)^2+(748)^2+\cdots+(701)^2+(771)^2]-\tfrac{1}{81}[(2258)^2+\cdots+(2274)^2]$$

The error s.s., with 70 d.f., is calculated by subtraction.

Step 4. In an experiment of this type, where each factor has equally spaced levels of an ingredient, it is usually advisable to examine the linear and quadratic components of the response curves. The contributions to the treatments s.s. are displayed in the lower section of table 6.12.

The method of calculation was described in section 5.26, where the data for N and P in this experiment were used as an example. All three fertilizers show significant linear responses, with no indication of any departure from linearity.

With regard to the two-factor interactions, the "linear by linear" components have been isolated for NP and NK (see section 5.26 for the method). For PK it was not thought worth while to do this, since the total s.s. for PK, 212.89, is not large enough to allow any single component to be significant. The "linear by linear" component is significant for NK but not for NP.

Where the "linear by linear" components of the two-factor interactions are large, it is desirable to isolate the "linear by linear by linear" component of the three-factor interaction. This will be done as an exercise. First we must define this component. With *two* factors, say N and P, the N_LP_L component has been defined as

$$(22) + (00) - (20) - (02)$$

where the numbers refer to the levels of N and P. The value of this quantity at the highest level of K, *minus* its value at the lowest level of K, measures the linear effect of K on N_LP_L, and is the $N_LP_LK_L$ component in question. Algebraically, it is

$$L = (222) + (200) + (020) + (002) - (022) - (202) - (220) - (000)$$

The estimate from the treatment totals will be found to be -27.

Since three-factor interactions are partially confounded with blocks, this component is also partially confounded and must be adjusted so as to remove block effects. The nature of the confounding is seen by writing down the expression above in the original table of plot yields (table 6.10). This has been done by means of the $+$ and $-$ signs that appear to the left of the seedling numbers in the table. It is evident that L is orthogonal to the totals of blocks $1C$, $2B$, $3C$, and $4C$, since each of these blocks has one $+$ and one $-$. In blocks $1B$, $2C$, $3B$, and $4B$, L has three $+$ signs, while in the remaining blocks it has three $-$ signs. Hence, a quantity that is free from block effects is

$$3L - (1B) - (2C) - (3B) - (4B) + (1A) + (2A) + (3A) + (4A)$$

or

$$3(-27) - 354 - 134 - 232 - 290 + 171 + 308 + 197 + 302 = -113$$

the block symbols denoting block totals. This expression, apart from a divisor, may be shown to be the least squares estimate of $N_LP_LK_L$, adjusted for block differences. This estimate requires the assumption that all other components of the NPK interaction are negligible.

If this quantity is expressed algebraically as a linear function of the plot yields, the sum of squares of the coefficients is found to be 216. The contribution of $N_L P_L K_L$ to the treatments s.s. is therefore $(113)^2/216$, or 59.12, which is practically the same as the error m.s. If the number of replicates differs from four, the procedure for isolating the "linear by linear by linear" component remains the same, except that the divisor for the square becomes $54r$ instead of 216.

Step 5. This concerns the presentation of the results. Usually it will be sufficient to show the three two-way tables of means, which are derivable from the two-way tables of totals (in table 6.11) on division by 12. Since all main effects and two-factor interactions are unconfounded, standard errors and t-tests for the 3×3 tables are obtained just as in a randomized blocks experiment. The principal results are that each fertilizer has produced a significant decrease in the numbers of seedlings that emerged, the decrease being substantially proportional to the amount of dressing. The significant $N_L K_L$ interaction represents the fact that the decrease in emergence from n_0 to n_2 was smaller at the k_2 level than at the k_0 level. There is some indication of a similar effect with N and P, though this did not attain significance.

The individual 27 treatment totals or means cannot be used as they stand for interpretative purposes, since they contain some block effects. A table of these totals or means will probably be unnecessary unless some aspect of the three-factor interaction requires study. To obtain such a table, we adjust each total so as to remove block effects. Each block effect is first estimated. For this purpose we do not use simply the observed block mean, since that in turn contains treatment effects. The least squares estimate of any block effect is

$$\frac{1}{27}[4(\text{block total}) - (\text{total of treatments appearing in the block})]$$

The basic data needed are available in the section headed "*NPK* components" in table 6.11. Thus, for block 1*A*, the estimated effect is

$$\frac{1}{27}[4(171) - (944)] = -9.6$$

The block effects are given below.

	Block		
Replication	*A*	*B*	*C*
1	−9.6	+11.6	+16.5
2	+4.2	−4.0	−15.1
3	−8.8	−5.4	−5.4
4	+3.9	+6.3	+6.0

To adjust any treatment total, we note from the plan the 4 blocks in which it appears, and compute the sum of the effects for these 4 blocks.

This quantity is *subtracted* from the unadjusted total to give the adjusted total. Thus for $n_1p_1k_0$, which appears in blocks 1*A*, 2*A*, 3*B*, and 4*B*, the adjusted total is

$$118 - [-9.6 + 4.2 - 5.4 + 6.3] = 122.5$$

To obtain the adjusted mean we divide by 4 as usual.

6.18 Mixed Series: the 3 × 3 × 2 Factorial. In the designs discussed in previous sections, all factors have the same number of levels. When the number of levels differs from one factor to another, the utility of confounding is limited. Usually some two-factor interactions must be confounded, and the computations become more laborious. Only five designs of this type are given at the end of this chapter, though a number of others are available in the literature. As an example, we will consider the confounding of a $3 \times 3 \times 2$ factorial, which has 18 treatment combinations.

The main effects of a factor will be kept clear of block effects if every block contains each level of the factor the same number of times. Thus the factor A, which occurs at 3 levels (0, 1, 2), is unconfounded if every block contains an equal number of 0's, 1's, and 2's. Consequently, block size has to be a multiple of 3. Similarly we can keep the main effects of B (0, 1, 2) unconfounded if the block size is a multiple of 3. For the factor C at 2 levels (0, 1), the block size must be a multiple of 2. Hence, in order to keep all main effects clear of blocks, the only feasible block size is 6.

Further, with 6 units in a block, every possible combination of the levels of A and those of C can appear once in every block, so that we may expect to be able to keep AC, and likewise BC, unconfounded. We cannot place all the 9 combinations for AB in a block, so that AB will be partially confounded.

From this approach the plan can be constructed rather easily. Every block is to contain the 6 combinations (00), (10), (20), (01), (11), (21) of B and C. The only question is the manner in which the 3 levels of A are combined with the 6 pairs above. This allocation is to be such that all 6 combinations of AC appear in the block. That is, the 0, 1, and 2 levels of A must each appear with the 0 level of C and each with the 1 level of C. We may impose one additional rule, designed to confound AB as little as possible. Although we cannot represent all 9 combinations of AB in a block, it will be well to represent as many as possible. Thus we introduce the restriction that any AB combination (e.g., 12) must not appear more than once in a block.

Under these rules only four types of replicate can be made up, as shown in table 6.13.

Consider, for instance, in how many ways the block containing (000) can be constructed. The first three A levels (reading from the top in table 6.13) must be either 0, 1, 2 or 0, 2, 1, so that each A level shall appear with the 0 level of C. If 0, 1, 2 is chosen, the remaining three A

TABLE 6.13 Possible blocks for a 3 × 3 × 2 factorial

		Level of A											
B	C	I*a*	I*b*	I*c*	II*a*	II*b*	II*c*	III*a*	III*b*	III*c*	IV*a*	IV*b*	IV*c*
0	0	1	2	0	2	0	1	1	2	0	2	0	1
1	0	2	0	1	0	1	2	0	1	2	1	2	0
2	0	0	1	2	1	2	0	2	0	1	0	1	2
0	1	2	0	1	1	2	0	2	0	1	1	2	0
1	1	0	1	2	2	0	1	1	2	0	0	1	2
2	1	1	2	0	0	1	2	0	1	2	2	0	1

levels must be either 1, 2, 0 or 2, 0, 1, these being the only sequences that make no AB combination appear twice. These sequences give blocks I*c* and II*b* respectively. Similarly, the choice of 0, 2, 1 for the first three A's leads only to blocks III*c* and IV*b*.

In the same way, we find that there are only four possible blocks containing the treatment combination (001) and four containing (002.) The three sets of 4 blocks can be grouped into 4 separate replications (I, II, III, and IV in table 6.13).

As Yates (6.1) has shown, the plan can be written in a more condensed form (table 6.14) which exhibits the nature of the confounding.

TABLE 6.14 Condensed form of the plan for a 3 × 3 × 2 confounded factorial

Level of C	Blocks											
	I_a	I_b	I_c	II_a	II_b	II_c	III_a	III_b	III_c	IV_a	IV_b	IV_c
0	I_2	I_3	I_1	I_3	I_1	I_2	J_2	J_3	J_1	J_3	J_1	J_2
1	I_3	I_1	I_2	I_2	I_3	I_1	J_3	J_1	J_2	J_2	J_3	J_1

For example, in block I_a the 3 combinations that appear with the zero level of C are (10), (21), and (02). This is the group that forms the

I_2 component of *AB* (see section 6.15). It follows that in the first 2 replicates the *I* component of *AB* is partially confounded; the *J* component can be verified to be unconfounded. In replicates III and IV the *J* component is partially confounded, while the *I* component is clear of blocks. *ABC* is partially confounded in all replicates. The relative information on *AB* has been shown to be 7/8 and that on *ABC*, 5/8.

6.19 Numerical Example of the 3 × 3 × 2 Design. Table 6.15 shows the plan and analysis of variance for an experiment on the response of young tung trees to fertilizers. The results of this experiment have been reported by Merrill, Kilby, and Greer (6.17). The general object was to

TABLE 6.15 DATA FOR 3 × 3 × 2 FACTORIAL EXPERIMENT IN BLOCKS OF 6 UNITS

I*a*	I*b*	I*c*		II*a*	II*b*	II*c*	
100: 80	200: 78	000: 38		200:136	000: 43	100: 89	
210: 86	010: 55	110: 73		010: 56	110: 81	210: 87	
020: 70	120: 82	220: 75		120: 64	220: 90	020: 66	
201: 74	001: 67	101: 78		101: 95	201: 81	001: 91	
011: 82	111: 67	211: 51		211: 76	011: 61	111: 97	
121: 86	221: 57	021: 66		021: 71	121: 65	221: 60	
478	406	381	1,265	498	421	490	1,409

III*a*	III*b*	III*c*		IV*a*	IV*b*	IV*c*	
100: 86	200: 73	000: 66		200: 88	000: 53	100: 81	
010: 79	110: 76	210: 85		110:107	210: 66	010: 58	
220: 73	020: 97	120:101		020: 70	120: 92	220: 56	
201: 97	001:116	101:117		101: 79	201: 88	001: 90	
111: 79	211: 86	011:106		011: 92	111:109	211: 68	
021:113	121: 81	221:102		221: 96	021: 95	121: 67	
527	529	577	1,633	532	503	420	1,455

Analysis of variance

Source	d.f.	s.s.	m.s.
Replications	3	3,837	1,279
Blocks within replications	8	2,836	354
Levels of 8-8-6 in row (*A*)	2	1,116	558
Meals (*B*)	2	254	127
Levels of 8-8-6 side-dressed (*C*)	1	868	868 *
8-8-6 in row × meals (*AB*)	4′	1,129	282
8-8-6 in row × 8-8-6 side-dressed (*AC*)	2	2,995	1,498 **
Meals by 8-8-6 side-dressed (*BC*)	2	424	212
8-8-6 in row × meals × 8-8-6 side-dressed (*ABC*)	4′	1,016	254
Error	43	8,909	207
Total	71	23,384	

discover the type of fertilizer application that would stimulate the early growth so that the young trees would be ready for transplanting to commercial orchards by early autumn. The data shown are the mean heights in centimeters of the 12 tung trees on each plot. For convenience in following the computations, the plot yields follow the same order as in table 6.13. Note that factor A appears last in table 6.13 and first in table 6.15.

The three factors were as follows, all dressings being *per acre.*

A	B	C
8-8-6 fertilizer applied in the row		8-8-6 fertilizer applied as side-dressing
0 None	0 None	0 None
1 200 lb.	1 Tung meal (650 lb.)	1 200 lb.
2 400 lb.	2 Cottonseed meal (650 lb.)	

The side-dressing was placed in a shallow furrow on one side of the row about 8 inches from the tung seedlings. Factors A and B were applied at planting (March 13), factor C on June 3, and the height measurements were made on October 23. It should be noted that all three factors supply each of the common plant nutrients (N, P, and K).

In an experiment of this type, one might anticipate that two-factor interactions will be present. If, for example, 400 lb. of fertilizer applied in the row supply all the nutrients that the young tree can utilize, plots that receive this dressing will show no response to either factor B or C. On the other hand, plots that receive the zero level of A may respond to the dressings supplied in factors B and C. It would not be wise to confound any two-factor interactions heavily. Since, however, the soils available for experiments had been found to be very variable, it was decided to arrange the experiment in blocks of 6 plots. As we have seen, this will partially confound AB, though as the relative information is high ($7/8$), not much information is lost on this comparison even if the incomplete blocks prove ineffective.

Table 6.16 shows the computation sheet leading to the analysis of variance. The steps are as follows.

i. Calculate the block totals (denoted by Ia, Ib, etc., and shown in table 6.15), and the treatment totals (table 6.16). Calculate the total s.s. and the blocks s.s. The latter has been divided into two components—replications (3 d.f.) and blocks within replications (8 d.f.). This separation enables us to estimate the decrease in error variance due to the reduction in block size from 18 to 6 plots. The analysis of variance appears at the foot of table 6.15.

ii. Form the two three-way totals as shown. From the margins of the table for $(c_1 + c_0)$, calculate the sum of squares for the main effects

TABLE 6.16 Computation sheet for 3 × 3 × 2 experiment, blocks of 6 units

(i) Treatment totals

	c_0			c_1		
	b_0	b_1	b_2	b_0	b_1	b_2
a_0	200	248	303	364	341	345
a_1	336	337	339	369	352	299
a_2	375	324	294	340	281	315

(ii) Two-way tables

	$c_1 + c_0$				$c_1 - c_0$			
	b_0	b_1	b_2	Totals	b_0	b_1	b_2	Totals
a_0	564(1)	589(4)	648(7)	1,801	+164(1)	+93(4)	+42(7)	299
a_1	705(2)	689(5)	638(8)	2,032	+ 33(2)	+15(5)	−40(8)	8
a_2	715(3)	605(6)	609(9)	1,929	− 35(3)	−43(6)	+21(9)	−57
Totals	1,984	1,883	1,895	5,762	162	65	23	250

(iii) From $(c_1 + c_0)$

$$I_1 = (1) + (5) + (9) = 1{,}862 \qquad J_1 = (1) + (6) + (8) = 1{,}807$$
$$I_2 = (2) + (6) + (7) = 1{,}958 \qquad J_2 = (2) + (4) + (9) = 1{,}903$$
$$I_3 = (3) + (4) + (8) = 1{,}942 \qquad J_3 = (3) + (5) + (7) = 2{,}052$$
$$5{,}762 \qquad\qquad 5{,}762$$

$$2I_1' = 2I_1 + \mathrm{I}_a + \mathrm{II}_a = 4{,}700 \qquad 2J_1' = 2J_1 + \mathrm{III}_a + \mathrm{IV}_a = 4{,}673$$
$$2I_2' = 2I_2 + \mathrm{I}_b + \mathrm{II}_b = 4{,}743 \qquad 2J_2' = 2J_2 + \mathrm{III}_b + \mathrm{IV}_b = 4{,}838$$
$$2I_3' = 2I_3 + \mathrm{I}_c + \mathrm{II}_c = 4{,}755 \qquad 2J_3' = 2J_3 + \mathrm{III}_c + \mathrm{IV}_c = 5{,}101$$
$$14{,}198 \qquad\qquad 14{,}612$$

$$\text{s.s. for } AB = \frac{(4{,}700)^2 + (4{,}743)^2 + \cdots + (5{,}101)^2}{84} - \frac{(14{,}198)^2 + (14{,}612)^2}{252} = 1{,}129$$

(iv) From $(c_1 - c_0)$

$R_1 = +200$, $R_2 = +32$, $R_3 = +18$; $S_1 = +81$, $S_2 = +147$, $S_3 = +22$

where, e.g., $R_1 = (1) + (5) + (9), \cdots, S_3 = (3) + (5) + (7)$

$$2R_1' = 2R_1 - \mathrm{I}_b + \mathrm{I}_c + \mathrm{II}_b - \mathrm{II}_c = +306$$
$$2R_2' = 2R_2 + \mathrm{I}_a - \mathrm{I}_c - \mathrm{II}_a + \mathrm{II}_c = +153$$
$$2R_3' = 2R_3 - \mathrm{I}_a + \mathrm{I}_b + \mathrm{II}_a - \mathrm{II}_b = +\ 41$$
$$+500$$

$$2S_1' = 2S_1 - \mathrm{III}_b + \mathrm{III}_c + \mathrm{IV}_b - \mathrm{IV}_c = +203$$
$$2S_2' = 2S_2 + \mathrm{III}_a - \mathrm{III}_c - \mathrm{IV}_a + \mathrm{IV}_c = +132$$
$$2S_3' = 2S_3 - \mathrm{III}_a + \mathrm{III}_b + \mathrm{IV}_a - \mathrm{IV}_b = +\ 75$$
$$+500$$

$$\text{s.s. for } ABC = \frac{(+306)^2 + (+153)^2 + \cdots + (+75)^2}{60} - \frac{(+500)^2}{90} = 1{,}016$$

of A and B. From the *total* of the table for $(c_1 - c_0)$, we calculate the sum of squares for the main effect of C: this is $(250)^2/72$, since there are 72 plots. Similarly, the margins of the table provide the sum of squares for the AC and BC interactions, which being unconfounded are obtained in the usual way.

iii. These steps lead to the sum of squares for AB. First calculate the I and J components of AB, which are obtained by summation from the $(c_1 + c_0)$ table in step ii, as shown in table 6.16 (iii). Then apply the adjustments for block effects. Since I_1, for instance, does not appear in blocks I_a or II_a, the adjustments involve the totals for these two blocks. The adjustments lead to the quantities $2I_1'$, etc., from which the sum of squares for AB follows as shown.

iv. These steps lead to the sum of squares for ABC. From the table for $(c_1 - c_0)$, step ii, form totals R_i and S_i which are calculated by the same rules as for the I's and J's, respectively. These totals are then adjusted for block effects, yielding quantities $2R_1'$, $2S_1'$, etc., from which the ABC s.s. is easily obtained. The error s.s. is then computed by subtraction.

v. Summary and presentation of results. Consider first the conclusions indicated by the analysis of variance. The only significant mean square for a main effect is that for C (side-dressing). The AC interaction is significant at the 1% level, while the mean square for A, though not significant, is substantially above the error m.s. Neither B (meals) nor any of its interactions approaches significance. These results suggest that the AC two-way table should be examined.

Since table 6.17 contains no confounded effects, the means per plot are found simply by addition and division from table 6.16, (i). It is

TABLE 6.17 AC TWO-WAY TABLE

Row dressing	Mean heights (centimeters) ± 4.15	
	Side dressing	
	c_0 None	c_1 200 lb.
a_0 None	62.6	87.5
a_1 200 lb.	84.3	85.0
a_2 400 lb.	82.8	78.0

evident that there are large and significant responses to the 200 lb. dressings, both when applied in the row and as side-dressing. The means, 84.3 for the row application and 87.5 for the side-dressing, do not differ significantly, so that the applications appear to have been about equally effective. There is no further increase to the 400 lb. row dressing, or to mixtures containing both types of dressing. Thus the large AC interaction arises primarily because each dressing is effective only in the absence of the other.

The analysis of variance would indicate that no definite effects of the meals (B) can be established. As an exercise, the AB two-way table, which involves partially confounded effects, will be examined. The means must be adjusted for block effects. This could be done by first calculating a mean for each incomplete block, adjusted for treatment effects, and then adjusting each AB mean for the 4 blocks in which it lies. It is slightly quicker to perform the adjustments by means of the I and J effects, as follows.

In table 6.16, (iii), divide the quantities $2I_1'$, etc., by 42, and calculate the deviations from their mean. The quantities $2J_1'$, etc., are treated similarly. The same procedure is carried out for the I's and J's, except that the divisor is 24. This gives

$$i_1' = -0.8, \quad i_2' = 0.2, \quad i_3' = 0.5 \qquad j_1' = -4.7, \quad j_2' = -0.8, \quad j_3' = 5.4$$
$$i_1 = -2.4, \quad i_2 = 1.6, \quad i_3 = 0.9 \qquad j_1 = -4.7, \quad j_2 = -0.7, \quad j_3 = 5.5$$
$$\delta i_1 = +1.6, \quad \delta i_2 = -1.4, \quad \delta i_3 = -0.4 \qquad \delta j_1 = 0.0, \quad \delta j_2 = -0.1, \quad \delta j_3 = -0.1$$

The unadjusted mean of (a_0b_0) is 70.5. From table 6.16, (ii) and (iii), we see that (a_0b_0) belongs to I_1 and J_1. Hence its adjusted value is

$$(a_0b_0)' = (a_0b_0) + \delta i_1 + \delta j_1 = 70.5 + 1.6 + 0.0 = 72.1$$

By the same principle, the adjustment to (a_2b_1), for instance, is $(\delta i_2 + \delta j_1)$. The complete two-way table is shown below.

TABLE 6.18 AB TWO-WAY TABLE, ADJUSTED FOR BLOCK EFFECTS

Row dressing	Mean heights (centimeters)		
	b_0 None	b_1 Tung	b_2 Cottonseed
a_0 None	72.1	73.1	79.5
a_1 200 lb.	86.6	87.6	79.4
a_2 400 lb.	88.9	74.2	77.6

The standard error of a single figure in the table is $\sqrt{207/8}$, or 5.09, since each figure is the mean of 8 plots. This standard error applies to t-tests of unconfounded effects; i.e., of the main effects of A and B. For t-tests of components of the AB interaction, the figure must be multiplied by $\sqrt{8/7}$, or 1.069, the value 8/7 being the inverse of the relative efficiency on AB.

For t-tests of quantities that are a mixture of main effects and interactions, the multiplier lies between 1.069 and 1. It will usually be sufficiently accurate to use the multiplier 1.069. If desired, the correct standard error may be obtained by means of a rule which will be given without proof. The quantity whose standard error is wanted is expressed in terms of components of main effects and interactions. Suppose that we wish to test the response to cottonseed when no row fertilizer is applied. The response is (79.5 − 72.1), or 7.4 cm. The approximate standard error is $5.09 \times 1.069 \times \sqrt{2}$, or 7.69. To obtain the correct value, we write symbolically,

$$(b_2 - b_0)a_0 = \frac{(b_2 - b_0)(a_0 + a_1 + a_2)}{3} + \frac{(b_2 - b_0)(2a_0 - a_1 - a_2)}{3}$$

If all means are unadjusted means, this is an algebraic identity. The first term on the right is a component of the main effect of B, while the second is a component of the AB interaction. Further, by a property of least squares, the equation remains valid if means adjusted for block effects are substituted for unadjusted means. By the rule for the variance of a linear function, the variance of the first term on the right is $6\sigma^2/9$, where σ^2 is the variance of a single figure. By the same rule, the variance of the second term is

$$\left(\frac{12}{9}\right)\left(\frac{8}{7}\right)\sigma^2$$

the factor 8/7 being introduced because the second term is a component of AB. The two terms are orthogonal, so that the estimated standard error of $(b_2 - b_0)a_0$ is

$$5.09\sqrt{\tfrac{6}{9} + (\tfrac{12}{9})(\tfrac{8}{7})} = 5.09\sqrt{2.190} = 7.53$$

Adjustment of the 18 individual treatment means is more tedious. In table 6.16, (iv), we divide the quantities $2R_1'$, etc., by 30, and the quantities R_1, etc., by 24, thereafter taking deviations from the mean of each trio. This gives

$r_1' = +4.6,\quad r_2' = -0.5,\quad r_3' = -4.2 \qquad s_1' = +4.2,\quad s_2' = -1.2,\quad s_3' = -3.1$

$r_1 = +4.8,\quad r_2 = -2.2,\quad r_3 = -2.7 \qquad s_1 = -0.1,\quad s_2 = +2.6,\quad s_3 = -2.6$

$\delta r_1 = -0.2,\quad \delta r_2 = +1.7,\quad \delta r_3 = -1.5 \qquad \delta s_1 = +4.3,\quad \delta s_2 = -3.8,\quad \delta s_3 = -0.5$

In this experiment there is little point in adjusting all 18 treatment means. However, the unadjusted treatment totals, table 6.16, (i), suggest that there might be an effect of the meals (factor B) when applied alone. Hence it may be of interest to test the difference between the adjusted means of $a_0b_2c_0$ and $a_0b_0c_0$.

From table 6.16, (iii) and (iv), we see that $a_0b_0c_0$ belongs to i_1, j_1, r_1, and s_1. Since the unadjusted mean is 50.0, the adjusted mean is

$$50.0 + \delta i_1 + \delta j_1 - \delta r_1 - \delta s_1 = $$
$$50.0 + 1.6 + 0.0 + 0.2 - 4.3 = 47.5 \text{ cm.}$$

Note that the terms in δr and δs carry a *minus* sign whenever the treatment is at the c_0 level, and a *plus* sign when the treatment is at the c_1 level. For $a_0b_2c_0$, which belongs to i_2, j_3, r_2, and s_3, the adjusted mean will be found to be 73.1 cm.

In order to obtain the standard error of the difference (73.1 − 47.5) = 25.6, this must be expressed in terms of components of main effects and interactions. This requires some practice, but the method is not hard to master. Consider the following table of values of $(b_2 - b_0)$, averaged over the 4 replicates.

	$(b_2 - b_0)$ a_0	a_1	a_2	Total
c_0	d	e	f	l
c_1	g	h	k	m
Total	n	o	p	q

The quantity whose standard error we wish is d. Now it is easy to verify that

$$6d = q + (l - m) + (2n - o - p) + [(2d - e - f) - (2g - h - k)]$$

components of $\quad B \quad BC \quad AB \quad ABC$

As before, this is an identity for the unadjusted means and remains valid for adjusted means. Hence by the same rule as before

$$36\sigma^2{}_d = \frac{\sigma^2}{4}\left[12 + 12 + 24\left(\frac{8}{7}\right) + 24\left(\frac{8}{5}\right)\right] = 22.46\sigma^2$$

where σ^2 is the estimated variance per plot. The extra factor 8/7 is used for the AB component and the factor 8/5 for the ABC component, for which the relative information is 5/8. Consequently, the estimated standard error of d is

$$\sqrt{\frac{22.46 \times 207}{36}} = 11.4 \text{ cm.}$$

It follows that the t-value for the response to cottonseed when applied alone is 25.6/11.4, or 2.25, with 43 d.f., which is significant at the 5% level.

These results illustrate a feature that is common to the mixed series of confounded factorials. If the examination of the results is directed mostly at simple rather than factorial effects, the analysis is likely to be time-consuming because of the adjustments needed to remove block effects and to construct standard errors by the method given above. It will also be apparent that considerable experience in the analysis of variance, or access to expert advice on this topic, is advisable.

6.2 The Use of Confounded Designs

6.21 General Recommendations. In order to make efficient use of confounding, it is necessary to consider whether the advantage is likely to outweigh the disadvantages.

The advantage comes from the reduction in the experimental error by the use of a block which is more homogeneous, or which can be subjected to a more uniform technique, than the complete replicate. Without some idea of the amount of this reduction, a realistic decision cannot be made on the question of confounding. If uniformity data have been collected, the experimenter may compare the variability within the incomplete blocks with the variability within replicates. In addition, from the results of an experiment in which confounding has been employed, it is usually possible to estimate what the experimental error would have been if confounding had not been used. The method of calculation, which was first outlined by Yates (6.5, p. 215), is given in section 6.35. In many types of research the reduction in error from confounding varies greatly from experiment to experiment. In such cases little reliance can be placed on the results of a single trial; a summary of a moderately large group of experiments is necessary.

The disadvantages of confounding consist of (1) the reduction in replication on the confounded treatment comparisons, and (2) in some cases, a greater complexity in the calculations. With regard to the first, the experimenter should take into account the importance of the interactions that must be confounded and the extent to which they are confounded. No interaction should be *completely* confounded unless there is good reason to believe, either from previous experience or from the nature of the factors to be tested, that the interaction will be negligible.

As an illustration of the type of reasoning that might be employed in reaching a decision, consider the $3 \times 3 \times 2$ design discussed in the previous section. The incomplete blocks contain 6 units, with 7/8 relative information on AB and 5/8 relative information on ABC. ABC is expected to be negligible; AB, however, is of sufficient interest so that confounding is undesirable if it involves any loss of precision for this comparison. The issue resolves itself into the question: is the error variance per unit with blocks of 6 units likely to be less than 7/8 of the error variance with blocks of 18 units? If so, the AB interaction is estimated more precisely from the confounded design than from randomized blocks, because the reduction in variance per unit more than compensates for the loss of replication due to confounding. Since the relative information is useful for this type of argument, it is given in table 6.19 for all the plans presented at the end of this chapter.

The increase in the amount of computation varies both with the type of design and with the type of analysis needed to summarize the results. Where all the confounding is *complete*, the statistical analysis is practically the same as with randomized blocks. With all factors at two or three levels (i.e., in the 2^n and 3^n series), the extra computations are not formidable even when partial confounding is used; they are much more troublesome when the number of levels is not the same for all factors, as in the 3×2^2 and 3×2^3 designs. The extra work consists in the calculation of adjustments to the partially confounded comparisons so that their sums of squares in the analysis of variance are freed from block effects. If it is desired to present tables of mean yields which involve confounded comparisons, the entries in the tables must be adjusted so as to remove block effects. This point should be remembered, especially with designs where two-factor interactions are confounded. If t-tests of simple effects are needed, special computations must be made to obtain the standard errors for these tests. Moreover, if part of the data is missing, the analysis tends to become much more complicated than with randomized blocks.

The experimenter who has not previously used confounding and who does not have access to expert statistical advice is advised to confine himself at first to the simplest types of confounding. *He should make sure that he understands the method of analysis before the experiment is conducted.* After the experiment is completed, he should investigate the increase in precision that has been achieved by confounding, as a guide to the future use of the technique.

Yates (6.5) has pointed out that in certain circumstances a confounded design might give rise to a misleading interpretation of the results. This can be illustrated by means of the 2^3 factorial in table 6.1 (p. 183). This has three replicates, with ABC completely confounded in each replicate. Suppose that the response to A is much larger than any other effects, and that this response varies greatly from block to block. That is, the interaction of A with blocks is much larger than the interactions of the other factors with blocks. Now the factorial effect total for BC may be written

$$BC = (abc) + (bc) + (a) + (1) - (ab) - (ac) - (b) - (c)$$

$$= [(abc) + (a) - (b) - (c)] - [(ab) + (ac) - (bc) - (1)]$$

$$= \text{response to } A \text{ in blocks 1, 3, 5} - \text{response to } A \text{ in blocks 2, 4, 6}$$

The point is that the factorial effect of BC is a component of the interaction of A with blocks. When tested against the error m.s. (12 d.f.) this response might be significant, not because there is an effect of BC, but because A has large interactions with blocks.

Another way of describing the source of the trouble is to say that the error m.s. is not homogeneous, in that the interaction of A with blocks is larger than the interactions of other treatments. If this were suspected, the danger of a misleading interpretation could be avoided by subdividing the error m.s. We can isolate 4 d.f. (two from blocks 1, 3, and 5 and two from blocks 2, 4, and 6) that represent the interaction of A with blocks. The mean square for these four could be used to test A and BC, while the remaining factorial effects would be tested by the other 8 d.f. for error. This device can usually be applied in the 2^n series, though generally not in the mixed series. Further, the heterogeneity of the error variance may not be suspected.

How frequently this danger arises will depend on the nature of the experimentation. Investigations by Yates (6.5) and Kempthorne (6.9) suggest that there is little danger with fertilizer experiments on English farm crops. In lines of research where errors have often been found to be heterogeneous, one should be on the alert for this effect.

In some experiments the use of confounding is practically unavoidable. For instance, an experiment on methods of manufacturing ice cream, planned at North Carolina State College, contained 3 factors each at 3 levels. It was desired to discover whether ordinary consumers could detect differences in the palatability of the 27 different types of ice cream. From previous experience it was believed that a taster (who was not an expert judge) could compare at most 3 ice creams at any single test. Moreover, it did not seem feasible to ask the tasters to grade the palatabilities on any kind of standard scale: they could merely be asked to rank the 3 ice creams in order of preference, if they had any preferences. Consequently, a design for a 3^3 factorial in blocks of 3 units was required, even though this involved confounding of two-factor interactions. The design adopted also employed partial confounding of the main effects, so as to obtain approximately equal accuracy on main effects and two-factor interactions.

TABLE 6.19 INDEX TO PLANS OF FACTORIAL EXPERIMENTS CONFOUNDED IN RANDOMIZED INCOMPLETE BLOCKS

a. Designs with which any number of replicates may be used

Type	Number of treatments	Number of units per block	Interactions confounded in a single replicate *	Plan
2^3	8	4	*ABC*	6.1
2^4	16	8	*ABCD*	6.2
2^4	16	4	*AB, ACD, BCD*	6.4
2^5	32	8	*ABC, ADE, BCDE*	6.5
2^6	64	16	*ABCD, ABEF, CDEF*	6.3
2^6	64	8	*ABC, CDE, ADF, BEF, ABDE, BCDF, ACEF*	6.6
3^3	27	9	*ABC* (3/4)	6.7
3^4	81	9	*ABC, ABD, ACD, BCD*, all (3/4)	6.8
4^2	16	4	*AB* (2/3)	6.12
4×2^2	16	8	*ABC* (2/3)	6.13

* The fractions in parentheses give the relative information on the comparisons which are confounded. Where no fraction is given, the comparison is completely confounded.

b. Balanced designs

Type	Number of treatments	Units per block	Number of replicates for a balanced design	Interactions confounded and relative information (in parentheses) *	Plan
2^4	16	4	$6n$ †	All two-factor (5/6); all three-factor (1/2)	6.4
2^5	32	8	$5n$	All three-factor (4/5); all four-factor (4/5)	6.5
2^6	64	8	$10n$	All three-factor (4/5); all four-factor (4/5)	6.6
3^3	27	9	$4n$	All three-factor (3/4)	6.7
3^4	81	9	$4n$	All three-factor (3/4)	6.8
3×2^2	12	6	$3n$ ‡	*BC* (8/9), *ABC* (5/9)	6.9
3×2^3	24	6	$3n$ ‡	*BC, BD, CD*, all (8/9); *ABC, ABD, ACD*, all (5/9)	6.10
$3^2 \times 2$	18	6	$4n$	*AB* (7/8), *ABC* (5/8)	6.11
4^2	16	4	$3n$	*AB* (2/3)	6.12
4×2^2	16	8	$3n$	*ABC* (2/3)	6.13
$4 \times 3 \times 2$	24	12	$9n$ ‡	*AC* (26/27), *ABC* (23/27)	6.14

* The factors *ABC* ··· are read from the left; thus the *BC* interaction in a 3×2^2 design is the interaction between the 2 factors at 2 levels.

† The symbol "$6n$" denotes that the number of replicates should be a multiple of 6.

‡ In these cases only the balanced design is recommended.

6.22 Index to Experimental Plans. The opportunities for confounding vary with the number of factors and with the number of levels of each factor. The plans given at the end of this chapter are those that are likely to be most frequently used.

Table 6.19, which forms an index to the plans, is divided into two parts. The first part shows a group of designs in which any number of replicates may be used. In successive replicates, the experimenter may either (i) repeat the first replicate (with a new randomization), in which case the same comparisons are confounded in all replicates, or (ii) change the grouping into blocks in successive replicates, so as to spread the confounding over a greater number of comparisons. In the plans, the factorial effects that are confounded are shown separately for each replication. Consequently, the reader can readily see what possibilities are open in this direction. Where confounding can be restricted to a high-order interaction, as in the 2^4 design in blocks of 8 units, the first procedure is preferable. On the other hand, with the 2^4 design in blocks of 4 units, where one two-factor interaction must be confounded in every replicate, the confounding should be changed in successive replications, unless it is certain that one of the two-factor interactions will be negligible. More detailed recommendations for the individual designs are given in section 6.3.

The second part, table 6.19*b*, shows a group of *balanced* designs, in which all interactions of the same order and the same type are confounded to the same extent. Thus in the $3 \times 2 \times 2 \times 2$ (or 3×2^3) design, all first-order interactions between the factors which have 2 levels are equally confounded, the relative information being 8/9. The first-order interactions between the factor at 3 levels and any factor at 2 levels are unconfounded. Balanced designs are useful in cases where all interactions of the same order and type are of equal interest, so that it is undesirable to confound some more fully than others. Also, the computations are usually simpler. For this reason, only the balanced designs should be used in certain cases, as indicated in table 6.19. On the other hand, with balanced designs the available numbers of replications are severely limited.

Table 6.20 shows a number of designs for which plans are given in the references cited. These designs are mainly for mixed series, and will not be discussed here. The most recent addition to this list consists of a series of designs by Binet *et al.* (6.1a), which may be used either in single replication (see section 6.24) or in more than one replication. These designs confound linear and quadratic components as little as possible. Table 6.20a presents selected replicates which appear to be most useful.

6.23 Randomization. The procedure is the same for all designs.

1. Arrange the positions of the blocks at random within each replication, using a new randomization for each replication.

2. Allot the treatments in any block at random to the units in the block, using a new randomization for each block.

TABLE 6.20 CONFOUNDED DESIGNS FOR OTHER FACTORIAL EXPERIMENTS

Type	Units per block	Number of replicates	Interactions confounded	Reference
$3^2 \times 2^2$	12	$2n$	AB (7/8), $ABCD$ (5/8)	6.6
$3^3 \times 2$	18	$2n$	ABC, $ABCD$	6.6
$3^3 \times 2$	6	$2n$	AB, AC, BC (7/8), ABD, ACD, BCD, ABC	6.1
4×3^2	12	$2n$	BC (7/8), ABC	6.6
$4^2 \times 2$	16	Any	ABC	6.6
$4^2 \times 3$	12	$3n$	AB (26/27), ABC	6.6
4^3	16	Any	ABC	6.7
4×2^3	8	$3n$	ABC, ABD, ACD	6.6
4^4	16	Any	ABC, ABD, ACD, BCD	6.7
5^2	5	$4n$	AB (3/4)	6.8
5×2^2	10	$5n$	BC (24/25), ABC	6.6
5^3	25	Any	ABC	6.8

TABLE 6.20a DEGREE CONFOUNDING DESIGNS IN SINGLE REPLICATES *

Type	Units per block	Relative information on confounded first and second degree effects	Table*	Plans
3×2	3	B_1 (8/9)	6	1
3×3	4 and 5	A_2 (8/9), B_2 (8/9)	59	..
$3^2 \times 2$	9	C_1 (80/81)	9	..
$3^2 \times 2^2$	12	A_1B_1 (31/32)	15	1–6
	12	A_1C_1 (11/12)	15	9 and 10
	18	C_1D_1 (80/81)	13	7
4×2	4	None confounded	20	2
4×3	6	A_2 (8/9)	21	3
4^2	4	A_1B_1 (21/25)	29	2
	8	None confounded	27	1 and 2
5×2	5	B_1 (24/25)	36	1 and 3
5×3	7 and 8	A_2 and B_2 (>96%)	61	..
5×4	10	B_2 (24/25)	38	1 and 2
5×3^2	15	B_1C_1 (43/50)	40	1–4
$5 \times 3 \times 2$	15	C_1 (>99%)	42	7 and 8
5^2	12 and 13	A_2 and B_2 (>99%)	62	3

* See reference Binet *et al.* (6.1a). A_1 is a linear (first degree) component, while A_2 or A_1B_1 are quadratic (second degree) components.

TABLE 6.20a (*Continued*)

Type	Units per block	Relative information on confounded first and second degree effects	Table	Plans
6×2	6	A_1B_1 (>99%)	45	1
6×3	9	A_1 (>99%)	46	1
6×4	12	None confounded	47	2 and 5
6×5	15	A_1 (>99%)	49	1
6^2	18	A_1B_1 (>99%)	59	1
6×2^2	12	None confounded	51	1 and 2
$6 \times 3 \times 2$	18	A_1C_1 (>99%)	52	9
7×2	7	B_1 ($^{48}/_{49}$)	55	..
7×4	14	B_2 ($^{48}/_{49}$)	56	..
7×2^2	14	B_1C_1 ($^{48}/_{49}$)	57	..
7×3	10 and 11	A_2 and B_2 (>98%)	63	..
7×5	17 and 18	A_2 and B_2 (>99%)	64	..

6.24 Experiments in Single Replication. In most of the plans at the end of this chapter, the total number of treatment combinations is large. This fact limits the number of replications that can be employed. Sometimes the resources are sufficient for only a single replication unless the experimenter omits some factors which he originally intended to include.

With only one replication, we cannot derive an estimate of error from the interactions of treatments with blocks. If, however, certain high-order interactions are negligible, their mean squares in the analysis of variance will behave exactly like components of the error m.s., and therefore can be used to provide an estimate of error. In the 2^n series, the smallest design which provides an adequate number of error d.f. is the 2^5 in blocks of 8 units (plan 6.5). In replicate 1 of this plan, ABC, ADE, and $BCDE$ are completely confounded. If the remaining three-, four-, and five-factor interactions furnish the estimate of error, the analysis of variance appears as follows.

	d.f.
Blocks	3
Main effects	5
Two-factor interactions	10
Error (from high-order interactions)	13
Total	31

Other designs in which a sufficient number of error d.f. are available from interactions among three or more factors are the 2^6, the 3^3, and the 3^4. The interactions that constitute the estimate of error should be chosen before the results have been inspected. Recommendations for each of the designs above are given in the notes on the plans (section

6.3). It may sometimes be wise to deviate from these recommendations. For instance, in the 2^5 example above, the main effects of A, C, and D and the AC and AD interactions might all turn out to be rather large. This would suggest that the ACD interaction may not be negligible, and that it should not be included in the interactions used as error. A decision of this type should be made before examining ACD itself. The practice of examining the high-order interactions and using those that are small as error leads to a serious underestimation of the true error variance.

The estimation of error from high-order interactions is, of course, open to criticism. If some of these interactions happen to be large, the error m.s. that is used will overestimate the true error variance and the fact that the interactions are large will not be discovered. Where a considerable series of experiments of the same general type is being conducted, some safeguard is obtained both by examining the high-order interactions in experiments that are replicated and by watching the two-factor interactions in experiments with single replication. If most two-factor interactions are small, it seems very unlikely (though still possible) that interactions of higher order will be large. If many two-factor interactions are found to be large, this suggests that some three-factor interactions may also be substantial. See Yates (6.5) and Cornish (6.10).

6.3 Notes on the Plans and Statistical Analysis

6.31 2^n Series. In plans 6.1 to 6.6, one level of a factor is denoted by the corresponding letter; the other level by the absence of the letter. With designs that are not balanced, care must be taken to allocate the letters a, b, c, . . . to the factors in the experiment so that the interactions which are confounded are those that the experimenter has decided to confound. For instance, the first replication of plan 6.4 (2^4 design in blocks of 4) confounds the interaction AB, but no other two-factor interaction. The factors labelled A and B should be those whose interaction is of least interest.

The statistical analysis follows the same general method for all 2^n designs. The total s.s. and the sum of squares among blocks are obtained by the usual methods. The factorial effects may be computed by one of the methods outlined in section 6.14. When this calculation has been completed, the sums of squares are added for all unconfounded comparisons. Let this total be S_T. If all the confounding is complete, S_T is the treatment s.s., since the effects that are confounded appear already in the block s.s. For each comparison which is partially confounded, the sum of squares is calculated separately *from those replications in which the comparison is unconfounded.* The replications to be used are clear from the plans, which show the factorial effects that are confounded in each replication. The total of these sums of squares is added to S_T to give the treatment s.s. The error s.s. is found by subtraction.

2^3 factorial, blocks of 4 units. Plan 6.1, which confounds ABC completely, may be used for every replication. For a worked example, see reference (6.1), p. 19.

Alternatively, with 4, or a multiple of 4, replicates, we may confound partially all two-factor interactions and the three-factor interaction, the relative information on each comparison being 3/4. The plan and statistical analysis are described in reference (6.1), p. 21.

2^4 factorial, blocks of 8 units. Plan 6.2 should be used for every replication, $ABCD$ being completely confounded.

Note that the treatment s.s. contains only 14 d.f. since the single degree of freedom for $ABCD$ is included in the blocks.

2^4 factorial, blocks of 4 units. Plan 6.4 gives a balanced design with 6 replications. If a smaller number of replicates is used, the replicates chosen depend on the interactions which the experimenter decides to confound. Since one two-factor interaction is confounded in each replicate, it will usually be preferable to confound different comparisons in successive replications. For instance, if AB and CD are to be confounded, replicates 1 and 6 should be selected. This arrangement gives 1/2 relative information on AB, CD and the three-factor interactions ACD, BCD, ABC, ABD, all other comparisons being clear of blocks.

With a 2^4 experiment, the investigator may be uncertain whether to confound in blocks of 8 units or in blocks of 4 units. The former design is recommended unless (*a*) the interactions which are confounded with blocks of 4 units are known to be small or (*b*) the block of 4 units is much more homogeneous than the block of 8 units. The latter condition may apply where the experimental material is extremely variable, or falls naturally into groups of 4 units.

2^5 factorial, blocks of 8 units. The balanced design (plan 6.5) requires 5 replicates. When fewer than 5 replicates are used, it may be satisfactory to confound completely two three-factor interactions and one four-factor interaction. In this case replication 1 is repeated the required number of times.

If some information is desired on all treatment comparisons, replications 1, 2, $\cdots$ may be used up to the proposed number. With replications 1 and 2, for instance, the interactions ABC, ADE, ABD, BCE, $BCDE$, and $ACDE$ are partially confounded, each with relative information 1/2.

The statistical analysis is described in reference (6.1), p. 27, for the case in which the experiment contains only a single replication.

2^6 factorial, blocks of 16 units. Plan 6.3, which shows a single replication with $ABCD$, $ABEF$, and $CDEF$ confounded, may be used for all

replications. The 3 interactions above are completely confounded, and are omitted from the treatment s.s. in the analysis of variance.

If there is only 1 replication, an estimate of error may be obtained from the unconfounded four-, five-, and six-factor interactions. These comparisons supply 22 d.f., of which three are confounded with blocks, leaving 19 for the estimate of error. The separation of the 63 d.f. is then as follows:

	d.f.
Blocks	3
Main effects	6
Two-factor interactions	15
Three-factor interactions	20
Error (from high-order interactions)	19

2^6 factorial, blocks of 8 units. In plan 6.6, the size of block is reduced from 64 to 8 units without confounding any two-factor interactions. Four of the 20 three-factor interactions and 3 of the 15 four-factor interactions are confounded in each replication. The balanced design requires 10 replications, i.e., 640 experimental units, and necessitates a large experiment.

With a single replication, the 63 d.f. in the analysis of variance subdivide as follows:

	d.f.
Blocks	7
Main effects	6
Two-factor interactions	15
Three-factor interactions	16
Error (from high-order interactions)	19

The simplest procedure when there is more than one replication is to confound completely a single set of 7 factorial effects by the repeated use of replication 1. For partial confounding, the successive replicates in plan 6.6 may be taken. In the first 4 replicates no three-factor interaction is confounded more than once.

6.32 3^n Series. From plan 6.7 onwards, the levels or variations of each factor are denoted by the figures 0, 1, 2, $\cdots$, while the factors are read from left to right. The treatment 102 implies the middle level of the first factor, the lowest level of the second factor, and the highest level of the third factor. In an experiment the numbers 0, 1, 2, $\cdots$ may be assigned to the levels or variations of a factor according to any convenient scheme.

The testing of three instead of two variations of each factor makes possible a more thorough evaluation of the effects of the factors. In particular, the 3^3 factorial experiment has been much used in investigations of the responses to fertilizers, since the effects of each of 3 different plant nutrients may be explored at 3 levels of application, which provide some information on the shape of the response-curve.

3^3 factorial, blocks of 9 units. In plan 6.7, two of the 8 d.f. for ABC are completely confounded in each replication, so that the balanced design requires 4 replicates and provides 3/4 relative information on ABC. Unless the three-factor interactions are known to be negligible, in which case replication I may be used for all replicates of the experiment, it is preferable to select the required number of replications in succession from the plan. For example, an experiment with 3 replicates gives 2/3 relative information on 6 of the 8 interaction d.f., and full replication on the remaining 2 d.f. As usual, the rule in the analysis of variance is to estimate confounded comparisons from those replications in which they are not confounded. An example of the analysis with 4 replicates has already been given (section 6.17).

For an example of the analysis with a single replicate, see reference (6.1), p. 53. If the three-factor interactions are used as error, the separation of the degrees of freedom is as shown on the left in table 6.25.

TABLE 6.25 Subdivision of the degrees of freedom in a 3^3 experiment with single replication

I. Three-factor interactions used as error

	d.f.
Blocks	2
Main effects	6
Two-factor interactions	12
Error	6
Total	26

II. For experiments where all effects are approximately linear

	d.f.
Blocks	2
Main effects	6
AB: linear × linear	1
AC: linear × linear	1
BC: linear × linear	1
Error	15
Total	26

The number of error d.f., 6, is rather small. There is an alternative scheme for experiments where the effects of the factors are linear, or nearly so. Under these circumstances, the only components of the two-factor interactions that are likely to be substantial are the interactions of the linear responses to the factors. Hence, nine of the 12 d.f. for two-factor interactions are added to the material used for an estimate of error, as shown in table 6.25, II. If in addition the linear components of the two-factor interaction are large, a further variant is to remove from the error the ABC: linear × linear × linear term. This leaves 14 d.f.

for error. The calculation of these components should be fairly simple with the help of the example worked in section 6.17.

3^4 factorial, blocks of 9 units. With 4 factors each at 3 levels (plan 6.8) the block size is reduced from 81 to 9 units by confounding three-factor interactions. There are 4 three-factor interactions: *ABC*, *ABD*, *ACD*, and *BCD*, each with 8 d.f. Two components from each set of 8 are confounded in any replication. If the experiment has more than 1 replicate, partial confounding is generally preferable, the required number of replicates being selected from the balanced set. The balanced design, with 4 replicates, gives threefold replication on all three-factor interactions.

Statistical analysis with a single replication. An analysis may be calculated without difficulty from the following partition of the degrees of freedom.

	d.f.
Blocks	8
Main effects	8
Two-factor interactions	24
Error (from three- and four-factor interactions)	40

All sums of squares are calculated in the usual way. For the main effects and two-factor interactions, six 3×3 tables are required. The error, which is found by subtraction, contains 24 d.f. representing unconfounded three-factor interactions and the 16 d.f. for the four-factor interactions.

This analysis gives no test of the three-factor interactions. If a test of *ABC*, for example, is wanted, the first step is to calculate in the usual way the sum of squares for the 8 d.f. From this we must subtract the sum of squares for the 2 d.f. that are confounded with blocks.

In plan 6.8, replication I, the 2 d.f. are denoted by *ABC* I, where the numeral I refers to replication I of plan 6.7. If we ignore *D*, it will be found that blocks 1, 6, and 8 contain the same set of 9 treatments as block I*a* in plan 6.7. Calculate the total yield of these 3 blocks (27 units). Similarly, we calculate the total yield of blocks 3, 5, and 7, which correspond to block I*b* in plan 6.7, and of blocks 2, 4, and 9, which correspond to block I*c*. The sum of squares of deviations of the 3 totals, divided by 27, is subtracted from the sum of squares for the 8 d.f. The resulting 6 d.f. are removed from the error, which now has 34 d.f.

The reader may verify that the 2 confounded degrees of freedom for *BCD* (i.e., *BCD* II) are obtained from the totals of blocks 1, 2, 3, blocks 4, 5, 6, and blocks 7, 8, 9. In this case *A* is ignored when plan 6.8 is compared with plan 6.7.

Statistical analysis with more than one replication. If all three-factor interactions are presumed to be negligible, they may be combined with the error. The sums of squares for blocks, main effects, and two-factor interactions are found in the usual way, while the remainder constitutes the error.

All 16 components, each with 2d.f., of the three-factor interactions may, however, be calculated and tested. Partially confounded comparisons are taken only from replicates in which they are unconfounded. For example, if there are two replications of plan 6.8, all the data are used in the calculation of components *ABC* III and *ABC* IV, which are unconfounded. For *ABC* II we use only the data from the first replicate of the experiment, and for *ABC* I only the data from the second replicate.

6.33 Mixed Series. *3×2^2 factorial, blocks of 6 units.* Only the balanced plan (6.9) should be used; this may be repeated if extra replications are wanted. Thus the number of replicates must be 3, or some multiple of 3. Although the *BC* interaction is partially confounded, the relative information is high (8/9).

The statistical analysis is described, with a numerical example, in reference (6.1), p. 58.

$3^2 \times 2$ factorial, blocks of 6 units. In plan 6.11, *AB* and *ABC* are partially confounded. Either 2 or 4 replicates may be used, though 4 are required for complete balance. A numerical example for 4 replicates was given in section 6.19.

With 2 replicates (say I and II in plan 6.11) the I and R components are partially confounded, while the J and S components are unconfounded. The I and R totals are adjusted by the same formulae as is table 6.16, (iii) and (iv). The J and S totals need no adjustment. The interaction s.s. in the analysis of variance are computed as follows.

$$AB(I) = \tfrac{1}{36}[(2I_1')^2 + (2I_2')^2 + (2I_3')^2] - \tfrac{1}{108}[2I_1' + 2I_2' + 2I_3']^2$$

$$AB(J) = \tfrac{1}{12}[J_1^2 + J_2^2 + J_3^2] - \tfrac{1}{36}[J_1 + J_2 + J_3]^2$$

$$ABC(R) = \tfrac{1}{12}[(2R_1')^2 + (2R_2')^2 + (2R_3')^2] - \tfrac{1}{36}[2R_1' + 2R_2' + 2R_3']^2$$

$$ABC(S) = \tfrac{1}{12}[S_1^2 + S_2^2 + S_3^2] - \tfrac{1}{36}[S_1 + S_2 + S_3]^2$$

The *AB* two-way table and the table of individual treatment means are adjusted in the same way as with four replicates. The quantities $2I'$, $2R'$ are divided by 18 and 6, respectively, to obtain i' and r'. For i and r, we divide I and R each by 12. After taking deviations from the mean in each case, we calculate δi and δr. Thereafter the adjustment proceeds exactly as with 4 replicates, being somewhat simpler since J and S components do not enter.

3×2^3 factorial, blocks of 6 units. In plan 6.10 the interactions BC, BD, and CD, between pairs of factors that occur at 2 levels, are partially confounded with 8/9 relative information, while 5/9 relative information is retained on the interactions ABC, ABD, and ACD. Only the balanced design which requires 3 replicates is recommended.

The following computations lead to the sum of squares due to partially confounded effects. Let

$$g_1 = \mathrm{I}_a + \mathrm{I}_b - \mathrm{I}_c - \mathrm{I}_d; \qquad g_1' = \mathrm{I}_a - \mathrm{I}_b + \mathrm{I}_c - \mathrm{I}_d;$$

$$g_1'' = \mathrm{I}_a - \mathrm{I}_b - \mathrm{I}_c + \mathrm{I}_d$$

where I_a is the total of block I_a, etc. Similar definitions hold for $g_2, \cdots, g_3''$ in the other replications. Then find

$$3Q = 3[CD] + g_1 + g_2 + g_3; \qquad 3Q' = 3[BD] + g_1' + g_2' + g_3';$$

$$3Q'' = 3[BC] + g_1'' + g_2'' + g_3''$$

$$3R_0 = 3[CD_{a}] - g_1 + g_2 + g_3; \qquad 3R_1 = 3[CD_{a_1}] + g_1 + g_2 - g_3;$$

$$3R_2 = 3[CD_{a_2}] + g_1 - g_2 + g_3$$

$$3R_0' = 3[BD_{a_0}] + g_1' - g_2' + g_3'; \quad 3R_1' = 3[BD_{a_1}] - g_1' + g_2' + g_3';$$

$$3R_2' = 3[BD_{a_2}] + g_1' + g_2' - g_3'$$

$$3R_0'' = 3[BC_{a_0}] + g_1'' + g_2'' - g_3''; \quad 3R_1'' = 3[BC_{a_1}] + g_1'' - g_2'' + g_3'';$$

$$3R_2'' = 3[BC_{a_2}] - g_1'' + g_2'' + g_3''$$

The sum of squares for the two-factor interactions are as follows.

$$CD = \frac{(3Q)^2}{576}; \qquad BD = \frac{(3Q')^2}{576}; \qquad BC = \frac{(3Q'')^2}{576}$$

For ACD, divide the sum of squares of deviations of the quantities $3R_0$, $3R_1$, $3R_2$ by 120, with similar rules for ABD and ABC.

The CD, BD, and BC two-way tables must be adjusted for block effects. For the means *per unit* in the BC table, the adjustment is

$$\frac{3Q''}{192} - \frac{[BC]}{72}$$

The adjustment is added (algebraically) when b and c are both present or both absent; when b or c alone is present the adjustment is subtracted.

4^2 factorial, blocks of 4 units. In plan 6.12 each replication confounds completely a set of three of the 9 d.f. for AB. Since the relative informa-

tion on AB is 2/3, this plan gives less precise estimates of the interactions than randomized blocks unless the reduction in block size brings a substantial decrease in the error m.s. The balanced design, which requires a multiple of 3 replicates, is preferable.

With 3 replicates the analysis of variance is as follows.

	d.f.
Replicates	2
Blocks within replicates	9
Main effects	6
Two-factor interactions	9
Error	21
Total	47

To compute the sum of squares for AB, let B_{mn} be the total of the nth block in the mth replicate, and let T_{mn} be the total of all treatments that appear in this block (this will be a total over 12 units). Let $P_{mn} = T_{mn} - B_{mn}$, so that there are 12 quantities P_{mn}. Then

$$AB = \tfrac{1}{8}(P_{11}^2 + P_{12}^2 + \cdots + P_{34}^2) - \tfrac{1}{32}(G^2 + R_{\mathrm{I}}^2 + R_{\mathrm{II}}^2 + R_{\mathrm{III}}^2)$$

where G is the grand total and the R's are replication totals.

The estimated block mean (adjusted for treatment effects) is

$$b_{mn} = \frac{3B_{mn} - T_{mn}}{8}$$

To present a two-way table, the adjusted mean of treatment (00) is, for example,

$$(00)' = (00) - \tfrac{1}{3}(b_{\mathrm{I}b} + b_{\mathrm{II}a} + b_{\mathrm{III}b})$$

since this treatment occurs in blocks Ib, IIa, and IIIb.

4×2^2 factorial, blocks of 8 units. In plan 6.13, one of the three components of ABC is confounded in each replicate. In other words, the difference between the treatments in block Ia and those in block Ib is one component of ABC. The plan gives 2/3 relative information on ABC. Any number of replicates may be used, though three are needed for balanced confounding.

To obtain the sum of squares for ABC, calculate the total for each component separately, taking it only from those replicates in which it is not confounded. The divisor for the square of this total is obtained by the usual rules. Thus with 2 replicates the first component is taken from replicate II and has divisor 16, the second component is taken from replicate I and has divisor 16, while the third component comes from both replicates and has divisor 32.

If there are r replicates, the adjusted block mean is

$$b_{mn} = \frac{rB_{mn} - T_{mn}}{8(r-1)}$$

where B_{mn} is the block total and T_{mn} that of all treatments appearing in the block. With 2 replicates, the adjusted mean of treatment 111 is

$$(111)' = (111) - \frac{b_{\text{I}a} + b_{\text{II}b}}{2}$$

since this treatment is found in blocks Ia and IIb.

4 × 3 × 2 factorial, blocks of 12 units. The 3 d.f. for the factor A at 4 levels may be divided into the following components.

$$A' = a_3 + a_2 - a_1 - a_0$$

$$A'' = a_3 - a_2 - a_1 + a_0$$

$$A''' = a_3 - a_2 + a_1 - a_0$$

In the first 3 replicates of plan 6.14, the interactions $A'C$ and $A'BC$ are partially confounded, with 8/9 and 5/9 relative information, respectively, the interactions of A'' and A''' being unconfounded; $A''C$ and $A''BC$ are partially confounded in replicates 4 to 6 and $A'''C$ and $A'''BC$ in replicates 7 to 9. Although nine replicates are required for complete balance, the design may be used with three or six replicates. In these cases the last three or the last six replications should be taken from plan 6.14. The AC interaction suffers only a trivial loss of replication from the confounding. The statistical analysis is described in reference (6.6).

6.34 Missing Values. When factorial experiments are arranged in randomized blocks or latin squares with no confounding, the formulae previously given (sections 4.25, 4.36) are used for the estimation of missing values. With confounding, the formulae are changed, becoming in general more complicated. The different cases are discussed below.

i. *Complete confounding of the 2^3, 2^4, 2^5, 2^6, 3^3, 3^4 factorials.* When the effects that are confounded are unimportant, we recommended that the first replication of the plan be used for all replications of the experiment, with complete confounding of these effects. In this case the estimation formula is simple. Let B be the total yield of all other units in the same block as the missing unit, and T be the total yield of all other units which have the same treatment as the missing unit. Every replication contains a block which has the same set of treatments as the block with the missing value. Let B' be the total yield of all such blocks (including the block with the missing value).

The value for the missing unit is estimated by the formula:

$$y = \frac{rB + kT - B'}{(r-1)(k-1)}$$

where k is the number of units per block, and r is the number of replications.

It will be realized that this is the standard formula for randomized blocks (section 4.25), applied only to those blocks which contain the treatment with the missing value.

ii. *Partial confounding of the* 2^3, 2^4, 2^5, 2^6, 3^3, 3^4, 4^2, 4×2^2 *factorials.* The formula given later is valid when a different replication of the plan is used for each replication of the experiment. A further essential condition is that no treatment comparison be confounded in more than one replication. This condition is satisfied in all plans for the designs above, except possibly plan 6.4 (2^4 in blocks of 4) and plan 6.6 (2^6 in blocks of 8). For instance, if the first 2 replications of plan 6.6 are used, no interaction is confounded more than once; but if the first 3 replications are used, $BCDE$ is confounded twice and the formula does not apply.

The formula requires some preliminary calculations, most of which can be used in the subsequent analysis of variance. In addition to B and T, as defined in case i, we require first the total R of all other units in the same replicate as the missing unit and the grand total G of all other units in the experiment. In the other replicates, calculate the total S_b of all blocks which contain the treatment with the missing unit.

Next form the table of treatment totals. From this table compute the sum U of the observations for all other treatments which appear in the block which has the missing unit. In the other replications, calculate the corresponding sum for each block which contains the treatment with the missing unit. Let the total of these $(r - 1)$ sums be V, where r is the number of replicates. Note that in the calculation of U and V the treatment with the missing value is not included.

The missing yield is estimated by the equation

$$y = \frac{kt(r-1)T + tr(r-1)B + kG + tV - krR - t(r-1)U - trS_b}{(r-1)[t(r-1)(k-1) - (t-k)]}$$

The table on p. 229 which shows the number of units whose observations are added to obtain the various totals T, B, $\cdots$, may be helpful in checking that the totals have been correctly identified.

Consider U, for example. There are $(k - 1)$ other treatments in the block with the missing value, and each treatment total is the sum of r observations. Consequently U is a total of $r(k - 1)$ individual observations.

Total	T	B	G	V	R	U	S_b
Number of units	$(r - 1)$	$(k - 1)$	$(tr - 1)$	$r(r - 1)(k - 1)$	$(t - 1)$	$r(k - 1)$	$k(r - 1)$

iii. *Partial confounding of the* 3×2^2, 3×2^3, $3^2 \times 2$, $4 \times 3 \times 2$ *factorials.* For these designs the general method given by Yates (6.16) should be followed.

iv. *Single replication of the* 2^5, 2^6, 3^3, 3^4 *factorials.* A missing value is estimated by minimizing the sum of squares for the interactions that are used as error.

The estimation of missing values in cases iii and iv requires a knowledge of the algebra of the analysis of variance. The expressions for the missing value are usually fairly simple in case iv but are complicated in case iii.

When the estimated value has been substituted, the analysis of variance is calculated by the appropriate method for the design. As usual, the error d.f. are reduced by one for each estimated value. Variance-ratio tests of the treatment effects are slightly disturbed. For t-tests, an approximate rule is to decrease the number of replicates ascribed to any mean by $1\frac{1}{2}$ for each estimated value that the mean contains. Suppose, for instance, that (abc) is missing in a 2^3 experiment with 3 replications. The mean of the units which receive a is taken over 12 observations, one of which is an estimated value. By the rule, the standard error of this mean is computed as $s/\sqrt{10\frac{1}{2}}$ instead of $s/\sqrt{12}$. Consequently the standard error of the mean response to A is taken as

$$\sqrt{s^2\left(\frac{1}{10\frac{1}{2}} + \frac{1}{12}\right)}$$

instead of

$$\sqrt{\frac{2s^2}{12}}$$

6.35 Estimation of the Gain in Precision from Confounding.

As mentioned previously, the advisability of confounding depends on the amount by which the experimental error is decreased when the block size is reduced. It is therefore worth while, especially where confounding has not been previously used, to estimate the gain in precision which has been obtained in each experiment.

The technique is essentially the same for all types of design. From the results of the analysis of variance of a confounded experiment, an estimate E_r is made of the experimental error which would have been present if the experiment had been laid out in randomized complete blocks.

Single replication. Let E_b be the mean square for blocks, and E_e the mean square for error. If there is only 1 replication in the experiment, the estimate E_r is

$$E_r = \frac{n_b E_b + n_e E_e}{n_b + n_e} \qquad (6.1)$$

where n_b = number of degrees of freedom for blocks.
n_e = total number of degrees of freedom *minus* n_b.

For example, with a 2^6 experiment in blocks of 8 units (single replication)

$$n_b = 7, \qquad n_e = 63 - 7 = 56$$

The estimate E_r is directly comparable with the estimated error E_e. As shown in section 2.31, the comparison may be expressed in terms of the relative amounts of replication required to obtain equal accuracy with the two types of layout.

The estimate E_r is subject to the assumption that the interactions which are confounded with blocks are negligible. Since confounding is complete, no method of estimation avoids this assumption. If the confounded interactions are *not* negligible, the calculation overestimates the gain from confounding.

More than one replication. If the confounding is complete, formula 6.1 holds for E_r, subject again to the assumption that the confounded comparisons are negligible. In this case

E_b = mean square for blocks *within replications.*
E_e = error mean square.
n_b = number of degrees of freedom for blocks *within replications.*
n_e = total number of degrees of freedom *minus* number of degrees of freedom for replications *minus* n_b.

Thus with 2 replicates of a 2^6 design in blocks of 8 units, the same interactions being confounded in both replicates, we have

$$n_b = 14, \qquad n_e = 127 - 1 - 14 = 112$$

With partial confounding, the same formula may be used if all confounded comparisons are assumed to be small. Evidence on the validity of this assumption is provided by the experimental results, since partially confounded comparisons can be tested for significance.

It is possible to avoid the assumption that confounded comparisons are negligible, though some extra computations are necessary. We first adjust the mean square among blocks within replicates so as to remove treatment effects. The following identity holds for the *sums of squares*.

Blocks within replicates (adj.) + treatments (unadj.) =
blocks within replicates (unadj.) + treatments (adj.)

The two quantities on the right of the equation appear in the analysis of variance of the experimental results. Consequently the *adjusted* blocks s.s. is found from this equation by calculating the unadjusted treatments s.s. From the adjusted blocks s.s. we obtain, as shown below, a new estimate E_b' for use in formula (6.1).

These procedures may be illustrated from the results of the 3 × 3 × 2 experiment in 4 replicates (section 6.19), in which AB and ABC are partially confounded. From table 6.15 the following summar"' of the analysis of variance is obtained.

	d.f.	s.s.	m.s.
Replications	3	3,837	1,279
Blocks within replications	8	2,836	354
Treatments	17	7,802	459
Error	43	8,909	207
Total	71	23,384	

If AB and ABC are assumed negligible, the estimate E_r is found by formula (6.1) for the case where there is more than one replication.

$$E_r = \frac{8 \times 354 + 60 \times 207}{68} = 224$$

In order to avoid the assumption, we calculate the unadjusted treatments s.s., which is found to be 8328. Since the adjusted treatments s.s. is 7802, we have

$$\text{Blocks s.s. (adj.)} = 2836 + 7802 - 8328 = 2310$$

corresponding to a mean square E_{b_a} of 289.

An unbiased estimate E_b' of the variance between blocks within replicates is

$$E_b' = \frac{4E_{b_a} - E_e}{3} = \frac{4 \times 289 - 207}{3} = 316$$

This value is inserted in formula (6.1) to give E_r':

$$E_r' = \frac{8 \times 316 + 60 \times 207}{68} = 220$$

The estimated gain in efficiency over randomized blocks is 8% when AB and ABC are assumed negligible and 6% without this assumption. The two results agree closely, as is usually the case unless the confounded interactions are large.

The method above applies to all designs with partial confounding. For the designs discussed in this chapter, the formula for E_b' in terms of E_{b_a} and E_e is

$$E_b' = \frac{rE_{b_a} - E_e}{r - 1}$$

where r is the number of replicates.

REFERENCES

6.1 Yates, F. The design and analysis of factorial experiments. *Imp. Bur. Soil Sci. Tech. Comm.* 35, 1937.

6.2 Barnard, M. M. An enumeration of the confounded arrangements in the 2^n factorial designs. *Jour. Roy. Stat. Soc. Suppl.* 3, 195–202, 1936.

6.3 Fisher, R. A. The theory of confounding in factorial experiments in relation to the theory of groups. *Ann. Eugen.* 11, 341–353, 1942.

6.4 Finney, D. J. The construction of confounded arrangements. *Emp. Jour. Exp. Agr.* 15, 107–112, 1947.

6.5 Yates, F. Complex experiments. *Jour. Roy. Stat. Soc. Suppl.* 2, 181–247, 1935.

6.6 Li, J. C. R. Design and statistical analysis of some confounded factorial experiments. *Iowa Agr. Exp. Sta. Res. Bull.* 333, 1944.

6.7 Nair, K. R. On a method of getting confounded arrangements in the general symmetrical type of experiments. *Sankhya* 4, 121–138, 1938.

6.8 Nair, K. R. Balanced confounded arrangements for the 5^n type of experiments. *Sankhya* 5, 57–70, 1940.

6.9 Kempthorne, O. A note on differential responses in blocks. *Jour. Agr. Sci.* 37, 245–48, 1947.

6.10 Cornish, E. A. Non-replicated factorial experiments. *Jour. Australian Inst. Agr. Sci.* 2, 79–82, 1936.

6.11 Finney, D. J. The fractional replication of factorial arrangements. *Ann. Eugen.* 12, 291–301, 1945.

6.12 Finney, D. J. Recent developments in the design of field experiments. III. Fractional replication. *Jour. Agr. Sci.* 36, 184–191, 1946.

6.13 Plackett, R. L., and Burman, J. P. The design of optimum multifactorial experiments. *Biometrika* 33, 305–325, 1946.

6 14 Kempthorne, O. A simple approach to confounding and fractional replication in factorial experiments. *Biometrika* 34, 255–272, 1947.

6.15 Mood, A. M. On Hotelling's weighing problem. *Ann. Math. Stat.* 17, 432–446, 1946.

6.16 YATES, F. The analysis of replicated experiments when the field results are incomplete. *Emp. Jour. Exp. Agr.* 1, 129–142, 1933.

6.17 MERRILL, S., KILBY, W. W., and GREER, S. R. Fertilization of tung seedlings in the nursery. *Proc. Amer. Soc. Hort. Sci.* 41, 167–170, 1942.

6.1a BINET, F. E., *et al.* Analysis of confounded factorial experiments in single replication. *North Carolina Agr. Exp. Sta. Tech. Bull.* 113, 1955.

ADDITIONAL READING

BOSE, R. C. Mathematical theory of the symmetrical factorial design. *Sankhya* 8, 107–166, 1947.

BOSE, R. C., and KISHEN, K. On the problem of confounding in the general symmetrical factorial design. *Sankhya* 5, 21–36, 1940.

KITAGAWA, R., and MITOME, M. *Tables for the design of factorial experiments.* Baifukan Co., Tokyo, 1953.

NAIR, K. R., and RAO, C. R. Confounded designs for asymmetrical factorial experiments. *Sci. and Cul.* 7, 313–314, 1941.

RAO, C. R. Factorial experiments derivable from combinatorial arrangements of arrays. *Jour. Roy. Stat. Soc. Suppl.* 9, 128–139, 1947.

STEVENS, W. L. Statistical analysis of a non-orthogonal tri-factorial experiment. *Biometrika* 35, 346–367, 1948.

PLANS

Plan 6.1

2^3 factorial, blocks of 4 units

Rep. I, *ABC* confounded

abc	*ab*
a	*ac*
b	*bc*
c	(1)

Plan 6.2

2^4 factorial, blocks of 8 units

Rep. I, *ABCD* confounded

a	(1)
b	*ab*
c	*ac*
d	*bc*
abc	*ad*
abd	*bd*
acd	*cd*
bcd	*abcd*

Plan 6.3

2^6 factorial, blocks of 16 units

Rep. I, *ABCD*, *ABEF*, *CDEF* confounded

a	*c*	*ab*	*ac*
b	*d*	*cd*	*ad*
acd	*abc*	(1)	*bc*
bcd	*abd*	*abcd*	*bd*
ce	*ae*	*ace*	*abe*
de	*be*	*ade*	*cde*
abce	*acde*	*bce*	*e*
abde	*bcde*	*bde*	*abcde*
cf	*af*	*acf*	*abf*
df	*bf*	*adf*	*cdf*
abcf	*acdf*	*bcf*	*f*
abdf	*bcdf*	*bdf*	*abcdf*
aef	*cef*	*abef*	*acef*
bef	*def*	*cdef*	*adef*
acdef	*abcef*	*ef*	*bcef*
bcdef	*abdef*	*abcdef*	*bdef*

Plan 6.4 Balanced group of sets for 2^4 factorial, blocks of 4 units

Two-factor interactions are confounded in 1 replication and three-factor interactions are confounded in 3 replications. The columns are the blocks.

Rep. I, *AB, ACD, BCD* confounded

(1)	ab	a	b
abc	c	bc	ac
abd	d	bd	ad
cd	abcd	acd	bcd

Rep. II, *AC, ABD, BCD*

(1)	ac	a	c
abc	b	bc	ab
acd	d	cd	ad
bd	abcd	abd	bcd

Rep. III, *AD, ABC, BCD*

(1)	ad	a	d
abd	b	bd	ab
acd	c	cd	ac
bc	abcd	abc	bcd

Rep. IV, *BC, ABD, ACD*

(1)	bc	b	c
abc	a	ac	ab
bcd	d	cd	bd
ad	abcd	abd	acd

Rep. V, *BD, ABC, ACD*

(1)	bd	b	d
abd	a	ad	ab
bcd	c	cd	bc
ac	abcd	abc	acd

Rep. VI, *CD, ABC, ABD*

(1)	cd	c	d
acd	a	ad	ac
bcd	b	bd	bc
ab	abcd	abc	abd

Plan 6.5 Balanced group of sets for 2^5 factorial, blocks of 8 units

Three- and four-factor interactions are confounded in 1 replication.

Rep. I, *ABC, ADE, BCDE* confounded

(1)	ab	a	b
bc	ac	abc	c
abd	d	bd	ad
acd	bcd	be	abcd
abe	e	ce	ae
ace	bce	ade	abce
de	abde	abcde	bde
bcde	acde	cd	cde

Rep. II, *ABD, BCE, ACDE*

(1)	ab	a	b
ad	bd	d	abd
abc	c	bc	ac
bcd	acd	abcd	cd
abe	e	be	ae
bde	ade	abde	de
ce	abce	ace	bce
acde	bcde	cde	abcde

BL 3
main 5
2 factor 10
Error 13
31

Plan 6.5 (Continued)

Balanced group of sets for 2^5 factorial, blocks of 8 units

Rep. III, *ACE, BCD, ABDE*

(1)	*ac*	*a*	*c*
ae	*ce*	*e*	*ace*
abc	*b*	*bc*	*ab*
bce	*abe*	*abce*	*be*
acd	*d*	*cd*	*ad*
cde	*ade*	*acde*	*de*
bd	*abcd*	*abd*	*bcd*
abde	*bcde*	*bde*	*abcde*

Rep. IV, *ACD, BDE, ABCE*

(1)	*ad*	*a*	*d*
ac	*cd*	*c*	*acd*
abd	*b*	*bd*	*ab*
bcd	*abc*	*abcd*	*bc*
ade	*e*	*de*	*ae*
cde	*ace*	*acde*	*ce*
be	*abde*	*abe*	*bde*
abce	*bcde*	*bce*	*abcde*

Rep. V, *ABE, CDE, ABCD*

(1)	*ae*	*a*	*e*
ab	*be*	*b*	*abe*
ace	*c*	*ce*	*ac*
bce	*abc*	*abce*	*bc*
ade	*d*	*de*	*ad*
bde	*abd*	*abde*	*bd*
cd	*acde*	*acd*	*cde*
abcd	*bcde*	*bcd*	*abcde*

Plan 6.6 Balanced group of sets for 2^6 factorial, blocks of 8 units

All three- and four-factor interactions are confounded in 2 replications.

Rep. I, *ABC, CDE, ADF, BEF, ABDE, BCDF, ACEF* confounded

abc	*a*	*b*	(1)	*bc*	*ac*	*c*	*ab*
bd	*cd*	*abcd*	*acd*	*abd*	*d*	*ad*	*bcd*
ae	*abce*	*ce*	*bce*	*e*	*abe*	*be*	*ace*
cde	*bde*	*ade*	*abde*	*acde*	*bcde*	*abcde*	*de*
cf	*bf*	*af*	*abf*	*acf*	*bcf*	*abcf*	*f*
adf	*abcdf*	*cdf*	*bcdf*	*df*	*abdf*	*bdf*	*acdf*
bef	*cef*	*abcef*	*acef*	*abef*	*ef*	*aef*	*bcef*
abcdef	*adef*	*bdef*	*def*	*bcdef*	*acdef*	*cdef*	*abdef*

Rep. II, *ABD, DEF, BCF, ACE, ABEF, ACDF, BCDE*

abd	*b*	*a*	(1)	*ad*	*bd*	*d*	*ab*
cd	*ac*	*bc*	*abc*	*bcd*	*acd*	*abcd*	*c*
be	*abde*	*de*	*ade*	*e*	*abe*	*ae*	*bde*
ace	*cde*	*abcde*	*bcde*	*abce*	*ce*	*bce*	*acde*
af	*df*	*abdf*	*bdf*	*abf*	*f*	*bf*	*adf*
bcf	*abcdf*	*cdf*	*acdf*	*cf*	*abcf*	*acf*	*bcdf*
def	*aef*	*bef*	*abef*	*bdef*	*adef*	*abdef*	*ef*
abcdef	*bcef*	*acef*	*cef*	*acdef*	*bcdef*	*cdef*	*abcef*

Plan 6.6 (*Continued*)

Balanced group of sets for 2^6 factorial, blocks of 8 units

Rep. III, *ABE*, *BDF*, *ACD*, *CEF*, *ADEF*, *BCDE*, *ABCF*

bc	*a*	*ac*	(1)	*abc*	*ab*	*b*	*c*
acd	*bd*	*bcd*	*abd*	*cd*	*d*	*ad*	*abcd*
abe	*ce*	*e*	*ace*	*be*	*bce*	*abce*	*ae*
de	*abcde*	*abde*	*bcde*	*ade*	*acde*	*cde*	*bde*
af	*bcf*	*bf*	*abcf*	*f*	*cf*	*acf*	*abf*
bdf	*acdf*	*adf*	*cdf*	*abdf*	*abcdf*	*bcdf*	*df*
cef	*abef*	*abcef*	*bef*	*acef*	*aef*	*ef*	*bcef*
abcdef	*def*	*cdef*	*adef*	*bcdef*	*bdef*	*abdef*	*acdef*

Rep. IV, *ABF*, *CDF*, *ADE*, *BCE*, *ABCD*, *BDEF*, *ACEF*

ac	*a*	*b*	(1)	*c*	*abc*	*bc*	*ab*
bd	*bcd*	*acd*	*abcd*	*abd*	*d*	*ad*	*cd*
bce	*be*	*ae*	*abe*	*abce*	*ce*	*ace*	*e*
ade	*acde*	*bcde*	*cde*	*de*	*abde*	*bde*	*abcde*
abf	*abcf*	*cf*	*bcf*	*bf*	*af*	*f*	*acf*
cdf	*df*	*abdf*	*adf*	*acdf*	*bcdf*	*abcdf*	*bdf*
ef	*cef*	*abcef*	*acef*	*aef*	*bef*	*abef*	*bcef*
abcdef	*abdef*	*def*	*bdef*	*bcdef*	*acdef*	*cdef*	*adef*

Rep. V, *ACF*, *BCD*, *ADE*, *BEF*, *ABDF*, *CDEF*, *ABCE*

ab	*a*	*bc*	(1)	*b*	*ac*	*c*	*abc*
bcd	*cd*	*abd*	*acd*	*abcd*	*d*	*ad*	*bd*
ce	*bce*	*ae*	*abce*	*ace*	*be*	*abe*	*e*
ade	*abde*	*cde*	*bde*	*de*	*abcde*	*bcde*	*acde*
acf	*abcf*	*f*	*bcf*	*cf*	*abf*	*bf*	*af*
df	*bdf*	*acdf*	*abdf*	*adf*	*bcdf*	*abcdf*	*cdf*
bef	*ef*	*abcef*	*aef*	*abef*	*cef*	*acef*	*bcef*
abcdef	*acdef*	*bdef*	*cdef*	*bcdef*	*adef*	*def*	*abdef*

Rep. VI, *ABC*, *BDE*, *ADF*, *CEF*, *ACDE*, *BCDF*, *ABEF*
Interchange *B* and *C* in replication I
Rep. VII, *ABF*, *DEF*, *BCD*, *ACE*, *ABDE*, *ACDF*, *BCEF*
Interchange *F* and *D* in replication II
Rep. VIII, *ABE*, *BDF*, *CDE*, *ACF*, *ADEF*, *ABCD*, *BCEF*
Interchange *A* and *E* in replication III
Rep. IX, *ABD*, *CDF*, *AEF*, *BCE*, *ABCF*, *BDEF*, *ACDE*
Interchange *F* and *D* in replication IV
Rep. X, *AEF*, *BDE*, *ACD*, *BCF*, *ABDF*, *CDEF*, *ABCE*
Interchange *E* and *C* in replication V

Plan 6.7 Balanced group of sets for 3^3 factorial, blocks of 9 units

ABC confounded

Rep. I, *ABC*(*W*) *

a	*b*	*c*
000	100	200
110	210	010
220	020	120
101	201	001
211	011	111
021	121	221
202	002	102
012	112	212
122	222	022

Rep. II, *ABC*(*X*)

a	*b*	*c*
000	100	200
110	210	010
220	020	120
201	001	101
011	111	211
121	221	021
102	202	002
212	012	112
022	122	222

Rep. III, *ABC*(*Y*)

a	*b*	*c*
000	100	200
210	010	110
120	220	020
101	201	001
011	111	211
221	021	121
202	002	102
112	212	012
022	122	222

Rep. IV, *ABC*(*Z*)

a	*b*	*c*
000	100	200
210	010	110
120	220	020
201	001	101
111	211	011
021	121	221
102	202	002
012	112	212
222	022	122

* Yates' notation.

Plan 6.8 Balanced group of sets for 3^4 factorial, blocks of 9 units

Three-factor interactions confounded

Rep. I, *ABC* I, *ABD* III, *ACD* IV, *BCD* II confounded

0000	0011	0022	0012	0020	0001	0021	0002	0010
1011	1022	1000	1020	1001	1012	1002	1010	1021
2022	2000	2011	2001	2012	2020	2010	2021	2002
0121	0102	0110	0100	0111	0122	0112	0120	0101
1102	1110	1121	1111	1122	1100	1120	1101	1112
2110	2121	2102	2122	2100	2111	2101	2112	2120
0212	0220	0201	0221	0202	0210	0200	0211	0222
1220	1201	1212	1202	1210	1221	1211	1222	1200
2201	2212	2220	2210	2221	2202	2222	2200	2211

Plan 6.8 (*Continued*)

Balanced group of sets for 3^4 factorial, blocks of 9 units

Rep. II, *ABC* II, *ABD* IV, *ACD* I, *BCD* III

0000	0022	0011	0021	0010	0002	0012	0001	0020
1022	1011	1000	1010	1002	1021	1001	1020	1012
2011	2000	2022	2002	2021	2010	2020	2012	2001
0112	0101	0120	0100	0122	0111	0121	0110	0102
1101	1120	1112	1122	1111	1100	1110	1102	1121
2120	2112	2101	2111	2100	2122	2102	2121	2110
0221	0210	0202	0212	0201	0220	0200	0222	0211
1210	1202	1221	1201	1220	1212	1222	1211	1200
2202	2221	2210	2220	2212	2201	2211	2200	2222

Rep. III, *ABC* IV, *ABD* I, *ACD* II, *BCD* I

0000	0021	0012	0022	0010	0001	0011	0002	0020
1021	1012	1000	1010	1001	1022	1002	1020	1011
2012	2000	2021	2001	2022	2010	2020	2011	2002
0122	0110	0101	0111	0102	0120	0100	0121	0112
1110	1101	1122	1102	1120	1111	1121	1112	1100
2101	2122	2110	2120	2111	2102	2112	2100	2121
0211	0202	0220	0200	0221	0212	0222	0210	0201
1202	1220	1211	1221	1212	1200	1210	1201	1222
2220	2211	2202	2212	2200	2221	2201	2222	2210

Rep. IV, *ABC* III, *ABD* II, *ACD* III, *BCD* IV

0000	0012	0021	0011	0020	0002	0022	0001	0010
1012	1021	1000	1020	1002	1011	1001	1010	1022
2021	2000	2012	2002	2011	2020	2010	2022	2001
0111	0120	0102	0122	0101	0110	0100	0112	0121
1120	1102	1111	1101	1110	1122	1112	1121	1100
2102	2111	2120	2110	2122	2101	2121	2100	2112
0222	0201	0210	0200	0212	0221	0211	0220	0202
1201	1210	1222	1212	1221	1200	1220	1202	1211
2210	2222	2201	2221	2200	2212	2202	2211	2220

Plan 6.9 Balanced group of sets for 3×2^2 factorial, blocks of 6 units

BC, *ABC* confounded

Rep. I		Rep. II		Rep. III	
a	*b*	*a*	*b*	*a*	*b*
001	000	000	001	000	001
010	011	011	010	011	010
100	101	101	100	100	101
111	110	110	111	111	110
200	201	200	201	201	200
211	210	211	210	210	211

Plan 6.10 Balanced group of sets for 3×2^3 factorial, blocks of 6 units

BC, *BD*, *CD*
ABC, *ABD*, *ACD* confounded

Rep. I

a	*b*	*c*	*d*
0100	0000	0001	0010
0011	0111	0110	0101
1010	1001	1000	1100
1101	1110	1111	1011
2001	2010	2100	2000
2110	2101	2011	2111

Rep. II

a	*b*	*c*	*d*
0010	0001	0000	0100
0101	0110	0111	0011
1001	1010	1100	1000
1110	1101	1011	1111
2100	2000	2001	2010
2011	2111	2110	2101

Rep. III

a	*b*	*c*	*d*
0001	0010	0100	0000
0110	0101	0011	0111
1100	1000	1001	1010
1011	1111	1110	1101
2010	2001	2000	2100
2101	2110	2111	2011

Plan 6.11 Balanced group of sets for $3^2 \times 2$ factorial, blocks of 6 units

AB, *ABC* confounded

Rep. I

a	*b*	*c*
100	200	000
210	010	110
020	120	220
201	001	101
011	111	211
121	221	021

Rep. II

a	*b*	*c*
200	000	100
010	110	210
120	220	020
101	201	001
211	011	111
021	121	221

Rep. III

a	*b*	*c*
100	200	000
010	110	210
220	020	120
201	001	101
111	211	011
021	121	221

Rep. IV

a	*b*	*c*
200	000	100
110	210	010
020	120	220
101	201	001
011	111	211
221	021	121

Plan 6.12 Balanced group of sets for 4^2 factorial, blocks of 4 units

AB confounded

Rep. I

a	*b*	*c*	*d*
33	32	31	30
22	23	20	21
10	11	12	13
01	00	03	02

Rep. II

a	*b*	*c*	*d*
33	30	32	31
21	22	20	23
12	11	13	10
00	03	01	02

Rep. III

a	*b*	*c*	*d*
33	31	32	30
20	22	21	23
11	13	10	12
02	00	03	01

Plan 6.13 **Balanced group of sets for 4 × 2^2 factorial, blocks of 8 units**

ABC confounded

Rep. I		Rep. II		Rep. III	
a	*b*	*a*	*b*	*a*	*b*
000	001	000	001	000	001
011	010	011	010	011	010
100	101	101	100	101	100
111	110	110	111	110	111
201	200	201	200	200	201
210	211	210	211	211	210
301	300	300	301	301	300
310	311	311	310	310	311

Plan 6.14

Balanced group of sets for 4 × 3 × 2 factorial, blocks of 12 units

A'C, *A'BC* confounded

Rep. I		Rep. II		Rep. III	
a	*b*	*a*	*b*	*a*	*b*
000	001	001	000	001	000
011	010	010	011	011	010
021	020	021	020	020	021
100	101	101	100	101	100
111	110	110	111	111	110
121	120	121	120	120	121
201	200	200	201	200	201
210	211	211	210	210	211
220	221	220	221	221	220
301	300	300	301	300	301
310	311	311	310	310	311
320	321	320	321	321	320

Plan 6.14 (Continued)

Balanced group of sets for 4 × 3 × 2 factorial, blocks of 12 units

$A''C$, $A''BC$

Rep. IV		Rep. V		Rep. VI	
a	*b*	*a*	*b*	*a*	*b*
001	000	000	001	000	001
010	011	011	010	010	011
020	021	020	021	021	020
100	101	101	100	101	100
111	110	110	111	111	110
121	120	121	120	120	121
200	201	201	200	201	200
211	210	210	211	211	210
221	220	221	220	220	221
301	300	300	301	300	301
310	311	311	310	310	311
320	321	320	321	321	320

$A' = a_3 + a_2 - a_1 - a_0$
$A'' = a_3 - a_2 - a_1 + a_0$
$A''' = a_3 - a_2 + a_1 - a_0$

$A'''C$, $A'''BC$

Rep. VII		Rep. VIII		Rep. IX	
a	*b*	*a*	*b*	*a*	*b*
000	001	001	000	001	000
011	010	010	011	011	010
021	020	021	020	020	021
101	100	100	101	100	101
110	111	111	110	110	111
120	121	120	121	121	120
200	201	201	200	201	200
211	210	210	211	211	210
221	220	221	220	220	221
301	300	300	301	300	301
310	311	311	310	310	311
320	321	320	321	321	320

CHAPTER 6A

Factorial Experiments in Fractional Replication

6A.1 Construction and Properties of Fractionally Replicated Designs

6A.11 Introduction. Often, even a single replication of a factorial experiment is beyond the resources of the investigator, or it gives more precision in the estimates of the main effects than is needed. In a single replication of a 2^6 factorial, each main effect is an average over 32 combinations of the other factors, and hence in effect has 32-fold replication. Perhaps 4- or 8-fold replication would suffice. In situations like this, an experiment which consists of only part of a complete replication is worth considering.

The use of experiments in fractional replication was proposed in 1945 by Finney (6A.1). He outlined methods of construction for 2^n and 3^n factorials, and described a half-replicate of a 4×2^4 agricultural field experiment that had been conducted in 1942. In 1946, Plackett and Burman (6A.2, 6A.3) gave designs for the minimum possible size of experiment with p^n factorials ($p = 2, 3, \ldots 7$) and pointed out their utility in physical and industrial research. Since then, the designs have found many applications, particularly in industrial research and development. Their chief appeal is that they enable 5 or more factors to be included simultaneously in an experiment of a practicable size, so that the investigator can discover quickly which factors have an important effect on the product.

This welcome reduction in the size of the experiment is not obtained without paying a price. As will be seen, the results of an experiment in fractional replication are open to misinterpretation in a way that does not arise with fully replicated designs. Before using these designs, it is important to understand the nature of this confusion.

6A.12 Aliases. In order to see what happens when the experiment contains part of a replication, consider a 2^3 factorial in which only the 4 treatment combinations a, b, c, and abc are tested. This is half of the complete replicate. From the experimental results the main effects and interactions would be estimated as follows (apart from a divisor 2).

FACTORIAL EFFECTS

Treatment combination	A	B	C	AB	AC	BC	ABC
a	+	−	−	−	−	+	+
b	−	+	−	−	+	−	+
c	−	−	+	+	−	−	+
abc	+	+	+	+	+	+	+

For the main effect of A, we add the yields for units that contain a and subtract the yields for units that do not, i.e.,

$$A = (abc) + (a) - (b) - (c)$$

and similarly for the main effects of B and C.. It will be seen from the table that the A, B, and C comparisons are mutually orthogonal. Thus the estimates of the three main effects are independent.

To compute the AB interaction we will follow the "Evens versus odds" rule (section 5.24). Add the yields for units that contain an even number of the letters a, b, and subtract the yields for units that contain an odd number. This gives

$$AB = (abc) + (c) - (a) - (b)$$

But this is the same quantity that is used to estimate the main effect of C. The name *aliases* was given by Finney (6A.1) to two factorial effects that are represented by the same comparison. Thus C and AB are aliases. We write $C = AB$. The table above also shows that

$$AC = B\colon BC = A$$

The three-factor interaction, ABC, cannot be estimated at all. If all 8 treatment combinations were available, ABC would be computed from

$$ABC = (abc) + (a) + (b) + (c) - (ab) - (ac) - (bc) - (1)$$

We have, however, chosen the 4 treatment combinations which carry a + sign in this expression to be our half replicate, so that no comparison is possible. ABC is called the *defining contrast,* since it was the contrast used to split the factorial into 2 half-replicates.

In summary, the outcome of using a half-replicate is to lose one factorial effect, ABC, entirely and to leave each main effect inextricably mixed with one of the two-factor interactions.

If the experiment shows an apparent effect of A, there is no way of knowing from the results whether the effect is really due to A, to the BC interaction, or to a mixture of the two. This kind of confusion is always present in experiments with fractional replication, since every factorial effect always has one or more aliases. The problem of deciding to which alias an effect is to be attributed is one that the experimenter must face when he comes to interpret the results.

In some types of research, previous experience with the same factors, or a knowledge of the nature of their actions, may lead the investigator to predict confidently that their interactions will be negligible. The ambiguity then disappears: we conclude that an apparent effect of A is in fact due to A, and likewise for B and C. With these assumptions, the half-replicate provides independent estimates of the three main effects.

In other cases, the nature of the results makes one interpretation of the data more plausible than another. Suppose that the comparison A shows a large effect, but B and C have no apparent effects. It could be argued, at least in some types of experimentation, that factors B and C are unlikely to have a large interaction, since their main effects are apparently absent. The large effect of A would therefore be ascribed to A and not be BC. This conclusion could be wrong: some risk of misinterpretation cannot be avoided.

On the other hand, the experiment might show large effects of the B and C comparisons and a much smaller effect of the A comparison. Since $A = BC$, we would have to consider the possibility that the apparent A effect was wholly or partly due to an interaction between the large effects of B and C. If an experiment takes only a short time to complete, this uncertainty can be resolved by running a new experiment consisting of the four combinations, (ab), (ac), (bc), and (1), that were omitted in the first experiment. By putting the results of the two experiments together we have a complete replicate, which gives independent estimates of A and BC. This device is, of course, less practicable in lines of work where an experiment takes a long time to produce its results. It follows that more risks can be taken with fractional replication when experiments are short than when they are long.

In order to use fractional replication with the minimum risk of misinterpretation, it is important to know what are the aliases of the factorial effects in which we are interested. The alias of any factorial effect can be found by a simple rule, which for the moment we will state for half-replicates.

In the 2^n system, the alias of any factorial effect is its generalized interaction with the defining contrast.

To illustrate, the defining contrast in this example is ABC. The alias of A is therefore the "interaction" $AABC$, or A^2BC. In interpreting a generalized interaction, any squared terms are canceled (as mentioned previously in section 6.13). Thus A^2BC is read as BC. Similarly, the alias of B is AB^2C or AC.

Does it make any difference which half-replicate is chosen? This choice affects the alias relationships in a way that is worth noting. Suppose that instead of (abc), (a), (b), and (c) we take the other half: (ab), (ac), (bc), and (1). The main effect of A (apart from the divisor) is

$$A = (ab) + (ac) - (bc) - (1)$$

In computing the BC interaction, we give a $+$ sign to combinations having both or none of the letters b and c; i.e.,

$$BC = (bc) + (1) - (ab) - (ac)$$

Note that A is now equal to the BC interaction *with its sign changed:* we write $A = -BC$. With the original half-replicate, on the other hand, A and BC are identical; i.e., $A = +BC$.

To pursue the issue further, suppose that B and C are factors whose interaction is positive, in the sense that the combination (bc) gives a higher response than would be expected from the sum of the responses to (b) and (c) alone. With the half-replicate (ab), (ac), (bc), and (1), the presence of this interaction will make the observed effect of A smaller (algebraically) than the true effect of A. With the original half-replicate, on the other hand, the true effect of A is overestimated if a positive BC interaction exists.

This issue should seldom be of practical importance, because it is not advisable to conduct an experiment in which A and BC are aliases if both effects may be of the same order of magnitude. The distinction between the $+$ and $-$ signs in an interaction alias is worth remembering, however, when we are trying to make sense of an experiment whose results are different from our anticipations.

In the next two sections we discuss a half-replicate of the 2^5 factorial and a half- and quarter-replicate of the 2^6 factorial, in order to become familiar with the properties of fractional replications in larger experiments.

6A.13 A Half-replicate of a 2^5 Factorial. The experiment is to consist of 16 of the 32 possible treatment combinations. Since all information is lost on the factorial effect that is used to divide the replication into halves, the obvious choice for the defining contrast is the five-factor interaction *ABCDE*. Having made this choice, we can write down the alias of any other factorial effect by the rule just given. These aliases can be examined to see if they are suitable for our purpose before going to the trouble of writing down the plan itself. The alias of A, for instance, is A^2BCDE, i.e., the four-factor interaction *BCDE*. The complete set of aliases appears in table 6A.1.

TABLE 6A.1 ALIASES IN A HALF-REPLICATE OF A 2^5 FACTORIAL

Defining contrast: *ABCDE*

Main effects	Alias	Two-factor interactions	Alias
A	*BCDE*	*AB*	*CDE*
B	*ACDE*	*AC*	*BDE*
C	*ABDE*	*AD*	*BCE*
D	*ABCE*	*AE*	*BCD*
E	*ABCD*	*BC*	*ADE*
		BD	*ACE*
		BE	*ACD*
		CD	*ABE*
		CE	*ABD*
		DE	*ABC*

Every main effect has a four-factor interaction as alias, and every two-factor interaction has a three-factor interaction as alias. No main effect is mixed up with a two-factor interaction as occurred in the 2^3 design. In practice, this design can be used in at least three different ways.

(i) If all interactions among 3 or more factors can be assumed negligible, the design gives independent estimates of all main effects and all two-factor interactions. Its chief defect is that it does not furnish any d.f. for the estimation of error. Since there are 16 observations in the whole experiment, the total sum of squares has 15 d.f., and of these, 5 are allotted to main effects and the remaining 10 to the two-factor interactions. If estimates of the error variance have been made in previous experiments, and these have shown stability from one experiment to another, they might be pooled to provide rough standard errors and tests of significance.

(ii) If all two-factor interactions are considered likely to be absent or small, the experiment can be used to estimate only the 5 main effects. The 10 d.f. attributed to the two-factor interactions provide the estimate of error.

(iii) If only a few of the two-factor interactions appear likely to be substantial, say *AB*, *AC*, and *BC*, the scheme of analysis may be to estimate all main effects plus these 3 interactions. An estimate of error is made from the remaining 7 d.f.

In situations (ii) and (iii), the experimenter should avoid any temptation to pick out for estimation those interactions that appear to be large and to use the remainder to estimate the error variance. This practice can produce a serious underestimation of the true error variance.

To write down the plan, we take either the 16 treatment combinations that contain 5, 3, or 1 of the letters *a*, *b*, *c*, *d*, *e*, or the 16 that contain 4, 2, or none of these letters.

TABLE 6A.2 THE TWO HALF-REPLICATES OF A 2^5 FACTORIAL

Defining contrast: *ABCDE*

+*ABCDE*	−*ABCDE*
abcde	*abcd*
abc	*abce*
abd	*abde*
abe	*acde*
acd	*bcde*
ace	*ab*
ade	*ac*
bcd	*ad*
bce	*ae*
bde	*bc*
cde	*bd*
a	*be*
b	*cd*
c	*ce*
d	*de*
e	(1)

The rule for finding aliases is easily extended to give the sign of the alias. One of the half-replicates contains the treatment combination in which all the letters appear (i.e., *abcde* in the 2^5 factorial). If this half-replicate is selected, call the defining contrast +. If the other half is chosen, the defining contrast is given a − sign. In finding aliases, the signs are combined by the usual laws of algebra. The interaction of *A* with $-ABCDE$ is $-BCDE$: the interaction of $-BC$ with $-CDE$ is $+BDE$.

6A.14 Half- and Quarter-replicates of a 2^6 Experiment. The half-replicate, which requires 32 units, will be discussed only briefly, since the principles are the same as in the 2^5 factorial. The best defining contrast is *ABCDEF*. Every main effect then has a five-factor interaction as alias (e.g., $A = BCDEF$), and every two-factor interaction has a four-factor interaction as alias (e.g., $AB = CDEF$). There are 20 three-factor interactions, which divide themselves into 10 pairs of aliases (e.g., $ABC = DEF$). Consequently, in order to obtain valid estimates of all main effects and two-factor interactions, we need only assume that four- and five-factor interactions can be neglected. If the three-factor interactions are also negligible, the 10 d.f. for the 10 alias pairs can be used as an estimate of error. In this scheme the partition of d.f. is as follows.

	d.f.
Main effects	6
Two-factor interactions	15
Error (from three-factor interactions)	10
Total	31

If the half-replicate corresponding to $+ABCDEF$ is selected, the plan consists of all treatment combinations that contain 6, 4, 2, or none of the letters *abcdef* (table 6A.3).

TABLE 6A.3 A HALF-REPLICATE OF A 2^6 FACTORIAL

abcdef	*abcd*	*abce*	*abcf*	*abde*	*abdf*	*abef*	*acde*
acdf	*acef*	*adef*	*bcde*	*bcdf*	*bcef*	*bdef*	*cdef*
ab	*ac*	*ad*	*ae*	*af*	*bc*	*bd*	*be*
bf	*cd*	*ce*	*cf*	*de*	*df*	*ef*	(1)

To obtain a quarter-replicate, another factorial effect is used as a defining contrast to divide the 32 combinations in table 6A.3 into 2 sets of 16. Since this factorial effect will not be estimable, it is natural to try a five-factor interaction, such as $ABCDE$. Taking $+ABCDE$, we select from the 32 combinations the 16 that contain 5, 3, or 1 of the letters *abcde*. It may be verified that we obtain the following 16 combinations:

abcdef	*abcf*	*abdf*	*abef*	*acdf*	*acef*	*adef*	*bcdf*
bcef	*bdef*	*cdef*	*af*	*bf*	*cf*	*df*	*ef*

These are the 16 combinations that have a + sign in the expression for $ABCDEF$ and also in the expression for $ABCDE$. Note, however, that all 16 combinations contain the letter *f*. We have inadvertently produced a design from which the main effect of F cannot be estimated, there being no tests made at the lower level of F. In other words, F has become a defining contrast: if F had been used instead of $ABCDE$

to construct the quarter-replicate, the same set of 16 treatments would have been obtained.

This example is an illustration of a general rule.

In the 2^n system, any two factorial effects may be used as defining contrasts to divide a factorial experiment into quarter-replicates. Their generalized interaction also acts as a defining contrast, and cannot be estimated from the quarter-replicate.

The generalized interaction of $ABCDEF$ and $ABCDE$ is $A^2B^2C^2D^2E^2F$, or F. This rule is reminiscent of the rule that applies (section 6.13) when a replicate of a factorial experiment is divided into 4 blocks by confounding factorial effects.

Before trying to improve on this design, we will state the rule governing aliases:

In a quarter-replicate in the 2^n system, any factorial effect that is not a defining contrast has 3 aliases. These are its generalized interactions with the 3 defining contrasts.

For instance, the aliases of A are its interactions with $ABCDEF$, $ABCDE$, and F, i.e., $BCDEF$, $BCDE$, and AF. To verify this, note that from the 16 treatment combinations that are available, A is estimated (apart from a divisor) as

$$A = (abcdef) + (abcf) + (abdf) + (abef) + (acdf) + (acef) + (adef) + (af) - (bcdf) - (bcef) - (bdef) - (cdef) - (bf) - (cf) - (df) - (ef)$$

This is also the expression for AF, since all the + signs contain both the letters a and f, while the − signs contain only one of these letters. The same check may be applied to $BCDEF$ and $BCDE$.

In seeking a better design for the 2^6 factorial in a quarter-replicate, we would like the defining contrasts, and all aliases of main effects and of two-factor interactions, to be interactions of high order that can be assumed negligible. Trial and error shows that this cannot be achieved. The best procedure is to select as defining contrasts 2 four-factor interactions like $ABCE$ and $ABDF$ that have two letters in common. Their generalized interaction, the third defining contrast, is another four-factor interaction $CDEF$.

With this design it is clear that the alias of any main effect is an interaction among at least three factors. Some of the aliases of two-factor interactions, however, are themselves two-factor interactions. If we proceed systematically to write down the aliases of each of the 15 two-factor interactions, we find that when we have done this for 7 of the two-factor interactions, the remaining 8 have already turned up as aliases. The 7 sets are shown in table 6A.4.

TABLE 6A.4 ALIASES OF TWO-FACTOR INTERACTIONS IN A 2^6 FACTORIAL (¼ replicate)

Defining contrasts: *ABCE*, *ABDF*, *CDEF*

Two-factor interaction	Aliases
AB	*CE*, *DF*, *ABCDEF*
AC	*BE*, *BCDF*, *ADEF*
AD	*BF*, *BDCE*, *ACEF*
AE	*BC*, *BDEF*, *ACDF*
AF	*BD*, *BCEF*, *ACDE*
CD	*EF*, *ABDE*, *ABCF*
CF	*DE*, *ABEF*, *ABCD*

To summarize, this arrangement furnishes valid estimates of the main effects, provided that three-factor interactions can be ignored. Two-factor interactions are badly mixed up.

There are several situations in which this design can be used. If all two-factor interactions are thought to be relatively small, we estimate only the main effects, the remaining comparisons providing an estimate of error. The analysis of variance is as follows:

	d.f.
Main effects	6
Error (from interactions)	9
Total	15

The 9 d.f. for error are composed of the 7 alias sets given above that involve two-factor interactions, plus 2 alias sets that contain only three-factor interactions, namely:

$$ACF = BCD = BEF = ADE \quad \text{and} \quad ACD = AEF = BCF = BDE$$

Alternatively, some of the two-factor interactions can be estimated, provided that their aliases are negligible. Suppose that *E* and *F* do not interact with any of the other factors or with each other. In the 7 alias sets given in table 6A.4, the first 6 sets can then be regarded as estimates of *AB*, *AC*, *AD*, *BC*, *BD*, and *CD*, respectively. The 7th set, which contains *CF* and *DE*, contributes to the estimate of error. This appears to be as far as we can go in disentangling the two-factor interactions. If this analysis is adopted, the d.f. subdivide as follows:

	d.f.
Main effects	6
Two-factor interactions of A, B, C, D	6
Error (from remaining interactions)	3
Total	15

The d.f. for error are scanty.

The plan for this quarter-replicate can be written down by using $ABCE$ to divide the 64 combinations into halves, and then using $ABDF$ to divide the chosen half into quarters. If the selected quarter is $+ABCE$, $+ABDF$, the following 16 treatment combinations are tested.

abcdef	*abdf*	*cdef*	*abce*	*acd*	*ade*	*bcd*	*bde*
acf	*aef*	*bcf*	*bef*	*ab*	*ce*	*df*	(1)

This account of fractional factorial designs is intended only as an introduction sufficient to enable the plans at the end of this chapter to be used with understanding. Further discussions are given in references (6A.1), (6A.4), and (6A.5).

6A.15 Example of the Statistical Analysis. The experiment is a half-replicate of a 2^6 factorial. Factors A, B, C, D, and E are different constituents that go into the commercial manufacture of icing for cakes, and factor F is a variable that the housewife may introduce in using the icing. The defining contrast is $ABCDEF$, and the half-replicate used is the same as that in table 6A.3. The variable to be analyzed is the texture of the icing. Table 6A.5 shows the texture readings for the 32 treatment combinations. The analysis proceeds as in table 6A.5.

1. Decide which factorial effects are to be isolated and tested, and which are to constitute the estimate of error. (If some of the factors to be isolated have aliases of the same order, e.g., if $CD = EF$, it is necessary to decide also which alias is to be regarded as producing any effect that is found.) In the example, all 6 d.f. for main effects and all 15 d.f. for two-factor interactions have higher-order interactions as aliases, and all will be estimated. The remaining 10 d.f. supply the error mean square.

2. There are several ways of computing the sum of squares for main effects and two-factor interactions. One is to construct the 15 two-way tables for every pair of variables. Each table gives the sum of squares for one interaction and two main effects: these sums of squares are calculated by the usual rules of the analysis of variance. Another method is to write down the combination of signs for each of the 21 effects that are to be estimated, as was done in table 6.4, p. 190, for a 2^4 factorial.

TABLE 6A.5 STATISTICAL ANALYSIS OF ½-REPLICATE OF A 2^6 FACTORIAL

Treatment combination	Texture	(1)	(2)	(3)	(4)	(5) Effect total	Identification
(1)	233	450	1034	2200	4634	8985	*G*
a(*f*)	217	584	1166	2434	4351	151	*A*
b(*f*)	267	483	1184	2416	−198	317	*B*
ab	317	683	1250	1935	349	−49	*AB*
c(*f*)	250	517	983	34	434	649	*C*
ac	233	667	1433	−232	−117	15	*AC*
bc	333	650	967	16	134	−83	*BC*
abc(*f*)	350	600	968	333	−183	151	*ABC* (Error)
d(*f*)	250	550	34	334	198	−247	*D*
ad	267	433	0	100	451	51	*AD*
bd	400	733	−116	−150	−34	−51	*BD*
abd(*f*)	267	700	−116	33	49	51	*ABD* (Error)
cd	400	467	17	100	−134	−515	*CD*
acd(*f*)	250	500	−1	34	51	119	*ACD* (Error)
bcd(*f*)	283	484	133	−150	302	−383	*BCD* (Error)
abcd	317	484	200	−33	−151	−617	*EF* = *ABCD*
		8985	9136	9404	10136	8544	
e(*f*)	200	−16	134	132	234	−283	*E*
ae	350	50	200	66	−481	547	*AE*
be	283	−17	150	450	−266	−551	*BE*
abe(*f*)	150	17	−50	1	317	−317	*ABE* (Error)
ce	400	17	−117	−34	−234	253	*CE*
ace(*f*)	333	−133	−33	0	183	83	*ACE* (Error)
bce(*f*)	317	−150	33	−18	−66	185	*BCE* (Error)
abce	383	34	0	67	117	−453	*DF* = *ABCE*
de	267	150	66	66	−66	−715	*DE*
ade(*f*)	200	−133	34	−200	−449	583	*ADE* (Error)
bde(*f*)	150	−67	−150	84	34	417	*BDE* (Error)
abde	350	66	184	−33	85	183	*CF* = *ABDE*
cde(*f*)	117	−67	−283	−32	−266	−383	*CDE* (Error)
acde	367	200	133	334	−117	51	*BF* = *ACDE*
bcde	267	250	267	416	366	149	*AF* = *BCDE*
abcde(*f*)	217	−50	−300	−567	−983	−1349	*F* = *ABCDE*
Odds	4417	4434	4336	5864	5072		
Evens	4568	4702	5068	4272	3472		
	8985	9136	9404	10136	8544	6944	

A third method, to be illustrated here, is an adaptation of Yates' automatic method (section 5.24a) of finding all factorial effect totals in a 2^n factorial. This is the quickest method for the larger fractional factorial experiments. The adaptation proceeds as follows.

Since the experiment has 32 treatment combinations, it contains a complete replication of the first 5 factors if F is ignored. Arrange these 5 factors in the systematic order required for the application of Yates' method, as shown in table 6A.5. In this table the treatment symbol f has been placed in ( ) as a reminder that f plays no part in the systematic order.

Next apply Yates' method as if the experiment were a complete replicate of a 2^5 experiment. The factorial effect totals appear in column (5). The Odds and Evens totals at the foot of each column and the totals at the foot of the first half of each column are check data as described in section 5.24a.

If the factor F were not present, the figures in column (5) would represent the factorial effect totals for all main effects and interactions of the factors A, B, C, D, and E. Now reintroduce the factor F. When this is done, each effect will have an alias involving F. Since the defining contrast is $ABCDEF$, we have $A = BCDEF$; $B = ACDEF$; and so on. It is necessary to identify the effects involving F that are to be estimated. These are the main effect, F, and the interactions AF, BF, CF, DF, and EF. The aliases of these effects are, respectively,

$$F = ABCDE; \qquad AF = BCDE; \qquad BF = ACDE;$$

$$CF = ABDE; \qquad DF = ABCE; \qquad EF = ABCD$$

Consequently, we identify the interaction $ABCDE$ as F, the interaction $BCDE$ as AF, and so on. The identification assigned to each of the 32 factorial effect totals appears on the right of column (5).

From the 10 factorial effects that furnish the estimate of error, the error mean square per unit can be computed directly as follows, without constructing an analysis of variance table.

$$\text{Error m.s.} = \frac{(151)^2 + (51)^2 + \cdots + (383)^2}{(32)(10)} = 3089$$

The divisor is the product of the number of units in the experiment (32) and the number of error d.f. (10). The standard error of a single unit is 55.6, and the standard error of a factorial effect total is $\sqrt{32}(55.6) = 309.6$. For 10 d.f., the 5% and 1% t-values are 2.228 and 3.169, respectively. Hence the two numbers

$$(2.228)(309.6) = 690; \qquad (3.169)(309.6) = 981$$

give the values required for statistical significance at the 5% and 1% levels, respectively. The main effect of F is significant at the 1% level, as appears obvious from inspection of the original data. Of the remaining 20 comparisons that are tested, only the DE interaction reaches the

5% level. This result may well be due to chance fluctuations, since one apparently significant result out of 20 would be expected.

For further study of the results, tables of the mean yields for certain groups of treatment combinations, e.g., for the lower and upper levels of F, will be wanted. These means can be computed directly from the original yields, or by expressing them as linear functions of the factorial effect totals. For example, the four mean yields in the DE two-way table are obtained as in table 6A.6.

TABLE 6A.6 Computation of mean yields in the DE two-way table from factorial effect totals

Mean for Combination	Factorial effect total G 8985	D −247	E −283	DE −715	Divisor	Mean yield
(−)	+	−	−	+	32	275.0
(d)	+	+	−	−	32	304.2
(e)	+	−	+	−	32	302.0
(de)	+	+	+	+	32	241.9

The divisor (32) is the total number of units in the experiment. The reader may verify that these results agree with those found by adding the appropriate groups of original yields and dividing by 8.

Yates' method is applicable to all fractional factorials of the 2^n series, including quarter- and eighth-replicate designs. The general approach is to select a sub-set of the factors for which the experiment is a complete replicate. With a quarter-replicate, two factors will be ignored temporarily: with an eighth-replicate, three factors. The factorial effect totals for the sub-set are computed by the Yates' method. Then the other factors are introduced by writing down the alias sets and making the necessary identifications in the list of factorial effects found for the sub-set.

6A.16 The Reduction of Block Size by Confounding. The designs discussed thus far assume that the experiment is to be completely randomized. Since arrangements in 16 or 32 units are rather imprecise in many types of experimentation, designs in blocks of smaller sizes may be wanted. These can be obtained by confounding certain factorial effects with blocks. The principles will be illustrated by the 2^6 factorial in 32 units (½ replicate), with $ABCDEF$ as defining contrast. As we have seen (section 6A.14), all other factorial effects group themselves into pairs of aliases: $A = BCDEF$, $AB = CDEF$, $ABC = DEF$, and so on.

If blocks of 16 units are wanted, any factorial effect other than *ABCDEF* may be used to divide the 32 units into two blocks of 16 units. All information on this factorial effect and on its alias will be lost, since both are confounded with blocks. The problem is therefore to find a pair of aliases both of which can be sacrificed. In the present instance, the best choice is clearly a pair of three-factor interactions like *ABC* and *DEF*. As with fully replicated designs, *this confounding does not disturb any of the other factorial effects.*

The plan (table 6A.7) is constructed by first taking all treatment combinations having an even number of the letters *a, b, c, d, e, f* to form the half-replicate. This set of 32 treatments is divided into two blocks such that block 1 contains only even numbers of the letters *a, b,* or *c.* (It also contains only even numbers of the letters *d, e,* and *f,* since *ABC* = *DEF*.) The partition of the d.f. is shown at the side of the plan.

TABLE 6A.7 A ½ REPLICATE OF A 2^6 FACTORIAL IN BLOCKS OF 16 UNITS

Defining contrast: *ABCDEF* Confounded with blocks: *ABC* = *DEF*

Block 1	Block 2
(1)	*ad*
de	*ae*
df	*af*
ef	*bd*
ab	*be*
ac	*bf*
bc	*cd*
abde	*ce*
abdf	*cf*
abef	*adef*
acde	*bdef*
acdf	*cdef*
acef	*abcd*
bcde	*abce*
bcdf	*abcf*
bcef	*abcdef*

	d.f.
Blocks	1
Main effects	6
Two-factor interactions	15
Error (from three-factor interactions)	9
Total	31

For a design in blocks of 8 units, another alias pair must also be confounded with blocks. Suppose that *ABD* and its alias *CEF* are selected. It will be recalled (section 6.13) that the generalized interaction of this alias pair with *ABC* and *DEF* is also confounded. This interaction is the alias pair *CD* and *ABEF*. It is unfortunate that a two-factor interaction is confounded, but further trial shows that this is unavoidable in blocks of 8 units. To construct the plan (table 6A.8), divide each block of table 6A.7 into halves that contain respectively an odd and even number of the letters *c* and *d*.

TABLE 6A.8 A HALF-REPLICATE OF A 2^6 FACTORIAL IN BLOCKS OF 8 UNITS

Confounded with blocks: *ABC*, *DEF*, *ABD*, *CEF*, *CD*, *ABEF*

Block			
1	2	3	4
(1)	*de*	*ae*	*ad*
ef	*df*	*af*	*bd*
ab	*ac*	*be*	*ce*
abef	*bc*	*bf*	*cf*
acde	*abde*	*cd*	*abce*
acdf	*abdf*	*abcd*	*abcf*
bcde	*acef*	*cdef*	*adef*
bcdf	*bcef*	*abcdef*	*bdef*

In practice, the letters should be allocated to the factors so that *CD* is the least important two-factor interaction. In the analysis of variance, the sums of squares for blocks, main effects, and the unconfounded two-factor interactions are all computed in the usual way. If all available two-factor interactions are to be estimated, the d.f. subdivide as follows:

	d.f.
Blocks (includes *CD*)	3
Main effects	6
Two-factor interactions (except *CD*)	14
Error	8
Total	31

With blocks of 4 units there are 7 d.f. for blocks, so that 7 alias pairs must be thrown away. The best solution is to choose *AE* to form blocks of 4 units from the blocks of 8 units. It will be found that one additional two-factor interaction, *BF*, is also lost (this being the interaction of *AE* with *ABEF*). This arrangement (shown in plan 6A.6) gives estimates of all main effects and 12 of the 15 two-factor interactions, but sacrifices *CD*, *AE*, and *BF*.

For blocks of 2 units, investigation shows that all 15 two-factor interactions must be confounded with blocks if main effects are to be kept clear.

To summarize, the reduction of block size by confounding has the same general consequences in fractional factorials as in ordinary factorials. It becomes impossible to estimate the factorial effects (and their aliases) that are confounded with blocks. The rest of the factorial effects are undisturbed and are estimated in the usual way (except those that serve as defining contrasts).

6A.2 The Use of Fractional Factorial Designs in Practice

6A.21 Designs in Which All Main Effects and Two-factor Interactions Can Be Estimated. These designs are useful in the following circumstances:

1. The factors are such that all two-factor interactions must be examined.
2. Experimental errors are substantial, requiring a relatively large amount of internal replication to provide satisfactory estimates of the main effects and two-factor interactions.
3. A complete replication is considered too large or is not feasible.

These conditions sometimes apply when numerous factors are being investigated in agricultural field experiments. Most of the uses of fractional factorials in this area have involved designs that give all, or nearly all, the two-factor interactions.

Table 6A.9 is an index to the plans of this type for experiments with up to 8 factors at 2 levels each. In these plans, no main effect or two-factor interaction is confounded with blocks, and all aliases of any main effect or any two-factor interaction are interactions among at least 3 factors.

TABLE 6A.9 Fractional factorial designs in which all main effects and two-factor interactions can be estimated

(All factors at 2 levels)

No. of factors	Size of experiment (units)	Fraction of a rep.	Block sizes	Plan no.
5	16	½	16 *	6A.3
6	32	½	32, 16	6A.6
7	64	½	64, 32, 16, 8	6A.13
8	64	¼	64, 32, 16	6A.16
8	128	½	128, 64, 32, 16	6A.17

* Provides no error d.f. if two-factor interactions are estimated.

All the experiments are half-replicates except one. The allowable block sizes are on the large side. If the size of the experiment is satisfactory but the block size is too large, the reader should consult the full index (table 6A.10) to see which two-factor interactions must be sacrificed to permit a further reduction in block size. Sometimes only a few are lost in this way.

In a number of these plans, a large majority of the three-factor interactions are also estimable. Where this is so, a note about the status of the three-factor interactions is given with the plans. If the investigator anticipates that he may wish to examine some of the three-factor interactions, the plan should be checked in advance to see whether this can be done.

With 9, 10, or 11 factors, 128 units are necessary in order to keep all main effects and two-factor interactions estimable. Plans for various block sizes are given by Connor (6A.6).

6A.22 Designs in Which Some Two-factor Interactions Are Sacrificed. If a more drastic reduction is made either in the size of the experiment or in the size of block, at least some of the two-factor interactions are lost, because they are either aliases of other two-factor interactions or are confounded with blocks. The number of d.f. available for an estimate of error also declines, particularly with a reduction in the size of the experiment. The situation with regard to two-factor interactions and error d.f. varies from plan to plan, and is best appreciated by examining some of the plans.

The designs discussed in this section are especially appropriate when the following conditions apply:

1. Most of the two-factor interactions are likely to be negligible.
2. Experimental errors are small, making a large experiment unnecessary from the viewpoint of precision.
3. A good estimate of the standard deviation of the experimental errors is available from previous experiments. This condition is not essential in all the plans, since some of them furnish enough d.f. for an internal estimate of error.

The attractiveness of a severe reduction in the size of the experiment is increased if experiments can be done quickly. In this situation the investigator often prefers to keep his experiments small, revising his strategy at the end of each experiment in order to avoid doing a lot of work on treatments that turn out to have only minor effects.

The plans cover from 4 to 8 factors, in experiments of sizes from 8 to 64 units (i.e., treatment combinations to be tested). Block sizes range from 4 to 64 units. Table 6A.10 is an index to the plans (including those discussed in the previous section).

Before the table is entered to see if a suitable plan exists, the user must have decided upon:

(i) the number of factors,
(ii) the size of the experiment,
(iii) the size of block,
(iv) the two-factor interactions that can be assumed negligible and the two-factor interactions that it is desired to estimate,
(v) whether an estimate of error must be made from the results of the experiment, or whether the error variance is well enough known from previous work.

TABLE 6A.10 Index to plans for 2^n factorials in fractional replication

No. of factors	Fraction of a rep.	Size of expt.	Size of block	Two-factor interactions: Total	Two-factor interactions: Max. no. estimable	Error d.f.: 2-factors used as error	Error d.f.: 2-factors estimated	Plan no.
4	1/2	8	8	6	3 *	3	0	6A.1 ‡
5	1/4	8	8	10	2 *	2	0	6A.2 ‡
	1/2	16	16	10	10	10	0	6A.3 ‡
			8	10	9	9	0	6A.3 ‡
			4	10	7	7	0	6A.3 ‡
6	1/8	8	8	15	1 *	1	0	6A.4 ‡
	1/4	16	16	15	7 *	9	2	6A.5 ‡
			8	15	7 *	8	1	6A.5 ‡
			4	15	6 *	6	0	6A.5 ‡
	1/2	32	32	15	15	25	10	6A.6
			16	15	15	24	9	6A.6
			8	15	14	22	8	6A.6
			4	15	12	18	6	6A.6
7	1/16	8	8	21	0	0	0	6A.7 ‡
	1/8	16	16	21	7 *	8	1	6A.9 ‡
			8	21	7 *	7	0	6A.9 ‡
			4	21	4 *	5	1	6A.8 ‡
	1/4	32	32	21	18 †	24	6	6A.11
			16	21	18 †	23	5	6A.11
			8	21	17 †	21	4	6A.11
			4	21	14 †	17	3	6A.10
	1/2	64	64	21	21	56	35	6A.13
			32	21	21	55	34	6A.13
			16	21	21	53	32	6A.13
			8	21	21	49	28	6A.13
			4	21	15	41	26	6A.12
8	1/16	16	16	28	7 *	7	0	6A.14 ‡
			8	28	6 *	6	0	6A.14 ‡
			4	28	4 *	4	0	6A.14 ‡
	1/8	32	32	28	20 †	23	3	6A.15 ‡
			16	28	19 †	22	3	6A.15 ‡
			8	28	17 †	20	3	6A.15 ‡
			4	28	13	16	3	6A.15 ‡
	1/4	64	64	28	28	55	27	6A.16
			32	28	28	54	26	6A.16
			16	28-	28	52	24	6A.16
			8	28	26	48	22	6A.16
			4	28	21	40	19	6A.16
	1/2	128	128	28	28	119	91	6A.17
			64	28	28	118	90	6A.17
			32	28	28	116	88	6A.17
			16	28	28	112	84	6A.17
			8	28	26	104	78	6A.17

* All these interactions have other 2-factor interactions as aliases, and are estimable only if the aliases can be considered negligible. See plans for details.

† Some of the interactions have other 2-factor interactions as aliases. See plans for details.

‡ Except in unusual circumstances, only main effects can be estimated with this design.

As regards (iv), informed guesses about the importance of the two-factor interactions can frequently be made from previous experimentation or from knowledge of the way in which the factors produce their effects. The greater the number of two-factor interactions that are unimportant, the better the chance that a suitable plan exists. It is worth noting that interactions can sometimes be reduced in magnitude by a skillful choice of the factors or of the scale in which results are analyzed and interpreted. The sizes of the differences between the two levels of the factors are also relevant. With small changes in levels that produce moderate or small main effects, the assumption that the effects of the factors are additive is less likely to be seriously wrong than when the experiment explores a wide range of variations in levels.

As regards the error variance, the provision of a self-contained estimate of error and of valid tests of significance is one of the principal strengths of modern experimental design. If the use of an estimate of error from previous experiments is contemplated, some preliminary questions should be answered. Were the previous experiments conducted on similar experimental material, under similar conditions of experimentation and with treatment effects of about the same order of magnitude? Were the previous error variances homogeneous from one experiment to another, as indicated by a test of homogeneity of variances? If these questions are answered in the affirmative, the risk in taking an external estimate of error is decreased.

The use of table 6A.10 will be illustrated by a few examples. To start with an experiment that has already been briefly discussed, suppose that there are 6 factors, to be tested in 32 units (½ replicate) and that a block size of 4 units is advisable for precise comparisons. The experiment will then consist of 8 blocks of 4 units each. Preferably, all the two-factor interactions are to be estimable, although a few of them may be non-existent. The experiment is to furnish its own estimate of error.

Entering table 6A.10 for 6 factors, size of experiment = 32, and block size = 4, we see that 12 of the 15 two-factor interactions can be estimated. To discover which interactions must be sacrificed we refer to the plan (plan 6A.6). The 3 two-factor interactions that are lost are *AE*, *BF*, and *CD*, which are confounded with blocks. This plan is practicable with respect to two-factor interactions if the letters can be assigned to factors so that these three interactions are unlikely to be appreciable. It is unfortunate that the three interactions involve all 6 factors.

The situation with regard to error d.f. is shown in the two columns of table 6A.10 headed "Error d.f." The right hand column of this pair gives the number of error d.f. that the experiment provides when all available two-factor interactions are estimated. In this case there are 6 d.f. The left hand column shows the number of error d.f. when all two-factor interactions are assumed small and are pooled with the error. For this plan we would have 18, i.e., (6 + 12) error d.f. in that event. An intermediate course may be adopted, some of the 12 d.f. for two-factor interactions being estimated and the rest pooled with the error.

To summarize, this plan gives independent estimates of all main effects and nearly all two-factor interactions, plus a small but tolerable number of d.f. for the estimate of error.

To take an example with a larger reduction in the size of the experiment, suppose that 7 factors are to be investigated in 16 tests (1/8 replicate), with block size 8. Table 6A.10 shows that at most 7 of the 21 two-factor interactions are estimable. Moreover, the asterisk attached to the number 7 is a warning that these 7 are all aliases of other two-factor interactions, and are estimable only if their two-factor aliases can be assumed negligible. Detailed scrutiny of the aliases (given with plan 6A.9) is needed to reach a verdict on this point.

The "Error d.f." columns of table 6A.10 reveal that 7 error d.f. are obtainable if all two-factor interactions are included with the error, but none if two-factor interactions are isolated.

We conclude that if the experiment is to provide its own estimate of error, this plan is good only for estimating main effects under conditions in which all interactions are negligible.

For 5, 6, and 7 factors the plans with the greatest reductions in the size of the experiment are those in experiments that include only 8 treatment combinations. In these plans all two-factor interactions must be negligible (since two-factor interactions are aliases of main effects) and an external estimate of error must be available. If all two-factor interactions are unimportant but an internal estimate of error is a necessity, the plans with size of experiment equal to 16 units may be suitable. Additional plans of this type in 12, 20, and 24 units in the experiment can be obtained in (6A.2).

6A.23 The Hazards of Fractional Replication. As we have seen, the results of an experiment with fractional replication may be misinterpreted if some of the interactions that have been assumed negligible are not so. Since the extent of this danger will vary from case to case, it is hard to present an overall picture of it. The following example, although it over-simplifies the issue, exhibits the nature of the problem.

There are 3 factors. A has no effect, while B and C have positive effects and a positive interaction. We shall suppose that B gives an increase of 40 when applied at the lower level of C and of 60 at the upper level of C, while C gives increases of 20 and 40 at the lower and upper levels of B, respectively. With a basal level of 100 for the (1) combination and no experimental error, the results for the 8 treatment combinations are found to be as follows:

TABLE 6A.11 Hypothetical results in a 2^3 experiment

(a)	100	(1)	100
(b)	140	(ab)	140
(c)	120	(ac)	120
(abc)	180	(bc)	180

The experimenter, of course, does not know these results. He runs a half-replicate, with ABC as defining contrast. The results that he obtains consist of those in the left column of table 6A.11 if $+ABC$ is chosen, and those in the right column if $-ABC$ is chosen.

The two sets of main effects are estimated as follows:

From $+ABC$:

$$A = \frac{[(abc) + (a) - (b) - (c)]}{2} = +10; \qquad B = +50; \qquad C = +30$$

From $-ABC$:

$$A = \frac{[(ab) + (ac) - (bc) - (1)]}{2} = -10; \qquad B = +50; \qquad C = +30$$

The two half-replicates give the same results for B and for C: these are in fact the correct responses to B and C over all 8 treatment combinations. The first half-replicate shows an erroneous increase of 10 to A, while the second half shows an equal decrease to A. Neither set of results gives a warning about the presence of a BC interaction (unless the experimenter speculates that this might be one interpretation of the results).

How much harm is done depends on the decisions made from these results. Three situations will be considered briefly.

(i) If the experiment is primarily of the "screening" type, designed to pick out the most important factors, the conclusion may be drawn that B and C are important but A is relatively unimportant. Further experimentation is confined to B and C and presumably reveals the BC interaction. Little or no harm has been done.

(ii) If the experiment is part of a program of fundamental research, the investigator wrongly concludes either that A has a beneficial effect or that it has a deleterious effect. In the absence of further experiments, energy may be dissipated in creating incorrect theories to explain the "effect" of A or in disputes with colleagues.

(iii) In a program of applied research, the object may be to select the best combination of levels of the factors. In the first half-replicate, (abc) gives the highest observed yield. If the results of this half-replicate are used to speculate about the response from the combinations (ab), (ac), and (bc) that were not tested, all these treatment combinations will be predicted to be inferior to (abc). For instance, (bc) is predicted as 10 units poorer than (abc), since A gives an increase of 10 in this half-replicate. Hence, (abc) is the recommended optimum.

If the second half-replicate is chosen, (bc) gives the best observed yield. It will be predicted to be superior to a combination like (abc) that was not tested, since A produces a decrease of 10 in the second half-replicate.

Actually, both choices (abc) and (bc) give the correct highest yield, 180. The confusion is about the level at which A must be set. If the level at which A is put involves a substantial difference in cost, and if the investigator argues as above without further experimentation, he stands a chance of recommending an unnecessarily expensive "optimum."

From the foregoing it is clear that one cannot make flat statements about the danger of misinterpreting the results of a fractional factorial. The penalties may be trivial, or they may be more serious. The example discussed above is not necessarily typical: examples in which more difficulty is created by the use of fractional replication can easily be constructed. In general, it seems wise not to lean heavily on fractional replication as a weapon for investigation unless the risk of confusion from the presence of factor interactions is judged to be small.

6A.24 Fractional Replication in Time. Davies and Hay (6A.7) have pointed out that fractional factorials may save effort in lines of work where an experiment can be conducted one part at a time. Suppose that a complete replication of a 2^4 experiment is planned, and that 8 treatment combinations can be processed at one time, say in a week. The obvious procedure is to divide the replicate into two blocks of 8, confounding $ABCD$ between blocks. In the fractional factorial approach, we stop when the first block (half-replicate) has been completed, and decide from the results whether to continue with the second block or to change the direction of experimentation. If A alone shows a definite response, we may decide that further experimentation with these 4 factors is unjustified. On the other hand, if A, B, and the alias pair $AD = BC$ all give marked effects, we complete the replicate in order to obtain independent estimates of AD and BC so as to elucidate the nature of these interactions. Similarly, if a half-replicate of a large factorial experiment is envisioned, it may be time-saving to start with a quarter-replicate and decide on the next step when the results of this part are known.

To save space, the plans in this book present only one specific fraction of the replicate. If the experimenter decides to follow one quarter-replicate with another, a rule is needed for writing down the plan for the second quarter-replicate. The rule is simple. To obtain another fractional replicate, multiply all the treatment combinations in the plan by any combination of letters that does not appear in the plan, with the usual convention that the square of any letter is replaced by 1. For example, the following are the 8 treatment combinations in plan 6A.2 for a 2^5 factorial in an experiment of size 8:

$$(1) \quad ab \quad cd \quad ace \quad bce \quad ade \quad bde \quad abcd$$

Multiplication by a produces a second quarter-replicate

$$a \quad b \quad acd \quad ce \quad abce \quad de \quad abde \quad bcd$$

To generate a third quarter, multiply the first set by any combination of letters that does not appear in either of the first two sets. Thus e will do, but not b or acd.

If the second quarter-replicate is being run in order to disentangle some aliases, it must be chosen with care. The following example illustrates the general procedure. In the first quarter-replicate above, the defining contrasts are ABE, CDE, and $ABCD$. When we create another quarter-replicate, the half-replicate so formed will have as its defining contrast *one* of the three effects ABE, CDE, or $ABCD$. By an appropriate choice of the second quarter, we can make any specified one of these three effects the defining contrast.

In the first quarter-replicate, main effects are entangled with two-factor interactions (e.g., $A = BE$, $C = DE$). If the second quarter is undertaken in order to free the main effects from this entanglement, we must have $ABCD$ as the defining contrast for the half-replicate, because the choice of ABE or CDE still leaves some main effects with two-factor interactions as aliases. To form the correct second experiment, multiply the treatment combinations in the first by any combination of letters that

(i) does not appear in the first fraction and
(ii) has an *even* number of the letters in the defining contrast that we want to have.

For instance, e is satisfactory as a multiplier, having no letters in common with a, b, c, d, but a is not.

The reasoning behind this rule is as follows. In the plans we have always shown the fractional replicate in which all treatment combinations have an *even* number of letters in common with each of the defining contrasts. (This fraction is called the *principal block.*) Consequently, if $ABCD$ is to remain a defining contrast, all treatment combinations in the half-replicate must have an even number of the letters a, b, c, and d. Multiplying the principal block by an even number of these letters, as in condition (ii) above, achieves this end.

6A.25 Randomization. When the plan has been selected, the steps are:

1. Randomize the blocks.
2. Randomize the treatment combinations within each block.

It is also preferable to select the fractional replication at random from those that can be generated by the set of defining contrasts that are being used. With a half-replicate, for instance, we decide at random which of the two halves is used. This step increases the validity of tests of significance made on the results, and if similar experiments are being conducted in different places it provides a wider sampling of the possible treatment combinations.

As we have mentioned, the plans present only one specific fraction of the replicate. Although the other fractions can be written down by the rules given in the preceding section, the following alternative method avoids this step, using the plan as given.

With a half-replicate, decide at random for *one* of the factors whether the presence of the letter is to represent the first or the second level. With factors that are quantitative, it is natural for the experimenter to let the letter, say a, denote the *higher* level. Consequently, this random choice should be made for a qualitative factor, if there is one in the experiment, or more generally for the factor that will cause the least confusion to the experimenter. The reader may verify, for any of the half-replicate plans, that this choice represents a random selection of one of the two possible halves.

For a quarter-replicate, assign the letter to the levels at random for any *two* of the factors, provided that this pair of letters does not occur as a treatment combination in the plan. To illustrate, in plan 6A.5 for 6 factors in a ¼ replicate, blocks of 4 units, the only pairs of letters that occur are *ab*, *ce*, and *df*. Any pair other than this trio (e.g., *ae*, *bc*) will do. For an eighth-replicate, any 3 of the factors whose letter combination does not appear in the plan may have their levels assigned at random to letters.

6A.26 Statistical Analysis. If there is no confounding with blocks, the procedure in section 6A.15 should be followed. When the experiment is arranged in blocks, the factorial effects divide themselves into three groups which must be clearly distinguished.

(i) The effects that are to be estimated. Each plan shows which two-factor interactions are estimable: this information is also given for three-factor interactions in the larger designs.

(ii) The effects that are confounded with blocks. These effects are also given, either under the plan or above the analysis of variance in cases where a plan is constructed by combining the blocks from a previous plan.

(iii) The effects that are to be used for an estimate of experimental error.

The example in section 6A.15 on the texture of cake icing will be used to provide an illustration for a half-replicate of a 2^6 factorial, confounded in four blocks of 8 units. In table 6A.12 the data for this experiment are arranged in the blocks given by plan 6A.6. Artificial block effects of -40, $+10$, 0, and $+30$ per observation have been superimposed on the data.

TABLE 6A.12 A 2^6 FACTORIAL (½ REPLICATE) IN BLOCKS OF 8 UNITS

Texture of icing after 72 hours

(1)		(2)		(3)		(4)	
(1)	193	*ac*	243	*ae*	350	*ad*	297
ab	277	*bc*	343	*af*	217	*bd*	430
ef	160	*de*	277	*be*	283	*ce*	430
abef	110	*df*	260	*bf*	267	*cf*	280
acde	327	*abde*	360	*cd*	400	*abce*	413
acdf	210	*abdf*	277	*abcd*	317	*abcf*	380
bcde	227	*acef*	343	*cdef*	117	*adef*	230
bcdf	243	*bcef*	327	*abcdef*	217	*bdef*	180

CD, *ABC*, *ABD* confounded.

The analysis proceeds as in section 6A.15. The factor F is temporarily suppressed. The remaining factors, which form a complete replicate, are arranged in systematic order for the application of Yates' method (table 6A.13). When the factorial effect totals have been obtained by this method (column 5), it remains to reintroduce F and identify each one appropriately.

We shall suppose that all main effects and the 14 available 2-factor interactions are to be estimated (CD is lost by confounding). The identifications are given on the right of table 6A.13. Be sure that the blocks

TABLE 6A.13 COMPUTATION OF FACTORIAL EFFECT TOTALS

Treatment combination	Texture	(1)	(2)	(3)	(4)	(5)	Identification
(1)	193	410	954	2200	4634	8985	G
$a(f)$	217	544	1246	2434	4351	151	A
$b(f)$	267	523	1264	2416	−198	317	B
ab	277	723	1170	1935	349	−49	AB
$c(f)$	280	557	903	34	434	649	C
ac	243	707	1513	−232	−117	15	AC
bc	343	610	1047	16	134	−83	BC
$abc(f)$	380	560	888	333	−183	631	ABC (Blocks)
$d(f)$	260	510	34	334	198	−247	D
ad	297	393	0	100	451	51	AD
bd	430	773	−116	−150	−34	−51	BD
$abd(f)$	277	740	−116	33	49	211	ABD (Blocks)
cd	400	507	17	60	−134	−1155	CD (Blocks)
$acd(f)$	210	540	−1	74	51	119	ACD (Error)
$bcd(f)$	243	444	133	−190	542	−383	BCD (Error)
$abcd$	317	444	200	7	89	−617	$EF = ABCD$
		8985	9136	9404	10616	8544	
$e(f)$	160	24	134	292	234	−283	E
ae	350	10	200	−94	−481	547	AE
be	283	−37	150	610	−266	−551	BE
$abe(f)$	110	37	−50	−159	317	−317	ABE (Error)
ce	430	37	−117	−34	−234	253	CE
$ace(f)$	343	−153	−33	0	183	83	ACE (Error)
$bce(f)$	327	−190	33	−18	14	185	BCE (Error)
$abce$	413	74	0	67	197	−453	$DF = ABCE$
de	277	190	−14	66	−386	−715	DE
$ade(f)$	230	−173	74	−200	−769	583	ADE (Error)
$bde(f)$	180	−87	−190	84	34	417	BDE (Error)
$abde$	360	86	264	−33	85	183	$CF = ABDE$
$cde(f)$	117	−47	−363	88	−266	−383	CDE (Error)
$acde$	327	180	173	454	−117	51	$BF = ACDE$
$bcde$	227	210	227	536	366	149	$AF = BCDE$
$abcde(f)$	217	−10	−220	−447	−983	−1349	$F = ABCDE$
Odds	4417	4434	4096	6344	5072		
Evens	4568	4702	5308	4272	3472		
	8985	9136	9404	10616	8544	6944	

are correctly identified. It may be noted that the numerical values of all factorial effect totals are the same in table 6A.13 as in table 6A.5, except for the three effects *CD*, *ABC*, and *ABD* that are confounded with blocks. This is as it should be.

In the analysis of variance (table 6A.14), the sums of squares for blocks, main effects, estimable two-factor interactions, and error are all obtained directly by adding the squares of the appropriate factorial effect totals, and dividing by the number of units in the experiment (32). Thus

$$\text{Blocks s.s.} = \frac{(631)^2 + (211)^2 + (1155)^2}{32} = 55{,}522$$

TABLE 6A.14 CONDENSED ANALYSIS OF VARIANCE

	d.f.	s.s.	m.s.
Blocks	3	55,522	18,507
Main effects	6	78,293	13,049
2-factor interactions	14	57,405	4,100
Error (from 3-factors)	8	30,091	3,761
Total	31	221,311	

Alternatively, the analysis of variance can be dispensed with by finding the error mean square directly. From this, the standard error of a factorial effect total is computed, and tests of significance are performed directly on the effect totals in table 6A.13.

6A.3 Designs with Factors at More Than Two Levels

6A.31 Index to Plans. When some or all of the factors contain more than 2 levels, most of the fractional replication designs have two-factor interactions heavily mixed up with one another. We shall confine ourselves to a few designs in which all or nearly all of the two-factor interactions can be estimated. Table 6A.15 gives an index to the designs for which plans are presented.

TABLE 6A.15 DESIGNS WITH FACTORS AT MORE THAN 2 LEVELS

Type	Fraction of a rep.	Size of experiment	Block sizes	Plan
3^4	⅓	27	27, 9	6A.18
3^5	⅓	81	81, 27, 9	6A.19
3×2^5	½	48	48, 24, 12	Section 6A.35
4×2^4	½	32	32, 16, 8	6A.20

A number of these arrangements have been used in agricultural research. Chinloy *et al.* (6A.8) describe the application of a 3^5 design, a 1/3 replicate, in blocks of 9 plots, in an experiment on the manuring of sugar cane, the factors being nitrogen, phosphorus, potash, filter press mud, and bagasse (the crushed stalks minus the cane juice), each at 3 levels. A 1/9 replicate of a 3^7 factorial was used by Tischer and Kempthorne (6A.4) to study 7 factors that might affect the performance of the Adams consistometer, an instrument for measuring the consistency of canned food. This experiment required $3^7/9$, or 243, different treatment combinations to be tested. Since 27 combinations could be handled in a day, the experiment was arranged in blocks of 27 units.

The following sections give notes on the plans and mention some other possibilities.

6A.32 The 3^n Series. The problems of construction can be illustrated from a 1/3 replicate of a 3^3 factorial. It will be recalled (sections 6.15, 6.16) that in a 3^3 factorial the 26 d.f. for the treatments sum of squares can be split into 13 pairs of d.f., as follows:

Main effects: A, B, C.
Two-factor interactions: $AB(I)$, $AB(J)$, $AC(I)$, $AC(J)$, $BC(I)$, $BC(J)$.
Three-factor interactions: $ABC(W)$, $ABC(X)$, $ABC(Y)$, $ABC(Z)$.

To obtain a 1/3 replicate it is natural to choose a component of ABC as the defining contrast. This is easily done with the aid of plan 6.7 (p. 238), in which four replicates of a 3^3 design are shown in blocks of 9 units. In each replicate, a different component of ABC is confounded between blocks. Replicate I confounds $ABC(W)$. If any single block of this replicate is selected, we obtain a 1/3 replicate with $ABC(W)$ as the defining contrast. Table 6A.16 shows the plan given by the first block.

TABLE 6A.16 A 1/3 REPLICATE OF A 3^3 DESIGN

Defining contrast: $ABC(W)$

000
110
220
101
211
021
202
012
122

For this plan, the alias system works out as shown below.

$$A = BC(J) = ABC(Z)$$

$$B = AC(I) = ABC(Y)$$

$$C = AB(I) = ABC(X)$$

$$AB(J) = AC(J) = BC(I)$$

To verify the aliases of A, note that the main effect of A is computed by comparing the totals for the 3 levels of A. These totals are:

$$A_0 = 000 + 021 + 012$$

$$A_1 = 101 + 110 + 122$$

$$A_2 = 220 + 211 + 202$$

Now consider the experiment with reference to B and C, ignoring A. From section 6.15, where the I and J components of a two-factor interaction are defined, it is seen that the three totals above represent respectively the J_1, J_2, and J_3 components of BC. Thus

$$A = BC(J)$$

The three-factor interaction $ABC(Z)$ is shown in replication IV of plan 6.7, where it is confounded with blocks. If we compute $ABC(Z)$ from the 9 treatment combinations that are available in the ⅓ replicate, it will again be found that this is the main effect of A. The other alias relationships can be checked in the same way.

Owing to the confusion between main effects and two-factor interactions, this plan is not of practical use unless all two-factor interactions are negligible relative to main effects.

For a 3^4 factorial, plan 6A.18 gives a ⅓ replicate in 27 observations, with a component of $ABCD$ as defining contrast. Main effects are clear of two-factor interactions, but some confusion still exists among the latter, i.e.,

$$AB(J) = CD(J); \qquad AC(I) = BD(I); \qquad BC(I) = AD(I)$$

This plan may serve the purpose if one of the factors, say D, does not interact with any of the others. The 3 alias pairs above are then regarded as estimating $AB(J)$, $AC(I)$, and $BC(I)$, respectively. Under this assumption, full information on the two-factor interactions of A, B, and C is obtained. The analysis of variance runs as follows:

Effects	d.f.
Main	8
AB, *AC*, *BC*	12
Error (from *AD*, *BD*, and *CD*)	6
Total	26

If only the linear $\times$ linear components of the *AB*, *AC*, and *BC* interactions (section 5.26) appear likely to be appreciable, 9 additional d.f. can be included in the estimate of error.

To obtain a design in blocks of 9 units, 2 d.f. representing two-factor interactions must be confounded. The best pair to sacrifice depends on the nature of the experiment. In plan 6A.18 we have confounded *CD*(*I*) with blocks, on the grounds that this design seems most useful when *D* does not interact with the other factors. The separation of d.f. is shown with the plan.

With the 3^5 factorial (plan 6A.19), all main effects and two-factor interactions are estimable in a ⅓ replicate, with blocks of either 27 or 81 units. If the block size is reduced to 9 units, 2 d.f. from *AE* are lost by confounding with blocks.

In the analysis of variance for 3^4 and 3^5 designs, all main effects and unconfounded two-factor interactions are computed in the usual way by setting up the necessary two-way tables.

In the 3^n system, as in the 2^n system, there are rules by which the aliases of any factorial effect can be discovered when the defining contrasts have been selected, without having to construct the plan (6A.1, 6A.4).

6A.33 The 4^n Series. Since experiments with all factors at 4 levels do not appear to be common, we have not presented plans for the 4^n series. It is worth noting, however, that plans with some or all factors at 4 levels can be constructed quickly from the 2^n plans. Let the 4 levels of a factor *A* be denoted by a_0, a_1, a_2, and a_3. As mentioned on p. 227, the three d.f. for the main effect of *A* can be partitioned into three orthogonal single d.f.

$$A' = (a_3) + (a_2) - (a_1) - (a_0)$$

$$A'' = (a_3) - (a_2) - (a_1) + (a_0)$$

$$A''' = (a_3) - (a_2) + (a_1) - (a_0)$$

Suppose now that we start with two different factors *C* and *D*, each at two levels. Make the transformations

$$cd = a_3; \qquad c = a_2; \qquad (1) = a_1; \qquad d = a_0$$

Then the reader may verify that the main effect of C becomes identical with A' as just given. Similarly, $D = A''$, and the interaction $CD = A'''$. Thus C, D, and their interaction CD are transformed into the 3 d.f. for the main effect of A.

By this device, any pair of factors in a 2^n plan can be replaced by a single factor at 4 levels. Take, for instance, the half-replicate plan for a 2^8 factorial, with $ABCDEFGH$ as defining contrast, and replace the pair A, B by I; C, D by J; E, F by K; and G, H by L, where I, J, K, L are at 4 levels each. The plan becomes a half-replicate of a 4^4 factorial, with $I'''J'''K'''L'''$ as defining contrast. The alias relationships in the 4^4 design are easily found from these in the 2^8 design. The equality $ABCD = EFGH$, for example, becomes $I'''J''' = K'''L'''$. Note that $I'''J'''$ is a component of the *two*-factor interaction of I and J in the 4-level system, and likewise for $K'''L'''$, so that the 4^4 design retains some confusion among the two-factor interactions.

When using this method, remember that some two-factor interactions in the 2^n system become main effects in the 4^n system, while some three- and four-factor interactions become two-factor interactions. The alias system should be carefully checked before adopting a design.

6A.34 Hyper-graeco-latin Squares As Fractional Factorials. When all factors have 2 levels, we have seen (section 6A.22) that the main effects of as many as 7 factors can be estimated independently in an experiment with only 8 treatment combinations, provided that all interactions are negligible. When factors have more than 2 levels, designs that give the maximum reduction in size of experiment are obtained from graeco-latin squares (section 4.51), or more generally from sets of completely orthogonal latin squares. Table 6A.17 shows the 3×3 graeco-latin square, rewritten (on the right) as a plan for a 3^4 factorial in 9 units ($1/9$ replicate).

TABLE 6A.17 A $1/9$ REPLICATE OF A 3^4 FACTORIAL

Graeco-latin square			Fractional factorial		
A_1	B_3	C_2	0000	0112	0221
B_2	C_1	A_3	1011	1120	1202
C_3	A_2	B_1	2022	2101	2210

In the rewritten form, the rows of the square represent the first factor, columns the second, latin letters the third, and suffixes the fourth. All main effects of the 4 factors are orthogonal to one another. Since each main effect carries 2 d.f., the four main effects account for all the 8 d.f. in the total sum of squares among the 9 treatment combinations in the experiment. This result shows that 4 is the greatest number of factors at 3 levels each for which main effects can be estimated in 9 units. This estimation is possible only if all interactions are absent.

For factors at 4 levels, there exists a set of three completely orthogonal 4×4 latin squares (6A.9). An experiment with 16 units will therefore accommodate 5 factors each at 4 levels, this being 1/64 replicate. Similarly, there is a design with 25 units (1/625 replicate) which gives estimates of 6 factors each at 5 levels.

These results are presented for their mathematical interest, not because the designs are to be recommended. Since main effects have two-factor interactions as aliases, such designs can rarely be used with safety. Even in the less extreme case in which the rows, columns, and letters of an ordinary $p \times p$ latin square represent different factors (giving a $1/p$ fraction of a p^3 factorial), main effects still have two-factor interactions as aliases, as was verified in section 6A.32 for the 3^3 factorial. Youden and Hunter (6A.10) have warned against the misuse of this design and have shown how to include some genuine replication in it.

6A.35 3×2^n and 4×2^n Factorials. If one factor is to be tested at 3 levels and all others at 2 levels, a design can be obtained by an expansion of the corresponding 2^n design. Take, for instance, plan 6A.3 for a 2^5 factorial in 16 units (½ replicate). If we call the additional 3-level factor F, and test all these 16 treatment combinations at *each* level of F, we have a 3×2^5 factorial in 48 units (½ replicate). The block sizes, of course, become three times as large as those in plan 6A.3. The two-factor interactions that are estimable are exactly the same as in that plan, CD, CE, and DE being lost in blocks of 12 units and DE in blocks of 24 units. The same method gives a 3×2^6 design in 96 units.

Plan 6A.20 shows a 4×2^n factorial in 32 units (½ replicate). All two-factor interactions are estimable except in blocks of 8 units, in which 1 d.f. must be sacrificed. Plans for 4×2^5 and $4^2 \times 2^4$ factorials can be constructed from the 2^n series of plans as explained in section 6A.33, but they are not presented here.

REFERENCES

6A.1 Finney, D. J. The fractional replication of factorial arrangements. *Ann. Eugen.* 12, 291–301, 1945.

6A.2 Plackett, R. L., and Burman, J. P. The design of optimum multifactorial experiments. *Biometrika* 33, 305–325, 1946.

6A.3 Plackett, R. L. Some generalizations in the multifactorial design. *Biometrika* 33, 328–332, 1946.

6A.4 KEMPTHORNE, O. *The design and analysis of experiments.* John Wiley and Sons, New York, 1952.

6A.5 BROWNLEE, K. A., KELLY, B. K., and LORAINE, P. K. Fractional replication arrangements for factorial experiments with factors at two levels. *Biometrika* 35, 268–276, 1948.

6A.6 CONNOR, W. S., ZELEN, M., and DEMING, LOLA. Fractional factorial experiment design for factors at two levels. *Nat. Bur. of Standard Applied Math. Series* 48, 1956.

6A.7 DAVIES, O. L., and HAY, W. A. The construction and uses of fractional factorial designs in industrial research. *Biometrics* 6, 233–249, 1950.

6A.8 CHINLOY, T., INNES, R. F., and FINNEY, D. J. An example of fractional replication in an experiment on sugar cane manuring. *Jour. Agr. Sci.* 43, 1–11, 1953.

6A.9 FISHER, R. A., and YATES, F. *Statistical tables for biological, agricultural and medical research.* Oliver and Boyd, Edinburgh, 4th ed., 1953.

6A.10 YOUDEN, W. J., and HUNTER, J. S. Partially replicated latin squares. *Biometrics* 11, 399–405, 1955.

ADDITIONAL READING

KEMPTHORNE, O., and TISCHER, R. G. An example of the use of fractional replication. *Biometrics* 9, 295–303, 1953.

WILLIAMS, E. J. Confounding and fractional replication in factorial experiments. *Jour. Australian Inst. Agr. Sci.* 15, 145–153, 1949.

PLANS

Plan 6A.1 2^4 factorial in 8 units (½ replicate)

Defining contrast: $ABCD$

Estimable 2-factor interactions: $AB = CD$, $AC = BD$, $AD = BC$.

(1)
ab
ac
ad
bc
bd
cd
abcd

Effects	d.f.
Main	4
2-factor	3
	–
Total	7

Plan 6A.2 2^5 factorial in 8 units (¼ replicate)

Defining contrasts: *ABE, CDE, ABCD*

Main effects have 2-factors as aliases. The only estimable 2-factors are $AC = BD$ and $AD = BC$.

(1)
ab
cd
ace
bce
ade
bde
abcd

Effects	d.f.
Main	5
2-factor	2
Total	7

Plan 6A.3 2^5 factorial in 16 units (½ replicate)

Defining contrast: *ABCDE*

Blocks of 4 units

Estimable 2-factors: All except *CD, CE, DE* (confounded with blocks).

Blocks	(1)	(2)	(3)	(4)
	(1)	*ac*	*ae*	*ad*
	ab	*bc*	*be*	*bd*
	acde	*de*	*cd*	*ce*
	bcde	*abde*	*abcd*	*abce*

Effects	d.f.
Block	3
Main	5
2-factor	7
Total	15

CD, CE, DE confounded.

Blocks of 8 units

Estimable 2-factors: All except *DE*.

Combine blocks 1 and 2; and blocks 3 and 4. *DE* confounded.

Effects	d.f.
Block	1
Main	5
2-factor	9
Total	15

Blocks of 16 units

Estimable 2-factors: All.

Combine blocks 1–4.

Effects	d.f.
Main	5
2-factor	10
Total	15

Plan 6A.4 2^6 factorial in 8 units (1/8 replicate)

Defining contrasts: *ACE, ADF, BCF, BDE, ABCD, ABEF, CDEF*

Main effects have 2-factors as aliases. The only estimable 2-factor is the set $AB = CD = EF$.

(1)
acf
ade
bce
bdf
abcd
abef
cdef

Effects	d.f.
Main	6
2-factor ($AB = CD = EF$)	1
Total	7

Plan 6A.5 2^6 factorial in 16 units (1/4 replicate)

Defining contrasts: *ABCE, ABDF, CDEF*

Blocks of 4 units

Estimable 2-factors: The alias sets $AC = BE$, $AD = BF$, $AE = BC$, $AF = BD$, $CD = EF$, $CF = DE$.

Blocks	(1)	(2)	(3)	(4)
	(1)	*acd*	*ab*	*acf*
	abce	*aef*	*ce*	*ade*
	abdf	*bcf*	*df*	*bcd*
	cdef	*bde*	*abcdef*	*bef*

Effects	d.f.
Block	3
Main	6
2-factor	6
Total	15

AB, ACF, BCF confounded.

Blocks of 8 units

Estimable 2-factors: Same as in blocks of 4 units, plus the set $AB = CE = DF$.

Combine blocks 1 and 2; and blocks 3 and 4. *ACF* confounded.

Effects	d.f.
Block	1
Main	6
2-factor	7
3-factor	1
Total	15

Blocks of 16 units

Estimable 2-factors: Same as in blocks of 8 units.

Combine blocks 1–4.

Effects	d.f.
Main	6
2-factor	7
3-factor	2
Total	15

Plan 6A.6 2^6 factorial in 32 units (½ replicate)

Defining contrast: *ABCDEF*

Blocks of 4 units

Estimable 2-factors: All except *AE*, *BF*, and *CD* (confounded with blocks).

Blocks	(1)	(2)	(3)	(4)	(5)	(6)	(7)	(8)
	(1)	*ab*	*ac*	*bc*	*ae*	*af*	*ad*	*bd*
	abef	*ef*	*de*	*df*	*bf*	*be*	*ce*	*cf*
	acde	*acdf*	*abdf*	*acef*	*cd*	*abcd*	*abcf*	*abce*
	bcdf	*bcde*	*bcef*	*abde*	*abcdef*	*cdef*	*bdef*	*adef*

AE, *BF*, *CD*, *ABC*, *ABD*, *ACF*, *ADF* confounded.

Effects	d.f.
Block	7
Main	6
2-factor	12
Higher order	6
Total	31

Blocks of 8 units

Estimable 2-factors: All except *CD*.

Combine blocks 1 and 2; blocks 3 and 4; blocks 5 and 6; and blocks 7 and 8. *CD*, *ABC*, *ABD* confounded.

Effects	d.f.
Block	3
Main	6
2-factor	14
Higher order	8
Total	31

Blocks of 16 units

Estimable 2-factors: All.

Estimable 3-factors: *ABC* = *DEF* is lost by confounding. The others are in alias pairs, e.g., *ABD* = *CEF*.

Combine blocks 1–4; and blocks 5–8. *ABC* confounded.

Effects	d.f.
Block	1
Main	6
2-factor	15
3-factor	9
Total	31

Blocks of 32 units

Estimable 2-factors: All.

Estimable 3-factors: These are arranged in 10 alias pairs.

Combine blocks 1–8.

Effects	d.f.
Main	6
2-factor	15
3-factor	10
Total	31

Plan 6A.7 2^7 factorial in 8 units (1/16 replicate)

Defining contrasts: *ABG, ACE, ADF, BCF, BDE, CDG, EFG, ABCD, ABEF, ACFG, ADEG, BCEG, BDFG, CDEF, ABCDEFG*

Main effects have 2-factors as aliases. No 2-factors are estimable.

(1)
abcd
abef
acfg
adeg
bceg
bdfg
cdef

Effects	d.f.
Main	7
Total	7

Plan 6A.8 2^7 factorial in 16 units (1/8 replicate)

Defining contrasts: *ABCD, ABEF, ACEG, ADFG, BCFG, BDEG, CDEF*

Blocks of 4 units

Estimable 2-factors: Only the alias sets $AE = BF = CG$; $AF = BE = DG$; $AG = CE = DF$; $BG = DE = CF$.

Blocks	(1)	(2)	(3)	(4)
	(1)	*abg*	*acf*	*ade*
	efg	*cdg*	*bdf*	*bce*
	abcd	*abef*	*aceg*	*adfg*
	abcdefg	*cdef*	*bdeg*	*bcfg*

Effects	d.f.
Block	3
Main	7
2-factor	4
Higher order	1
Total	15

$AB = CD = EF$,
$AC = BD = EG$,
$AD = BC = FG$ confounded.

Plan 6A.9 2^7 factorial in 16 units (1/8 replicate)

Defining contrasts: *ABCD*, *ABEF*, *ACEG*, *ADFG*, *BCFG*, *BDEG*, *CDEF*

Blocks of 8 units

Estimable 2-factors: Same as in blocks of 4 units, plus the alias sets $AB = CD = EF$, $AC = BD = EG$, $AD = BC = FG$.

Blocks (1)	(2)
(1)	*abg*
abcd	*acf*
abef	*ade*
aceg	*bce*
adfg	*bdf*
bcfg	*cdg*
bdeg	*efg*
cdef	*abcdefg*

Effects	d.f.
Block	1
Main	7
2-factor	7
Total	15

ABG confounded.

Blocks of 16 units

Estimable 2-factors: Same as in blocks of 8 units.

Combine blocks 1 and 2 of the plan for blocks of 8 units.

Effects	d.f.
Main	7
2-factor	7
Higher order	1
Total	15

Plan 6A.10 2^7 factorial in 32 units (1/4 replicate)

Defining contrasts: *ABCDE*, *ABCFG*, *DEFG*

Blocks of 4 units

Estimable 2-factors: *AB*, *AC*, *BC*, and $DF = EG$ are lost by confounding. All other 2-factors are estimable, except that $DE = FG$ and $DG = EF$, so that members of these alias pairs cannot be separated.

Blocks (1)	(2)	(3)	(4)	(5)	(6)	(7)	(8)
(1)	*de*	*ab*	*cdg*	*ac*	*bdg*	*bc*	*adg*
defg	*fg*	*cdf*	*cef*	*bdf*	*bef*	*adf*	*aef*
abcdf	*abcdg*	*ceg*	*abde*	*beg*	*acde*	*aeg*	*bcfg*
abceg	*abcef*	*abdefg*	*abfg*	*acdefg*	*acfg*	*bcdefg*	*bcde*

AB, *AC*, *BC*, $DF = EG$, *ADG*, *BDG*, *CDG* confounded.

Effects	d.f.
Block	7
Main	7
2-factor	14
Higher order	3
Total	31

Plan 6A.11 2^7 factorial in 32 units (¼ replicate)

Defining contrasts: *ABCDE, ABCFG, DEFG*

Blocks of 8 units

Estimable 2-factors: All except *DF* = *EG* (confounded with blocks). However, *DE* = *FG* and *DG* = *EF* are alias pairs which cannot be separated.

Blocks (1)	(2)	(3)	(4)
(1)	*bdg*	*ab*	*de*
bc	*bef*	*ac*	*fg*
adf	*cef*	*bdf*	*adg*
aeg	*abfg*	*beg*	*aef*
defg	*acfg*	*cdf*	*bcde*
abcdf	*abde*	*ceg*	*bcfg*
abceg	*acde*	*acdefg*	*abcdg*
bcdefg	*cdg*	*abdefg*	*abcef*

Effects	d.f.
Block	3
Main	7
2-factor	17
Higher order	4
	—
Total	31

DF = *EG*, *ADE*, *AEF* confounded.

Blocks of 16 units

Estimable 2-factors: All, except that *DE* = *FG*, *DG* = *EF*, and *DF* = *EG* are alias pairs.

Combine blocks 1 and 2; and blocks 3 and 4. *AEF* confounded.

Effects	d.f.
Block	1
Main	7
2-factor	18
Higher order	5
	—
Total	31

Blocks of 32 units

Estimable 2-factors: Same as in blocks of 16 units.

Combine blocks 1–4.

Effects	d.f.
Main	7
2-factor	18
Higher order	6
	—
Total	31

Plan 6A.12 2^7 factorial in 64 units (½ replicate)

Defining contrast: *ABCDEFG*

Blocks of 4 units

Estimable 2-factors: All except *AB*, *AC*, *BC*, *EF*, *EG*, and *FG* (confounded with blocks).

Blocks	(1)	(2)	(3)	(4)	(5)	(6)	(7)	(8)
	(1)	*ab*	*ac*	*bc*	*ae*	*be*	*ce*	*abce*
	abcd	*cd*	*bd*	*ad*	*bcde*	*acde*	*abde*	*de*
	defg	*abdefg*	*acdefg*	*bcdefg*	*adfg*	*bdfg*	*cdfg*	*abcdfg*
	abcefg	*cefg*	*befg*	*aefg*	*bcfg*	*acfg*	*abfg*	*fg*

(9)	(10)	(11)	(12)	(13)	(14)	(15)	(16)
af	*bf*	*cf*	*abcf*	*ef*	*abef*	*acef*	*bcef*
bcdf	*acdf*	*abdf*	*df*	*abcdef*	*cdef*	*bdef*	*adef*
adeg	*bdeg*	*cdeg*	*abcdeg*	*dg*	*abdg*	*acdg*	*bcdg*
bceg	*aceg*	*abeg*	*eg*	*abcg*	*cg*	*bg*	*ag*

AB, *AC*, *BC*, *EF*, *EG*, *FG*, *ADE*, *ADF*, *ADG*, *BDE*, *BDF*, *BDG*, *CDE*, *CDF*, *CDG* confounded.

Effects	d.f.
Block	15
Main	7
2-factor	15
Higher order	26
Total	63

Plan 6A.13 2^7 factorial in 64 units (½ replicate)

Defining contrast: *ABCDEFG*

Blocks of 8 units

Estimable 2-factors: All.

Estimable 3-factors: All except *ABC, ADE, AFG, BDF, BEG, CDG, CEF* (confounded with blocks).

Blocks	(1)	(2)	(3)	(4)	(5)	(6)	(7)	(8)
	(1)	*bc*	*ac*	*ab*	*ag*	*af*	*ae*	*ad*
	abdg	*de*	*df*	*dg*	*bd*	*be*	*bf*	*bg*
	abef	*fg*	*eg*	*ef*	*ce*	*cd*	*cg*	*cf*
	acdf	*abdf*	*abde*	*acde*	*abcf*	*abcg*	*abcd*	*abce*
	aceg	*abeg*	*abfg*	*acfg*	*adef*	*adeg*	*adfg*	*aefg*
	bcde	*acdg*	*bcdg*	*bcdf*	*befg*	*bdfg*	*bdeg*	*bdef*
	bcfg	*acef*	*bcef*	*bceg*	*cdfg*	*cefg*	*cdef*	*cdeg*
	defg	*bcdefg*	*acdefg*	*abdefg*	*abcdeg*	*abcdef*	*abcefg*	*abcdfg*

Effects	d.f.
Block	7
Main	7
2-factor	21
3-factor	28
Total	63

Blocks of 16 units

Estimable 2-factors: All.

Estimable 3-factors: All except *ABC, ADE, AFG* (confounded).

Combine blocks 1 and 2; blocks 3 and 4; blocks 5 and 6; and blocks 7 and 8.

Effects	d.f.
Block	3
Main	7
2-factor	21
Higher order	32
Total	63

Blocks of 32 units

Estimable 2-factors: All.

Estimable 3-factors: All except *ABC* (confounded).

Combine blocks 1–4; and blocks 5–8.

Effects	d.f.
Block	1
Main	7
2-factor	21
3-factor	34
Total	63

Blocks of 64 units

Estimable 2-factors: All.

Estimable 3-factors: All.

Combine blocks 1–8.

Effects	d.f.
Main	7
2-factor	21
3-factor	35
Total	63

Plan 6A.14 2^8 factorial in 16 units (1/16 replicate)

Defining contrasts: *ABCD, ABEF, ABGH, ACEH, ACFG, ADEG, ADFH, BCEG, BCFH, BDEH, BDFG, CDEF, CDGH, EFGH, ABCDEFGH.*

Blocks of 4 units

Estimable 2-factors: Only the 4 alias sets $AE = BF = CH = DG$; $AF = BE = CG = DH$; $AG = BH = CF = DE$; $AH = BG = CE = DF$. Except in special circumstances, only main effects are estimable.

Blocks	(1)	(2)	(3)	(4)
	(1)	*abef*	*adeg*	*aceh*
	abcd	*abgh*	*adfh*	*acfg*
	efgh	*cdef*	*bceg*	*bdeh*
	abcdefgh	*cdgh*	*bcfh*	*bdfg*

AB, AC, AD confounded.

Effects	d.f.
Block	3
Main	8
2-factor	4
Total	15

Blocks of 8 units

Estimable 2-factors: As in blocks of 4 units, plus the alias sets $AC = BD = EH = FG$; $AD = BC = EG = FH$.

Combine blocks 1 and 2; and blocks 3 and 4. *AB* confounded.

Effects	d.f.
Block	1
Main	8
2-factor	6
Total	15

Blocks of 16 units

Estimable 2-factors: As in blocks of 8 units, plus the alias set $AB = CD = EF = GH$.

Combine blocks 1–4.

Effects	d.f.
Main	8
2-factor	7
Total	15

Plan 6A.15 2^8 factorial in 32 units (1/8 replicate)

Defining contrasts: *BCDH, BDFG, CFGH, ABCEF, ABEGH, ACDEG, ADEFH*

Blocks of 4 units

Estimable 2-factors: All interactions of *A* and of *E*. All other 2-factors are lost by confounding.

Blocks	(1)	(2)	(3)	(4)	(5)	(6)	(7)	(8)
	(1)	*abd*	*dgh*	*cdf*	*afg*	*ach*	*bcg*	*bfh*
	ae	*bde*	*abcf*	*abgh*	*efg*	*ceh*	*adfh*	*acdg*
	abcdfgh	*cfgh*	*bcef*	*begh*	*bcdh*	*bdfg*	*defh*	*cdeg*
	bcdefgh	*acefgh*	*adegh*	*acdef*	*abcdeh*	*abdefg*	*abceg*	*abefh*

BD, BF, BH, CF, DF, DH, FH and their aliases confounded.

Effects	d.f.
Block	7
Main	8
2-factor	13
Higher order	3
Total	31

Blocks of 8 units

Estimable 2-factors: All interactions of *A* and of *E*, and the alias pairs *BC* = *DH*, *BF* = *DG*, *BG* = *DF*, *BH* = *CD*.

Combine blocks 1 and 2; blocks 3 and 4; blocks 5 and 6; and blocks 7 and 8. *BD, CF, FH* confounded.

Effects	d.f.
Block	3
Main	8
2-factor	17
Higher order	3
Total	31

Blocks of 16 units

Estimable 2-factors: As in blocks of 8 units, plus the alias sets *CG* = *FH*, *BD* = *CH* = *FG*.

Combine blocks 1–4; and blocks 5–8. *CF* confounded.

Effects	d.f.
Block	1
Main	8
2-factor	19
Higher order	3
Total	31

Blocks of 32 units

Estimable 2-factors: All interactions of *A* and of *E*, plus the alias sets *BC* = *DH*, *BF* = *DG*, *BG* = *DF*, *BH* = *CD*, *CG* = *FH*, *CF* = *GH*, *BD* = *CH* = *FG*.

Combine blocks 1–8.

Effects	d.f.
Main	8
2-factor	20
Higher order	3
Total	31

Plan 6A.16 2^8 factorial in 64 units (¼ replicate)

Defining contrasts: *ABCEG, ABDFH, CDEFGH*

Blocks of 4 units

Estimable 2-factors: All except *AF, AH, BC, BG, CG, DE, FH* (confounded with blocks).

Blocks	(1)	(2)	(3)	(4)	(5)	(6)	(7)	(8)
	(1)	*adg*	*ach*	*beh*	*eg*	*fh*	*bef*	*acf*
	adefh	*abce*	*bfg*	*abdf*	*bcd*	*ade*	*abdh*	*bgh*
	bcdeg	*efgh*	*cdef*	*cdgh*	*adfgh*	*abcg*	*cdfg*	*cdeh*
	abcfgh	*bcdfh*	*abdegh*	*acefg*	*abcefh*	*bcdefgh*	*acegh*	*abdefg*

(9)	(10)	(11)	(12)	(13)	(14)	(15)	(16)
ab	*ce*	*afg*	*df*	*acd*	*cg*	*dh*	*agh*
cfgh	*bdg*	*bch*	*aeh*	*abeg*	*bde*	*aef*	*bcf*
acdeg	*acdfh*	*degh*	*bcefg*	*cefh*	*abfh*	*bcegh*	*defg*
bdefh	*abefgh*	*abcdef*	*abcdgh*	*bdfgh*	*acdefgh*	*abcdfg*	*abcdeh*

AF, AH, BC, BG, CG, DE, FH, ACD, ADG, BEF, BEH, CDF, CEF, DFG, EFG confounded.

Effects	d.f.
Block	15
Main	8
2-factor	21
Higher order	19
Total	63

Blocks of 8 units

Estimable 2-factors: All except *BC* and *FH* (confounded with blocks).

Combine blocks 1 and 2; blocks 3 and 4; blocks 5 and 6; blocks 7 and 8; blocks 9 and 10; blocks 11 and 12; blocks 13 and 14; and blocks 15 and 16. *BC, FH, ACD, BEF, BEH, CEF, DFG* confounded.

Effects	d.f.
Block	7
Main	8
2-factor	26
Higher order	22
Total	63

Blocks of 16 units

Estimable 2-factors: All.

Combine blocks 1–4; blocks 5–8; blocks 9–12; and blocks 13–16. *ACD, BEF, DFG* confounded.

Effects	d.f.
Block	3
Main	8
2-factor	28
Higher order	24
Total	63

Plan 6A.16 (Continued)

Blocks of 32 units

Estimable 2-factors: All.

Combine blocks 1–8; and blocks 9–16. *ACD* confounded.

Effects	d.f.
Block	1
Main	8
2-factor	28
Higher order	26
Total	63

Blocks of 64 units

Estimable 2-factors: All.

Combine blocks 1–16.

Effects	d.f.
Main	8
2-factor	28
Higher order	27
Total	63

Plan 6A.17 2^8 factorial in 128 units (½ replicate)

Defining contrast: *ABCDEFGH*

For blocks of 8 units, see end of plan.

Blocks of 16 units

Estimable 2-factors: All.

Estimable 3-factors: All.

Blocks	(1)	(2)	(3)	(4)	(5)	(6)	(7)	(8)
	(1)	*acfh*	*fh*	*ac*	*af*	*ch*	*ah*	*cf*
	abcd	*bdfh*	*abcdfh*	*bd*	*bcdf*	*abdh*	*bcdh*	*abdf*
	adeg	*abefgh*	*abefgh*	*abeg*	*defg*	*begh*	*degh*	*befg*
	bceg	*cdefgh*	*bcefgh*	*cdeg*	*abcefg*	*acdegh*	*abcegh*	*acdefg*
	adfh	*ab*	*ad*	*abfh*	*dh*	*bf*	*df*	*bh*
	bcfh	*cd*	*bc*	*cdfh*	*abch*	*acdf*	*abcf*	*acdh*
	efgh	*aceg*	*eg*	*acefgh*	*aegh*	*cefg*	*aefg*	*cegh*
	abcdefgh	*bdeg*	*abcdeg*	*bdefgh*	*bcdegh*	*abdefg*	*bcdefg*	*abdegh*
	abgh	*ef*	*abfg*	*eh*	*bfgh*	*ae*	*bg*	*aefh*
	aceh	*adfg*	*acef*	*adgh*	*cefh*	*dg*	*ce*	*dfgh*
	bdeh	*bcfg*	*bdef*	*bcgh*	*abdefh*	*abcg*	*abde*	*abcfgh*
	cdgh	*abcdef*	*cdfg*	*abcdeh*	*acdfgh*	*bcde*	*acdg*	*bcdefh*
	abef	*gh*	*abeh*	*fg*	*be*	*afgh*	*befh*	*ag*
	acfg	*adeh*	*acgh*	*adef*	*cg*	*defh*	*cfgh*	*de*
	bdfg	*bceh*	*bdgh*	*bcef*	*abdg*	*abcefh*	*abdfgh*	*abce*
	cdef	*abcdgh*	*cdeh*	*abcdfg*	*acde*	*bcdfgh*	*acdefh*	*bcdg*

ABCD, ABEF, ACFG, ADEG, BCEG, BDFG, CDEF confounded.

Effects	d.f.
Block	7
Main	8
2-factor	28
3-factor	56
Higher order	28
Total	127

Plan 6A.17 (Continued)

Blocks of 32 units

Estimable 2-factors: All.

Estimable 3-factors: All.

Combine blocks 1 and 2; blocks 3 and 4; blocks 5 and 6; and blocks 7 and 8. *ABCD, ABEF, CDEF* confounded.

Effects	d.f.
Block	3
Main	8
2-factor	28
3-factor	56
Higher order	32
Total	127

Blocks of 64 units

Estimable 2-factors: All.

Estimable 3-factors: All.

Combine blocks 1–4; and blocks 5–8. *ABCD* confounded.

Effects	d.f.
Block	1
Main	8
2-factor	28
3-factor	56
Higher order	34
Total	127

Blocks of 128 units

Estimable 2-factors: All.

Estimable 3-factors: All.

Combine blocks 1–8.

Effects	d.f.
Main	8
2-factor	28
3-factor	56
Higher order	35
Total	127

Blocks of 8 units

Estimable 2-factors. All except *EG* and *FH* (confounded). Start with the plan for blocks of 32 units. The first 4 rows of blocks 1 and 2 form the first block of 8 units: i.e., this block contains 1, *abcd, adeg, bceg, acfh, bdfh, abefgh, cdefgh*. Similarly, rows 5–8 of blocks 1 and 2 give the second block, rows 9–12 the third and rows 13–16 the fourth. The remaining 12 blocks are formed likewise from blocks 3 and 4; blocks 5 and 6; and blocks 7 and 8.

Effects	d.f.
Block	15
Main	8
2-factor	26
Higher order	78
Total	127

Plan 6A.18 3^4 factorial in 27 units (1/3 replicate)

Defining contrasts: 2 d.f. from *ABCD*, equivalent to putting $D = ABC(Y)$

Blocks of 9 units

Estimable 2-factors: 16 of the 24 d.f. are clear. $CD(I)$ is lost by confounding. Also $AB(J) = CD(J)$; $AC(I) = BD(I)$; $AD(I) = BC(I)$.

Blocks (1)	(2)	(3)
0000	0021	0012
0122	0110	0101
0211	0202	0220
1022	1010	1001
1111	1102	1120
1200	1221	1212
2011	2002	2020
2100	2121	2112
2222	2210	2201

Effects	d.f.
Block	2
Main	8
2-factor	16
Total	26

If all interactions of *D* are negligible, the analysis may be written:

Effects	d.f.
Block	2
Main	8
AB, AC, BC	12
Error (from interactions of *D*)	4
Total	26

Blocks of 27 units

Estimable 2-factors: As in blocks of 9 units, plus $CD(I)$.

Combine blocks 1, 2, and 3. See section 6A.32 for analysis of variance.

Plan 6A.19 3^5 factorial in 81 units (1/3 replicate)

Defining contrasts: 2 d.f. from *ABCDE*

Blocks of 9 units

Estimable 2-factors: All except *AE*(*J*), which is confounded with blocks.

Blocks	(1)	(2)	(3)	(4)	(5)	(6)	(7)	(8)	(9)
ab	*cde*	*cde*	*cde*	*cde*	*cde*	*cde*	*cde*	*cde*	*cde*
00	000	201	102	120	021	222	111	012	210
10	122	020	221	212	110	011	200	101	002
20	211	112	010	001	202	100	022	220	121
01	110	011	212	200	101	002	221	122	020
11	202	100	001	022	220	121	010	211	112
21	021	222	120	111	012	210	102	000	201
02	220	121	022	010	211	112	001	202	100
12	012	210	111	102	000	201	120	021	222
22	101	002	200	221	122	020	212	110	011

Effects	d.f.
Block	8
Main	10
2-factor	38
Higher order	24
Total	80

Blocks of 27 units

Estimable 2-factors: All.

Combine blocks 1–3; blocks 4–6; blocks 7–9.

Effects	d.f.
Block	2
Main	10
2-factor	40
Higher order	28
Total	80

Blocks of 81 units

Estimable 2-factors: All.

Combine blocks 1–9.

Effects	d.f.
Main	10
2-factor	40
Higher order	30
Total	80

Plan 6A.20 4×2^4 factorial in 32 units (½ replicate)

Defining contrast: $A''' BCDE$

Blocks of 8 units

Estimable 2-factors: All except *DE* (confounded with blocks).

Blocks	(1)	(2)	(3)	(4)
ab	*cde*	*cde*	*cde*	*cde*
00	100	010	111	001
01	011	101	000	110
10	011	101	000	110
11	100	010	111	001
20	111	001	100	010
21	000	110	011	101
30	000	110	011	101
31	111	001	100	010

Effects	d.f.
Block	3
Main	7
2-factor	17
Higher order	4
Total	31

Blocks of 16 units

Estimable 2-factors: All.

Combine blocks 1, 2; and blocks 3, 4.

Effects	d.f.
Block	1
Main	7
2-factor	18
Higher order	5
Total	31

Blocks of 32 units

Estimable 2-factors: All.

Combine blocks 1, 2, 3, and 4.

Effects	d.f.
Main	7
2-factor	18
Higher order	6
Total	31

CHAPTER 7

Factorial Experiments with Main Effects Confounded: Split-plot Designs

7.1 The Simple Split-plot Design

7.11 Description. In field experiments an extra factor is sometimes introduced into an experiment by dividing each plot into a number of parts. For example, if the experiment is planned originally to test a factor A with five levels, the division of each plot into halves permits the inclusion of an extra factor B at two levels. Within each plot the two levels of B are allotted at random to the two sub-plots. If the whole plots are in a randomized block design, the plan (after randomization) might appear as shown in table 7.1. It is worth noting the difference be-

TABLE 7.1 Example of a split-plot design

Rep. 1					Rep. 2					Rep. 3				
a_3	a_1	a_2	a_0	a_4	a_1	a_4	a_0	a_2	a_3	a_1	a_3	a_0	a_2	a_4
b_0	b_1	b_0	b_0	b_0	b_1	b_1	b_0	b_0	b_0	b_1	b_0	b_0	b_0	b_1
b_1	b_0	b_1	b_1	b_1	b_0	b_0	b_1	b_1	b_1	b_0	b_1	b_1	b_1	b_0

tween this arrangement and ordinary randomized blocks. In the latter the ten treatment combinations are assigned to the ten sub-plots in a replication completely at random. Here we have a more orderly assignment in which the two treatment combinations that have any given level of A always appear in the same whole-plot.

This type of arrangement is common in industrial experimentation, although the connection may not at first be realized. Frequently, one series of treatments requires rather a large bulk of experimental material, while another series can be compared with much smaller amounts. For instance, the comparison of different types of furnace for the preparation of an alloy would use much greater amounts of alloy than the comparison of different types of mould into which the alloy might be poured. In an experiment in which both factors are to be tested, the natural procedure is to take the material prepared in any furnace, and pour some of it into each mould. That is, material prepared in one furnace at one time provides a complete replication for the comparisons among moulds, just as the plot containing any level of A provides a complete replication of the factor B. Another instance is the comparison of different machines for milking dairy cows. Each machine would necessitate rather substantial amounts of milk, whereas other comparisons, for example on the best method of pasteurizing or of cooling, could be conducted with a much smaller amount of milk per treatment. The produce from any machine could be subdivided for these subsequent tests.

We may describe this design in another way that brings out more clearly its relation to the confounded factorial designs discussed in chapter 6. If the sub-units are regarded as the experimental units, it is seen that the treatments $a_0, a_1, \ldots, a_4$ are applied to groups or blocks of two units. Differences among these blocks are confounded with differences among the levels of A; i.e., the main effects of A are confounded. Accordingly, the split-plot design is sometimes considered as one in which certain *main* effects are confounded, as contrasted with the designs in chapter 6, where the confounding is restricted to interactions.

7.12 Nature of the Experimental Error. In the statistical analysis, account must be taken of the fact that the observations from different sub-units in the same unit may be correlated. In field experiments this correlation is just a reflection of the fact that neighboring pieces of land tend to be similar in fertility and in other agronomic properties. In industrial experiments the same correlation may be present, because any factor that affects the whole batch of alloy prepared in one furnace at one time will tend to create similarity among smaller batches poured from it.

The mathematical analysis used to examine the effects of this correlation will be illustrated for the experiment in table 7.1. Let i refer to the replication, j to the level of A, and k to that of B: for sub-units in the same unit, the i and j subscripts will be the same. The assumption is made that a correlation ρ exists between the experimental errors e_{ijk} and e_{iju} for any two sub-units in the same unit. Sub-units in different units are assumed to be uncorrelated. Mathematically, we have

$$E(e_{ijk}e_{iju}) = \rho\sigma^2; \qquad E(e_{ijk}e_{stu}) = 0$$

We now consider the effects of this model on the most important treatment comparisons. The main effects of A are calculated entirely from unit totals or means. With two sub-units per unit, the error variance of a unit total is

$$E(e_{ij1} + e_{ij2})^2 = E(e_{ij1}^2) + E(e_{ij2}^2) + 2E(e_{ij1}e_{ij2}) = \sigma^2 + \sigma^2 + 2\rho\sigma^2$$
$$= 2\sigma^2(1 + \rho)$$

The factor 2 may be regarded as representing the effect of adding over 2 sub-units. Consequently, for the main effects of A, the appropriate error variance per sub-unit is $\sigma^2(1 + \rho)$. If there are β sub-units per unit, the corresponding variance works out at $\sigma^2[1 + (\beta - 1)\rho]$.

The main effects of B, on the other hand, are derived from the difference between the two sub-units in a unit. For this we have

$$E(e_{ij1} - e_{ij2})^2 = 2\sigma^2(1 - \rho)$$

Thus the effective variance per sub-unit applicable to the main effects of B is $\sigma^2(1 - \rho)$. This expression remains unchanged when there are β sub-units per unit. This variance also applies to any component of the AB interaction, since such components involve comparisons of $(b_1 - b_0)$ at different levels of A.

For other treatment comparisons, the basic error variance may be different from either of the two expressions above. Consider, for instance, $(a_3b_0 - a_1b_0)$, a comparison of a_3 with a_1 at the zero level of B. In any replication the 2 sub-units involved are in different units, and are therefore independent. The variance of their difference is therefore $2\sigma^2$, and the basic variance per sub-unit is σ^2. However, the appropriate variance for all other comparisons of this type can be derived from the basic variances $\sigma^2[1 + (\beta - 1)\rho]$ and $\sigma^2(1 - \rho)$. As will be seen, the analysis of variance gives unbiased estimates of these two variances, and from these, unbiased estimates of any particular variance can be obtained.

In practice, ρ is nearly always positive. The result is that the main effects of A (the factor applied to the units) are less precisely estimated than those of B or than the AB interaction.

The analysis of variance is fairly easy. For the example in table 7.1 we first compute the 15 plot totals. Their sum of squares of deviations is partitioned in the usual way into 2 d.f. for replications, 4 for the main effects of A, and 8 for the experimental error applicable to a whole-plot. The mean square for the latter is an unbiased estimate of $\sigma^2(1 + \rho)$. All computations are divided by 2 to convert them to a sub-unit basis.

Next take the difference $(b_1 - b_0)$ on every unit. The 15 differences provide 1 d.f. which represents the main effect of B, 4 d.f. which represent the AB interactions, and the remaining 10 d.f. whose mean square gives an unbiased estimate of the sub-plot error variance $\sigma^2(1 - \rho)$. All sums of squares are again divided by 2. The complete separation of degrees of freedom is shown in table 7.2.

TABLE 7.2 ANALYSIS OF VARIANCE FOR THE SPLIT-PLOT EXPERIMENT IN TABLE 7.1

	d.f.
Whole plots	
Replications	2
A	4
Whole-plot error	8
Total	14
Sub-plots	
B	1
AB	4
Sub-plot error	10
Grand total	29

The point to be noted is that the whole-plot error is computed entirely from plot totals, and the sub-plot error entirely from the differences between sub-plots in the same plot. It will be seen that the grand total of the degrees of freedom is 29. As might be expected, the corresponding sum of squares is the sum of squares of deviations of the 30 observations from their mean. A computation of this quantity provides a check on all sums of squares. In practice, particularly with more than 2 sub-unit treatments, we usually find the sub-plot error by subtraction.

7.13 Comparison with Randomized Blocks. The experiment in table 7.1 might be arranged in ordinary randomized blocks, with 3 blocks of 10 treatments each. In a consideration of the relative merits of the two arrangements, the following points are relevant.

1. With the split-plot design, usually the B and the AB effects are estimated more precisely than the A effects. Moreover, the number of degrees of freedom available for the experimental error m.s. is smaller for whole-unit comparisons than for sub-unit comparisons.

2. It can be shown that the *average* experimental error over all treatment comparisons is the same for both designs. Consequently there is no *net* gain in precision resulting from the use of the split-plot design; the increased precision on B and AB is obtained by the sacrifice of precision on A. For tests of significance or the construction of confidence limits the randomized block design holds a slight advantage on the average since it provides more degrees of freedom for the estimate of the single error variance that it requires. For instance, the experiment cited has

8 d.f. for the whole-unit error and 10 d.f. for the sub-unit error, whereas randomized blocks would provide 18 d.f.

3. As we have indicated, the chief practical advantage of the split-plot arrangement is that it enables factors that require relatively large amounts of material and factors that require only small amounts to be combined in the same experiment. If the experiment is planned to investigate the first type of factor, so that large amounts of material are going to be used anyway, factors of the second type can often be included at very little extra cost, and some additional information obtained very cheaply.

To summarize, the split-plot design is advantageous if the B and AB effects are of greater interest than the A effects, or if the A effects cannot be tested on small amounts of material.

Two disadvantages have been mentioned by experimenters. Sometimes the whole-unit error is much larger than the sub-unit error. It may happen that the effects of A, though large and exciting, are not significant, whereas those of B, which are too small to be of practical interest, are statistically significant. The experimenter tends to be uncomfortable in reporting results of this type. Secondly, the fact that different treatment comparisons have different basic error variances makes the analysis more complex than with randomized blocks, especially if some unusual type of comparison is being made.

When the number of replications and the experimental conditions are suitable, the whole units may be arranged in a latin square. A split-plot latin square, which eliminates the error variation arising from two types of grouping, may be preferable to randomized blocks. Summarizing 22 field experiments in latin squares where the plots were split into halves, Yates (7.1) found a substantial net increase in precision over randomized blocks. The superiority of the latin square was so pronounced that even the whole-plot comparisons would have been less precisely determined in randomized block designs. Factorial combinations that lend themselves to the use of split-plot latin squares are the 5×2, 5×3, 5×4, 6×2, 6×3, 7×2, 7×3, 8×2.

7.14 Randomization. The treatments applied to the *units* are randomized according to the instructions for the design (e.g. randomized blocks, latin square, etc.) in which the units are arranged. The treatments applied to the *sub-units* are allotted at random within each unit. A separate randomization is carried out for each unit.

7.15 Statistical Analysis. If the factor A (applied to the units) contains α levels, and the factor B (applied to the sub-units) contains β

levels, the subdivision of degrees of freedom in the analysis of variance is shown in table 7.3. The details of the analysis are illustrated in section 7.17.

TABLE 7.3 PARTITION OF DEGREES OF FREEDOM FOR A SPLIT-PLOT DESIGN

Units arranged in randomized blocks (r replicates)		Units arranged in a latin square ($r = \alpha$ replicates)	
Units	d.f.	Units	d.f.
Blocks	$(r - 1)$	Rows	$(\alpha - 1)$
A	$(\alpha - 1)$	Columns	$(\alpha - 1)$
Error (a)	$(\alpha - 1)(r - 1)$	A	$(\alpha - 1)$
		Error (a)	$(\alpha - 1)(\alpha - 2)$
Total	$(r\alpha - 1)$	Total	$(\alpha^2 - 1)$
Sub-units		Sub-units	
B	$(\beta - 1)$	B	$(\beta - 1)$
AB	$(\alpha - 1)(\beta - 1)$	AB	$(\alpha - 1)(\beta - 1)$
Error (b)	$\alpha(r - 1)(\beta - 1)$	Error (b)	$\alpha(\alpha - 1)(\beta - 1)$
Total	$r\alpha(\beta - 1)$	Total	$\alpha^2(\beta - 1)$

7.16 Standard Errors. Let E_a and E_b be the error mean squares for error (a) and error (b), respectively, on a sub-unit basis. For the treatment means, also expressed on a sub-unit basis, the standard errors shown in table 7.4 apply. The final comparison in table 7.4 contains

TABLE 7.4 STANDARD ERRORS FOR THE SPLIT-PLOT DESIGN

Treatment comparison	s.e.
Difference between two A means: e.g., $[(a_1) - (a_0)]$	$\sqrt{2E_a/r\beta}$
Difference between two B means: e.g., $[(b_1) - (b_0)]$	$\sqrt{2E_b/r\alpha}$
Difference between two B means at the same level of A: e.g., $[(a_1b_1) - (a_1b_0)]$	$\sqrt{2E_b/r}$
Difference between two A means at the same level of B or at different levels of B: e.g., $[(a_1b_1) - (a_0b_1)]$ or $[(a_1b_1) - (a_0b_0)]$	$\sqrt{2[(\beta - 1)E_b + E_a]/r\beta}$

both the main effect of A and the AB interaction; consequently the appropriate error is a weighted mean of E_a and E_b. This error also applies to the difference between two A means which have *different* levels of B. In such cases the ratio of the treatment difference to its standard error does not follow the t-distribution. For practical purposes the approximate rule of section 4.14 may be used, though this method gives slightly too few significant results. Let t_a, t_b be the significance levels of t corresponding to the degrees of freedom in E_a and E_b, respectively. The significance level of t is taken as

$$t = \frac{(\beta - 1)E_b t_b + E_a t_a}{(\beta - 1)E_b + E_a}$$

For an application, see the next section.

7.17 Numerical Example. In an experiment on the preparation of chocolate cakes, conducted at Iowa State College (7.2), 3 recipes for preparing the batter were compared. Recipes I and II differed in that the chocolate was added at 40°C. and 60°C., respectively, while recipe III contained extra sugar. In addition, 6 different baking temperatures were tested: these ranged in 10° steps from 175° to 225°. Each time that a mix was made by any recipe, enough batter was prepared for 6 cakes, each of which was baked at a different temperature. Thus the recipes are the "whole-unit" treatments, while the baking temperatures are the "sub-unit" treatments. There were 15 replications, and it will be assumed that these were conducted serially according to a randomized blocks scheme: that is, one replication was completed before starting the next, so that differences among replicates represent time differences. (The notes suggest that this was done, though they are not quite explicit.)

A number of measurements were made on the cakes. The measurement presented here is the breaking angle. One half of a slab of cake is held fixed, while the other half is pivoted about the middle until breakage occurs. The angle through which the moving half has revolved is read on a circular scale. Since breakage is gradual, the reading tends to have a subjective element. The data are shown in table 7.5.

It is customary to compute the analysis of variance on a sub-unit basis. To avoid confusion, this should be clearly stated in the analysis of variance table itself. The calculations may be presented in three steps.

Step 1. Analyze the whole-unit totals by the method appropriate to the design in which they are arranged.

$$\text{Correction factor: } \frac{(8673)^2}{270} = 278{,}596$$

$$\text{Total: } \frac{(269)^2 + (260)^2 + \cdots + (155)^2}{6} - 278{,}596 = 11{,}538$$

$$\text{Replications: } \frac{(843)^2 + (820)^2 + \cdots + (479)^2}{18} - 278{,}596 = 10{,}204$$

$$\text{Recipes: } \frac{(2981)^2 + (2848)^2 + (2844)^2}{90} - 278{,}596 = 135$$

$$\text{Error } (a)\text{, by subtraction: } 11{,}538 - 10{,}204 - 135 = 1{,}199$$

TABLE 7.5 BREAKING ANGLES (DEGREES)

	Rep.	Temperature 175°	185°	195°	205°	215°	225°	Unit totals	Rep. totals
Recipe I	1	42	46	47	39	53	42	269	
	2	47	29	35	47	57	45	260	
	3	32	32	37	43	45	45	234	
	4	26	32	35	24	39	26	182	
	5	28	30	31	37	41	47	214	
	6	24	22	22	29	35	26	158	
	7	26	23	25	27	33	35	169	
	8	24	33	23	32	31	34	177	
	9	24	27	28	33	34	23	169	
	10	24	33	27	31	30	33	178	
	11	33	39	33	28	33	30	196	
	12	28	31	27	39	35	43	203	
	13	29	28	31	29	37	33	187	
	14	24	40	29	40	40	31	204	
	15	26	28	32	25	37	33	181	
Totals		437	473	462	503	580	526	2981	
Recipe II	1	39	46	51	49	55	42	282	843
	2	35	46	47	39	52	61	280	820
	3	34	30	42	35	42	35	218	665
	4	25	26	28	46	37	37	199	599
	5	31	30	29	35	40	36	201	583
	6	24	29	29	29	24	35	170	517
	7	22	25	26	26	29	36	164	492
	8	26	23	24	31	27	37	168	493
	9	27	26	32	28	32	33	178	492
	10	21	24	24	27	37	30	163	519
	11	20	27	33	31	28	33	172	538
	12	23	28	31	34	31	29	176	557
	13	32	35	30	27	35	30	189	574
	14	23	25	22	19	21	35	145	502
	15	21	21	28	26	27	20	143	479
Totals		403	441	476	482	517	529	2848	8673
Recipe III	1	46	44	45	46	48	63	292	
	2	43	43	43	46	47	58	280	
	3	33	24	40	37	41	38	213	
	4	38	41	38	30	36	35	218	
	5	21	25	31	35	33	23	168	
	6	24	33	30	30	37	35	189	
	7	20	21	31	24	30	33	159	
	8	24	23	21	24	21	35	148	
	9	24	18	21	26	28	28	145	
	10	26	28	27	27	35	35	178	
	11	28	25	26	25	38	28	170	
	12	24	30	28	35	33	28	178	
	13	28	29	43	28	33	37	198	
	14	19	22	27	25	25	35	153	
	15	21	28	25	25	31	25	155	
Totals		419	434	476	463	516	536	2844	
Temp. totals		1259	1348	1414	1448	1613	1591	8673	

Step 2. This concerns the sub-unit treatments. Their main effects are obtained directly.

$$\text{Temperatures: } \frac{(1259)^2 + (1348)^2 + \cdots + (1591)^2}{45} - 278{,}596 = 2100$$

The sum of squares for interactions between sub-unit and whole-unit treatments is found by subtraction. First calculate the total s.s. for the two-way table that shows both sets of treatments.

Total treatments:

$$\frac{(437)^2 + (473)^2 + \cdots + (516)^2 + (536)^2}{15} - 278{,}596 = 2441$$

Then,

$$\text{Recipes} \times \text{temperatures: } 2441 - 135 - 2100 = 206$$

Step 3. Compute the total s.s. among all sub-units.

$$\text{Total: } (42)^2 + (47)^2 + \cdots + (35)^2 + (25)^2 - 278{,}596 = 18{,}143$$

The sum of squares for error (b) is then found by subtraction in table 7.6.

TABLE 7.6 Analysis of variance of breaking angles (on a sub-unit basis) for the experiment on chocolate cakes

Source of variation	d.f.	s.s.	m.s.	F
Replications	14	10,204		
Recipes	2	135	67.5	1.58
Error (a)	28	1,199	42.8	
Temperatures	5	2,100	420.0	20.49
Linear regression	1	1,967	1,967.0	95.95
Deviations	4	133	33.2	1.62
Recipes × temperatures	10	206	20.6	1.00
Error (b)	210	4,299	20.5	
Total	269	18,143		

The mean square for recipes, while above that for error (a), does not approach the 5% level. Temperature effects are highly significant as compared with error (b). Since there is a fairly steady increase in the breaking angle with increased temperature, the sum of squares has been divided into the component due to a linear regression on temperature and that due to deviations from the straight line. The regression coefficient amounts to an increase of 1.6° in the breaking angle for each 10° rise in baking temperature. The mean square for deviations is not significant, though it is higher than expectation. There is no indication of any interaction. It will be observed that the error (a) m.s. is about twice as large as that for error (b).

In table 7.7 are shown the treatment means, with the principal standard errors as obtained from table 7.4.

TABLE 7.7 BREAKING ANGLE MEANS (DEGREES)

Recipe	175°	185°	195°	205°	215°	225°	Recipe means
I	29.1	31.5	30.8	33.5	38.7	35.1	33.1
II	26.9	29.4	31.7	32.1	34.5	35.3	31.6
III	27.9	28.9	31.7	30.9	34.4	35.7	31.6
Temp. means	28.0	30.0	31.4	32.2	35.9	35.4	32.1

Standard error of difference between

Two recipe means: $\sqrt{\dfrac{2(42.8)}{90}} = 0.98$ (28 d.f.)

Two temperature means: $\sqrt{\dfrac{2(20.5)}{45}} = 0.95$ (210 d.f.)

Two temperature means for one recipe: $\sqrt{\dfrac{2(20.5)}{15}} = 1.65$ (210 d.f.)

Two recipe means for a given temperature: $\sqrt{\dfrac{2[5(20.5) + 42.8]}{90}} = 1.80$

The 5% levels of t are 2.05 and 1.97, respectively, for 28 and 210 d.f. Consequently, the 5% level for the last standard error above is

$$\frac{(5)(20.5)(1.97) + (42.8)(2.05)}{(5)(20.5) + (42.8)} = 1.99$$

This value always lies between the two individual t-values. In practice it need rarely be calculated.

7.18 Missing Data. The formulae for inserting estimates of missing values have been developed by Anderson (7.3). Suppose first that the observation for a single sub-unit is missing, and that this sub-unit receives the treatment combination $(a_i b_j)$. Let

U = total for the unit containing the missing observations.
$(A_i B_j)$ = total of all sub-units that receive the treatment combination $(a_i b_j)$.
(A_i) = total of all observations that receive the ith level of A.

Then the estimate to be inserted for the missing value is

$$y = \frac{rU + \beta(A_iB_j) - (A_i)}{(r-1)(\beta-1)}$$

Example. Suppose that the observation for recipe II, 195° temperature, in replication 2 is missing. (Its actual value is 47.) Then

$$U = 280 - 47 = 233; \qquad (A_iB_j) = 476 - 47 = 429;$$

$$(A_i) = 2848 - 47 = 2801$$

$$y = \frac{15(233) + 6(429) - 2801}{(14)(5)} = \frac{3268}{70} = 47$$

By accident, the estimated value is the same as the actual value. If several observations are missing, repeated use is made of the formula as described in section 4.25.

In the analysis of variance 1 d.f. is subtracted from the error (*b*) d.f. for each missing observation. An unbiased estimate of E_b is obtained, but the treatment sums of squares and that for error (*a*) are biased upwards. If only a small fraction of the values is missing, it appears that these biases can be ignored; methods for obtaining unbiased estimates are given by Anderson (7.3).

Standard errors. When there are missing observations, the formulae in table 7.8 are suggested for the differences between 2 means.

TABLE 7.8 STANDARD ERRORS FOR SPLIT-PLOT EXPERIMENTS WITH MISSING DATA

Treatment comparison	s.e.
Difference between two A means:	$\sqrt{2[E_a + fE_b]/r\beta}$
Difference between two B means:	$\sqrt{2E_b[1 + (f\beta/\alpha)]/r\alpha}$
Difference between two B means at the same level of A:	$\sqrt{2E_b[1 + (f\beta/\alpha)]/r}$
Difference between two A means at the same level of B, or at different levels of B:	$\sqrt{(2E_a/r\beta) + (2E_b)[(\beta - 1) + f\beta^2]/r\beta}$

If only one value is missing in the experiment, and if a mean containing that value is compared with another mean, the factor f is $1/2(r-1)(\beta-1)$, and the formulae are exact, reference (7.3).

With more than 1 missing observation, the value of f depends on the locations of the missing observations. The following approximation, developed by G. S. Watson, is correct for a number of cases but otherwise tends to be slightly too high.

$$f = \frac{k}{2(r - d)(\beta - k + c - 1)}$$

When counting the values of k, c, and d, as defined below, be sure to ignore all missing observations except those that occur in the 2 means that are being compared.

k = number of missing observations.

c = number of replications which contain 1 or more missing observations.

d = number of missing observations in the sub-unit treatment (a_ib_j) that is most affected.

For example, suppose that in the cake experiment the values for recipe I, 175° temperature, are missing in replications 1, 2, and 3, while those for recipe I, 185° temperature, are missing in replications 3 and 4. We are comparing the means for the 2 temperatures, taken over all recipes. Then k is 5, and c is 4. The sub-unit treatment with the most missing values is recipe I, 175° temperature, which has 3 missing, so that d is 3.

The case where a complete unit is missing is discussed by Anderson (7.3).

7.2 Repeated Subdivision

7.21 Description. In order to include a new factor C at γ levels, the sub-units may each be divided into sub-sub-units. There are three experimental error variances. Errors (a) and (b) have the same functions as described previously and are calculated in the same manner, except that an extra divisor γ is introduced into all sums of squares in order to present the analysis on a sub-sub-unit basis. Error (c), which will usually be the smallest of the three, applies to the C, AC, BC, and ABC effects. The process of subdivision may be carried as far as is convenient.

The partition of degrees of freedom has already been given (table 7.3) for factors A and B. For C and its interactions. the partition is shown in table 7.9.

TABLE 7.9 Analysis of variance for C and its interactions

Effect	d.f.
C	$(\gamma - 1)$
AC	$(\alpha - 1)(\gamma - 1)$
BC	$(\beta - 1)(\gamma - 1)$
ABC	$(\alpha - 1)(\beta - 1)(\gamma - 1)$
Error (c)	$\alpha\beta(r - 1)(\gamma - 1)$

The computations in the analysis of variance are a straightforward extension of those for the simple split-plot design. The sum of squares for AC is obtained by calculating the total s.s. for the A by C two-way table, and subtracting the sums of squares due to A and C; and similarly for the other factorial effects in table 7.9. Finally, the sum of squares for error (c) is found by calculating the total s.s. among sub-sub-units, and subtracting the contributions from all other items in the analysis. Worked examples are given in references (7.4) and (7.5).

7.22 Standard Errors. The formulae in table 7.4 for standard errors applicable to A and B effects remain valid apart from division by an additional factor $\sqrt{\gamma}$. For the principal comparisons which involve C effects, standard errors are shown in table 7.10. All treatment means and error mean squares are assumed to be on a sub-sub-unit basis.

TABLE 7.10 STANDARD ERRORS FOR THE SPLIT-PLOT DESIGN WITH TWO SUBDIVISIONS

Treatment comparison	s.e.
$[(c_1) - (c_0)]$	$\sqrt{2E_c/r\alpha\beta}$
$[(a_1c_1) - (a_1c_0)]$	$\sqrt{2E_c/r\beta}$
$[(b_1c_1) - (b_1c_0)]$	$\sqrt{2E_c/r\alpha}$
$[(a_1b_1c_1) - (a_1b_1c_0)]$	$\sqrt{2E_c/r}$
$[(b_1c_1) - (b_0c_1)]$ or $[(b_1c_1) - (b_0c_0)]$	$\sqrt{2[(\gamma - 1)E_c + E_b]/r\alpha\gamma}$
$[(a_1b_1c_1) - (a_1b_0c_1)]$	$\sqrt{2[(\gamma - 1)E_c + E_b]/r\gamma}$
$[(a_1c_1) - (a_0c_1)]$ or $[(a_1c_1) - (a_0c_0)]$	$\sqrt{2[(\gamma - 1)E_c + E_a]/r\beta\gamma}$
$[(a_1b_1c_1) - (a_0b_1c_1)]$	$\sqrt{2[\beta(\gamma - 1)E_c + (\beta - 1)E_b + E_a]/r\beta\gamma}$

7.3 Some Variants of the Split-plot Design

7.31 Systematic Arrangement of the Treatments Applied to the Units. The device of subdividing each unit into a number of sub-units is very flexible, and can be used with any type of design in which the units are arranged. Some of the more useful variations that are possible are described in the remainder of this chapter. The first arises when the nature of the experiment makes it necessary or desirable to have the A treatments arranged in a systematic design. This arrangement has been used in experiments where the A treatments were varieties of wheat known to mature at different dates, and the B treatments were preparations applied to the seed before sowing. Since each variety was to be harvested when ripe, the field operations were made easier by planting the varieties in the order in which they would be ready for harvest. The field plan is illustrated in table 7.11, where a_1 is the earliest variety and a_5 the latest.

TABLE 7.11 EXAMPLE OF SYSTEMATIC ARRANGEMENT OF TREATMENTS APPLIED TO UNITS

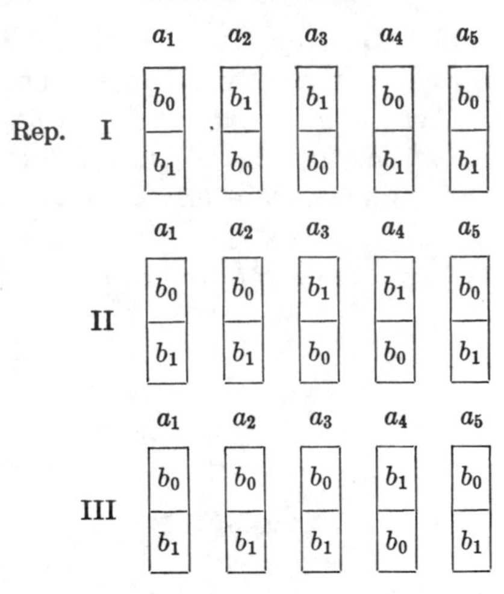

Rep.	a_1	a_2	a_3	a_4	a_5
I	b_0	b_1	b_1	b_0	b_0
	b_1	b_0	b_0	b_1	b_1
II	b_0	b_0	b_1	b_1	b_0
	b_1	b_1	b_0	b_0	b_1
III	b_0	b_0	b_0	b_1	b_0
	b_1	b_1	b_1	b_0	b_1

The sub-unit treatments are randomized within each unit; the unit treatments, however, appear in separate strips.

This arrangement provides no valid estimate of error for the A main effects, or for comparisons such as $(a_1b_0 - a_2b_0)$ that involve A effects. Thus the "whole-plot" analysis of variance is irrelevant. In the sub-plot analysis of variance, error (b) is still valid for testing the B effects and the AB interactions. One point should be watched. A comparison such as $(a_1b_1 - a_1b_0)$ is derived entirely from the first strip in the plan. Consequently, if the sub-plot error variance differs from strip to strip, the use of the error (b) m.s. for testing this quantity is not justified; instead, it is necessary to compute a separate error using only data from the first strip.

To summarize, systematic placement of the A treatments is advisable only where it is essential in order to conduct the experiment and where no test of the A main effects is required.

7.32 Sub-unit Treatments in Strips. In a further variant the sub-unit treatments, instead of being randomized independently within each unit, are arranged in strips across each replication. For a 5 × 3 design the appropriate rearrangement (after randomization) might be as shown in table 7.12.

This layout may be convenient for field experiments where it is necessary to test both factors on relatively large areas and to leave free access

TABLE 7.12 DESIGN WITH SUB-UNIT TREATMENTS IN STRIPS

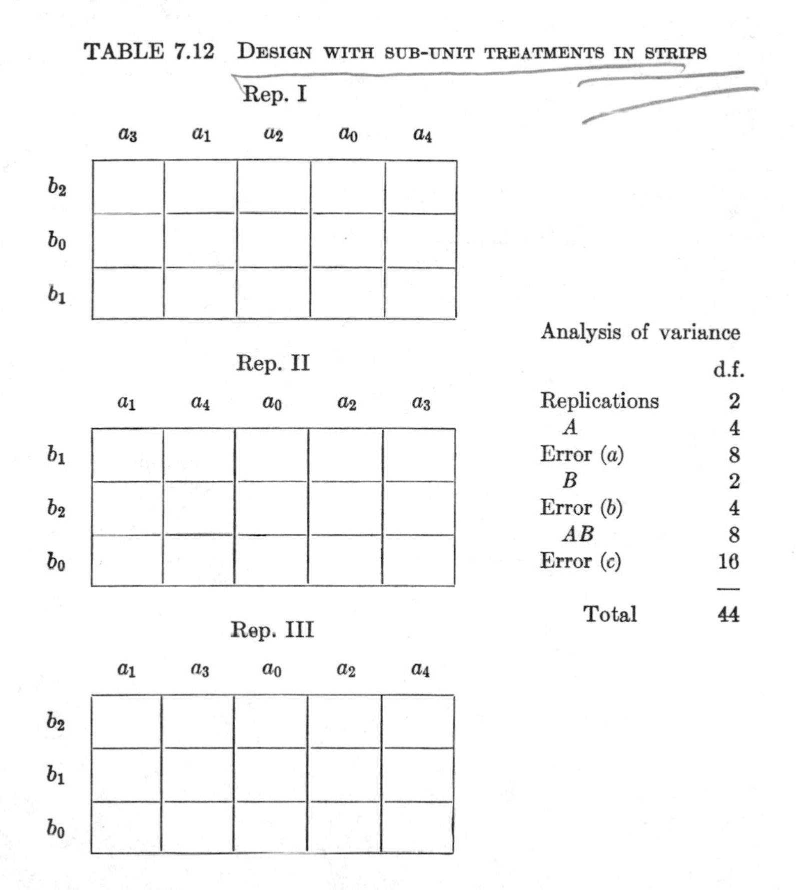

Analysis of variance

	d.f.
Replications	2
A	4
Error (a)	8
B	2
Error (b)	4
AB	8
Error (c)	16
Total	44

at both ends. Similar conditions may apply in other types of experimentation.

As with the ordinary split-plot design, the average precision over all treatment comparisons is the same as that of randomized blocks arranged on the sub-units. The present design sacrifices precision on the main effects of A and B in order to provide higher precision on the interactions, which will generally be more accurately determined than in either randomized blocks or the simple split-plot design. Since in addition the numbers of degrees of freedom for estimating the errors of the A and B

effects are likely to be small, the design is not recommended unless practical considerations necessitate its use or unless the interactions are the principal object of study. Both the A and B treatments should be randomized independently within each replication.

TABLE 7.13 LATIN SQUARE DESIGN WITH SUB-UNIT TREATMENTS IN STRIPS

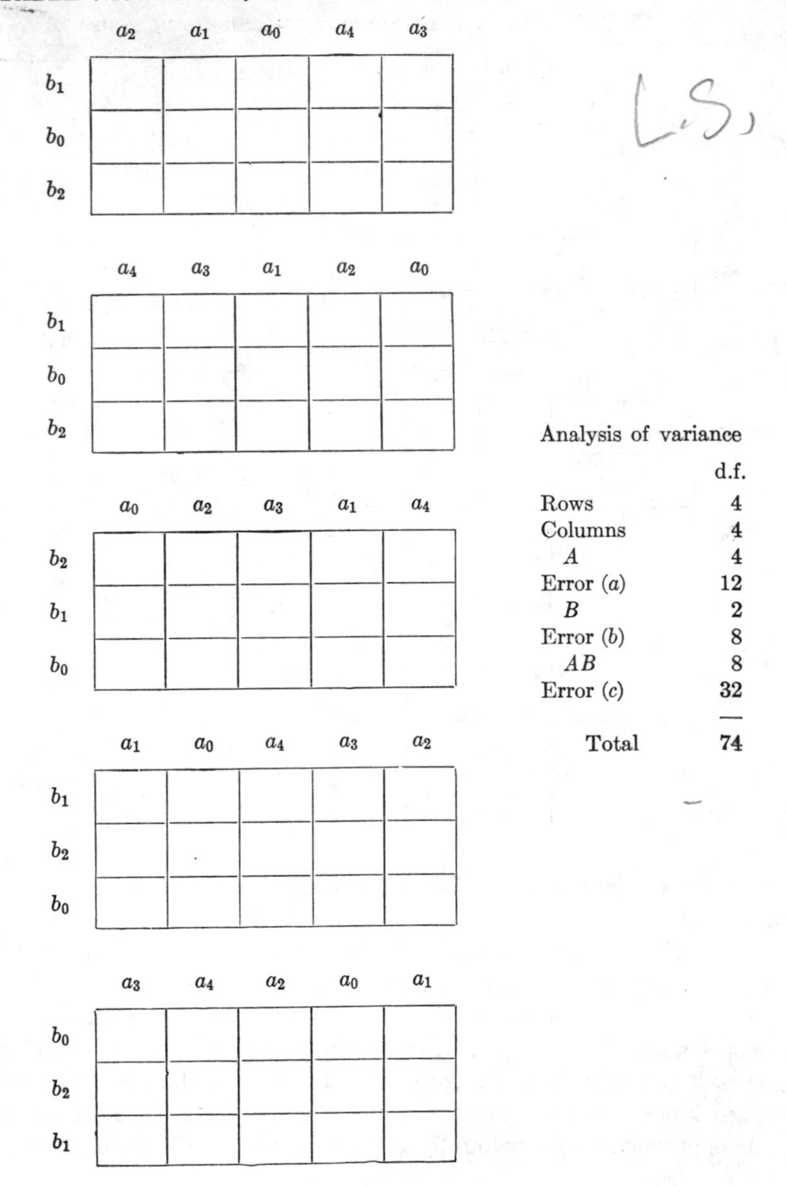

Analysis of variance

	d.f.
Rows	4
Columns	4
A	4
Error (a)	12
B	2
Error (b)	8
AB	8
Error (c)	32
Total	74

In the statistical analysis, separate estimates of error are obtained for A, B, and AB. Both sets of main effects may be regarded as tested in randomized block designs, in which the sum of squares for replications (blocks) is the same for both designs. A worked example (on uniformity data) is given in reference (7.6).

Formulae for the estimated standard errors of the differences between treatment means per *sub-unit* are given in table 7.14. The mean squares E_a, E_b, and E_c are assumed to be on a sub-unit basis.

TABLE 7.14 STANDARD ERRORS WHEN SUB-UNIT TREATMENTS ARE IN STRIPS

(α = number of levels of A; β = number of levels of B; r = number of replicates)

Treatment comparison	s.e.
$[(a_1) - (a_0)]$	$\sqrt{2E_a/r\beta}$
$[(b_1) - (b_0)]$	$\sqrt{2E_b/r\alpha}$
$[(a_1b_1) - (a_0b_1)]$	$\sqrt{2[(\beta - 1)E_c + E_a]/r\beta}$
$[(a_1b_1) - (a_1b_0)]$	$\sqrt{2[(\alpha - 1)E_c + E_b]/r\alpha}$

One of the factors may be arranged in a latin square. With a 5×3 design in 5 replicates, a layout of this type is shown in table 7.13.

The statistical analysis is straightforward. For the B treatments, the "blocks" of the randomized blocks design are the rows of the latin square. The formulae given above for the standard errors are applicable (with r equal to α).

7.33 Sub-unit Treatments in a Latin Square. In certain cases the sub-unit treatments may be arranged in a latin square, with the prospect of a further increase in accuracy on sub-unit comparisons. There are two types of layout, of which the first will be described in this section; the second, which is used when there are only a few whole-unit treatments whose main effects are not required with high accuracy, is discussed in section 8.6.

In the first type the number of replications must be a multiple of the number of sub-unit treatments. For each whole-unit treatment, the sub-unit treatments are arranged in one or more separate latin squares. An example with 4 whole-unit treatments, 3 sub-unit treatments, and 3 replications is shown in table 7.15. The latin squares are not very clearly displayed in the plan. If we examine the three units that receive a_3, it is seen that each b appears once in the top sub-unit, once in the middle, and once at the bottom. Thus the rows of the latin square are the positions of the sub-units within the unit. The only change from the ordinary split-plot arrangement is that differences among these rows are eliminated from the sub-unit errors. If these differences are large, sub-unit comparisons are more precisely determined than in the ordinary design. Whole-unit comparisons are unaffected.

The designs are useful where a fairly consistent gradient is expected within each unit. Examples have been quoted previously. In experiments on plant viruses, the sub-units may be the top, middle, and lower leaves of a plant, with a regular gradient of susceptibility down the plant. The sub-unit treatments may be operations performed in succession by the same person, or drugs injected in succession to the same animal, where a time-trend is anticipated.

TABLE 7.15 SPLIT-PLOT DESIGN WITH SUB-TREATMENTS IN A LATIN SQUARE

	a_3	a_2	a_1	a_0
	b_2	b_1	b_0	b_1
Rep. I	b_0	b_0	b_1	b_0
	b_1	b_2	b_2	b_2

	a_0	a_2	a_1	a_3
	b_0	b_0	b_1	b_1
II	b_2	b_2	b_2	b_2
	b_1	b_1	b_0	b_0

	a_3	a_0	a_2	a_1
	b_0	b_2	b_2	b_2
III	b_1	b_1	b_1	b_0
	b_2	b_0	b_0	b_1

Analysis of variance

Units	d.f.
Replications	2
A	3
Error (a)	6
Total	11
Sub-units	
Rows	8
B	2
AB	6
Error (b)	8
Total	24

The number of replicates may be any multiple of the number of sub-unit treatments. With 6 replicates of the design in table 7.15, the experimental material is divided into 2 groups of 3 replicates each. If practicable, the division should be made in such a way that the within-unit gradient is constant in a given group. Since the rows are eliminated separately for each group, changes in the gradient from group to group do not increase the error.

In the randomization the first step is to randomize the *units* according to the instructions for the design in which they are arranged. For each whole-unit treatment a latin square of the size required to accommodate the sub-treatments is then randomized as described in section 4.33. A new randomization of the latin square is taken for each successive whole-unit treatment. The first column of each square is assigned to the first replication of the experiment, and so on.

For extra replication the basic plan is repeated with a new randomization.

Since the whole-unit analysis of variance is unaffected, it is necessary to show only the subdivision of the degrees of freedom among sub-units. Table 7.16 applies to the case in which the basic plan is repeated k times.

TABLE 7.16 PARTITION OF DEGREES OF FREEDOM AMONG SUB-UNITS WHEN SUB-TREATMENTS FORM LATIN SQUARES

	d.f.
Rows	$k\alpha(\beta - 1)$
Sub-unit treatments	$(\beta - 1)$
Sub-unit × whole-unit treatments	$(\alpha - 1)(\beta - 1)$
Error (b)	$\alpha(\beta - 1)(r - 1 - k)$
Total	$r\alpha(\beta - 1)$

Notation: α = number of whole-unit treatments.
β = number of sub-unit treatments.
r = number of replications = $k\beta$.

The number of degrees of freedom for rows should be noted. Since each latin square supplies $(\beta - 1)$ degrees of freedom for rows, every whole-unit treatment contributes $k(\beta - 1)$ degrees of freedom, so that the total is $k\alpha(\beta - 1)$ degrees of freedom. The sum of squares for rows is calculated by addition of the sums of squares for each of the $k\alpha$ latin squares. The sum of squares among the columns of the latin squares constitutes a whole-unit comparison and does not appear in the portion of the analysis shown above.

Before using this design it is advisable to verify that the number of degrees of freedom for error (b) is adequate. In the example in table 7.15 there are only 8 d.f., as against 16 with an ordinary split-plot arrangement. From section 2.31 it appears that the sub-unit error variance must be reduced by at least 15% in order to compensate for the decrease in degrees of freedom.

7.34 Split-plot Technique with Confounded Designs. The treatments applied to the units may constitute a factorial system arranged in a design that confounds certain interactions completely or partially.

Since experimenters are sometimes uncertain how to compute the analysis of variance in such cases, the following notes indicate the method.

1. The totals for each unit are analyzed by the instructions for the confounded design in which the units are arranged.

2. In computing the sums of squares for the sub-unit treatments and for all interactions between sub-unit and whole-unit treatments, the confounding is ignored. Even if a whole-unit factorial effect is completely confounded, its interactions with sub-unit treatments are unconfounded and may be tested by means of error (*b*).

3. The sum of squares for error (*b*) is found as usual by subtracting the total s.s. on the whole units plus the total s.s. for sub-unit treatments and for interactions between sub-unit and whole-unit treatments from the total s.s. for sub-units.

A difficulty arises occasionally in the presentation of tables. Suppose that a $3 \times 3 \times 2$ design is applied to the units in blocks of 6 units, with AB and ABC partially confounded, and that a factor D at 2 levels is applied to the sub-units. In the A by B two-way table, the entries must be adjusted to eliminate block effects. The problem is how to present the ABD three-way table. The observed means $(a_0b_0d_0)$ and $(a_0b_0d_1)$ will average to the *unadjusted* mean for (a_0b_0), which contains block effects. The simplest process is to calculate the difference between the adjusted and the unadjusted means of (a_0b_0). This difference is then applied to $(a_0b_0d_0)$ and $(a_0b_0d_1)$. Notice that this adjustment does not change the difference between these two entries and brings their mean to the correct value.

7.35 Confounding of Comparisons among Sub-unit Treatments.

When the sub-unit treatments are factorial, it may be advantageous to confound certain interactions among these factors with the units. The technique, first presented by Yates (7.7), is illustrated in table 7.17, which shows a single replication of the design. The factor A applied to the units has 2 levels, while the sub-unit treatments are the 8 combinations of factors B, C, and D at 2 levels each. The ordinary split-plot arrangement is shown in (i), while (ii) and (iii) are two alternatives for the new method.

Consider first (i) and (ii). In the ordinary arrangement the unit contains 8 sub-units, to which the 8 sub-treatments are allotted at random. With the new arrangement the sub-units are in groups of 4, chosen so that BCD is confounded with group totals. The advantage of the new layout is that the size of the whole-unit has been reduced from 8 to 4 sub-units. This will in general lead to more accurate estimates both of the A effects and of the sub-unit treatments. The only exceptions are the high-order interactions BCD and $ABCD$. In the ordinary arrangement these interactions are derived from comparisons among sub-units in the same unit, whereas in the new design they are estimated from comparisons among the units, as the reader may verify.

Arrangement (iii) goes a step further. The a_0 and a_1 treatments which have *opposite* sets of sub-unit treatments are placed in the same sub-

TABLE 7.17 CONFOUNDING OF BCD IN A SPLIT-PLOT EXPERIMENT

i. Ordinary arrangement

a_0		a_1	
bd	*d*	*bcd*	*b*
c	*cd*	*d*	(1)
bcd	*b*	*bd*	*bc*
(1)	*bc*	*c*	*cd*

ii. New arrangement

a_0	a_0	a_1	a_1
cd	*b*	(1)	*c*
bd	*bcd*	*bc*	*d*
(1)	*d*	*cd*	*bcd*
bc	*c*	*bd*	*b*

iii. New arrangement in sub-blocks

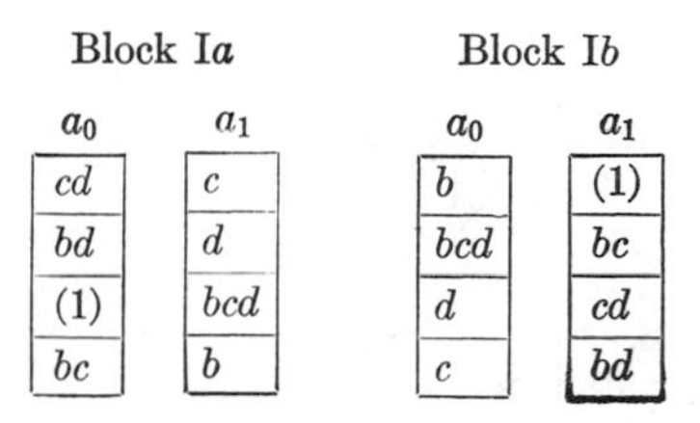

Block I*a*		Block I*b*	
a_0	a_1	a_0	a_1
cd	*c*	*b*	(1)
bd	*d*	*bcd*	*bc*
(1)	*bcd*	*d*	*cd*
bc	*b*	*c*	*bd*

block. The result is that $ABCD$ is completely confounded between sub-blocks, while BCD, like A, is orthogonal with sub-blocks. The aim is to sacrifice information on $ABCD$ in the hope of obtaining more precise estimates of A and BCD. Consequently, alternative (iii) is usually preferable to (ii).

With 5 replications of arrangement (iii), the partition of degrees of freedom is set out in table 7.18 in condensed form. There are 80 sub-units and 10 sub-blocks.

TABLE 7.18 PARTITION OF DEGREES OF FREEDOM IN A 2×2^3 DESIGN WITH SPLIT-PLOT CONFOUNDING

Units	d.f.	Sub-units	d.f.
Between sub-blocks	9	B, C, D, BC, BD, CD	6
A	1	$AB, AC, AD, ABC, ABD, ACD$	6
BCD	1	Error (b)	48
Error (a)	8		—
	—	Grand total between sub-units	79
Total between units	19		

All sums of squares are calculated in the usual way. Note that $ABCD$ does not appear, since it is a component of the sum of squares between sub-blocks.

When the sub-unit treatments form a 2^3 system, a similar technique may be used for any number of levels of the factor A. Single replications of the layouts are shown in table 7.19 for A at 2, 3, 4, 5, and 6 levels.

TABLE 7.19 SPLIT-PLOT DESIGN FOR AN $\alpha \times 2^3$ FACTORIAL SYSTEM

Arrangement of whole-units in a single replication

$\alpha = 2;(2 \times 2^3)$

I_a		I_b	
a_0	a_1'	a_0'	a_1

$\alpha = 3;(3 \times 2^3)$

I_a			I_b		
a_0	a_1	a_2	a_0'	a_1'	a_2'

$\alpha = 4;(4 \times 2^3)$

I_a				I_b			
a_0	a_1'	a_2	a_3'	a_0'	a_1	a_2'	a_3

$\alpha = 5;(5 \times 2^3)$

I_a					I_b				
a_0	a_1	a_2	a_3	a_4	a_0'	a_1'	a_2'	a_3'	a_4'

$\alpha = 6;(6 \times 2^3)$

I_a						I_b					
a_0	a_1'	a_2	a_3'	a_4	a_5'	a_0'	a_1	a_2'	a_3	a_4'	a_5

Each unit contains 4 sub-units. A prime (′) implies that the sub-unit treatments are (1), *bc*, *bd*, and *cd*, and an unaccented *a* that they are *b*, *c*, *d*, and *bcd*.

The arrangements differ slightly according as the number of levels of A is even or odd. With an even level, one component of $ABCD$ is confounded between sub-blocks. For example, with A at 4 levels, this component is $(a_3 + a_1 - a_2 - a_0)\ BCD$. When A has an odd number of levels, we are forced to confound BCD at least partially with sub-blocks, and for simplicity it has been completely confounded.

The randomization for a single replicate is as follows. (i) Allot the sub-blocks *a* and *b* at random to the 2 halves into which the replicate has been divided. (ii) Allot the levels of A at random to units within each sub-block. (iii) Allot the appropriate set of 4 sub-treatments at random within each unit. A separate randomization is used in each replication.

The partition of degrees of freedom in the analysis of variance is shown in table 7.20 for r replications of the basic design.

TABLE 7.20 PARTITION OF DEGREES OF FREEDOM FOR AN $\alpha \times 2^3$ SPLIT-PLOT DESIGN

	Units		Sub-units	
	α even d.f.	α odd d.f.		d.f.
Sub-blocks	$(2r - 1)$	$(2r - 1)$	Sub-treatments	6
A	$(\alpha - 1)$	$(\alpha - 1)$	$A \times$ sub-treatments	$6(\alpha - 1)$
BCD	1		Error (b)	$6\alpha(r - 1)$
$ABCD$	$(\alpha - 2)$	$(\alpha - 1)$	Total	$6\alpha r$
Error (a)	$2(\alpha - 1)(r - 1)$	$2(\alpha - 1)(r - 1)$		
Total	$(2\alpha r - 1)$	$(2\alpha r - 1)$		

When α is odd, the sum of squares for BCD is omitted from the treatments; otherwise all sums of squares are obtained in the usual way. With α even, 1 d.f. is missing from $ABCD$, since it is confounded with sub-blocks. Since these designs are unlikely to be used unless $ABCD$ is small, it will usually be satisfactory to combine the $ABCD$ sum of squares with error (a). The pooled sum of squares can be found by subtraction without calculating that for $ABCD$.

A considerable number of arrangements of this type are possible, though some care is required in their construction in order to avoid confounding important comparisons. Finney (7.8) gives an account of the principles of construction and designs with worked examples, for a 4×2^2 and a 6×2^2 factorial, where the units are in a 4×4 and 6×6 latin square, respectively. Yates (7.7) gives a similar design for 6×2^3 factorial.

REFERENCES

7.1 YATES, F. Complex experiments. *Jour. Roy. Stat. Soc. Suppl.* 2, 181–247, 1935.

7.2 COOK, FRANCES E. Chocolate cake, I. Optimum baking temperature. Master's thesis, Iowa State College, 1938.

7.3 ANDERSON, R. L. Missing-plot techniques. *Biom. Bull.* 2, 41–47, 1946.

7.4 GOULDEN, C. H. *Methods of statistical analysis.* John Wiley & Sons, New York, 151, 1939.

7.5 CHAKRAVERTI, S. S., BOSE, S. S., and MAHALANOBIS, P. C. A complex experiment on rice. *Indian Jour. Agr. Sci.* 6, 34–51, 1936.

7.6 LEONARD, W. H., and CLARK, A. G. *Field plot technique.* Burgess Publishing Company, Minneapolis, 214, 1939.

7.7 YATES, F. The design and analysis of factorial experiments. *Imp. Bur. Soil Sci. Tech. Comm.* 35, 77, 1937.

7.8 FINNEY, D. J. Split-plot confounding. *Jour. Agr. Sci.* 36, 56–62, 1946.

ADDITIONAL READING

FINNEY, D. J. Recent developments in the design of field experiments. II. Unbalanced split-plot confounding. *Jour. Agr. Sci.* 36, 63–68, 1946.

JACOB, W. C. Split-plot half-plaid squares for irrigation experiments. *Biometrics* 9, 157–175, 1953.

SMITH, H. F., and TRUETT, JEANNE TITUS. Adjustment by covariance and consequent tests of significance in split-plot experiments. *Biometrics* 12, 23–39, 1956.

TAYLOR, J. The comparison of pairs of treatments in split-plot experiments. *Biometrika* 37, 443–444, 1950.

CHAPTER 8

Factorial Experiments Confounded in Quasi-latin Squares

8.1 Introduction

Yates (8.1) has constructed a number of confounded arrangements for factorial designs in latin squares. In the first group of designs (plans 8.1–8.7), some of the interactions among the factors are confounded with the rows and columns of the squares. These designs resemble ordinary latin squares in that the differences among the rows and columns of the squares are eliminated from the experimental errors of the unconfounded treatment effects; however, the designs do not possess the typical latin square property that each treatment appears once in every row and every column. Yates (8.1) has proposed the name *quasi-latin squares.*

Before one of these plans is used, it is advisable to note the interactions that are confounded and the extent to which they are confounded, particularly since more treatment effects must be confounded with latin squares than with randomized blocks. Of the eight plans shown, only one involves the confounding of two-factor interactions.

In the statistical analysis, the sums of squares for rows, columns, and all unconfounded treatment effects are calculated in the usual way. Partially confounded effects, if considered negligible, may be combined with the error, which is found by subtraction; or they may be tested for significance by calculating their sums of squares from those portions of the experiment in which the effects are unconfounded.

In a further group of designs (plans 8.8–8.14), certain treatments are applied to complete rows or columns of a latin square. These designs have been called *half-plaid* squares or *plaid* squares, from the resemblance of the plan to a Scotch plaid.

8.2 Randomization of Quasi-latin Squares

The rows and the columns of each latin square are rearranged at random. In certain plans (e.g., 8.2) each square contains several replications. The randomization must *not* be restricted in such cases with the object of keeping the replications separate.

8.3 Notes on the Plans and Statistical Analysis

2^3 factorial in a 4×4 square. There are a number of alternative designs, of which two are given in plan 8.1*a* and 8.1*b*. In the former, which requires 4 replications, ½ relative information is retained on AB, AC, BC, and ABC, main effects being unconfounded. On account of the high degree of confounding of the two-factor interactions, this plan is not recommended unless the factors are presumed to act independently or unless the 4×4 square is considered much more accurate than alternative designs.

A numerical example illustrates the general method of analysis for all designs presented here. The data in table 8.1 represent twice the yields

TABLE 8.1 Yields and statistical analysis of a 2^3 factorial in two 4×4 quasi-Latin squares

Square I (columns confounded: ABC, AC)

					Totals	
	c 57.9	(1) 62.3	*abc* 60.2	*ab* 58.6	239.0	AB
	a 66.2	*ac* 60.5	*b* 60.0	*bc* 53.4	240.1	
	abc 64.1	*bc* 61.3	(1) 63.2	*a* 59.2	247.8	BC
	b 58.4	*ab* 61.8	*ac* 67.0	*c* 61.9	249.1	
Totals	246.6	245.9	250.4	233.1	976.0	

Square II (columns confounded: AB, BC)

					Totals	
	c 62.4	*a* 64.6	*abc* 68.5	*b* 63.1	258.6	ABC
	(1) 61.5	*ac* 62.7	*bc* 58.6	*ab* 69.3	252.1	
	abc 66.0	*b* 56.5	(1) 63.5	*ac* 67.3	253.3	AC
	ab 65.9	*bc* 52.6	*a* 58.1	*c* 58.1	234.7	
Totals	255.8	236.4	248.7	257.8	998.7	

	d.f.	s.s.	m.s.
Squares	1	16.10	
Rows	6	100.90	16.82
Columns	6	112.71	18.78
A, B, C	3	146.43	
AB	1′	18.92	18.92
AC	1′	0.06	0.06
BC	1′	0.33	0.33
ABC	1′	8.27	8.27
Error	11	101.86	9.26
Total	31	505.58	

in bushels per acre from a soil-treatment test on soybeans, at Muscatine, Iowa. The plots were split for a varietal comparison which has been omitted. For convenience, rows and columns were rearranged so that the field plan corresponds to plan 8.1*a*. The factors were limestone (A), phosphorus (B), and potash (C); in each case the lower level implies no application of the treatment.

To obtain the sums of squares for the treatment effects, first calculate the factorial effect totals in the usual way. These are shown in the first line of table 8.2. For the partially confounded effects, these totals must

TABLE 8.2 TREATMENT EFFECTS FOR A 2^3 FACTORIAL IN 4×4 QUASI-LATIN SQUARES

	[A]	[B]	[C]	[AB]	[AC]	[BC]	[ABC]
Totals	+65.3	−18.1	−9.7	+35.7	+34.9	− 8.1	− 4.3
From rows and columns				+18.3	+35.9	−10.4	+ 7.2
Adjusted totals				+17.4	− 1.0	+ 2.3	−11.5

be adjusted. Now AB is completely confounded with the first 2 rows of square I and the first 2 columns of square II. Calculate AB as obtained from these rows and columns, i.e.,

$$239.0 - 240.1 + 255.8 - 236.4 = +18.3$$

This figure is placed in the second row of table 8.2, with corresponding figures for AC, BC, and ABC. The adjusted totals are found by subtracting the second from the first row. Of course, these represent just the totals for each effect as found from that part of the experiment in which the effect is unconfounded. The divisors for the squares are 32 for main effects and 16 for the other effects.

The two-way tables of means require adjustment. For the AB table the adjustment is

$$\frac{(+17.4)}{16} - \frac{(+35.7)}{32} = -0.03$$

This value is added (algebraically) to the (1) and (ab) means and subtracted from the (a) and (b) means.

Plan 8.1*b* completely confounds ABC with squares; all other effects are unconfounded. The arrangement has the peculiarity that half the treatments appear only in square I, while the other half appear only in square II. Although this layout avoids the confounding of any two-factor interactions, Yates has pointed out that any interaction of the A

effect with squares will appear in the analysis as a *BC* interaction. The results are therefore likely to be misleading if the experimental material in square I differs in responsiveness from that in square II. Where this danger seems likely, Yates suggests that the columns (or rows) of the two squares may be interlaced, so that each square comprizes a randomly chosen half of the experimental material. A numerical example of the statistical analysis is given by Yates (8.1, p. 33), using uniformity data.

2^4 factorial in an 8 × 8 square. Four replications are required for plan 8.2. *ABCD* is completely confounded, and threefold replication is obtained on all three-factor interactions. The degrees of freedom subdivide as follows.

	d.f.
Rows	7
Columns	7
Treatments	14
Error	35

In the computation for *ABC, ABD, ACD,* or *BCD,* the appropriate pair of rows is deleted.

2^5 factorial in an 8 × 8 square. The confounding in plan 8.3 (for 2 replications) is not completely balanced. Eight of the 10 three-factor interactions and four of the 5 four-factor interactions receive only single replication. The remaining high-order interactions, *ABE, CDE, ABCD,* and *ABCDE,* are unconfounded. By a proper assignment of the letters *a, b,* ··· to the factors, any single three-factor interaction may be left unconfounded, though not any desired pair of interactions.

The partition of degrees of freedom is:

	d.f.
Rows	7
Columns	7
Treatments	31
Error	18

In the computation of the sum of squares for *ABC, ADE,* or *BCDE,* the 4 rows in the experiment which correspond to the first 4 rcws of plan 8.3 are omitted. The 4 rows or columns to be omitted for any other partially confounded comparison are found similarly from plan 8.3. If the high-order interaction effects are likely to be negligible, all partially confounded effects and *ABCD* and *ABCDE* may be combined with the error. This procedure gives 17 d.f. for treatments and 32 for error.

2⁶ factorial in an 8 × 8 square. In each square of plan 8.4, 8 three-factor and 6 four-factor interactions are completely confounded. If the experiment has a single replication and if only the main effects and two-factor interactions are isolated for tests of significance, the degrees of freedom are allotted as follows.

	d.f.
Rows	7
Columns	7
Main effects	6
Two-factor interactions	15
Error (estimated from high-order interactions)	28

Any unconfounded three-factor interaction may also be tested.

When there are several replications in the experiment, the squares should be chosen so that the most important three-factor interactions are free from confounding. The balanced design, which requires 5 replications, gives 3/5 relative information on all three- and four-factor interactions.

3³ factorial in a 9 × 9 square. Three or six replicates may be used (plan 8.5). Four of the 8 d.f. for *ABC* are completely confounded in each square. The 2 squares together give balanced confounding of *ABC*, with 1/2 relative information.

When square I alone is used, the components of the analysis of variance are:

	d.f.
Rows	8
Columns	8
A, B, C	6
AB, AC, BC	12
ABC	4
Error	42

The 4 d.f. for *ABC* are evaluated as follows. Compute the total yields of the 3 groups of 9 treatments which are described as set II (*a*), (*c*), and (*b*) in square 2. The sum of squares of deviations of these 3 totals, with divisor 27, constitutes 2 of the 4 d.f. The remaining 2 are found similarly from set IV (*a*), (*b*), and (*c*) in square 2.

With both squares, all 8 d.f. for *ABC* may be computed, sets I and III being taken from the experimental data for square II, and sets II and IV from the data for square I.

3⁴ factorial in a 9 × 9 square. Each square in plan 8.6 provides 1/2 relative information on *ABC*, *ABD*, *ACD*, and *BCD*. For an experiment with a single replication, the simplest subdivision of the degrees of freedom is:

	d.f.
Rows	8
Columns	8
Main effects	8
Two-factor interactions	24
Error (estimated from high-order interactions)	32

When 2 replications are available, both squares should be used. All 8 d.f. for any three-factor interaction may be tested, though each is based on only a single replication. The calculation of the sums of squares for the three-factor interactions follows the method outlined for the 3^4 factorial in blocks of 9 units (section 6.32). Plan 8.6 indicates the replication from which any component must be taken. Thus, if ABC is being calculated, we use the data from square II for ABC I and ABC IV and the data from square I for ABC II and ABC III.

4×2^2 factorial in an 8×8 square. Plan 8.7 requires 4 replications. The symbols A', A'', A''' refer to components of the A effects.

$$A' = a_3 + a_2 - a_1 - a_0; \qquad A'' = a_3 - a_2 - a_1 + a_0;$$

$$A''' = a_3 - a_2 + a_1 - a_0$$

No two-factor interaction is confounded.

In view of the small relative information on ABC (only 1/3 on the average), the sum of squares for this term will usually be combined with the error. In this event the analysis of variance reads as follows.

	d.f.
Rows	7
Columns	7
Main effects and two-factor interactions	12
Error (by subtraction)	37

8.4 Other Quasi-latin Squares

Yates (8.1) gives experimental plans which can be used for the following factorial combinations: $3^2 \times 2$ (6×6 square), 4×2^3 (8×8 square), and 9^2 (9×9 square). Analogous to the last design are designs for 7^2 (7×7 square) and 5^2 (5×5 square). In a general discussion of methods for constructing such plans, Rao (8.3) presents designs for a 2^4 factorial in 4×4 squares.

For certain factorial combinations of treatments, the *lattice squares* (chapter 12) and the incomplete latin squares (chapter 13) confound all treatment effects to the same extent. These arrangements may be useful in cases where the experimenter does not wish to sacrifice information on the interactions.

8.5 Estimation of the Efficiency of Quasi-latin Squares

In many cases the precision of a quasi-latin square can be compared with that of some other design which might have been used. The data from the example in table 8.1 will illustrate two types of comparison.

In this experiment a 2^3 factorial system applied to plots of soybeans was arranged in two 4×4 squares, with 1/2 relative information on *AB*, *AC*, *BC*, and *ABC*. As an alternative, the experiment might have been laid out in randomized incomplete blocks of 4 plots each, with complete confounding of *ABC* (plan 6.1). Since the columns were more compact in shape than the rows, we will suppose that the columns would have been chosen as blocks.

We first estimate the error m.s. that would have been obtained with incomplete blocks. It is helpful to construct analyses of variance for the 2 designs (table 8.3) for the case where there are no treatments. The

TABLE 8.3 Analyses of variance (applicable to uniformity data)

Quasi-latin square			Incomplete blocks		
	d.f.	m.s.		d.f.	m.s.
Columns	6	18.78	Blocks (columns)	6	18.78
Rows	6	16.82	Intra-blocks	24	(11.15)
Intra-row-and-column	18	9.26			

mean squares to be inserted in the table are explained in the next paragraph.

With incomplete blocks, the degrees of freedom to be divided between treatments and error are the 24 d.f. *within columns* (3 from each of the 8 columns). Since the randomization ascribes to treatments a random selection from these 24 d.f., the mean square for the 24 d.f. provides an unbiased estimate of the error appropriate to the incomplete block design. The 24 d.f. subdivide into 6 representing variations among rows, and 18 representing intra-row-and-column variation. If the interactions partially confounded with the rows are regarded as negligible, the rows m.s. (16.82) from the quasi-latin square analysis provides an estimate of the 6 d.f. The quasi-latin square analysis does not supply information on the whole of the remaining 18 d.f., since in the experiment 7 of these were ascribed to treatments. By virtue of the randomization for the quasi-latin square, however, the error m.s. (9.26), with 11 d.f., may be taken as an estimate of the mean square for the whole 18 d.f.

Accordingly, an unbiased estimate of the error m.s. for the incomplete block design is

$$\frac{18 \times 9.26 + 6 \times 16.82}{24} = 11.15$$

For the estimation of main effects, the efficiency of the quasi-latin square relative to incomplete blocks is estimated as 11.15/9.26, or 120%. Owing to the confounding in the quasi-latin square, the relative efficiency on the interactions is only 60%.

The quasi-latin square may also be compared with ordinary randomized blocks of 8 plots, where each replication is a pair of columns. In table 8.1 the column totals are as follows.

Rep. I		Rep. II		Rep. III		Rep. IV	
246.6	245.9	250.4	233.1	255.8	236.4	248.7	257.8

With uniformity data, the randomized blocks error contains 28 d.f. (7 from each replicate). To obtain this error, the incomplete blocks error (24 d.f.) must be pooled with the 4 d.f. which measure differences between the 2 columns in each pair. Since the differences between the 2 column totals in each replication are 0.7, 17.3, 19.4, and 9.1, respectively, the *sum of squares* for the 4 d.f. is

$$\frac{(0.7)^2 + (17.3)^2 + (19.4)^2 + (9.1)^2}{8} = 94.87$$

This gives $[24(11.15) + 94.87]/28 = 12.95$ as an estimate of the randomized blocks error. The relative efficiency of the quasi-latin square is 140% on main effects and 70% on interactions.

The comparisons above require the assumption that treatment effects which are confounded with rows or columns are non-existent. By more complicated procedures, estimates can be made which avoid those assumptions. The relative efficiencies quoted above are slightly too high, because we neglected the effect of the additional error d.f. that would be available in the alternative designs (section 2.31).

8.6 Treatments Applied to Complete Rows of a Latin Square

A simple example (taken from plan 8.8) is shown below. The latin square treatments form a 2×2 factorial for factors B and C. An extra factor A has been superimposed on the rows. The first 2 rows receive the lower level of A, and the last 2 rows the higher level.

ABC

(–)	*b*	*c*	(1)	*bc*	
(–)	*c*	*b*	*bc*	(1)	
					A
a	(1)	*bc*	*b*	*c*	
a	*bc*	(1)	*c*	*b*	

This plan might be tried in a case where the extra factor was not suitable for application to individual units. No precise estimate of the effect of A would be expected, but some useful information on the interactions between A and the other factors might be hoped for. However, this device has one property that necessitates caution in its use. Whenever an extra factor is superimposed on the rows, certain interactions between that factor and the latin square treatments are automatically confounded with columns, as noted by Yates (8.2). In the plan above, the ABC interaction

$$(abc) + (a) + (b) + (c) - (ab) - (ac) - (bc) - (1)$$

is the difference between the first 2 columns and the last 2 columns. On the other hand, AB and AC are free of column effects. In the plans presented here, which are due to Yates (8.1), such confounding is confined as far as possible to high-order interactions. The reader who constructs his own designs should verify, before using them, what interactions have been confounded in this way.

As indicated above, these designs are useful in experiments where (i) it is not easy to apply factor A to individual units, and (ii) accurate estimates of the interactions of A, but not of its main effects, are wanted. For instance, in an experiment on an irrigated crop, we might wish to find out how the responses to plant nutrients B and C are affected by the level of the water supply. Factor A could then represent a restricted and an abundant amount of irrigation water, applied to whole rows of the square.

These *half-plaid squares* are related both to quasi-latin squares and to split-plot designs. The difference from quasi-latin squares is that in the latter the confounding is confined to interactions, whereas main effects also are confounded in half-plaid squares. The connection with split-plot designs is seen when the rows are regarded as the experimental units, and the units (or plots) as sub-units. From this point of view the plan above may be considered as a split-plot design with the sub-units arranged in latin squares.

A moderately large number of useful half-plaid squares can be constructed; only five of the simplest are given here. In the notation employed, the factorial system for the latin square treatments is placed in parentheses ( ), while the number of treatments imposed on the rows is shown before the parentheses. Thus plan 8.8 shows a $2 \times (2^2)$ factorial.

As with an ordinary latin square, rows and columns must both be permuted at random. The whole-row treatments cannot be arranged in separate blocks.

The statistical analysis may be followed from the partition of the degrees of freedom, which is given in condensed form with each plan. As in the split-plot design, there are two errors. The error for the factor A applied to the rows is derived from the sum of squares among rows minus the sum of squares for A.

Most of the other treatment effects are unconfounded. Their sums of squares are calculated in the usual way and tested against the latin square error. In plans 8.9 and 8.10 certain interactions are *partially* confounded with columns; the calculations for these effects are described in the notes below. In the remainder of the plans, some interactions are *completely* confounded with rows or columns and consequently do not appear in the treatments s.s. The error s.s. is found by subtraction.

8.61 Notes on the Plans and Statistical Analysis. *$2 \times (2^2)$ factorial in a 4×4 square* (plan 8.8). At least two squares (4 replications) should be used.

$2 \times (3 \times 2)$ factorial in a 6×6 square (plan 8.9). Each square contains 3 replicates. BC and ABC are partially confounded with columns, the relative information being 8/9 and 5/9, respectively. The sums of squares for these effects are calculated by the procedure for the 3×2^2 factorial in incomplete blocks of 6 units, where the columns constitute the blocks. A numerical example is given by Yates (8.1, p. 58). Columns Ia ··· IIIb are identical, respectively, with Yates' blocks Ia ··· IIIb. Notice that the factor applied to the rows is C.

$3 \times (3 \times 2)$ factorial in a 6×6 square (plan 8.10). Each square requires 2 replicates, while the 2 squares comprize a balanced set. Column effects are partially confounded with AB and ABC, the average relative information being 7/8 and 5/8, respectively. If the columns are regarded as incomplete blocks, the numerical example in section 6.19 may be followed for the computation of the sums of squares for AB and ABC.

The statistical analyses for the $2 \times (2^3)$ factorial (plan 8.11) and the $2 \times (2^4)$ factorial (plan 8.12) in 8×8 squares are straightforward. In the latter, the *BCDE* and *ABCDE* interactions have been combined with the error for *A*.

Other designs. Yates (8.1, p. 79) presents the plans for a $4 \times (2^3)$ factorial in an 8×8 square and (8.1, p. 80) for a $3 \times (3^3)$ factorial in a 9×9 square. Plans can also be constructed for a $2 \times (2^5)$ system in a 8×8 square and for a $3 \times (3^2)$ system in a 9×9 square.

8.7 Treatments Applied to Complete Rows and Columns of a Latin Square

In these designs one set of treatments *A* is applied to complete rows of the latin square and another set *B* to complete columns. Although relatively low precision is obtained on the row and column treatments, the latin square treatments and their more important interactions with the *A* and *B* treatments are subject only to the latin square error. The number of useful designs appears to be limited.

Two examples are shown in plans 8.13 and 8.14. The latin square treatments form a 2^3 and 2^4 system, respectively; the additional treatments are both at 2 levels. No two-factor interactions are confounded. The rows and columns of each square should be completely randomized.

The subdivision of degrees of freedom is indicated on each plan. Interactions which are confounded with rows and columns must be omitted from the treatments s.s.: all other treatment effects are calculated in the usual way. In plan 8.14, where the square contains a single replication, the only available estimate of error comes from the high-order interactions, unless two or more squares are used.

A 9×9 square for a 3^2 system of treatments, with row and column treatments also at 3 levels each, is given in reference (8.1), p. 81.

REFERENCES

8.1 Yates, F. The design and analysis of factorial experiments. *Imp. Bur. Soil Sci. Tech. Comm.* 35, 1937.

8.2 Yates, F. The principles of orthogonality and confounding in replicated experiments. *Jour. Agr. Sci.* 23, 108–145, 1933.

8.3 Rao, C. R. Confounded factorial designs in quasi-latin squares. *Sankhya* 7, 295–304, 1946.

ADDITIONAL READING

Grundy, P. M., and Healy, M. J. R. Restricted randomization—quasi-latin squares. *Jour. Roy. Stat. Soc. B* 11, 286–291, 1950.

Healy, M. J. R. Latin rectangle designs for 2^n factorial experiments on 32 plots. *Jour. Agr. Sci.* 41, 315–316, 1951.

PLANS

Plan 8.1a 2^3 factorial in two 4 × 4 quasi-latin squares

Square I

ABC		*AC*		
c	(1)	*abc*	*ab*	*AB*
a	*ac*	*b*	*bc*	
abc	*bc*	(1)	*a*	*BC*
b	*ab*	*ac*	*c*	

Square II

AB		*BC*		
c	*a*	*abc*	*b*	*ABC*
(1)	*ac*	*bc*	*ab*	
abc	*b*	(1)	*ac*	*AC*
ab	*bc*	*a*	*c*	

Plan 8.1b 2^3 factorial in two 4 × 4 latin squares

ABC completely confounded with squares

Square I

(1)	*bc*	*ac*	*ab*
ac	*ab*	(1)	*bc*
bc	(1)	*ab*	*ac*
ab	*ac*	*bc*	(1)

Square II

b	*a*	*abc*	*c*
abc	*c*	*a*	*b*
a	*b*	*c*	*abc*
c	*abc*	*b*	*a*

Plan 8.2 2^4 factorial in an 8 × 8 quasi-latin square

ABCD

c	*abcd*	*b*	*ad*	*a*	*bd*	*abc*	*cd*	*ABC*
abd	(1)	*bcd*	*bc*	*acd*	*ac*	*d*	*ab*	
d	*bc*	*a*	*abcd*	*b*	*cd*	*abd*	*ac*	*ABD*
bcd	*ad*	*acd*	*bd*	*abc*	*ab*	*c*	(1)	
a	*bd*	*c*	*ab*	*d*	*abcd*	*acd*	*bc*	*ACD*
abc	*ac*	*abd*	*cd*	*bcd*	(1)	*b*	*ad*	
b	*ab*	*d*	*ac*	*c*	*ad*	*bcd*	*abcd*	*BCD*
acd	*cd*	*abc*	(1)	*abd*	*bc*	*a*	*bd*	

Plan 8.3 2^5 factorial in an 8×8 quasi-latin square

ACE, BCD, ABDE				*ACD, BDE, ABCE*				
(1)	*abe*	*bc*	*ace*	*abd*	*acd*	*bcde*	*de*	*ABC, ADE, BCDE*
bce	*ac*	*e*	*ab*	*bcd*	*d*	*abde*	*acde*	
cde	*abcd*	*bde*	*ad*	*abce*	*ae*	*b*	*c*	
bd	*ade*	*cd*	*abcde*	*be*	*ce*	*abc*	*a*	
abc	*ce*	*acde*	*bcd*	(1)	*bde*	*ad*	*abe*	*ABD, BCE, ACDE*
acd	*bcde*	*abce*	*c*	*ade*	*ab*	*e*	*bd*	
abde	*d*	*a*	*be*	*cde*	*bc*	*ace*	*abcd*	
ae	*b*	*abd*	*de*	*ac*	*abcde*	*cd*	*bce*	

Plan 8.4 2^6 factorial in 8×8 quasi-latin squares

Square I

Columns confounded with *CDE, BDF, ABE, ACF, BCEF, ABCD, ADEF*

								Rows confounded with
abcdef	*cef*	*bf*	*bde*	*ae*	*abc*	*adf*	*cd*	*ABC*
cde	*abce*	*a*	*adef*	*bef*	*cf*	*bd*	*abcdf*	*BDE*
bdf	*af*	*abcef*	*abcd*	*c*	*be*	*cdef*	*ade*	*ADF*
acf	*bcdf*	*def*	(1)	*abd*	*acde*	*abef*	*bce*	*CEF*
bc	*acd*	*abde*	*abf*	*df*	*bcdef*	*e*	*acef*	*ACDE*
ef	*abdef*	*acdf*	*ace*	*bcde*	*d*	*bcf*	*ab*	*BCDF*
ad	*b*	*ce*	*cdf*	*abcf*	*aef*	*abcde*	*bdef*	*ABEF*
abe	*de*	*bcd*	*bcef*	*acdef*	*abdf*	*ac*	*f*	

Square II

ABC, CDE, ADF, BEF, ABDE, BCDF, ACEF

abcdef	*bf*	*cdf*	*bce*	*abd*	*ac*	*aef*	*de*	*BCE*
ae	*cd*	*b*	*def*	*acf*	*abdf*	*abcde*	*bcef*	*CDF*
cde	*a*	*abcd*	*acef*	*df*	*bcf*	*be*	*abdef*	*ABD*
adf	*cef*	*bdef*	(1)	*acde*	*abe*	*abcf*	*bcd*	*AEF*
abc	*bde*	*ce*	*bcdf*	*abef*	*acdef*	*ad*	*f*	*BDEF*
bd	*abce*	*ade*	*abf*	*bcdef*	*ef*	*c*	*acdf*	*ACDE*
bef	*abcdf*	*af*	*abde*	*bc*	*d*	*cdef*	*ace*	*ABCF*
cf	*adef*	*abcef*	*acd*	*e*	*bcde*	*bdf*	*ab*	

Square III

ABF, CDF, ADE, BCE, ABCD, BDEF, ACEF

abcdef	*be*	*def*	*bcf*	*abd*	*af*	*ace*	*cd*	*ABD*
ac	*df*	*b*	*cde*	*aef*	*abde*	*abcdf*	*bcef*	*DEF*
cdf	*a*	*abdf*	*acef*	*de*	*bef*	*bc*	*abcde*	*BCF*
ade	*cef*	*bcde*	(1)	*acdf*	*abc*	*abef*	*bdf*	*ACE*
abf	*bcd*	*cf*	*bdef*	*abce*	*acdef*	*ad*	*e*	*ABEF*
bd	*abcf*	*acd*	*abe*	*bcdef*	*ce*	*f*	*adef*	*ACDF*
bce	*abdef*	*ae*	*abcd*	*bf*	*d*	*cdef*	*acf*	*BCDE*
ef	*acde*	*abcef*	*adf*	*c*	*bcdf*	*bde*	*ab*	

Plan 8.4 (Continued) 2^6 factorial in 8 × 8 quasi-latin squares

Square IV

BCD, ACE, ABF, DEF, ABDE, ACDF, BCEF

ab	*acd*	*ce*	*bde*	*bcf*	*df*	*aef*	*abcdef*	*AEF*
de	*bce*	*abcd*	*a*	*acdef*	*abef*	*bdf*	*cf*	*BDE*
c	*bd*	*abe*	*acde*	*af*	*abcdf*	*bcef*	*def*	*ACD*
bef	*cdef*	*acf*	*abdf*	*abce*	*ade*	(1)	*bcd*	*BCF*
adf	*abcf*	*bcdef*	*ef*	*cd*	*b*	*abde*	*ace*	*ABDF*
acef	*abdef*	*bf*	*cdf*	*e*	*bcde*	*abc*	*ad*	*CDEF*
abcde	*ae*	*d*	*bc*	*bdef*	*cef*	*acdf*	*abf*	*ABCE*
bcdf	*f*	*adef*	*abcef*	*abd*	*ac*	*cde*	*be*	

Square V

ACF, BCD, ADE, BEF, ABDF, CDEF, ABCE

abcdef	*bdf*	*bc*	*acd*	*de*	*cef*	*abe*	*af*	*ABE*
ce	*a*	*acdf*	*bcf*	*abef*	*abcde*	*def*	*bd*	*BDF*
bcd	*abde*	*abcef*	*cdef*	*adf*	*ac*	*bf*	*e*	*ACD*
bef	*abcf*	*abd*	(1)	*ace*	*adef*	*bcde*	*cdf*	*CEF*
ade	*cd*	*f*	*abdf*	*bcdef*	*be*	*acef*	*abc*	*ADEF*
ab	*bce*	*bdef*	*aef*	*cf*	*d*	*abcdf*	*acde*	*BCDE*
acf	*ef*	*cde*	*abce*	*b*	*bcdf*	*ad*	*abdef*	*ABCF*
df	*acdef*	*ae*	*bde*	*abcd*	*abf*	*c*	*bcef*	

Plan 8.5 3^3 factorial in 9 × 9 quasi-latin squares

Square I

ABC III confounded

Set III	(*a*)	000	102	201	011	110	212	022	121	220	
	(*c*)	101	200	002	112	211	010	120	222	021	
	(*b*)	202	001	100	210	012	111	221	020	122	
	(*a*)	011	110	212	022	121	220	000	102	201	
	(*c*)	112	211	010	120	222	021	101	200	002	*ABC* II
	(*b*)	210	012	111	221	020	122	202	001	100	
	(*a*)	022	121	220	000	102	201	011	110	212	
	(*c*)	120	222	021	101	200	002	112	211	010	
	(*b*)	221	020	122	202	001	100	210	012	111	
	Set I	(*a*)	(*b*)	(*c*)	(*a*)	(*b*)	(*c*)	(*a*)	(*b*)	(*c*)	

Plan 8.5 (Continued)

Square II

ABC IV

Set IV	(*a*)	000	101	202	012	110	211	021	122	220	
	(*b*)	102	200	001	111	212	010	120	221	022	
	(*c*)	201	002	100	210	011	112	222	020	121	
	(*a*)	012	110	211	021	122	220	000	101	202	
	(*b*)	111	212	010	120	221	022	102	200	001	*ABC* I
	(*c*)	210	011	112	222	020	121	201	002	100	
	(*a*)	021	122	220	000	101	202	012	110	211	
	(*b*)	120	221	022	102	200	001	111	212	010	
	(*c*)	222	020	121	201	002	100	210	011	112	
Set II		(*a*)	(*c*)	(*b*)	(*a*)	(*c*)	(*b*)	(*a*)	(*c*)	(*b*)	

Plan 8.6 3^4 factorial in 9×9 quasi-latin squares

Square I

Columns confounded with *ABC* IV, *ABD* I, *ACD* II, *BCD* I

Rows confounded with *ABC* I, *ABD* III, *ACD* IV, *BCD* II

0000	1011	2022	0121	1102	2110	0212	1220	2201
1021	2002	0010	1112	2120	0101	1200	2211	0222
2012	0020	1001	2100	0111	1122	2221	0202	1210
0122	1100	2111	0210	1221	2202	0001	1012	2020
1110	2121	0102	1201	2212	0220	1022	2000	0011
2101	0112	1120	2222	0200	1211	2010	0021	1002
0211	1222	2200	0002	1010	2021	0120	1101	2112
1202	2210	0221	1020	2001	0012	1111	2122	0100
2220	0201	1212	2011	0022	1000	2102	0110	1121

Square II

Columns confounded with *ABC* III, *ABD* II, *ACD* III, *BCD* IV

Rows confounded with *ABC* II, *ABD* IV, *ACD* I, *BCD* III

0000	1022	2011	0112	1101	2120	0221	1210	2202
1012	2001	0020	1121	2110	0102	1200	2222	0211
2021	0010	1002	2100	0122	1111	2212	0201	1220
0111	1100	2122	0220	1212	2201	0002	1021	2010
1120	2112	0101	1202	2221	0210	1011	2000	0022
2102	0121	1110	2211	0200	1222	2020	0012	1001
0222	1211	2200	0001	1020	2012	0110	1102	2121
1201	2220	0212	1010	2002	0021	1122	2111	0100
2210	0202	1221	2022	0011	1000	2101	0120	1112

Plan 8.7 4×2^2 factorial in an 8×8 quasi-latin square

$A'''BC$

110	311	211	001	200	010	101	300	$A'BC$
011	100	000	210	301	201	310	111	
101	210	011	311	000	300	110	201	$A''BC$
310	001	200	010	211	111	301	100	
200	010	101	300	110	311	211	001	$A'BC$
301	201	310	111	011	100	000	210	
000	300	110	201	101	210	011	311	$A''BC$
211	111	301	100	310	001	200	010	

Plan 8.8 $2 \times (2^2)$ factorial in a 4×4 half-plaid square

ABC

–	b	c	(1)	bc	A
–	c	b	bc	(1)	
a	(1)	bc	b	c	
a	bc	(1)	c	b	

k squares ($2k$ replicates)

Squares		$(k-1)$
Rows		$3k$
A	1	
Error	$(3k-1)$	
Columns		$3k$
B, C, BC, AB, AC		5
Error		$(9k-5)$
Total		$(16k-1)$

Plan 8.9 $2 \times (3 \times 2)$ factorial in a 6×6 half-plaid square

BC, ABC partially confounded

c	ab						
0	01	00	11	10	21	20	C
0	10	11	20	21	00	01	
0	20	21	00	01	10	11	
1	00	01	10	11	20	21	
1	11	10	21	20	01	00	
1	21	20	01	00	11	10	
	Ia	Ib	IIa	IIb	IIIa	IIIb	

Rows		5
C	1	
Error	4	
Columns		5
A, B, AB, AC		7
BC		1′
ABC		2′
Error		15
Total		35

Plan 8.10 3 × (3 × 2) factorial in a 6 × 6 half-plaid square

Square I

AB, ABC partially confounded

a	*bc*					
0	20	10	00	21	11	01
0	11	01	21	10	00	20
1	00	20	10	01	21	11
1	21	11	01	20	10	00
2	01	21	11	00	20	10
2	10	00	20	11	01	21
	I*a*	I*b*	I*c*	II*a*	II*b*	II*c*

A

Square II

AB, ABC partially confounded

a	*bc*					
0	10	20	00	11	21	01
0	21	01	11	20	00	10
1	00	10	20	01	11	21
1	11	21	01	10	20	00
2	20	00	10	21	01	11
2	01	11	21	00	10	20
	III*a*	III*b*	III*c*	IV*a*	IV*b*	IV*c*

Square I alone (2 replicates)

Rows		5
A	2	
Error	3	
Columns		5
B, C, BC, AC		7
AB		4′
ABC		4′
Error		10
Total		35

Both squares (4 replicates)

Squares		1
Rows		10
A	2	
Error	8	
Columns		10
B, C, BC, AC		7
AB		4′
ABC		4′
Error		35
Total		71

Plan 8.11 2 × (2^3) factorial in an 8 × 8 half-plaid square

ABCD confounded

A								
–	(1)	*bc*	*bd*	*cd*	*b*	*c*	*d*	*bcd*
–	*bc*	(1)	*cd*	*bd*	*d*	*bcd*	*b*	*c*
–	*bd*	*cd*	(1)	*bc*	*bcd*	*d*	*c*	*b*
–	*cd*	*bd*	*bc*	(1)	*c*	*b*	*bcd*	*d*
a	*b*	*c*	*d*	*bcd*	(1)	*bd*	*cd*	*bc*
a	*c*	*b*	*bcd*	*d*	*bc*	*cd*	*bd*	(1)
a	*d*	*bcd*	*b*	*c*	*cd*	(1)	*bc*	*bd*
a	*bcd*	*d*	*c*	*b*	*bd*	*bc*	(1)	*cd*

Rows		7
A	1	
Error	6	
Columns		7
B, C, D		3
AB, AC, AD, BC, BD, CD		6
ABC, ABD, ACD, BCD		4
Error		36
Total		63

Plan 8.12 $2 \times (2^4)$ factorial in an 8×8 half-plaid square

ABD, BCE, ACDE confounded

	–	(1)	*bd*	*bc*	*cd*	*ce*	*bcde*	*be*	*de*
	–	*bcd*	*c*	*d*	*bce*	*bde*	*e*	*cde*	*b*
	–	*ce*	*bcde*	*be*	*de*	(1)	*bd*	*bc*	*cd*
A	–	*bde*	*e*	*cde*	*b*	*bcd*	*c*	*d*	*bce*
BCDE									
ABCDE	*a*	*bc*	*cd*	*ce*	*bcde*	*be*	*de*	(1)	*bd*
	a	*d*	*b*	*bcd*	*c*	*cde*	*bce*	*bde*	*e*
	a	*be*	*de*	(1)	*bd*	*bc*	*cd*	*ce*	*bcde*
	a	*cde*	*bce*	*bde*	*e*	*d*	*b*	*bcd*	*c*

Rows		7
A	1	
Error	6	
Columns		7
Main effects		4
Two-factor interactions		10
Three-factor "		8
Four-factor "		3
Error		24
Total		63

Plan 8.13 $2 \times 2 \times (2^3)$ factorial in an 8×8 plaid square

B, ACDE, ABCDE confounded

		–	–	–	–	*b*	*b*	*b*	*b*
	–	(1)	*e*	*cd*	*cde*	*ce*	*c*	*de*	*d*
	–	*ce*	*c*	*de*	*d*	*cd*	*e*	(1)	*cde*
	–	*cd*	*cde*	(1)	*e*	*de*	*d*	*ce*	*c*
A	–	*de*	*d*	*ce*	*c*	(1)	*cde*	*cd*	*e*
BCD									
ABCD	*a*	*e*	(1)	*cde*	*cd*	*c*	*ce*	*d*	*de*
	a	*c*	*de*	*d*	*ce*	*e*	(1)	*cde*	*cd*
	a	*cde*	*cd*	*e*	(1)	*d*	*de*	*c*	*ce*
	a	*d*	*ce*	*c*	*de*	*cde*	*cd*	*e*	(1)

Rows		7
A	1	
Error	6	
Columns		7
B	1	
Error	6	
Main effects		3
Two-factor interactions		10
Three-factor "		9
Four-factor "		3
Error		24
Total		63

Plan 8.14 $2 \times 2 \times (2^4)$ factorial in an 8×8 plaid square

B, ACD, CEF, ABCD, BCEF,
ADEF, ABDEF

		–	–	–	–	*b*	*b*	*b*	*b*
	–	*e*	*cde*	*df*	*cf*	*cd*	(1)	*cef*	*def*
A	–	*f*	*cdf*	*de*	*ce*	*cdef*	*ef*	*c*	*d*
BEF	–	*cd*	(1)	*cef*	*def*	*e*	*cde*	*df*	*cf*
CDF	–	*cdef*	*ef*	*c*	*d*	*f*	*cdf*	*de*	*ce*
ABEF									
ACDF	*a*	*df*	*cf*	*e*	*cde*	*cef*	*def*	*cd*	(1)
BCDE	*a*	*de*	*ce*	*f*	*cdf*	*c*	*d*	*cdef*	*ef*
ABCDE	*a*	*cef*	*def*	*cd*	(1)	*df*	*cf*	*e*	*cde*
	a	*c*	*d*	*cdef*	*ef*	*de*	*ce*	*f*	*cdf*

Rows		7
A	1	
Error	6	
Columns		7
B	1	
Error	6	
Main effects		4
Two-factor interactions		15
Error (from high-order interactions)		30
Total		63

CHAPTER 8A

Some Methods for the Study of Response Surfaces

8A.1 First Order Designs

8A.11 Introduction. In chapter 6 it was pointed out that the most informative method of analysis of the results of a factorial experiment depends on the nature of the factors. If all the factors represent quantitative variables like time, temperature, amount of nitrogen, it is natural to think of the yield or response y as a function of the levels of these variables. We may write

$$y_u = \phi(x_{1u}, x_{2u}, \cdots, x_{ku}) + e_u$$

where $u = 1, 2, \ldots N$ represents the N observations in the factorial experiment and x_{iu} represents the level of the ith factor in the uth observation. The function ϕ is called the *response surface*. The residual e_u measures the experimental error of the uth observation. A knowledge of the function ϕ gives a complete summary of the results of the experiment and also enables us to predict the response for values of the x_{iu} that were not tested in the experiment.

When the mathematical form of ϕ is not known, this function can sometimes be approximated satisfactorily, within the experimental region, by a polynomial in the variables x_{iu}. In this chapter we describe some experimental designs and methods of analysis that have been developed for fitting polynomials of the first and second degree. Box (8A.1) has called these designs *first order* designs and *second order* designs, respectively. The second part of this chapter deals with an important problem in which these designs can be used—namely, that of finding the level at which each of the x-variables should be set in order to maximize ϕ.

First, a word of caution. Polynomial response surfaces have the great advantage that they are easy to fit. With a suitable choice of design, even a quadratic surface in 6 variables, which contains 28 coefficients to be estimated, is not too formidable a task. On the other hand, polynomials are notoriously untrustworthy when extrapolated. A polynomial surface should be regarded only as an approximation to ϕ *within the region covered by the experiment.* Any prediction made from the polynomial about the response outside the region should be verified by experiments before putting reliance on it.

8A.12 Calculations for a Multiple Linear Regression. Since the fitting of a polynomial can be treated as a particular case of multiple linear regression, we shall review the calculations required to fit a multiple linear regression of y_u on the k variables $x_{iu} (i = 1, 2, \cdots k;\ u = 1, 2, \cdots N)$. The relation between y_u and the x_{iu} is of the form

$$y_u = \beta_0 + \beta_1 x_{1u} + \beta_2 x_{2u} + \cdots + \beta_k x_{ku} + e_u$$

Since β_0 occurs in every equation, it is customary, for the sake of uniformity, to introduce a dummy variable x_{0u} which has the value $+1$ for every observation in the sample. The equation can then be written

$$y_u = \beta_0 x_{0u} + \beta_1 x_{1u} + \beta_2 x_{2u} + \cdots + \beta_k x_{ku} + e_u$$

Step 1. List the sample values of the x_{iu} and the y_u as in table 8A.1. The two-way array of the x_{iu} values is called the X matrix, and the column of y_u values is called the Y vector.

Step 2. The least squares estimates b_i of the β_i are chosen so as to minimize the sum of squares of deviations

$$\sum_{u=1}^{N} (y_u - b_0 x_{0u} - b_1 x_{1u} - \cdots - b_k x_{ku})^2$$

The values of the b_i which minimize this expression satisfy the *normal equations*

$$\begin{aligned} b_0(00) + b_1(01) + \cdots + b_k(0k) &= (0y) \\ b_0(10) + b_1(11) + \cdots + b_k(1k) &= (1y) \\ \cdots\cdots\cdots\cdots\cdots\cdots & \\ b_0(k0) + b_1(k1) + \cdots + b_k(kk) &= (ky) \end{aligned}$$

where

$$(ij) = (ji) = \sum_{u=1}^{N} x_{iu} x_{ju} = \text{Sum of products of } i\text{th and } j\text{th columns in } X.$$

$$(ii) = \sum_{u=1}^{N} x_{iu}^2 = \text{Sum of squares of } i\text{th column in } X.$$

$$(iy) = \sum_{u=1}^{N} x_{iu} y_u = \text{Sum of products of } i\text{th column in } X \text{ with } Y.$$

It is sufficient to list the sums of squares and products (ij) and (iy), as shown in table 8A.1, without explicitly writing out the equations. The square array (ij) is called the $X'X$ matrix, and the column (iy) is called the $X'Y$ vector.

Step 3. Although the normal equations can be solved directly, it is useful to find a set of intermediate quantities c_{ij}, known as the inverse of the matrix (ij). The equations satisfied by $c_{i0}, c_{i1}, \cdots c_{ik}$ have the same left-hand side as those for the b_i, but the right side consists of zeros except in the ith equation, where the right side is unity. Various methods for computing the inverse matrix are available (8A.2, 8A.3, 8A.4).

Step 4. The regression coefficients b_i are obtained as follows.

$$b_i = \sum_{j=0}^{k} c_{ji}(jy) = \text{Sum of products of } i\text{th column of } c_{ij} \text{ with the column } (jy).$$

The formulae for obtaining the residual mean square and the standard errors of the b_i are given at the foot of table 8A.1.

TABLE 8A.1 CALCULATIONS FOR A MULTIPLE LINEAR REGRESSION

Data

	X			Y
x_{01}	x_{11}	$\cdots$	x_{k1}	y_1
x_{02}	x_{12}	$\cdots$	x_{k2}	y_2
$\cdot$	$\cdot$	$\cdots$	$\cdot$	$\cdot$
x_{0N}	x_{1N}	$\cdots$	x_{kN}	y_N

Normal equations

	$(ij) = X'X$			$(iy) = X'Y$
(00)	(01)	$\cdots$	$(0k)$	$(0y)$
(10)	(11)	$\cdots$	$(1k)$	$(1y)$
$\cdot$	$\cdot$	$\cdots$	$\cdot$	$\cdot$
$(k0)$	$(k1)$	$\cdots$	(kk)	(ky)

Inverse matrix

	$c_{ij} = (X'X)^{-1}$		
c_{00}	c_{01}	$\cdots$	c_{0k}
c_{10}	c_{11}	$\cdots$	c_{1k}
$\cdot$	$\cdot$	$\cdots$	$\cdot$
c_{k0}	c_{k1}	$\cdots$	c_{kk}

Solutions

$$b_i = \sum_{j=0}^{k} c_{ji}(jy)$$

$$\text{Residual sum of squares} = R_k = (yy) - \sum_{i=0}^{k} b_i(iy)$$

$$\text{Residual mean square} = s^2 = R_k/(N - k - 1)$$

$$V(b_i) = c_{ii}s^2 \qquad Cov(b_i b_j) = c_{ij}s^2$$

The normal equations are particularly easy to solve when the variables x_{iu} are mutually orthogonal. In this case all sums of products (ij) vanish $(i \neq j)$, and the normal equation for b_i reduces to

$$(ii)b_i = (iy)$$

so that

$$b_i = \frac{(iy)}{(ii)}$$

The values of the inverse matrix become $c_{ii} = 1/(ii)$; $c_{ij} = 0$, $i \neq j$. Since $c_{ij} = 0$, the b_i are uncorrelated.

In constructing experimental designs from which regression equations are to be computed, it is desirable to obtain normal equations that present little difficulty in solution. Fortunately, the designs that are constructed so as to give the most precise estimates in a polynomial regression also turn out to be fairly simple to handle computationally.

8A.13 Numerical Example for a 2^3 Factorial. As an illustration of the general approach, a linear equation will be fitted to the results of a 2^3 factorial. The data are artificial, and deal with three factors that influence the yield of a chemical reaction.

$A = x_1 =$ Time of reaction (5.5 and 6.5 hr.)

$B = x_2 =$ Temperature (260° and 340°)

$C = x_3 =$ Amount of an ingredient C (1.4 and 1.8 gm.)

The equation to be fitted is

$$\hat{y}_u = b_0 + b_1x_{1u} + b_2x_{2u} + b_3x_{3u}$$

where $\hat{y}_u$ is the predicted value of y, $(u = 1, 2, \cdots 8)$.

Since each x-variable takes only two different values, the calculations are simplified by coding each x-scale so that the upper level of x is $+1$ and the lower level is -1. The relations between the coded and the original scales are as follows:

$$x_1 = 2(\text{Time} - 6); \quad x_2 = \frac{\text{Temp.} - 300}{40}; \quad x_3 = \frac{\text{Amount of C} - 1.6}{0.2}$$

The coded values of the x's and the yields appear in table 8A.2. The computations follow the steps outlined in the preceding section. In the X matrix (table 8A.2) the sum of squares of any column is 8 and the sum of products of any two columns is 0. The (ij) matrix therefore assumes the simple orthogonal form, as does its inverse c_{ij}. From the

TABLE 8A.2 Fitting a linear equation to a 2^3 factorial

Data

		X			Y
		Coded scale			
Treatment	x_0	x_1	x_2	x_3	y
(1)	1	−1	−1	−1	161
a	1	1	−1	−1	183
b	1	−1	1	−1	151
ab	1	1	1	−1	170
c	1	−1	−1	1	166
ac	1	1	−1	1	192
bc	1	−1	1	1	156
abc	1	1	1	1	183

Normal equations

$(ij) = X'X$				$(iy) = X'Y$
8				$1362 = G$
	8			$94 = 4A$
		8		$-42 = 4B$
			8	$32 = 4C$

Inverse matrix

$c_{ij} = (X'X)^{-1}$			
1/8			
	1/8		
		1/8	
			1/8

Solutions

$$b_0 = \frac{1362}{8} = 170.25; \quad b_1 = \frac{94}{8} = 11.75; \quad b_2 = \frac{-42}{8} = -5.25; \quad b_3 = \frac{32}{8} = 4.00$$

TABLE 8A.3 Analysis of variance of yield

	d.f.	s.s.	m.s.
Linear model	3	1453	484.3
Lack of fit	4	23	5.8

solutions (foot of table 8A.2), the fitted equation (rounded slightly) is

$$\hat{y} = 170.2 + 11.8x_1 - 5.2x_2 + 4.0x_3$$

In the analysis of variance, the reduction in sum of squares of y due to the regression separates into the usual reduction due to the mean

$$\frac{(1362)^2}{8} = 231,880$$

and the reduction due to the linear terms

$$\frac{(94)^2 + (42)^2 + (32)^2}{8} = 1{,}453$$

The sum of squares of deviations from the fitted regression, i.e., $\sum(y_u - \hat{y}_u)^2$ is found by the short-cut method as

$$\sum_u y_u^2 - \sum_{i=0}^{4} b_i(iy) = 233{,}356 - 231{,}880 - 1{,}453 = 23$$

This sum of squares contains two sources of variation. It receives a contribution due to the experimental errors, which make the values y_u deviate from the true response surface ϕ. It is inflated also by any failure of the linear equation to represent the correct shape of the response surface ϕ. If the corresponding mean square is substantially larger than the mean square for experimental error, this is a warning that the linear equation does not adequately represent the true response surface. Consequently, the mean square of the deviations is sometimes described as measuring the "lack of fit."

These data, chosen for simplicity of illustration, do not furnish any estimate of the experimental error mean square with which the mean square for lack of fit can be compared. In practice, provision for an estimate of error is essential when first or second order designs are employed, since otherwise the experimenter has no idea whether the equation is an adequate approximation to the true surface. Some methods for obtaining an estimate of error for first order designs are discussed in section 8A.15.

8A.14 Relation to the Standard Factorial Analysis. If we had applied the standard method of analysis for a 2^3 factorial experiment (chapter 5), the data in table 8A.2 would have been analyzed into the main effects A, B, and C, the two-factor interactions AB, AC, and BC, and the three-factor interaction ABC. In table 8A.2 the treatment combinations have been written in standard factorial notation on the left of the matrix X. It is easily verified that the right sides of the normal equations are, respectively, the grand total G and the factorial effect *totals* of A, B, and C. Since the main effects are $\frac{1}{4}$ of the corresponding factorial effect totals, the coefficients b_1, b_2, and b_3 are $\frac{1}{2}$ of the main effects of A, B, and C, respectively.

It can be shown, as might be expected, that the 4 d.f. for "lack of fit" in table 8A.3 represent the contributions due to the interactions AB, AC, BC, and ABC. In fact, if we fit a more complex regression model of the form

$$\hat{y} = b_0x_0 + b_1x_1 + b_2x_2 + b_3x_3 + b_{12}x_1x_2 + b_{13}x_1x_3 + b_{23}x_2x_3 + b_{123}x_1x_2x_3$$

it will be found that

$$b_{12} = \frac{1}{2}AB; \qquad b_{13} = \frac{1}{2}AC; \qquad b_{23} = \frac{1}{2}BC; \qquad b_{123} = \frac{1}{2}ABC$$

A response surface of this type is unlikely to be considered, however, since there seems little reason to include cross-product terms like x_1x_2 without including terms in x_1^2 and x_2^2. For squared terms of this type, more than two levels of each factor must be introduced.

8A.15 Some Useful First Order Designs. The 2^k factorial designs, in single or fractional replication, are convenient in exploratory work for fitting a linear relation between the response y and the k factors or x-variables. The designs given in table 8A.4 also supply some d.f. for

TABLE 8A.4 SOME USEFUL 2^k FACTORIAL DESIGNS FOR FITTING LINEAR EQUATIONS

No. of factors	Size of experiment (units)	Fraction of a rep.	No. of d.f. for "lack of fit"	Plan no.
3	8	1	4	....
4	8	½	3	6A.1
5	8	¼	2	6A.2
5	16	½	10	6A.3
6	8	⅛	1	6A.4
6	16	¼	9	6A.5

"lack of fit" by which the experimenter can judge whether the linear equation fits adequately, provided that he has a measure of the experimental error variance at his disposal.

Plans for the fractional factorial designs in table 8A.4 are obtained from the plans in chapter 6A. The notation is easily translated into that of this chapter. For instance, the treatment combination *ac* in a 2^4 factorial corresponds to $x_1 = 1$, $x_2 = -1$, $x_3 = 1$, $x_4 = -1$.

As given, these designs do not provide any estimate of the experimental error variance. This can be obtained (i) by replication of the whole experiment, (ii) in some situations, by the use of an estimate from previous experimentation, if there is convincing evidence that the error variance remains stable through time, (iii) by adding to the 2^k factorial a number of tests made at the point at which all x's have the value 0 in the coded scale. Since the levels of every x in the 2^k factorial are -1 and $+1$, this point may be called the *center* of the design.

Suppose that the center point is replicated n_1 times. As can be seen by writing down the normal equations, these extra points do not change the values of any of the regression coefficients b_i, except that b_0 becomes the mean of the whole experiment. The extra points serve two purposes. The sum of squares of deviations of the n_1 responses at the center from their mean furnishes $(n_1 - 1)$ d.f. for measuring the experimental error. Further, if $\bar{y}_1$ is the mean response at the central points and $\bar{y}_2$ is the mean response at the exterior points (i.e., the points in the 2^k factorial), the comparison $(\bar{y}_2 - \bar{y}_1)$ gives an additional d.f. for measuring the lack of fit. (If the true response surface is quadratic rather than linear, this comparison estimates the sum of the coefficients of the terms in x_1^2, $x_2^2 \cdots$ in the quadratic surface.) The sum of squares for this single d.f. is computed as

$$\frac{n_1 n_2}{n_1 + n_2} (\bar{y}_1 - \bar{y}_2)^2$$

where n_2 is the number of exterior points.

Since the linear equation in k x-variables contains $(k + 1)$ regression coefficients that must be estimated, the smallest experiment to which a linear equation can be fitted is one that has $(k + 1)$ observations. Plackett and Burman (8A.5) and Box (8A.1) have shown how to construct the most efficient designs of this size. For practical purposes the use of these optimum designs is likely to be limited, since they supply no d.f. for measuring lack of fit and since most of the designs are less convenient to handle than the 2^k factorials.

8A.2 Second Order Designs

8A.21 The Quadratic Response Surface. The general form of a quadratic (second degree) polynomial is illustrated by the equation for two x-variables

$$y_u = \beta_0 + \beta_1 x_{1u} + \beta_2 x_{2u} + \beta_{11} x_{1u}^2 + \beta_{22} x_{2u}^2 + \beta_{12} x_{1u} x_{2u} + e_u$$

The surface contains linear terms in x_{1u}, x_{2u}, squared terms in x_{1u}^2, x_{2u}^2, and the cross-product term $x_{1u}x_{2u}$.

In order to estimate the regression coefficients in this model, each variable x_{iu} must take at least three different levels. This suggests the use of factorial designs of the 3^k series. If the three levels of any x are coded as -1, 0, 1, the second order surface is easy to fit to the results of a 3^k factorial. One disadvantage of the 3^k series, however, is that with more than three x-variables the experiments become large, although this difficulty can be reduced by the use of fractional replication. Further, the coefficients β_{11}, β_{22} of the squared terms are estimated with relatively low precision from a 3^k factorial, as Box and Wilson (8A.6) have pointed out.

These authors (8A.6) developed new designs specifically for fitting second order response surfaces. Their earliest designs, called *composite designs,* are constructed by adding further treatment combinations to those obtained from a 2^k factorial. If the coded levels of each x-variable are -1 and $+1$ in the 2^k factorial, the new designs are as follows.

Central composite designs. Test the $(2k + 1)$ additional factor combinations

$$
\begin{array}{llll}
(0, 0, \cdots, 0); & (-\alpha, 0, \cdots, 0); & (\alpha, 0, \cdots, 0); & \\
 & (0, -\alpha, \cdots, 0); & (0, \alpha, \cdots, 0); & \\
 & \cdots & (0, 0, \cdots, -\alpha); & (0, 0, \cdots, \alpha)
\end{array}
$$

The total number of treatment combinations is $(2^k + 2k + 1)$. For 2, 3, and 4 x-variables, the experiment requires 9, 15, and 25 units respectively, as compared with 9, 27, and 81 in the 3^k series. The value of α can be chosen to make the regression coefficients orthogonal to one another, or to minimize the bias that is created if the true form of the response surface is not quadratic, or to give the design the property of being rotatable (section 8A.22).

The central composite design can be fitted into a sequential program of experimentation. The program starts with an exploratory 2^k factorial to which a linear response surface is fitted. Suppose, however, that the center of the first experiment is actually close to a point of maximum response. The results of the first experiment may indicate, by the "lack of fit" terms, that the response surface is curved, and may suggest that the center of the experiment is close to a maximum. By adding the $(2k + 1)$ treatment combinations needed to create a central composite design, the experimenter utilizes the results of the initial 2^k experiment over again when he fits the quadratic surface. This procedure requires the assumption that there are no time trends which change the level of y between the first and the second experiment. When time trends are present, the designs in section 8A.27 are preferable.

The results of the 2^k factorial may suggest that the point of maximum response is nearer to one of the other factor combinations than to the center. In this event a non-central composite design is likely to give a more precise location of the maximum.

Non-central composite designs. This design has k extra points, one for each factor. The level of any factor is moved to $(1 + \alpha)$ if the optimum level seems nearer to 1 than to -1, and to $(-1 - \alpha)$ if the optimum level seems nearer to -1 than to 1. For instance, if the optimum is thought to be near $(1, -1, 1)$ with three factors, the extra combinations are as follows:

$$(1 + \alpha, -1, 1); \qquad (1, -1 - \alpha, 1); \qquad (1, -1, 1 + \alpha)$$

Figure 8A.1 illustrates the two types of composite design for three factors.

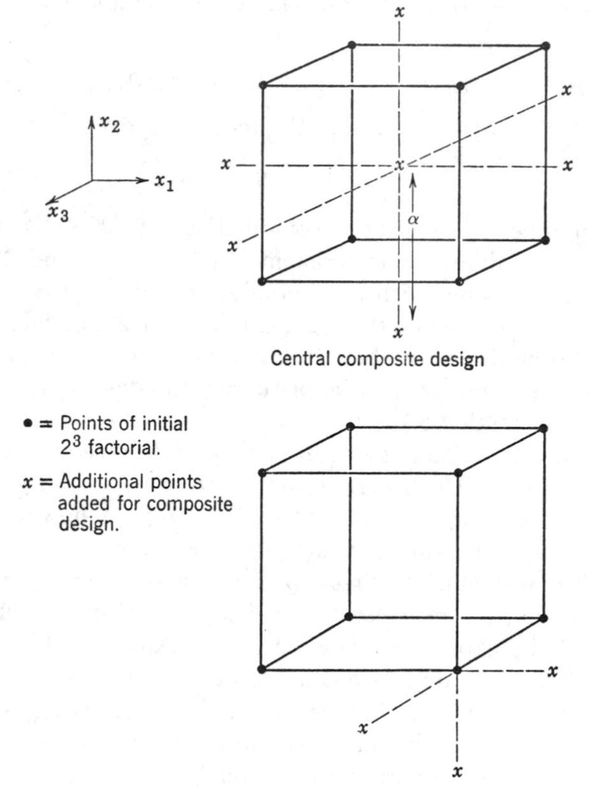

FIG. 8A.1

8A.22 Rotatable Second Order Designs. In order to appraise different designs that may be proposed for fitting a second order response surface, some criteria are needed as to what constitutes a good design. One useful property is, of course, that the computations should not be too difficult. In a more intensive consideration of desirable properties, Box and Hunter (8A.7) proposed the criterion of *rotatability*.

To explain this concept, suppose that the point $(0, 0, \cdots, 0)$ represents the center of the region in which the relation between y and the x's is under investigation. Let

$$\hat{y}_u = b_0 + \sum_{i=1}^{k} b_i x_{iu} + \sum_{i=1}^{k} b_{ii} x_{iu}^2 + \sum_{i<j} b_{ij} x_{iu} x_{ju}$$

be the estimated response at a point on the fitted second order surface. From the results of any experiment we can compute $s(\hat{y}_u)$, the standard error of $\hat{y}_u$, at any point on the fitted surface. This standard error will be a function of the coordinates x_{iu} of the point. In a rotatable design, this standard error is the same for all points that are at the same distance ρ from the center of the region, i.e., for all points for which

$$x_{1u}^2 + x_{2u}^2 + \cdots + x_{ku}^2 = \rho^2 = \text{constant}$$

This property is a reasonable one to adopt for exploratory work, in which the experimenter does not know in advance how the response surface will orient itself with respect to the x-axes. Consequently he has no rational basis for specifying that the standard error of $\hat{y}_u$ should be smaller in some directions than in others.

With two x-variables, Box and Hunter showed that a rotatable design is obtained by making tests at n_2 points equally spaced around the circumference of a circle in the x_1, x_2 plane with center $(0, 0)$, plus one or more tests at the center itself. The points on the circumference lie at the vertices of a regular polygon inscribed in the circle. Since there are six regression coefficients to be determined when $k = 2$, the smallest design consists of a pentagon, i.e., five points around the circumference, plus one point at the center. The general formulae for the actual values of x_1 and x_2 may be written as in table 8A.5. The factor α is the radius of the circle on which the points lie.

The replicated points at the center have two purposes. They provide $(n_1 - 1)$ d.f. for estimating the experimental error, and they determine the precision of $\hat{y}$ at and near the center. If there are many replications of the center point, the standard error of $\hat{y}$ is low at the center and increases rapidly as we move away from the center. With only one or two center points, on the other hand, the standard error of $\hat{y}$ may be greater in the center than at points like $(1, 0)$ or $(0, 1)$. As a compromise, Box and Hunter suggest that the number of center points be chosen so that the standard error of $\hat{y}$ is approximately the same at the

TABLE 8A.5 ROTATABLE SECOND ORDER DESIGNS IN TWO x-VARIABLES

x_1	x_2	
$\alpha \cos \theta$	$\alpha \sin \theta$	
$\alpha \cos 2\theta$	$\alpha \sin 2\theta$	
.	.	n_2 points on the circumference
.	.	$\theta = 2\pi/n_2$
.	.	
$\alpha \cos n_2\theta$	$\alpha \sin n_2\theta$	
0	0	
0	0	
.	.	n_1 points at the center
.	.	
.	.	
0	0	

center as at all points on the circle with radius 1. If this choice is made, the standard error remains roughly the same at all points within the circle of radius 1.

With three or more x-variables, rotatable designs are obtained by placing the points $(x_1, x_2, \cdots x_k)$ at the vertices of certain of the regular solid figures, or combinations of two solid figures, plus points at the center. The non-central points do not always lie on a sphere.

8A.23 Central Composite Rotatable Designs. The rotatable designs most likely to be useful in practice belong to a series that are also central

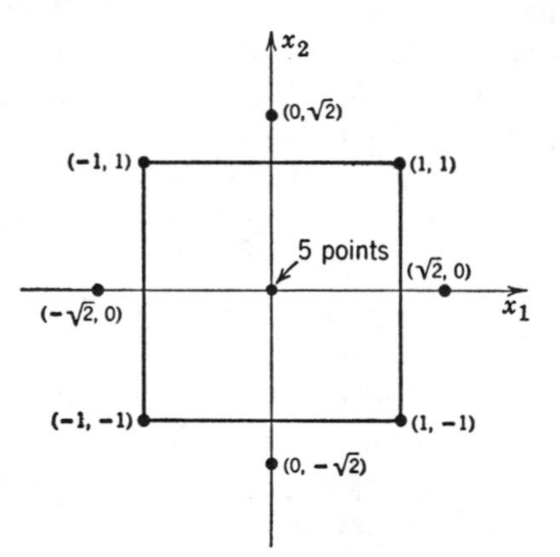

FIG. 8A.2 Central composite rotatable design in two x-variables.

composite designs. Figure 8A.2 shows the design in two x-variables, obtained from table 8A.5 by putting $n_2 = 8$, $n_1 = 5$, $\alpha = \sqrt{2}$. The design may be subdivided into three parts.

(i) The four points $(-1, -1)$, $(1, -1)$, $(-1, 1)$, and $(1, 1)$ constitute a 2^2 factorial.

(ii) The four points $(-\sqrt{2}, 0)$, $(\sqrt{2}, 0)$, $(0, -\sqrt{2})$, $(0, \sqrt{2})$ are the extra points included to form a central composite design with $\alpha = \sqrt{2}$. The figure formed by these points is called a *star*.

(iii) Five points are added at the center to give roughly equal precision for $\hat{y}$ within a circle of radius 1.

Rotatable designs for any number k of x-variables can be built up from these three components. The value of α must be $2^{k/4}$ in order to make the design rotatable. Table 8A.6 shows the components of the

TABLE 8A.6 COMPONENTS OF CENTRAL COMPOSITE SECOND ORDER ROTATABLE DESIGNS

No. of x-variables k	Number of points in 2^k factorial	Star	Center	Total N	Value of α
3	8	6	6	20	1.682
4	16	8	7	31	2.000
5	16	10	6	32	2.000
6	32	12	9	53	2.378

design for $k = 3, 4, 5, 6$. Note that with 5 and 6 x-variables, the size of the experiment is reduced by using a half-replicate of the 2^k factorial. With a half-replicate, α becomes $2^{(k-1)/4}$. Plans 8A.1 to 8A.3 at the end of this chapter give the designs for $k = 4, 5, 6$. The plans for $k = 2, 3$ are presented in the following sections. These plans assume that the experiment is to be completely randomized. If the different treatment combinations are applied one after another, the order in this sequence should be randomized.

8A.24 Statistical Analysis. The design for two x-variables is shown in table 8A.7. The columns headed x_1 and x_2, which specify the actual combinations to be used, constitute the plan for the experiment. Subsequent steps are as follows.

1. Complete the columns headed x_0, ${x_1}^2$, ${x_2}^2$ and x_1x_2 as shown. The two-way array, with 6 columns and 13 rows, comprises the X matrix of the x-variables. The corresponding values of the response y are placed on the right.

2. Form the sum of products of each column in the X matrix with the column of y values. These sums of products are denoted by $(0y)$, $(1y)$, $(2y)$ and so on.

(It is not necessary to compute the $X'X$ matrix of sums of squares and products of the x-variables. However, this matrix is shown in table 8A.7 in order to exhibit its structure, which remains the same for any value of k. Note that the coefficients b_1, b_2 of the linear terms and the coefficient b_{12} of the cross-product term are orthogonal to all the others. The only coefficients which are entangled are b_0 and the coefficients b_{11}, b_{22} of the squared terms.)

3. From the values of $(0y)$, $(1y)$, etc., the regression coefficients are computed directly by the equations given at the foot of table 8A.7. The only auxiliary quantity needed is $\sum(iiy)$, found by adding the cross-products of all the squared terms with y.

TABLE 8A.7 CENTRAL COMPOSITE ROTATABLE DESIGN IN TWO x-VARIABLES

	X = Matrix of x-variables Design					Y
x_0	x_1	x_2	x_1^2	x_2^2	x_1x_2	y
1	−1	−1	1	1	1	
1	1	−1	1	1	−1	
1	−1	1	1	1	−1	
1	1	1	1	1	1	
1	−1.414	0	2	0	0	
1	1.414	0	2	0	0	
1	0	−1.414	0	2	0	
1	0	1.414	0	2	0	
1	0	0	0	0	0	
1	0	0	0	0	0	
1	0	0	0	0	0	
1	0	0	0	0	0	
1	0	0	0	0	0	

$X'X$						$X'Y$
13	0	0	8	8	0	$(0y)$
0	8	0	0	0	0	$(1y)$
0	0	8	0	0	0	$(2y)$
8	0	0	12	4	0	$(11y)$
8	0	0	4	12	0	$(22y)$
0	0	0	0	0	4	$(12y)$

Solutions

$$b_0 = 0.2(0y) - 0.1 \sum (iiy);\ \text{where} \sum (iiy) = (11y) + (22y)$$

$$b_i = 0.125(iy)$$

$$b_{ii} = 0.125(iiy) + 0.01875 \sum (iiy) - 0.1\,(0y)$$

$$b_{ij} = 0.25(ijy)$$

4. In the analysis of variance, it is usually of interest to partition the sum of squares of the y's into the contribution due to a first order (linear) equation, the additional contribution due to the second order (quadratic) terms, a 'lack of fit' component which measures the deviations of the responses from the fitted surface, and finally a measure of the experimental errors, obtained from the replicated points at the center. General formulae for the sums of squares are as follows:

	Sums of squares	d.f.
First order terms:	$\sum_{i=1}^{k} b_i(iy)$	k
Second order terms:	$b_0(0y) + \sum_{i=1}^{k} b_{ii}(iiy) + \sum_{i<j} b_{ij}(ijy) - G^2/N$	$\frac{k(k+1)}{2}$
Lack of fit:	Found by subtraction	$n_2 - \frac{k(k+3)}{2}$
Experimental error:	$\sum (y_{1u} - \bar{y}_1)^2$	$n_1 - 1$
Total:	$\sum_{u=1}^{N} y_u^2 - G^2/N$	$n_1 + n_2 - 1$

In these formulae G is the grand total, N the number of points, and the y_{1u} represent the responses at the central points, with mean $\bar{y}_1$.

The multiplying factors which give the regression coefficients (foot of table 8A.7) also provide their standard errors. For $k = 2$

$$\text{s.e. }(b_i) = s\sqrt{0.125} = 0.354s$$

$$\text{s.e. }(b_{ii}) = s\sqrt{0.125 + 0.01875} = 0.379s$$

$$\text{s.e. }(b_{ij}) = s\sqrt{0.25} = 0.5s$$

where s is the standard error per observation. Note that in the case of b_{ii}, two factors are added to give the multiplier for the standard error. These relations between the standard errors and the multipliers hold for all the solutions given in this book.

8A.25 Numerical Example of the Analysis for 3 x-variables. The data come from a larger investigation by Moore *et al.* (8A.8) on the effects of the minor elements copper (Cu), molybdenum (Mo), and iron (Fe) on the growth of lettuce in water culture. The statistical analysis has been described by Hader *et al.* (8A.9). Table 8A.8 shows the design, the X matrix, and the lettuce yields y.

A preliminary step in any experiment of this type is to set up the relations between the coded x-scales and the original scales in which the levels are recorded. For Cu and Mo, the experiment was intended to explore the range of concentrations between 0.0002 and 2 parts per million (ppm). In the design scale (table 8A.8), the lowest and highest

TABLE 8A.8 CENTRAL COMPOSITE ROTATABLE DESIGN FOR $k = 3$

X = Matrix of x-variables; Y

	Design									
x_0	x_1	x_2	x_3	x_1^2	x_2^2	x_3^2	x_1x_2	x_1x_3	x_2x_3	y
1	−1	−1	−1	1	1	1	1	1	1	16.44
1	1	−1	−1	1	1	1	−1	−1	1	12.50
1	−1	1	−1	1	1	1	−1	1	−1	16.10
1	1	1	−1	1	1	1	1	−1	−1	6.92
1	−1	−1	1	1	1	1	1	−1	−1	14.90
1	1	−1	1	1	1	1	−1	1	−1	7.83
1	−1	1	1	1	1	1	−1	−1	1	19.90
1	1	1	1	1	1	1	1	1	1	4.68
1	−1.682	0	0	2.828	0	0	0	0	0	17.65
1	1.682	0	0	2.828	0	0	0	0	0	0.20
1	0	−1.682	0	0	2.828	0	0	0	0	25.39
1	0	1.682	0	0	2.828	0	0	0	0	18.16
1	0	0	−1.682	0	0	2.828	0	0	0	7.37
1	0	0	1.682	0	0	2.828	0	0	0	11.99
1	0	0	0	0	0	0	0	0	0	22.22
1	0	0	0	0	0	0	0	0	0	19.49
1	0	0	0	0	0	0	0	0	0	22.76
1	0	0	0	0	0	0	0	0	0	24.27
1	0	0	0	0	0	0	0	0	0	27.88
1	0	0	0	0	0	0	0	0	0	27.53
										324.18

Solutions

$$b_0 = 0.166338(0y) - 0.056791 \sum (iiy)$$

$$b_i = 0.073224(iy)$$

$$b_{ii} = 0.062500(iiy) + 0.006889 \sum (iiy) - 0.056791(0y)$$

$$b_{ij} = 0.125000(ijy)$$

values of x are -1.682 and 1.682. Rounding slightly, we take

$$x = -1.68 \text{ when conc.} = 0.0002; \qquad x = 1.68 \text{ when conc.} = 2.$$

Further, it was decided to express y as a quadratic function of the *log* concentration rather than of the concentration itself. Write

$$x = a + b \log (\text{conc.})$$

where a and b are chosen to satisfy the desired conditions at the ends of the scale. We find, for Cu and Mo,

$$x = 1.4272 + 0.84 \log \text{(conc.)}$$

From this equation the concentration levels corresponding to $x = -1$, 0, and 1 are determined. For Fe the range in concentration was from 0.0025 to 25 ppm, and the equation is

$$x = 0.5057 + 0.84 \log \text{(conc.)}$$

In the original experiment, the Fe concentration for $x = -1$ was set wrongly owing to a computational error, so that the actual design was not quite rotatable and required a more complex analysis. To avoid this complication, estimated yields for the correct concentration have been used here for the four observations that were affected by this error.

The statistical analysis follows the procedure outlined in the previous section. Each column of the X matrix is multiplied in turn by the Y column, giving the sums of products $(0y)$, $(1y)$, etc., shown in table 8A.9. The auxiliary total $\sum(iiy)$ is computed, and the regression co-

TABLE 8A.9 REGRESSION COEFFICIENTS AND ANALYSIS OF VARIANCE

$(0y)$	324.180	b_0	24.0401
$(1y)$	−64.761	b_1	−4.7420
$(2y)$	−16.231	b_2	−1.1885
$(3y)$	3.121	b_3	0.2285
$(11y)$	149.750	b_{11}	−5.4262
$(22y)$	222.429	b_{22}	−0.8837
$(33y)$	154.020	b_{33}	−5.1593
$(12y)$	−13.390	b_{12}	−1.6738
$(13y)$	−9.170	b_{13}	−1.1462
$(23y)$	7.770	b_{23}	−0.9712
$\sum (iiy)$	526.199		

	d.f.	s.s.	m.s.
First order terms	3	327.1	109.0
Second order terms	6	760.3	126.7
Lack of fit	5	40.1	8.0
Error	5	52.6	10.5
Total	19	1180.1	

efficients b_0, b_1, etc. are found by the equations given in the solutions at the foot of table 8A.8.

The sums of squares for the first and second order terms in the analysis of variance (foot of table 8A.9) are obtained from the general formulae in section 8A.24 as follows:

First order terms:

$$\sum_{i=1}^{k} b_i(iy) = (-64.761)(-4.7420) + (-16.231)(-1.1885) + (3.121)(0.2285) = 327.1$$

Second order terms:

$$b_0(0y) + \sum_{i=1}^{k} b_{ii}(iiy) + \sum_{i<j} b_{ij}(ijy) - G^2/N$$

$$= (324.180)(24.0401) + (149.750)(-5.4262) + \cdots + (7.770)(-0.9712) - (324.18)^2/20 = 760.3$$

The total sum of squares and the error sum of squares (from the 6 central points) are computed in the usual way. The sum of squares for lack of fit is obtained by subtraction. Since the mean square for lack of fit is about the same size as the error mean square, a second order surface appears to be adequate. Both the first and second order terms give significant mean squares.

8A.26 Examination of the Fitted Surface. If there are only two x-variables, the interpretation of the results is facilitated by plotting the contours of equal response. To draw a given contour, $\hat{y}$ is set equal to the appropriate response level. For each of a set of values of x_1, the fitted equation is solved as a quadratic in x_2. This gives a series of points (x_1, x_2) that lie on the contour. Alternatively, $\hat{y}$ can be calculated from the fitted equation for each of a grid of values of x_1 and x_2 which cover the experimental region. The series of contours are then drawn in freehand.

With three x-variables, a three-dimensional representation of the contours is illuminating. Choose one of the variates, say x_3, as the "vertical" variate. For each of a series of equally spaced values of x_3, draw the x_1, x_2 contours on transparent paper, using a separate sheet of paper for each value of x_3. These sheets are then placed one above the other on a vertical frame, at heights above the base which represent the values of x_3. A fuller description of this type of model, with a photograph, is given by Box (8A.13).

If there are more than three x-variables, geometrical representation can be used only partially The most helpful device is to transform the response surface to its canonical form (see section 8A.55). The interpretation of the canonical form is discussed, with examples, in references (8A.10), (8A.13), and (8A.14). Reference (8A.13) contains an experiment with five x-variables.

8A.27 Designs in Incomplete Blocks. Arrangements of the central composite second order designs in incomplete blocks have been given by DeBaun (8A.18) and Box and Hunter (8A.7). The design for two x-variables (table 8A.10) shows the typical features. The 2^k factorial,

TABLE 8A.10 CENTRAL COMPOSITE SECOND ORDER DESIGNS IN INCOMPLETE BLOCKS

2 x-variables $N = 12$ treatment combinations

Block I		Block II	
x_1	x_2	x_1	x_2
−1	−1	−1.414	0
1	−1	1.414	0
−1	1	0	−1.414
1	1	0	1.414
0	0	0	0
0	0	0	0

Solutions

$$b_0 = 0.25(0y) - 0.125 \sum (iiy)$$

$$b_i = 0.125(iy)$$

$$b_{ii} = 0.125(iiy) + 0.03125 \sum (iiy) - 0.125(0y)$$

$$b_{ij} = 0.25(ijy)$$

plus some central points, forms one block, or sometimes two blocks in the larger designs. The star part of the design, plus some central points, forms an additional block. Plans 8A.4 to 8A.7 give designs for 3 to 6 x-variables.

As usual, the blocking provides the opportunity for obtaining increased precision. If the treatment combinations are tested in sequence, blocking enables part of the effect of time trends to be eliminated from the experimental errors. In particular, if the initial experiment is a first order design, the 2^k factorial should be set out as in the blocked plans. The augmented part required to fit a second order surface can be added later without worrying about whether time trends have changed the level of y.

In the designs shown, blocks are orthogonal to the coefficients in the fitted quadratic surface. This simplifies the analysis. In order that the orthogonality condition hold, the average value of $\hat{y}$ must be the same in every block. In table 8A.10, for instance, it may be verified that in both blocks I and II the average value of $\hat{y}$ is

$$b_0 + \tfrac{2}{3}b_{11} + \tfrac{2}{3}b_{22}$$

The orthogonality condition determines the value of α in the star part of the design. Thus, in table 8A.10, if the non-zero values of x_1 and x_2 are taken as $-\alpha$ and α in block II, the mean value of $\hat{y}$ in this block works out as

$$b_0 + \frac{\alpha^2}{3}b_{11} + \frac{\alpha^2}{3}b_{22}$$

In order to agree with the mean value of block I, we must take $\alpha^2 = 2$, or $\alpha = 1.414$. In some of the designs, the value of α required for orthogonality differs slightly from that required for rotatability, so that the rotatable feature must be sacrificed to some extent.

The computation of the coefficients in the quadratic surface remains essentially the same as in section 8A.24. Write down the X matrix and the Y vector, ignoring the blocks. The sums of products $(0y)$, $(1y)$, etc. are then formed and the regression coefficients are calculated from the equations given at the foot of each plan.

In the analysis of variance, the sums of squares for the first order and second order terms are found by the formulae in section 8A.24. The blocks sum of squares is computed in the usual way as

$$\sum \frac{B_i^2}{\gamma_i} - \frac{G^2}{N}$$

where γ_i is the number of units in the ith block. (In some plans, γ_i differs from block to block.) The sum of squares for experimental error is the pooled sum of squares between central points *in the same block.* In table 8A.10, for instance, this sum of squares has 2 d.f., one from each block. The sum of squares for lack of fit is found by subtraction.

8A.3 Methods for Determining the Optimum Combination of Factor Levels

8A.31 Nature of the Problem. In recent years, progress has been made on a problem of great importance in applied research. A response y is thought to be affected by a number of quantitative factors x_1, $x_2 \cdots x_k$. An experimental program is undertaken so as to discover the level at which each of these factors must be set in order to maximize the response. How should the program be conducted? This problem is common in research and developmental work in industry, where the investigator is seeking ways to manufacture a new product with certain desired characteristics, or to obtain an old product more economically. Examples are found in other fields also, as for instance the search for the best nutrient medium in which an organism can multiply in the biological laboratory.

The response may be the actual amount of product or a measure of the quality of the product. Sometimes the object is to minimize y, as when y represents the amount of an undesirable by-product or the cost of manufacture per unit of output. The problem of locating a minimum does not require separate discussion, since a change in the sign of the response turns the problem into one of maximization.

In addition to locating the maximum of y, it is of value to learn something about how y varies, in the neighborhood of the maximum, when the levels of the factors are changed from their optimum values. This is so for several reasons.

(*i*) In large-scale application of the results, it may not be feasible to set every factor at exactly its optimum level. Some combination of levels different from the optimum may be cheaper to maintain.

(ii) Frequently, more than one characteristic of the product is important. A shift in the x_i away from the optimum may be preferable because of its effect on one of these other characteristics.

(iii) The shape of the response surface near the optimum may provide clues to the nature of the underlying process, as illustrated by Box and Voule (8A.10).

(iv) The surface may not have a true maximum in the region of experimentation (or in the region in which the process can be carried on in practice). In this event we want to know the nature of the surface in regions of relatively high response.

The location of an optimum usually requires a coordinated series of experiments. In the following sections, some methods for planning the series are described. In practical applications, each method demands considerable exercise of judgment by the investigator, and the methods cannot be presented as a set of inflexible rules. Moreover, since the problem has been investigated only recently, future work may result in modification of the methods or in the development of new methods.

At the outset, the experimenter must have decided which factors are to be included in the program. Sometimes there are initially as many as a dozen or more factors that might influence the response. The methods to be described have been applied mostly for numbers of factors between two and six, and become cumbersome for larger numbers of factors. Some preliminary weeding out of factors that seem likely to be of minor importance is therefore advisable before starting the methods described here. The range within which the level of each factor is to be varied must also have been selected.

8A.4 The Single-factor Method

8A.41 Description. A good account of this method has been given by Friedman and Savage (8A.11). The experimenter first makes an advance estimate of the optimum combination of factor levels, which will be denoted by $x_{11}, x_{21}, \cdots x_{k1}$. Since each experiment deals with only one factor, the factors must be arranged in the order in which they are to be experimented upon. In general, a good plan is to start with the factor that is expected to give the largest response.

In the first experiment all factors except the first are held constant at their initial levels $x_{21}, x_{31}, \cdots x_{k1}$. The purpose of this experiment is to find the level of x_1 which maximizes y for this particular combination of levels of the other factors. There are various ways in which this experiment can be carried out. In order to establish a maximum, at least three levels of x_1 must be compared. Four or five levels may be advisable if the range of x_1 is wide and the position of its optimum is poorly known; or perhaps an initial experiment with five widely spaced levels, followed by a second experiment with three levels at a narrower spacing. Hotelling (8A.12) has shown that equal spacing of the levels is not in general the most efficient procedure for locating the optimum. For experiments with three levels he has worked out the most efficient spacing in situations where the position of the optimum x_1 is fairly well delimited in advance.

When it is clear that the experimental points straddle a maximum of y, the position of the maximum may be estimated by fitting a parabola to the observed responses.

If the equation of the fitted parabola is

$$Y = b_0 + b_1x_1 + b_2x_1^2$$

with b_2 negative, the optimum x_1 is known from calculus to be

$$x_{12} = -\frac{b_1}{2b_2}$$

Sometimes the value of x_1 that gives the highest observed y is close enough to the estimated optimum. The optimum, however estimated, will be denoted by x_{12}.

In the second experiment, x_1 is fixed at the level x_{12} and the values of $x_3 \cdots x_k$ remain fixed at their initial levels. Several levels of the second factor are compared in order to find the optimum level of x_2, say x_{22}. The third experiment deals similarly with x_3, the other factors being held at the levels x_{12}, x_{22}, x_{41}, $\cdots$ x_{k1}, respectively. The process continues until an optimum has been estimated in turn for the k factors.

This ends the first round of experiments, the estimated optimum factor combination at this point being the set of levels x_{12}, x_{22}, $\cdots$ x_{k2}. If this new set of levels is close to the initial set x_{11}, x_{21}, $\cdots$ x_{k1}, and if the values of y have improved little during the first round, the experimenter may decide to terminate the experiments, concluding that no appreciable gain over the initial estimate of the optimum is possible. If more substantial changes have occurred, a second round of the whole process is undertaken, starting with fixed levels x_{22}, x_{32}, $\cdots$ x_{k2} and testing several levels of x_1 to determine whether the position of the optimum has changed from the level x_{12}. At the end of this round a new set of estimates x_{13}, x_{23}, $\cdots$ x_{k3} has been obtained. The experimenter then surveys the situation once again, and decides whether to terminate or continue to a third round.

Several variations may be introduced into the method. At the end of the second round, Friedman and Savage suggest moving along the vector defined by the ends of the first two rounds. This means that at the beginning of the third round all x-variables are changed simultaneously, the change in x_i being proportional to $(x_{i3} - x_{i2})$. This suggestion, which departs from the "single-factor" approach, resembles the "steepest ascent" method to be described next. It may be decided to omit some of the factors in later rounds, on the grounds that they have shown no effects in earlier rounds.

This method does not provide an estimate of the shape of the response curve. When the process has terminated, it will be advisable to conduct additional experiments for this purpose.

8A.5 The Method of Steepest Ascent

8A.51 The First Experiment. This method was proposed in 1951 by Box and Wilson (8A.6) and has been further developed and expounded by Box and others (8A.10, 8A.13, 8A.14, 8A.15). As in the single-factor approach, the maximum is located by means of a series of experiments, each planned from the results of the preceding ones. The results of any experiment must become known quickly if the method is to be expeditious. Further, like the single-factor approach, the method works best when experimental errors are small and a previous estimate of the error variance is known, because under these circumstances the individual experiments in the series can be kept moderate in size.

All factors are included in the initial experiments. At the end of each experiment a polynomial approximation to the response surface $\phi(x_1, x_2, \cdots x_k)$ is fitted to the results and is used to determine the nature of the next experiment. The first and second order designs described in sections 8A.15 and 8A.2 were developed in connection with this method.

The first experiment has two purposes.

(i) To fit a linear equation

$$\hat{y} = b_0 + b_1x_1 + b_2x_2 + \cdots + b_kx_k$$

as an approximation to ϕ in the vicinity of the starting point.

(ii) To test whether the linear approximation fits within the limits of experimental errors.

The 2^k factorial or fractional factorial designs described in section 8A.15 are useful for this purpose. Note that the experiment must provide some d.f. to measure the lack of fit, and that an estimate of experimental error is required. Unless there is a trustworthy external estimate of experimental error, the experiment itself must supply this, either by the addition of points at the center as mentioned in section 8A.15 or by replication of all points.

The artificial data to which a linear equation was fitted in section 8A.13 will serve as an example. There are three factors, the time during which a reaction is run, the temperature at which the reaction is maintained, and the amount C of a chemical substance. The best initial estimate of the optimum combination is, say,

Time: 6 hours; Temperature: 300°; Amount of C: 1.6 gm.

The next step is to decide the amount by which each of these factors will be varied in the first 2^3 experiment. A rough guiding rule is to change the level of a factor by an amount that (i) allows the effect of the factor a chance to show up, and (ii) is not large enough to cause doubts that the response will be curved in the interval between the two levels. Errors in these choices can be corrected by later experiments: a good choice means the optimum is attained more quickly.

In the example, the investigator decides that the time of reaction can be changed by 1 hour, the temperature by 80°, and the amount of C by 0.4 gm. Taking the initial factor combination as the *center* of the set of levels to be tested, the first experiment consists of the 8 combinations of the following factor levels:

x_1	Time	x_2	Temp.	x_3	Amt. of C
-1	5.5 hr.	-1	260°	-1	1.4 gm.
$+1$	6.5 hr.	$+1$	340°	$+1$	1.8 gm.

Notice that coded levels -1 and $+1$ have been assigned to each of the three x-variables. In section 8A.13 the fitted linear equation in the coded scale was found to be

$$\hat{y} = 170.2 + 11.8x_1 - 5.2x_2 + 4.0x_3$$

The experimental data did not provide any estimate of the experimental error variance. We shall suppose that a good external estimate of this variance was available, and that the mean square for lack of fit appeared to be well within the limits of experimental error.

8A.52 The Direction of Steepest Ascent. When the first experiment has been completed, the region of experimentation is shifted to another set of levels of the x's. This set is to be chosen so that the maximum expected increase in y occurs. If the center of the first experiment is taken as the origin, the problem is to move from the origin, with x-coordinates $(0, 0, \cdots 0)$, to the position P, with coordinates $(x'_1, x'_2, \cdots x'_k)$, so that the response $\phi(x'_1, x'_2, \cdots x'_k)$ is maximized.

The change in ϕ clearly depends on the size of the jump that is made from 0 to the point P, i.e., on whether we change the x's by small or by large amounts. By geometrical analogy, the "distance" r from 0 to P is defined as

$$r = \sqrt{x'^2_1 + x'^2_2 + \cdots + x'^2_k} \tag{1}$$

We suppose that the value of r has been chosen by the experimenter: the issues affecting this choice will be discussed later. For a given value of r, i.e., a given size of jump, we determine the values of the x_i that maximize $\phi(x'_1, x'_2, \cdots x'_k)$. The path of steepest ascent is the path from 0 to the point P whose coordinates are these maximizing x_i'.

By calculus, the maximizing values of the x_i are given by the equations

$$x_i = \mu \frac{\partial \phi}{\partial x_i}$$

where $\partial\phi/\partial x_i$ is the partial derivative of ϕ with respect to x_i, taken at the point P, and μ is a multiplier calculated so as to satisfy equation (1). The actual value of μ is $r \Big/ \sqrt{\Sigma \left(\dfrac{\partial \phi}{\partial x_i}\right)^2}$.

In order to apply this method, we must estimate the partial derivatives $\partial\phi/\partial x_i$ from the results of the first experiment. This task is simplified if ϕ can be satisfactorily represented by a linear equation

$$\phi = \phi_0 + b_1x_1 + b_2x_2 + \cdots + b_kx_k$$

where ϕ_0 is the value of ϕ at the origin.

From the linear function, it follows that

$$\frac{\partial \phi}{\partial x_i} = b_i$$

Thus the change in x_i in the path of steepest ascent is proportional to b_i, provided that the linear equation holds.

The calculation of the path of steepest ascent is set out for the numerical example in table 8A.11. From the linear equation we have

$$b_1 = +11.8; \qquad b_2 = -5.2; \qquad b_3 = +4.0$$

in coded levels.

TABLE 8A.11 CALCULATION OF PATH OF STEEPEST ASCENT

	x_1 Time (hr.)	x_2 Temp. (°)	x_3 C (gm.)	
(1) Relative change in design units = b_i	+11.8	−5.2	+4.0	
(2) No. of original units = 1 design unit	0.5	40.0	0.2	
(3) Relative change in original units	+5.9	−208.0	+0.8	
(4) Change per 1 hr. change in Time	1.0	−35.3	+0.14	
Path of steepest ascent				Predicted $\hat{y}$
(5) Initial levels	6.0	300	1.60	
(6) Levels on path	7.0	265	1.74	201
(7) " " "	8.0	229	1.88	232
(8) " " "	9.0	194	2.02	263

Row (1) shows the relative changes required in coded units. In rows (2) and (3), these are converted into relative changes in the *original* units. For Time, one coded unit represents 0.5 hr., so that 11.8 coded units represent

$$(11.8)(0.5) = 5.9 \text{ hr.}$$

as in row (3). At this point it is convenient to select one of the variables as a standard, say Time, and to compute the changes in the other variables that correspond to 1 hr. change in Time, as shown in row (4).

To obtain the path of steepest ascent, start with the initial levels which formed the center of the first experiment. For each hour by which Time is increased, the Temperature should be decreased by 35.3°, and the amount of C increased by 0.14 gm. Rows (6) to (8) show the factor combinations on the path of steepest ascent for Times of 7, 8, and 9 hr.

The expected yield $\hat{y}$ at each point on the path of steepest ascent can be predicted from the linear equation, by expressing the x-levels on the path in coded units and substituting in the equation. These yields are shown in table 8A.11.

8A.53 Subsequent Steps. The next step is to test a single factorial combination at some point on the path of steepest ascent, in order to discover whether the predicted increase in y actually takes place. The size of the jump that is made is a matter of judgment. In the example, a jump to the point (7, 265, 1.74) in the original units would probably be regarded as too small, since the first experiment contained a test at the point (6.5, 260, 1.8). The new test could be made either at the point (8, 229, 1.88) or at the point (9, 194, 2.02). The danger in making too large a jump is that it may pass beyond the optimum factor combination. In cases of doubt it is best to choose the smaller of two proposed jumps, since at the worst this involves merely testing an extra factor combination that could have been omitted.

If the actual yield in the new experiment is close to the predicted yield, a further jump is made along the same path. The process is continued until the actual yield is substantially different from the predicted yield.

At this point a new 2^k factorial experiment is conducted, with its center at the last point on the path. If a linear equation is still a satisfactory fit to the new data, a new path of steepest ascent is calculated, and tests are made along this new path.

In the course of time a situation is reached in which the 2^k factorial gives one of the following results:

(i) The linear equation still appears to fit, but all coefficients b_i are small. This suggests that we have reached a plateau.

(ii) The lack of fit terms show that the linear approximation is inadequate. This suggests that we are in a region in which the curvature of the surface must be taken into account.

For further exploration of cases (i) and (ii), second order designs in which ϕ is approximated by a quadratic function of the x's are used. Frequently the second order design can be constructed by adding additional points to the last 2^k factorial that was conducted, as in the rotatable second order designs (blocked or completely randomized) described in sections 8A.27 and 8A.23. Note that these designs provide a measure of the lack of fit of the quadratic surface and an estimate of the experimental error variance.

Situations (i) and (ii) can, of course, arise in the first experiment. In fact, situation (i) is to be anticipated in the first experiment if the initial estimate of the optimum is a good one and the optimum is flat. In this event we proceed at once to a design that permits the calculation of quadratic effects, without attempting to compute a path of steepest ascent from the first experiment.

8A.54 Correction for Errors in Level in the First Experiment. The importance of an appropriate choice of levels in the first experiment has been mentioned. Before discussing the interpretation of a quadratic response surface, it is worth noting how to correct for a poor choice of levels of a factor.

If the levels of a factor are chosen too close together, the factor is likely to show a negligible or small effect in the first experiment, and the factor levels will be changed very slowly along the path of steepest ascent. Suppose that in table 8A.11 the levels of factor C in the first experiment had differed by 0.2 instead of by 0.4. The regression coefficient b_3 in table 8A.11 would have been about $+2.0$ instead of $+4.0$. The relative change in x_3 in original units, in row (3) of this table, would have been $(+2.0)(0.1)$, or $+0.2$ instead of 0.8. A halving of the interval between levels reduces the relative change along the path of steepest ascent to $\frac{1}{4}$ of its original value. Thus too small an interval between levels may lead the experimenter to conclude that the factor is unimportant and to omit it from consideration.

When a factor shows little effect in the first experiment and it is suspected that the change in levels was too small, a much greater change in levels should be made in subsequent trials than the path of steepest ascent allows. If the factor actually produces no effect, this should become clear in later experiments, and the factor can then be dropped. Alternatively, a single confirmatory test can be run at the end of the first experiment, keeping the other factors fixed at their best levels in this experiment, and changing the level of the suspect factor by a substantial amount.

If the distance between the levels of a factor is too large in the first experiment, so that the response to this factor is markedly curved in the interval between the two levels, the failure of the simple linear equation should be revealed by the lack of fit terms, with the consequence that the investigator proceeds quickly to experiments from which quadratic effects can be estimated. The effect on overall rate of progress should not be drastic.

8A.55 Simplification of the Quadratic Surface. When the quadratic surface has been fitted, the position of the maximum value of $\hat{y}$, if one exists, is found by differentiating with respect to each x in turn. With three factors, the surface is

$$\hat{y} = b_0 + b_1x_1 + b_2x_2 + b_3x_3 + b_{11}x_1^2 + b_{22}x_2^2 + b_{33}x_3^2 + b_{12}x_1x_2 + b_{13}x_1x_3 + b_{23}x_2x_3$$

$$\frac{\partial \hat{y}}{\partial x_1} = b_1 + 2b_{11}x_1 + b_{12}x_2 + b_{13}x_3 = 0$$

$$\frac{\partial \hat{y}}{\partial x_2} = b_2 + b_{12}x_1 + 2b_{22}x_2 + b_{23}x_3 = 0$$

$$\frac{\partial \hat{y}}{\partial x_3} = b_3 + b_{13}x_1 + b_{23}x_2 + 2b_{33}x_3 = 0$$

Denote the solutions of these equations by x_{1S}, x_{2S}, x_{3S}, respectively. These give the factor combination for which $\hat{y}$ has a maximum, or more generally a stationary, value.

For further study of the response surface, some transformations of the x-variables simplify the task. The origin of the x-coordinates can be transferred to the stationary point by making the transformations

$$x_1' = x_1 - x_{1S}; \qquad x_2' = x_2 - x_{2S}; \qquad x_3' = x_3 - x_{3S}$$

With this substitution, the linear terms disappear from the quadratic expression, which becomes, for three factors,

$$\hat{y} = \hat{y}_S + b_{11}x_1'^2 + b_{22}x_2'^2 + b_{33}x_3'^2 + b_{12}x_1'x_2' + b_{13}x_1'x_3' + b_{23}x_2'x_3'$$

where

$$\hat{y}_S = b_0 + \tfrac{1}{2}(b_1x_{1S} + b_2x_{2S} + b_3x_{3S})$$

is the estimated stationary value of $\hat{y}$.

The cross-product terms can also be made to vanish by transforming to new variables X_i that are orthogonal linear functions of the x_i. This transformation is familiar in both geometry and algebra. In geometry, the transformation amounts to rotating the coordinate axes so that the principal axes of the quadratic surface become the coordinate axes. In algebra, the transformation reduces the quadratic expression to its *canonical form*. Methods for making the transformation are described in (8A.14).

In the new variables, the response surface becomes

$$\hat{y} = \hat{y}_S + \lambda_1X_1^2 + \lambda_2X_2^2 + \lambda_3X_3^2$$

where the coefficients λ_i are functions of the b's, and the X_i are linear combinations of the x_i.

8A.56 Interpretation and Further Experimentation. The next steps in the experimental program are determined by examining the form of the quadratic surface. From experience in the chemical industry, several different situations may be encountered.

(i) The λ_i are all negative. In this event the quadratic surface has a maximum at the point determined by the equations $X_1 = X_2 = X_3 = 0$. If this point is inside the region covered by the second order experiment, we have succeeded in locating a maximum. Further work will probably be confined to a confirmatory test at the estimated maximum and perhaps some additional experiments designed to determine the function ϕ more precisely.

A pictorial representation of the quadratic surface aids in interpreting the results. If there are only two factors, a series of contour lines can be drawn on graph paper as shown in Fig. 8A.3. A given contour shows all pairs of values of x_1 and x_2 for which the response has a specified value. When λ_1 and λ_2 are negative, these contours are ellipses: the point S is the optimum combination of x_1 and x_2. Figure 8A.3(a) shows a surface in which λ_1 and λ_2 are nearly equal: in Fig. 8A.3(b) λ_2 is much smaller than λ_1. The limiting form of Fig. 8A.3(b), in which λ_2 becomes zero, is shown in Fig. 8A.3(c). In this case the response remains the same as we move along any line X_1 = constant. In particular, the maximum response can be obtained by any combination of values of x_1 and x_2 that makes $X_1 = 0$. Box has called this type of surface a *stationary ridge*.

(ii) The λ_i are all negative, but the estimated maximum lies *outside* the region covered by the last experiment. In this event the estimated maximum may not be at all close to a true maximum, because the fitted quadratic can give very misleading results outside the region from which the data came. It is advisable to proceed cautiously in the general direction of the estimated maximum. If this path seems to lead toward a turning value, another experiment to which a new quadratic surface can be fitted is conducted further along the path.

With two variables, the contours obtained from the first quadratic surface may appear as in Fig. 8A.3(d) or (e). Case (e), in which λ_2 is near zero, has been called a *rising ridge* by Box, who recommends a move along the X_2 axis in the direction in which the predicted response increases.

(iii) Some λ_i are positive and some are negative. This turning value, illustrated for two factors in Fig. 8A.3(f), is called a minimax, being a minimum for some of the x-variables and a maximum for others. Should this case arise, the response is expected to increase most rapidly as we increase the X_i that has the highest *positive* λ_i. Factor combinations which give several increasing values of this X_i may be tried.

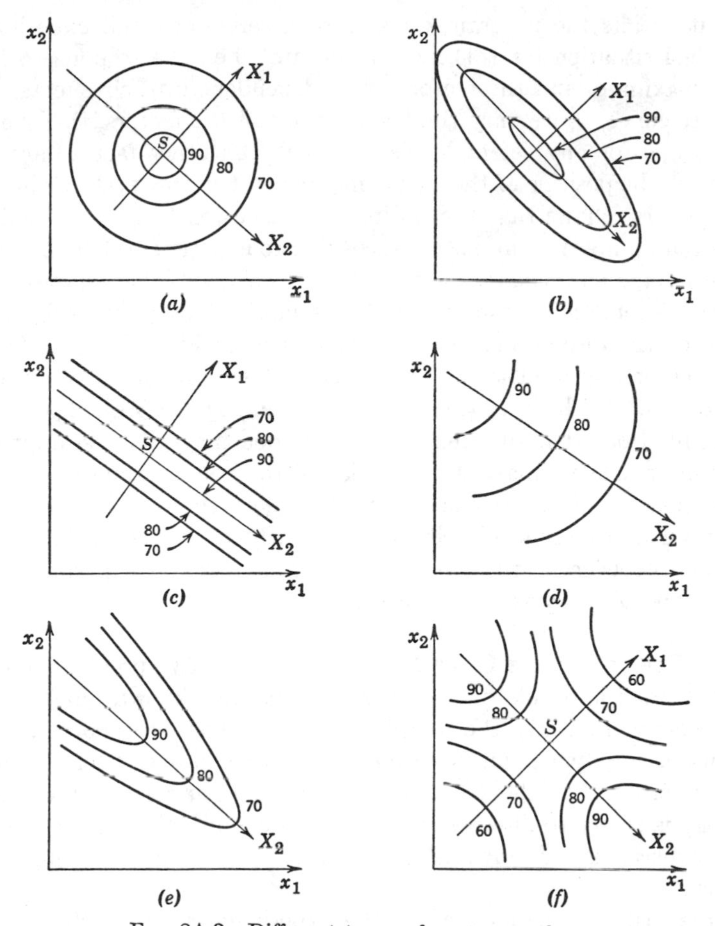

FIG. 8A.3 Different types of contour surface.

(iv) The quadratic surface does not fit. The best course of action varies from case to case. Addition of points in order that a cubic surface can be fitted may account for the discrepancy. In other cases it will be evident that no polynomial of low order provides a satisfactory approximation. Further experiments in the neighborhood of a point of relatively high response, with smaller intervals between the levels of the factors, may then be advisable.

8A.6 Summary Comments

8A.61 The Single-factor and Steepest Ascent Methods. On the score of simplicity the single-factor method has much to recommend it, since it breaks the program down into a series of small experiments. The ideal situation for this method occurs when the response y has a single maximum and the factors are independent in their effects. The response curve, as we vary the levels of one of the factors, then has the same *shape* for any set of levels at which the other factors are fixed (although the position of the curve may depend on the levels of the other factors). It follows that the optimum can be reached at the end of a single round, apart from disturbances due to experimental errors and to the fact that a parabola may not be the exact shape of the response curve.

When interactions among the effects of the factors are substantial, i.e., when the response surface is oriented as in Fig. 8A.3, (b)–(e), the single-factor approach tends to converge slowly and may require several rounds. With rising ridge surfaces as in Fig. 8A.3(e), Box (8A.13) has pointed out that the single-factor method may give the impression of having reached a maximum that is not the true maximum.

The method of steepest ascent probes more thoroughly and is safer and more informative. It should frequently reach the optimum with less total experimentation, although it requires larger experiments and the method of analysis is more involved.

8A.62 Use of a Single Large Experiment. The two previous methods require that the results of an experiment become known quickly, since each experiment awaits the completion and analysis of the previous one. In lines of work where an experiment must extend over a long period of time in order that the treatments produce their effects, the natural strategy is to try to discover the optimum combination at the end of a single experiment. Rotatable second order designs may be useful in this problem.

If the region to be covered in the experiment is large, the response surface may be too complex to be approximated by a quadratic polynomial. Use of polynomials of higher order, or further exploration in a second experiment, as previously suggested, should be considered.

8A.63 Selection of Test Points at Random. The possibility of using random combinations of the levels of the factors in exploratory work has been suggested, e.g., by Anderson (8A.16). If the factor x_i is to be studied between the levels 0 and 4, say, we choose a random number lying between these limits as the level of x_i. An independent choice is made for each factor and for each successive factor combination that is tested. Alternatively, a number of equally spaced points, 0, ½, 1, 1½, $\cdots$, 4 could be chosen, and a random choice of one of these levels made for each trial. At the end of n trials, the factor combination giving the highest observed response is regarded as an estimate of the optimum combination.

Although this method has no planned strategy, it may have practical possibilities when the number of factors is large and experimental errors are small. Two goals which the experimenter might hope to accomplish at the end of a moderate series of preliminary trials are (i) to reach a region of relatively high response and (ii) to learn something about the x-variables that have the greatest influence on y, in order that future experiments can be confined to a smaller number of factors. Suppose that the sub-region in the x-space in which the response is satisfactorily high has a volume which is a fraction a of the volume of the whole x-region that is being explored. If experimental error is ignored, the probability that at least one of the random combinations out of n trials lies in the desirable sub-region is

$$P = 1 - (1 - a)^n$$

This equation holds irrespective of the number of factors. For instance, if $a = 0.1$, P exceeds 0.8 after 16 trials have been completed, 0.9 after 22 trials, and 0.95 after 29 trials.*

Since the levels of the different factors are independently assigned, methods of regression analysis can be applied to provide clues to the factors exerting most influence. This approach must be developed further and compared with other methods before its utility can be assayed.

8A.64 An Experimental Comparison of the Four Methods. Direct comparisons of the performances of these methods in practical application are unlikely to be made for reasons of expense. A mathematical comparison which simulates a practical test has been conducted by Brooks (8A.17). He constructed four surfaces, each with well-defined maxima in two variables: typical contours are shown in Fig. 8A.4. For each surface, 9 different starting positions were tried. In his first series of artificial experiments, each method was allotted a total of 16 trials (i.e., 16 factor combinations) from which to locate the optimum; in his second, 30 trials were allowed. A further variation was that the series were carried through both with no experimental error present and with a moderate amount of error. Several different methods of estimating the optimum for each method were investigated.

The methods compared were the single-factor, the steepest ascent, the factorial, the selection of test points at random, and a stratified random selection of test points (not reported here). The factorial method represented the situation in which a single large experiment

* S. Brooks, personal communication.

must be used. A 4×4 factorial was employed with 16 trials and a 5×6 factorial with 30 trials.

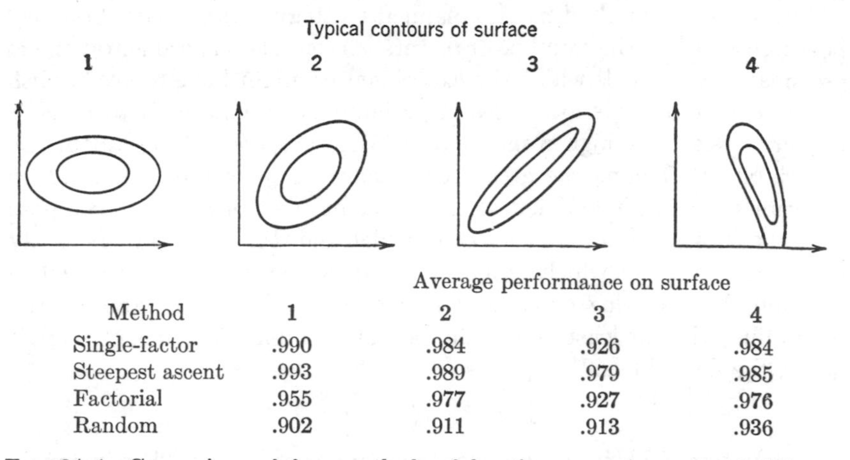

Method	Average performance on surface 1	2	3	4
Single-factor	.990	.984	.926	.984
Steepest ascent	.993	.989	.979	.985
Factorial	.955	.977	.927	.976
Random	.902	.911	.913	.936

FIG. 8A.4 Comparison of four methods of locating a maximum on 4 different response surfaces.

The performance of any method was gauged by the true response $\phi(x_1, x_2)$ at the optimum values of x_1 and x_2 as found by the method. The average results for each method over all trials are shown under Fig. 8A.4, using for each method the estimating procedure that turned out to be best.

Since the true maximum of each surface was 1, all methods did well. The steepest ascent method was best on all four surfaces. The single-factor method was almost as successful on all surfaces except surface 3. This is the type of surface for which one of the λ's in the canonical form of the quadratic approximation is small. It frequently brought about a "rising ridge" situation in which the single-factor method made little progress after the first round. The factorial method was consistently poorer than the two sequential methods except on surface 3, where it performed about the same as the single-factor method. A common difficulty was that a polynomial surface (quadratic or cubic) did not fit the data satisfactorily because of the relatively wide spacing of the factorial grid.

The use of random combinations of levels was distinctly inferior to the other methods. This method would not be expected to show promise when there are only two x-variables.

Needless to say, these conclusions may be altered if different types of response surface, different numbers of factors, or different amounts of experimental error are studied. For the surfaces covered, the results suggest an all-round reliability of the method of steepest ascent.

REFERENCES

8A.1 Box, G. E. P. Multifactor designs of first order. *Biometrika* 39, 49–57, 1952.

8A.2 Fisher, R. A. *Statistical methods for research workers.* Oliver & Boyd, Edinburgh, 12th ed. 1952.

8A.3 Dwyer, P. S. *Linear computations.* John Wiley & Sons, New York, 1951.

8A.4 Anderson, R. L., and Bancroft, T. A. *Statistical theory in research.* McGraw-Hill, New York, 1952.

8A.5 Plackett, R. L., and Burman, J. P. The design of optimum multifactorial experiments. *Biometrika* 33, 305–325, 1946.

8A.6 Box, G. E. P., and Wilson, K. B. On the experimental attainment of optimum conditions. *Jour. Roy. Stat. Soc. B*, 13, 1–45, 1951.

8A.7 Box, G. E. P., and Hunter, J. S. Multifactor experimental designs. *Ann. Math. Stat.* 28, 1957.

8A.8 Moore, D. P., *et al.* An investigation of some of the relationships of copper, iron, and molybdenum in the growth and nutrition of lettuce: II. *Proc. Soil Sci. Soc. Amer.* 21, 65–74, 1957.

8A.9 Hader, R. J., *et al.* An investigation of some of the relationships of copper, iron, and molybdenum in the growth and nutrition of lettuce: I. *Proc. Soil Sci. Soc. Amer.* 21, 59–64, 1957.

8A.10 Box, G. E. P., and Voule, P. V. The exploration and exploitation of response surfaces: An example of the link between the fitted surface and the basic mechanism of the system. *Biometrics* 11, 287–322, 1955.

8A.11 Friedman, M., and Savage, L. J. Planning experiments seeking maxima. In *Techniques of statistical analysis.* McGraw-Hill, New York, 1947.

8A.12 Hotelling, H. Experimental determination of the maximum of a function. *Ann. Math. Stat.* 12, 20–45, 1941.

8A.13 Box, G. E. P. The exploration and exploitation of response surfaces: some general considerations and examples. *Biometrics* 10, 16–60, 1954.

8A.14 Davies, O. L. (Ed.). *Design and analysis of industrial experiments.* Oliver & Boyd, Edinburgh. Chap. 11, 1954.

8A.15 Read, D. R. The design of chemical experiments. *Biometrics* 10, 1–15, 1954.

8A.16 Anderson, R. L. Recent advances in finding best operating conditions. *Jour. Amer. Stat. Assoc.* 48, 789–798, 1953.

8A.17 Brooks, S. *Comparison of methods for estimating the optimal factor combination.* Sc.D. thesis. Johns Hopkins University, 1955.

8A.18 DeBaun, R. M. Block effects in the determination of optimum conditions. *Biometrics* 12, 20–22, 1956.

PLANS

Central composite rotatable second order designs

Plan 8A.1 4 x-variables $N = 31$ treatment combinations

2^4 factorial + star design + 7 points in the center

x_1	x_2	x_3	x_4	x_1	x_2	x_3	x_4
−1	−1	−1	−1	−2	0	0	0
1	−1	−1	−1	2	0	0	0
−1	1	−1	−1	0	−2	0	0
1	1	−1	−1	0	2	0	0
−1	−1	1	−1	0	0	−2	0
1	−1	1	−1	0	0	2	0
−1	1	1	−1	0	0	0	−2
1	1	1	−1	0	0	0	2
−1	−1	−1	1	0	0	0	0
1	−1	−1	1	0	0	0	0
−1	1	−1	1	0	0	0	0
1	1	−1	1	0	0	0	0
−1	−1	1	1	0	0	0	0
1	−1	1	1	0	0	0	0
−1	1	1	1	0	0	0	0
1	1	1	1				

Solutions

$$b_0 = 0.142857(0y) - 0.035714 \sum (iiy)$$

$$b_i = 0.041667(iy)$$

$$b_{ii} = 0.031250(iiy) + 0.003720 \sum (iiy) - 0.035714(0y)$$

$$b_{ij} = 0.0625(ijy)$$

	b_i	b_{ii}	b_{ij}
s.e.	$0.204s$	$0.185s$	$0.250s$

Plan 8A.2 5 x-variables $N = 32$ treatment combinations

½ replicate of a 2^5 factorial + star design + 6 points in the center

x_1	x_2	x_3	x_4	x_5	x_1	x_2	x_3	x_4	x_5
−1	−1	−1	−1	1	−2	0	0	0	0
1	−1	−1	−1	−1	2	0	0	0	0
−1	1	−1	−1	−1	0	−2	0	0	0
1	1	−1	−1	1	0	2	0	0	0
−1	−1	1	−1	−1	0	0	−2	0	0
1	−1	1	−1	1	0	0	2	0	0
−1	1	1	−1	1	0	0	0	−2	0
1	1	1	−1	−1	0	0	0	2	0
−1	−1	−1	1	−1	0	0	0	0	−2
1	−1	−1	1	1	0	0	0	0	2
−1	1	−1	1	1	0	0	0	0	0
1	1	−1	1	−1	0	0	0	0	0
−1	−1	1	1	1	0	0	0	0	0
1	−1	1	1	−1	0	0	0	0	0
−1	1	1	1	−1	0	0	0	0	0
1	1	1	1	1	0	0	0	0	0

Solutions

$$b_0 = 0.159091(0y) - 0.034091 \sum (iiy)$$

$$b_i = 0.041667(iy)$$

$$b_{ii} = 0.031250(iiy) + 0.002841 \sum (iiy) - 0.034091(0y)$$

$$b_{ij} = 0.0625(ijy)$$

	b_i	b_{ii}	b_{ij}
s.e.	$0.204s$	$0.185s$	$0.250s$

Plan 8A.3 6 x-variables $N = 53$ treatment combinations

½ replicate of a 2^6 factorial + star design + 9 points in the center

x_1	x_2	x_3	x_4	x_5	x_6	x_1	x_2	x_3	x_4	x_5	x_6
−1	−1	−1	−1	−1	−1	−1	1	−1	1	1	1
1	−1	−1	−1	−1	1	1	1	−1	1	1	−1
−1	1	−1	−1	−1	1	−1	−1	1	1	1	1
1	1	−1	−1	−1	−1	1	−1	1	1	1	−1
−1	−1	1	−1	−1	1	−1	1	1	1	1	−1
1	−1	1	−1	−1	−1	1	1	1	1	1	1
−1	1	1	−1	−1	−1	−2.378	0	0	0	0	0
1	1	1	−1	−1	1	2.378	0	0	0	0	0
−1	−1	−1	1	−1	1	0	−2.378	0	0	0	0
1	−1	−1	1	−1	−1	0	2.378	0	0	0	0
−1	1	−1	1	−1	−1	0	0	−2.378	0	0	0
1	1	−1	1	−1	1	0	0	2.378	0	0	0
−1	−1	1	1	−1	−1	0	0	0	−2.378	0	0
1	−1	1	1	−1	1	0	0	0	2.378	0	0
−1	1	1	1	−1	1	0	0	0	0	−2.378	0
1	1	1	1	−1	−1	0	0	0	0	2.378	0
−1	−1	−1	−1	1	1	0	0	0	0	0	−2.378
1	−1	−1	−1	1	−1	0	0	0	0	0	2.378
−1	1	−1	−1	1	−1	0	0	0	0	0	0
1	1	−1	−1	1	1	0	0	0	0	0	0
−1	−1	1	−1	1	−1	0	0	0	0	0	0
1	−1	1	−1	1	1	0	0	0	0	0	0
−1	1	1	−1	1	1	0	0	0	0	0	0
1	1	1	−1	1	−1	0	0	0	0	0	0
−1	−1	−1	1	1	−1	0	0	0	0	0	0
1	−1	−1	1	1	1	0	0	0	0	0	0
						0	0	0	0	0	0

Solutions

$$b_0 = 0.110749(0y) - 0.018738 \sum (iiy)$$

$$b_i = 0.023087(iy)$$

$$b_{ii} = 0.015625(iiy) + 0.001217 \sum (iiy) - 0.018738(0y)$$

$$b_{ij} = 0.03125(ijy)$$

	b_i	b_{ii}	b_{ij}
s.e.	$0.152s$	$0.130s$	$0.177s$

Central composite second order designs in incomplete blocks

Plan 8A.4 3 x-variables $N = 20$ treatment combinations

Block I			Block II			Block III		
x_1	x_2	x_3	x_1	x_2	x_3	x_1	x_2	x_3
−1	−1	1	−1	−1	−1	−1.633	0	0
1	−1	−1	1	−1	1	1.633	0	0
−1	1	−1	−1	1	1	0	−1.633	0
1	1	1	1	1	−1	0	1.633	0
0	0	0	0	0	0	0	0	−1.633
0	0	0	0	0	0	0	0	1.633
						0	0	0
						0	0	0

Solutions

$$b_0 = 0.165385(0y) - 0.057692 \sum (iiy)$$

$$b_i = 0.075(iy)$$

$$b_{ii} = 0.070312(iiy) + 0.005409 \sum (iiy) - 0.057692(0y)$$

$$b_{ij} = 0.125(ijy)$$

	b_i	b_{ii}	b_{ij}
s.e.	$0.274s$	$0.275s$	$0.354s$

Plan 8A.5 4 x-variables $N = 30$ treatment combinations

Block I				Block II				Block III			
x_1	x_2	x_3	x_4	x_1	x_2	x_3	x_4	x_1	x_2	x_3	x_4
−1	−1	−1	−1	−1	−1	−1	1	−2	0	0	0
1	−1	−1	1	1	−1	−1	−1	2	0	0	0
−1	1	−1	1	−1	1	−1	−1	0	−2	0	0
1	1	−1	−1	1	1	−1	1	0	2	0	0
−1	−1	1	1	−1	−1	1	−1	0	0	−2	0
1	−1	1	−1	1	−1	1	1	0	0	2	0
−1	1	1	−1	−1	1	1	1	0	0	0	−2
1	1	1	1	1	1	1	−1	0	0	0	2
0	0	0	0	0	0	0	0	0	0	0	0
0	0	0	0	0	0	0	0	0	0	0	0

Solutions

$$b_0 = 0.166667(0y) - 0.041667 \sum (iiy)$$

$$b_i = 0.041667(iy)$$

$$b_{ii} = 0.031250(iiy) + 0.005208 \sum (iiy) - 0.041667(0y)$$

$$b_{ij} = 0.0625(ijy)$$

	b_i	b_{ii}	b_{ij}
s.e.	$0.204s$	$0.191s$	$0.250s$

Plan 8A.6 5 x-variables $N = 33$ treatment combinations

Block I

x_1	x_2	x_3	x_4	x_5
−1	−1	−1	−1	1
1	−1	−1	−1	−1
−1	1	−1	−1	−1
1	1	−1	−1	1
−1	−1	1	−1	−1
1	−1	1	−1	1
−1	1	1	−1	1
1	1	1	−1	−1
−1	−1	−1	1	−1
1	−1	−1	1	1
−1	1	−1	1	1
1	1	−1	1	−1
−1	−1	1	1	1
1	−1	1	1	−1
−1	1	1	1	−1
1	1	1	1	1
0	0	0	0	0
0	0	0	0	0
0	0	0	0	0
0	0	0	0	0
0	0	0	0	0
0	0	0	0	0

Block II

x_1	x_2	x_3	x_4	x_5
−2	0	0	0	0
2	0	0	0	0
0	−2	0	0	0
0	2	0	0	0
0	0	−2	0	0
0	0	2	0	0
0	0	0	−2	0
0	0	0	2	0
0	0	0	0	−2
0	0	0	0	2
0	0	0	0	0

Solutions

$$b_0 = 0.137255(0y) - 0.029412 \sum (iiy)$$

$$b_i = 0.041667(iy)$$

$$b_{ii} = 0.031250(iiy) + 0.001838 \sum (iiy) - 0.029412(0y)$$

$$b_{ij} = 0.0625(ijy)$$

	b_i	b_{ii}	b_{ij}
s.e.	$0.204s$	$0.182s$	$0.250s$

Plan 8A.7 6 x-variables N = 54 treatment combinations

Block I

x_1	x_2	x_3	x_4	x_5	x_6
−1	−1	−1	−1	−1	−1
1	−1	−1	−1	1	−1
−1	1	−1	−1	1	−1
1	1	−1	−1	−1	−1
−1	−1	1	−1	−1	1
1	−1	1	−1	1	1
−1	1	1	−1	1	1
1	1	1	−1	−1	1
−1	−1	−1	1	−1	1
1	−1	−1	1	1	1
−1	1	−1	1	1	1
1	1	−1	1	−1	1
−1	−1	1	1	−1	−1
1	−1	1	1	1	−1
−1	1	1	1	1	−1
1	1	1	1	−1	−1
0	0	0	0	0	0
0	0	0	0	0	0
0	0	0	0	0	0
0	0	0	0	0	0

Block II

x_1	x_2	x_3	x_4	x_5	x_6
−1	−1	−1	−1	1	1
1	−1	−1	−1	−1	1
−1	1	−1	−1	−1	1
1	1	−1	−1	1	1
−1	−1	1	−1	1	−1
1	−1	1	−1	−1	−1
−1	1	1	−1	−1	−1
1	1	1	−1	1	−1
−1	−1	−1	1	1	−1
1	−1	−1	1	−1	−1
−1	1	−1	1	−1	−1
1	1	−1	1	1	−1
−1	−1	1	1	1	1
1	−1	1	1	−1	1
−1	1	1	1	−1	1
1	1	1	1	1	1
0	0	0	0	0	0
0	0	0	0	0	0
0	0	0	0	0	0
0	0	0	0	0	0

Block III

x_1	x_2	x_3	x_4	x_5	x_6
−2.366	0	0	0	0	0
2.366	0	0	0	0	0
0	−2.366	0	0	0	0
0	2.366	0	0	0	0
0	0	−2.366	0	0	0
0	0	2.366	0	0	0
0	0	0	−2.366	0	0
0	0	0	2.366	0	0
0	0	0	0	−2.366	0
0	0	0	0	2.366	0
0	0	0	0	0	−2.366
0	0	0	0	0	2.366
0	0	0	0	0	0
0	0	0	0	0	0

Solutions

$$b_0 = 0.099600(0y) - 0.016892 \sum (iiy)$$

$$b_i = 0.023148(iy)$$

$$b_{ii} = 0.015944(iiy) + 0.000862 \sum (iiy) - 0.016892(0y)$$

$$b_{ij} = 0.03125(ijy)$$

	b_i	b_{ii}	b_{ij}
s.e.	$0.152s$	$0.130s$	$0.177s$

CHAPTER 9

Incomplete Block Designs

9.1 Balanced Designs

As their name implies, these designs, introduced by Yates (9.1), are arranged in blocks or groups that are smaller than a complete replication, in order to eliminate heterogeneity to a greater extent than is possible with randomized blocks and latin squares. In the designs described in chapters 6 to 8, this reduction in the size of block was achieved by sacrificing all or part of the information on certain treatment comparisons. The present designs, on the other hand, were developed for experiments in plant breeding and selection, where it is desired to make all comparisons among pairs of treatments with equal precision. Consequently, a different method for reducing the size of block is employed.

The designs may be arranged either in randomized incomplete blocks or in quasi-latin squares. They may be *balanced* or *partially balanced.* The balanced designs will be illustrated first by simple examples of the experimental plans.

Consider the plan in table 9.1 which compares 9 treatments in incomplete blocks of 3 experimental units with 4 replications.

Every pair of treatments will be found to occur once, and only once, in the same block. For instance, treatment 1 occupies the same block

TABLE 9.1 Balanced design for 9 treatments in blocks of 3 units

Block	Rep. I	Block	Rep. II	Block	Rep. III	Block	Rep. IV
(1)	1 2 3	(4)	1 4 7	(7)	1 5 9	(10)	1 8 6
(2)	4 5 6	(5)	2 5 8	(8)	7 2 6	(11)	4 2 9
(3)	7 8 9	(6)	3 6 9	(9)	4 8 3	(12)	7 5 3

with treatments 2 and 3 in the first replication, with treatments 4 and 7 in the second replication, with treatments 5 and 9 in the third replication, and with treatments 6 and 8 in the fourth replication. When the results are analyzed by the method of least squares, this property, to which the adjective *balanced* is applied, ensures that all pairs of treatments are compared with approximately the same precision, even though the differences among blocks may be large. This design belongs to the group known as *balanced lattices*, so-called because the plans are conveniently written down by drawing a square lattice, with the treatment numbers at the intersections of the lines. In the balanced lattices, the number of treatments must be an exact square while the number of units per block is the corresponding square root.

Balanced designs can be constructed for other numbers of treatments and of units per block. The plan in table 9.2 shows 7 treatments arranged in blocks of 3 units.

TABLE 9.2 BALANCED DESIGN FOR 7 TREATMENTS IN BLOCKS OF 3 UNITS

Block				Block				Block				Block			
(1)	1	2	4	(3)	3	4	6	(5)	1	5	6	(7)	1	3	7
(2)	2	3	5	(4)	4	5	7	(6)	2	6	7				

Again every pair of treatments occurs once within some block. In this case, however, the blocks cannot be grouped in separate replications, since 7 is not divisible by 3. Designs of this type are known as *balanced incomplete blocks.*

For certain numbers of treatments and units per block, both the types above can be laid out in a kind of latin square formation so as to allow the elimination of variation arising from two types of grouping. The appropriate rearrangement for the first example is shown in table 9.3.

TABLE 9.3 BALANCED DESIGN FOR 9 TREATMENTS IN 4 LATTICE SQUARES

Rep. I				Rep. II				Rep. III				Rep. IV			
Columns															
Rows	(1)	(2)	(3)		(4)	(5)	(6)		(7)	(8)	(9)		(10)	(11)	(12)
(1)	1	2	3	(4)	1	4	7	(7)	1	6	8	(10)	1	9	5
(2)	4	5	6	(5)	2	5	8	(8)	9	2	4	(11)	6	2	7
(3)	7	8	9	(6)	3	6	9	(9)	5	7	3	(12)	8	4	3

It may be verified by inspection that every pair of treatments now occurs once in the same row and also once in the same column. All comparisons between pairs of treatments are of nearly equal precision. This design is known as a *lattice square.*

The second example is rewritten somewhat differently.

Every treatment now appears in each of the 3 rows, and every pair of treatments appears together once in the same column. Since the plan

TABLE 9.4 BALANCED DESIGN FOR 7 TREATMENTS IN AN INCOMPLETE LATIN SQUARE

	Columns (Blocks)						
	(1)	(2)	(3)	(4)	(5)	(6)	(7)
Rows							
(1)	1	2	3	4	5	6	7
(2)	2	3	4	5	6	7	1
(3)	4	5	6	7	1	2	3

above could represent the first 3 rows of a 7 × 7 latin square, this type of design has been called an *incomplete latin square*, or alternatively a *Youden square*, after W. J. Youden, who developed these designs for greenhouse experiments.

9.2 Partially Balanced Designs

Although a balanced design can be constructed with any number of treatments and any number of units per block, the minimum number of replications is fixed by these 2 variables. In most cases this number is too large for the usual conditions of experimentation. In order to allow more freedom of choice in the number of replicates, designs which lack the complete symmetry of the balanced designs must be used.

Simple examples of such designs are the *lattices*. These are constructed in the same way as balanced lattices except that there are fewer replications. The design with 2 replications (e.g., the first 2 replications in table 9.1) is called a *simple lattice*, and that with 3 replications a *triple lattice*. Similarly, with a lattice square, as in table 9.3, we may use less than the full number of replicates necessary for balance. For all these designs the number of treatments must be a perfect square.

A further set, called *cubic lattices*, is useful when the number of treatments is very large. The number of treatments is the cube of the number of units per block, so that a drastic reduction in block size is obtained. The number of replicates is 3 or some multiple of 3.

Partially balanced designs are less suitable than balanced designs. The statistical analysis is more complicated. When the variation among blocks (or rows and columns) is large, some pairs of treatments are more precisely compared than others, and several different standard errors may have to be computed for tests of significance. These difficulties increase as the design departs more and more from the symmetry of the balanced design.

The first comprehensive examination of the construction and properties of designs that lack full balance was made by Bose and Nair (9.1a). They were particularly interested in developing designs for which the analysis should not be unduly difficult. For this to be so, the design must satisfy certain properties described in section 11.61a. Bose and Nair chose the name "partially balanced" for designs that satisfy these properties, although in the literature this phrase is sometimes found to denote any design that has some features of symmetry.

The simplest of the partially balanced designs are those with *two associate classes*. In these, some pairs of treatments occur in the same block λ_1 times, while other pairs occur in the same block λ_2 times, where λ_1 and λ_2 are whole numbers. In the lattice designs, which are of this type, some pairs of treatments never occur together in a block, i.e., $\lambda_1 = 0$; other pairs occur once in the same block ($\lambda_2 = 1$).

A catalogue of 376 designs with two associate classes has been published by Bose, Clatworthy, and Shrikhande (9.2a), including plans, instructions for analysis, and worked examples. All these designs have a block size greater than 2 units. Instructions for obtaining 92 plans with block size 2 have been given by Clatworthy (9.3a). These designs represent a great enlargement of the available selection of incomplete block arrangements. They should be particularly useful in experiments where the block size that permits precise comparisons is dictated by circumstances beyond the control of the experimenter.

Many of these designs can be arranged in a kind of latin square formation in which variations among rows and columns can both be eliminated from the experimental errors. Table 9.1a shows an example for 6 treatments in blocks of size 4. Each row forms a complete replication. With regard to the columns (blocks), a pair of treatments either occurs *twice* in the same block (e.g., treatments 1 and 2) or *four times* (e.g., treatments 1 and 4). This type of design is appropriate in the same circumstances as the Youden square.

TABLE 9.1a A PARTIALLY BALANCED INCOMPLETE BLOCK DESIGN ARRANGED IN LATIN SQUARE FORMATION

	Columns (blocks)					
Rows	(1)	(2)	(3)	(4)	(5)	(6)
(1)	1	2	3	4	5	6
(2)	4	5	6	1	2	3
(3)	2	3	1	5	6	4
(4)	5	6	4	2	3	1

9.3 Basis of the Statistical Analysis

9.31 Analysis without Recovery of Inter-block Information. To those who have previously used only the simplest designs, one feature of the incomplete block designs will be unfamiliar. The treatment averages must be adjusted in order to secure the full accuracy available in the design. That this is necessary can be seen by examining, for example, the balanced lattice in table 9.1. In three of the 4 replications, treatments 1 and 2 are in different blocks. Consequently, if the simple averages of the 2 treatments are compared, the effects of differences among blocks are eliminated only for the first replication.

Yates developed two methods of analysis for the designs. The first is an analysis by standard least squares, reference (9.1), sometimes referred to as the "intra-block" analysis. Later he showed that comparisons among the block totals also contained information about treatment effects that can be utilized in the larger experiments. This approach is analogous to that in split-plot experiments, where it will be recalled that use is made both of comparisons within whole-units (corresponding to incomplete blocks), and of comparisons among the totals of different whole-units. In this section the original method of analysis will be outlined for balanced lattice experiments. This method has not been entirely superseded by the newer analysis, because it must be used with small experiments. The newer analysis is sketched in section 9.32.

For the original analysis the mathematical model and the assumptions are essentially the same as those for previous designs in this book, as described in section 3.2. In the balanced lattice there are k^2 treatments in blocks of k units, with $(k + 1)$ replications. Let y_{ijq} be the observation for the qth treatment, which we suppose to be in the jth block within the ith replication. The model is

$$y_{ijq} = \mu + \pi_i + \beta_{ij} + \tau_q + e_{ijq} \tag{9.1}$$

where μ, π_i, β_{ij}, and τ_q represent the effects of the mean, the replicate, the incomplete block, and the treatment, respectively, and e_{ijq} is the intra-block residual or error, assumed to be normally and independently distributed with mean zero and variance σ_e^2. As with previous designs, we find estimates of the parameters by minimizing

$$\sum (y_{ijq} - m - p_i - b_{ij} - t_q)^2$$

subject to the usual relations

$$\sum_i p_i = 0, \qquad \sum_j b_{ij} = 0 \quad \text{(for any } i\text{)}, \qquad \sum_q t_q = 0 \tag{9.2}$$

The normal equation for any unknown is obtained as before by equating the observed total to the expected total over all units whose equation contains the constant. Thus, if T_q denotes the total for all $(k + 1)$ units that receive the qth treatment, the normal equation for t_q is

$$T_q = (k + 1)m + \sum b_{ij} + (k + 1)t_q \tag{9.3}$$

the terms in the p_i having disappeared in virtue of equations (9.2), since any treatment appears in *all* replicates.

This equation does not at once give the value of t_q, since it still contains the unknowns m and b_{ij}. From the normal equation for m, it is easy to see that m is the average over the whole experiment. The normal equation for any b_{ij} is

$$B_{ij} = km + kp_i + kb_{ij} + \sum t_q \tag{9.4}$$

where B_{ij} is the observed block total, and the treatments sum is over the k treatments that are in the block. If we add these equations for all blocks that contain the qth treatment, we obtain

$$B_t = k(k + 1)m + k\sum b_{ij} + \sum \sum t_q \tag{9.5}$$

where B_t is the total of all such blocks.

It is at this point that the structure of the design is important. In the blocks that contain t_q, every other treatment occurs once and only once, while t_q appears $(k + 1)$ times. Hence

$$\sum \sum t_q = (k + 1)t_q + \sum \text{(all other } t\text{'s)} = kt_q + \sum \text{(all } t\text{'s)} = kt_q$$

from (9.2).

This simplifies (9.5) to

$$B_t = k(k + 1)m + k\sum b_{ij} + kt_q \tag{9.6}$$

We may now eliminate the unwanted term $\sum b_{ij}$ from (9.3) and (9.6), giving

$$kT_q - B_t = k^2 t_q \tag{9.7}$$

a relatively simple solution. In effect, we simply adjust the treatment total for the total of the blocks in which it appears.

Since t_q represents the *deviation* of the estimated treatment mean from the mean for all treatments, it is customary in practice to compute $(t_q + m)$, which represents the estimated mean itself. A little algebraic manipulation from (9.7) shows that

$$t_q + m = \frac{T_q}{k+1} + \frac{[kT_q - (k+1)B_t + G]}{k^2(k+1)} \tag{9.8}$$

where $G = k^2(k+1)m$ is the grand total for the whole experiment. Note that $T_q/(k+1)$ is the simple or unadjusted treatment mean. The last term on the right therefore represents the adjustment that is applied to the ordinary mean. This result will be used in a later comparison with the newer analysis.

A continuation of the analysis shows that the variance of the difference between two t_q values is $2\sigma_e^2/k$. If the experiment had been in randomized blocks, with the same number of replications, $(k+1)$, the corresponding variance would have been $2\sigma^2/(k+1)$, where σ^2 is the error variance for randomized blocks. Hence the new design, with this method of analysis, gives a more accurate experiment than randomized blocks if and only if

$$\frac{\sigma_e^2}{\sigma^2} < \frac{k}{k+1}$$

The quantity $k/(k+1)$ is called the *efficiency factor* of the design. Note that the incomplete block design would be less accurate than randomized blocks if the variation among incomplete blocks were as great as that within complete replications. With the newer method of analysis this disadvantage is largely removed.

9.32 Analysis with Recovery of Inter-block Information. This analysis rests on more difficult theory and will be outlined only in part. Formally, the mathematical model is exactly the same:

$$y_{ijq} = \mu + \pi_i + \tau_q + \beta_{ij} + e_{ijq} \tag{9.9}$$

However, the additional assumptions are made that the block effects β_{ij} are normally and independently distributed with zero means and variance σ_b^2, and that they are independent of the e_{ijq}. It follows from the assumptions that for the difference between two observations in the *same* block, the residual variance is $2\sigma_e^2$, since the β_{ij}'s cancel, whereas for the difference between two observations in different blocks, the residual variance is $2(\sigma_e^2 + \sigma_b^2)$. The reader will note the analogy with split-plot experiments (section 7.12). With regard to the justification for the additional assumptions, it can be said that in agricultural field experiments, for which the designs were first developed, it is usually just as reasonable to make the assumptions for the β_{ij} as for the e_{ijq}. The assumptions cannot be taken for granted, and with certain types of data we might not wish to make them.

As a result of the assumptions, observations that are in the same block are positively correlated, so that the simplest type of least squares estimation cannot be applied. Instead, we may write down the joint frequency distribution of all the observations, and use the more general method of estimation known as maximum likelihood. By a well-known device in theory, maximum likelihood estimation may be shown to be equivalent to the minimization of a weighted sum of squares, consisting of two parts. The first part is the sum of squares of deviations of the residuals from their block means; the second is the sum of squares of the residuals of the block totals. The two parts receive weights w and w'/k, respectively, where

$$w = \frac{1}{\sigma_e^2}; \qquad w' = \frac{1}{\sigma_e^2 + k\sigma_b^2}$$

For the moment we suppose that w and w' are known. For the balanced lattice, the quantity to be minimized is

$$\sum w[y_{ijq} - y_{ij\cdot} - (t_q - t_{ij\cdot})]^2 + \sum \frac{w'}{k}[B_{ij} - km - kp_i - kt_{ij\cdot}]^2 \tag{9.10}$$

where $y_{ij\cdot}, t_{ij\cdot}$ denote, respectively, the observed mean of a block and the mean of the t's that occur in the block. The first sum is over all observations, the second over all blocks.

When we differentiate with respect to a given t_q, we must note that the term $t_{ij\cdot}$ will contain t_q, with coefficient $1/k$, whenever the block in question contains the qth treatment. The derivative, when equated to zero (omitting a factor -2), gives

$$\sum w[y_{ijq} - y_{ij\cdot} - t_q + t_{ij\cdot}] + \sum \frac{w'}{k}[B_{ij} - km - kp_i - kt_{ij\cdot}] = 0 \tag{9.11}$$

The first sum is over the observations that receive the qth treatment; the second over the blocks that contain this treatment. The derivative contains an additional part arising from the terms in $t_{ij\cdot}$ in the first bracket of (9.10), but this contribution will be found to vanish.

The next step is to simplify (9.11). Both brackets involve a sum of the terms $t_{ij\cdot}$. In each case, because of the symmetry of the balanced lattice, the sum contains all other t's once, and t_q $(k + 1)$ times. Hence

$$\sum t_{ij\cdot} = \frac{(k + 1)t_q + (\text{rest of } t\text{'s})}{k} = \frac{kt_q}{k} = t_q$$

since the total of all t's is zero. Further, we may note that the observed block mean $y_{ij.}$ equals B_{ij}/k. Writing as before T_q for the treatment total and B_t for the total of all blocks in which the treatment appears, we have from (9.11)

$$w\left[T_q - \frac{B_t}{k} - (k + 1)t_q + t_q\right] + \frac{w'}{k}[B_t - k(k + 1)m - kt_q] = 0 \qquad (9.12)$$

or

$$(kw + w')t_q + (k + 1)w'm = wT_q - (w - w')\frac{B_t}{k} \qquad (9.13)$$

The constant m may be shown to be the mean for the whole experiment, or $G/k^2(k + 1)$. After some rearrangement we may express (9.13) in a form comparable with the original estimate given in (9.8), as follows.

$$t_q + m = \frac{T_q}{k + 1} + \frac{(w - w')[kT_q - (k + 1)B_t + G]}{k(k + 1)(kw + w')} \qquad (9.14)$$

The adjustments are seen to be of the same form in the new as in the original analysis. The two are equal when w' is zero. This occurs when σ_b^2 is very large; that is, when differences among blocks are great. In all other cases the adjustments are smaller with the new than with the original analysis. They reduce to zero when $w' = w$, which happens only if σ_b^2 is zero, or in other words if there are no real differences among blocks. Thus the new analysis is a generalization of the original analysis, reducing to it in the extreme case where there are marked variations among incomplete blocks. At the other extreme, when the arrangement into blocks has been ineffective, the new analysis makes no adjustments for the non-existent block differences, and in fact reduces to an analysis by the method for randomized blocks.

In practice the weights w and w' are not known. Yates has shown that they can be estimated from the analysis of variance, in the same way as we estimate two separate errors in a split-plot experiment. The details will not be given. The fact that weights are estimated rather than exact introduces some additional sampling variation into the adjustments. This is unimportant in the larger experiments, but in certain of the smaller experiments the weights cannot be estimated accurately and the original analysis is recommended. For other accounts of the analysis by different approaches, see references (9.2), (9.3), (9.4), (9.5), and (9.4a).

9.4 Comparison of Incomplete Block and Randomized Block Designs

So far as the experimental operations are concerned, incomplete block designs are no more difficult than randomized blocks. Some extra planning is involved in drawing up and randomizing the experimental plan, especially if care is taken to make the best possible grouping of the experimental units. According to the computing equipment and experience available, the time required for the statistical analysis may exceed that for randomized blocks by 20 to 150%.

The gain in accuracy over randomized blocks depends on the type of experimental material and may be expected to increase as the number of treatments is increased. Most of the incomplete block designs cover the range from 6 to 200 treatments, while the cubic lattices extend this range to 1000 treatments without requiring a large incomplete block. The number of treatments for which a substantial increase in accuracy is attained must be determined by experience. If the experimental material is highly variable and yet lends itself to the formation of small groups which are homogeneous, the designs may be advantageous even with small numbers of treatments. From the results of varietal trials a number of comparisons with randomized blocks have been made: see, for example, references (9.6), (9.7), and (9.8). The experiments varied in size from the 3×3 to the 11×11, and included both lattices and lattice squares. They indicated an average gain in accuracy of the order of 25%. This means that 4 replications of an incomplete block design were about as accurate as 5 replications of randomized blocks.

The ease with which the number of replications can be increased is also a factor. The object in the new designs is to obtain the most accurate comparisons that are possible from a given number of experimental units. Accordingly, the designs are likely to be most helpful when the amount of experimental material or considerations of cost and labor force the experiment to be smaller than is desired. Where the number of replicates can be increased without difficulty, the experimenter may prefer some extra replication of a simpler design which avoids the calculation of adjustments.

There is one important property, possessed by many of the designs, which increases their attractiveness relative to randomized blocks. As the plans show, the lattice square and the lattice designs are arranged in complete replications as well as in incomplete blocks. A few of the balanced incomplete block designs can also be grouped into replications.

Such designs may be regarded as randomized block designs which have additional restrictions within each replication. It has been shown by Yates (9.2) that *these designs can be analyzed as if they were ordinary randomized blocks.* This implies that the unadjusted treatment means give unbiased estimates of the true treatment effects, and that the F- and t-tests do not lose their validity. Of course, this analysis will in general be less accurate than the complete analysis.

The question of the validity of t-tests requires discussion. If real block differences exist, the true error of the difference between two unadjusted treatment means is smaller for a pair of treatments that occur in the same block than for a pair that do not. Hence the randomized blocks analysis, which ascribes the same standard error to the difference between any pair of treatments, will overestimate the true error for some pairs and underestimate it for others. Yates (9.2) shows, however, that if an ordinary randomized blocks design is superimposed on the same experimental site with the same block effects, the true error of the difference still varies from one pair of treatments to another, because the randomization places some pairs of treatments in the same block and others not. This variation in accuracy appears to be of the same order of magnitude for an incomplete block design as for a randomized block design. This argument can break down for certain types of comparison. For instance, if we happened to compare the mean of all treatments that occur in one block with the mean of all those that occur in another block, the standard error of the difference could be substantially larger than the value obtained from the randomized blocks analysis.

Thus, if there is any criterion for forming incomplete blocks, an incomplete block design is worth a trial in preference to a randomized block design *which occupies the same set of replications.* When the data have been collected, the experimenter may choose whether to analyze them as randomized blocks or to complete the full analysis, with the adjustments for incomplete block variations. In fact, if the variation among incomplete blocks is no greater than that within blocks, the complete statistical analysis reduces automatically to that for randomized blocks, as mentioned in section 9.32.

The analysis as randomized blocks may be useful in experiments where several measurements are made on each experimental unit. In the formation of incomplete blocks the units are usually grouped with regard to the most important measurement. For certain other measurements from the same experiment, the grouping may be less effective; in such cases the "randomized blocks" analysis will be satisfactory, as also with measurements of subsidiary interest, where the greatest attainable accuracy is not required.

In some types of research an appreciable number of units are likely to be injured or destroyed in the course of the experiment, so that they must be omitted from the statistical analysis. With incomplete data (except where only a few units are missing) laborious computations are required to calculate the block (or row and column) adjustments. Consequently, in experiments where missing data are of frequent occurrence, incomplete block designs cannot be used to full advantage. Even in this case nothing is lost by using an incomplete block design which can be arranged in complete replicates. If when the experiment is completed it becomes evident that an appreciable number of units must be discarded, the experimenter may use the randomized blocks analysis in which the extra complication due to missing data is smaller. The same considerations apply in cases where certain treatments may have to be ignored in the final analysis.

9.5 Comparisons with Other Designs

Yates (9.1) has discussed three other designs that have been used when the number of treatments is large.

9.51 Systematic Controls. In agricultural field experiments, a control variety is sometimes placed at regular intervals over the site, the experimental varieties being arranged in complete blocks. From the yields of the control plots a fertility index may be calculated for every plot in the experiment. These indices are used to adjust the experimental plot yields for local variations in fertility.

The method of systematic controls is very flexible, since it can be used with any number of treatments and any number of replicates. Little evidence is available about the increase in accuracy obtained from the controls, though it seems probable that the increase is seldom large if the extra space occupied by the controls is taken into account. The calculation of the best adjustments, as described by Yates (9.1), is rather tedious, but if a crude type of adjustment is made most of the potential advantages of the method may be lost.

There may be additional reasons for the presence of extra controls, e.g., for their use in an observational scoring of the experimental material. In this connection it should be noted that extra controls can be included in an incomplete block design.

9.52 Random Controls. In another method the treatments are divided into groups, the grouping remaining the same in all replicates. Each group is arranged in a separate randomized block or latin square experiment. In order to obtain comparisons between treatments that are in different groups, one or more controls are included in each group and randomized along with the other treatments. Before comparing treatments that are in different groups, we subtract from each treatment mean the mean of the controls that are in the same group. Thus the controls serve to correct for differences in the fertility of the sites on which different groups are tested.

This design allows considerable flexibility in number of treatments and amount of replication and is simple to analyze. From theoretical considerations it is likely to be inferior in accuracy to a comparable incomplete block design, if one exists. Moreover the error variance is not the same for all types of comparison. If only one control is included, the variance is twice as great for the difference between two treatments that are in different groups as for two treatments in the same group.

9.53 Split-plot Designs. This arrangement is a variant of the previous design which avoids extra controls. As in section 9.52, the treatments are first divided into a number of groups of equal sizes. Instead of testing each group on a separate site, the groups are combined into a single experiment of the split-plot type. For example, with 25 treatments and 5 replications, we might first divide the treatments into 5 groups, A, B, C, D, and E. The groups are now regarded as whole-plot treatments, and could be arranged in a 5×5 latin square on plots 5 times as large as the basic plots. Within each group the treatments are arranged at random on the individual plots.

From the analysis for a split-plot design (section 7.15), it will be seen that we can test the difference between two treatments whether they are in the same group (i.e., whole-plot) or in different groups. Controls are no longer needed. The advantages and disadvantages of this arrangement are in general similar to those of the "random controls": the split-plot design is more accurate if the space that would have been allotted to the controls is utilized. Sometimes the split-plot design is really more appropriate than an incomplete block design because the treatments divide themselves naturally into groups, and comparisons between members of the same group are regarded as more important than comparisons between members of different groups.

9.6 Choice of Incomplete Block Design

Tables 9.5 (randomized incomplete blocks), 9.6 (lattice squares and incomplete and extended incomplete latin squares), and 9.2a (partially balanced incomplete block designs) form an index to the incomplete block designs. Each table is arranged by number of treatments (t) and number of units per block (k) and shows the numbers of replications for which designs are available, with references to the plans. It is hoped that the tables provide a rapid means of locating a suitable design, if one has been constructed.

Where the number of treatments and the size of block are fixed in advance by the conditions of the experiment, little choice is available to the experimenter. In certain types of research, however, both the number of treatments and the size of block (or row and column) can be varied to some extent without impairing the experiment.

Where more than one design appears appropriate, a design which can be arranged in separate replicates is preferable to one which cannot, and a balanced design is preferable to a partially balanced design. These recommendations usually narrow the choice to one or two designs. For example, with 25 treatments to be compared in 8 replications, we might use (see table 9.5) either a balanced incomplete block design in blocks of 4 units, or a lattice design (partially balanced) in blocks of 5 units. The latter design, though not the former, can be laid out in separate replicates.

The relative advantages of lattice and lattice square designs can be learned only by experience. From the results of a lattice square experiment we may estimate what the standard error would have been if either of the two types of grouping had not been used. Consequently the experimenter may test additional methods of grouping by using lattice squares. If the extra grouping turns out to be ineffective, the accuracy is only slightly less than that of the corresponding lattice design. The statistical computations are, however, more laborious for lattice squares, since correction terms must be obtained for both row and column effects. In general, the lattice squares may be expected to be successful in types of experimentation where the latin square has been found superior to randomized blocks.

TABLE 9.5 DESIGNS ARRANGED IN RANDOMIZED INCOMPLETE BLOCKS

Number of treatments, t	Number of units per block, k	Number of replications, r *		Plan number or chapter ‡	Number of treatments, t	Number of units per block, k	Number of replications, r *		Plan number or chapter ‡
		p.b.†	bal.†				p.b.†	bal.†	
4	2		3	11.1	16	4	2 ···	5	10.2
	3		3	11.		6		6	11.27
5	2		4	11.2		6		9	11.28
	3		6	11.1a		10		10	11.29
	4		4	11.	19	3		9	11.30
6	2		5	11.3		9		9	11.31
	3		5	11.4		10		10	11.32
	3		10	11.5	20	4	2, 3		10.11
	4		10	11.6	21	3		10	11.33
	5		5	11.		5		5	11.34
7	2		6	11.2a		7		10	11.35
	3		3	11.7	25	4		8	11.36
	4		4	11.8		5	2 ···	6	10.3(
	6		6	11.		9		9	11.37
8	2		7	11.9	27	3	3		10.
	4		7	11.10	28	4		9	11.38
	7		7	11.		7		9	11.39
9	2		8	11.3a	30	5	2, 3		10.12
	3	2, 3	4	10.1	31	6		6	11.40
	4		8	11.11		10		10	11.41
	5		10	11.12	36	6	2, 3		10.7
	6		8	11.13	37	9		9	11.42
	8		8	11.	41	5		10	11.43
10	2		9	11.14	42	6	2, 3		10.13
	3		9	11.15	49	7	2 ···	8	10.4
	4		6	11.16	56	7	2, 3		10.14
	5		9	11.17	57	8		8	11.44
	6		9	11.18	64	4	3		10.
	9		9	11.	64	8	2 ···	9	10.5
11	2		10	11.4a	72	8	2, 3		10.15
	5		5	11.19	73	9		9	11.45
	6		6	11.20	81	9	2 ···	10	10.6
	10		10	11.	90	9	2, 3		10.16
12	3	2, 3		10.10	91	10		10	11.46
13	3		6	11.21	100	10	2, 3		10.8
	4		4	11.22	121	11	2 ···		10.(12.7)
	9		9	11.23	125	5	3		10.
15	3		7	11.24	144	12	2, 3, 4		10.9
	7		7	11.25	169	13	2 ···		10.(12.8)
	8		8	11.26					

* Or any multiple of this number.
† p.b. = partially balanced designs.
bal. = balanced designs.
‡ References 10., 11., denote *chapters*.

TABLE 9.6 DESIGNS ARRANGED IN LATTICE SQUARES AND INCOMPLETE AND EXTENDED INCOMPLETE LATIN SQUARES

Number of treatments, t	Number of units per col. and row, k	Number of replications, r *		Plan number	Number of treatments, t	Number of units per col. and row, k	Number of replications, r *		Plan number
		p.b.†	bal.†				p.b.†	bal.†	
3	6, 3		6	2LS	9	9, 4		8	13.10a
	9, 3		9	3LS		9, 5		10	13.11a
	4, 4	5+		13R+C+		8, 8	7+		13R−C−
	3, 5	5		13.16		9, 8		8	13 ‡
	3, 7	7		13.17		8, 10	9−		13R−C+
	3, 8	8		13.18	10	10, 3		9	13.12a
	3, 10	10		13.19		9, 9	8+		13R−C−
4	4, 3		(3) §, 6	13 ‡		10, 9		9	13 ‡
	8, 4		8	2LS		9, 11	10−		13R−C+
	3, 5	4−		13R−C+	11	11, 2		10	13.13a
	4, 5	5		13.20		11, 5		5	13.3
	5, 5	6+		13R+C+		11, 6		6	13.4
	4, 7	7		13.21		10, 10	9+		13R−C−
	4, 9	9		13.22		11, 10		10	13 ‡
5	5, 2		4	13.6a	12	11, 11	10+		13R−C−
	5, 3		6	13.7a	13	13, 3		6	13.14a
	4, 4	3+		13R−C−		13, 4		4	13.5
	5, 4		4	13 ‡		13, 9		9	13.6
	10, 5		10	2LS	15	15, 7		7	13.7
	4, 6	5−		13R−C+		15, 8		8	13.8
	5, 6	6		13.23	16	4, 4		5	12.2
	6, 6	7+		13R+C+		16, 6		6	13.9
	5, 9	9		13.24		16, 10		10	13.10
6	5, 5	4+		13R−C−	19	19, 3		9	13.15a
	6, 5		5	13 ‡		19, 9		9	13.11
	5, 7	6−		13R−C+		19, 10		10	13.12
	6, 7	7		13.25	21	21, 5		5	13.13
	7, 7	8+		13R+C+	25	25, 4		8	13.16a
7	7, 2		6	13.8a		5, 5		3	12.3
	7, 3		3	13.1		25, 9		9	13.1a
	7, 4		4	13.2	31	31, 6		6	13.14
	6, 6	5+		13R−C−		31, 10		10	13.2a
	7, 6		6	13 ‡	37	37, 9		9	13.15
	6, 8	7−		13R−C+	41	41, 5		10	13.17a
	7, 8	8		13.26	49	7, 7	3	4	12.4
	8, 8	9+		13R+C+	57	57, 8		8	13.3a
8	7, 7	6+		13R−C−	64	8, 8	3 ···	9	12.5
	7, 8		7	13 ‡	73	73, 9		9	13.4a
	7, 9	8−		13R−C+	81	9, 9	3, 4	5	12.6
	9, 9	10+		13R+C+	91	91, 10		10	13.5a
9	9, 2		8	13.9a	121	11, 11	3 ···	6	12.7
	3, 3		(2) §, 4	12.1	169	13, 13	3 ···	7	12.8

\+ = One treatment has one more replicate.

− = One treatment has one less replicate.

R+C+ means one row and one col. added to a $t \times t$ latin square.

R−C+ means one row omitted and one col. added.

R−C− means one row and one col. omitted.

* Or any multiple of this number.

† p.b. = partially balanced designs.
bal. = balanced designs.

‡ Constructed from a $t \times t$ latin square by omission of last col.

( ) § Not enough degrees of freedom for error.

TABLE 9.2a PARTIALLY BALANCED INCOMPLETE BLOCK DESIGNS

t	k	r	Reference	Design number
5	3	3	11.6a	C12 *
	3	9	11.6a	C15 *
6	3	2	11.2a	SR1
	3	3	11.2a	R1 *
	3	4	11.2a	SR2
	3	6	11.2a	SR3,* R3 *
	3	7	11.6a	R44, R45
	3	8	11.6a	R46, R47
	4	2	11.2a	S1 *
	4	4	11.2a	S2,* R2 *
	4	6	11.2a	SR4 *
	4	8	11.2a	R4 *
	4	8	11.6a	R84 *
8	3	3	11.2a	R5 *
	3	9	11.2a	R7 *
	4	3	11.2a	S6
	4	4	11.2a	SR7 *
	4	6	11.2a	S8,* SR8
	4	6	11.2a	SR9 *
	4	10	11.2a	SR11
	5	5	11.6a	R108, R109
	6	3	11.2a	S7 *
	6	6	11.2a	S9 *
	6	9	11.6a	R131
9	3	3	11.2a	SR12 *
	3	5	11.2a	R9
	3	6	11.2a	R10,* LS3 *
	3	7	11.6a	R55
	3	7	11.2a	R11
	3	8	11.2a	LS4
	3	8	11.6a	R57
	3	9	11.2a	R12
	3	10	11.2a	R13, LS5
	4	4	11.2a	R8,* LS1 *
	5	5	11.2a	LS10,* R112
	6	2	11.2a	S12 *
	6	4	11.2a	LS8 *
	6	6	11.2a	S14,* SR14*
	7	7	11.6a	R134
10	3	3	11.2a	T6 *
	3	6	11.2a	T14 *
	4	2	11.2a	T1 *
	4	4	11.2a	T2,* T12 *
	4	4	11.2a	S17 *
	4	8	11.2a	R14 *
	5	3	11.2a	T9
	5	4	11.2a	SR16
	5	5	11.6a	R114
	5	6	11.2a	SR17

t	k	r	Reference	Design number
10	5	8	11.2a	SR18
	5	10	11.2a	SR19 *
	6	3	11.2a	T15 *
	6	6	11.2a	T16,* T18 *
	6	6	11.2a	S19 *
	6	6	11.6a	R132
	7	7	11.2a	T19 *
	8	4	11.2a	S18 *
	8	8	11.2a	S21 *
12	3	4	11.2a	SR21
	3	5	11.2a	R16
	3	5	11.6a	S1.5
	3	6	11.2a	R17 *
	3	7	11.2a	R18
	3	7	11.6a	R64
	3	8	11.2a	R19
	3	9	11.2a	R20 *
	3	10	11.2a	R22
	4	3	11.2a	SR20
	4	4	11.2a	R15 *
	4	5	11.2a	S26
	4	6	11.2a	SR23 *
	4	9	11.2a	SR29
	4	10	11.2a	R23 *
	5	5	11.6a	R116, R117
	5	5	11.6a	R118
	6	3	11.2a	S23 *
	6	4	11.2a	SR22 *
	6	5	11.2a	S27
	6	6	11.2a	S29 *
	6	6	11.2a	SR24,*SR25 *
	6	8	11.2a	SR28*
	6	10	11.2a	SR30, SR31 *
	7	7	11.6a	R135, R136
	7	7	11.6a	R137, R138
	8	2	11.2a	S22 *
	8	6	11.2a	SR26 *
	8	8	11.2a	S32 *
	8	10	11.2a	S38
	9	3	11.2a	S24 *
	9	9	11.2a	S34 *
	9	9	11.6a	R149
	10	5	11.2a	S28 *
	10	10	11.2a	S39 *
13	3	3	11.2a	C1 *
	3	9	11.2a	C3 *
	4	8	11.2a	C2 *
	6	6	11.6a	C23 *
	7	7	11.6a	C24 *
	10	10	11.2a	C4 *

t	k	r	Reference	Design number
14	3	6	11.2a	R25 *
	4	4	11.2a	R24 *
	4	6	11.2a	S42 *
	6	3	11.2a	S40 *
	6	6	11.2a	S43 *
	7	4	11.2a	SR32
	7	6	11.2a	SR33
	7	7	11.6a	R139
	7	8	11.2a	SR34
	7	10	11.2a	SR35
	8	4	11.2a	S41 *
	8	8	11.2a	S44 *
	8	8	11.6a	R145
15	3	3	11.2a	T28 *
	3	4	11.2a	T23
	3	5	11.2a	SR36
	3	6	11.2a	R28 *
	3	8	11.2a	R29
	3	9	11.2a	R32,* R33 *
	3	10	11.2a	T30
	4	4	11.2a	R27 *
	4	8	11.2a	R31 *
	5	2	11.2a	T20
	5	6	11.2a	SR37
	5	9	11.2a	SR38
	5	10	11.2a	R34 *
	6	4	11.2a	S47,* T22 *
	6	6	11.6a	R133
	9	6	11.2a	S49,* T27 *
	9	9	11.6a	R150
	10	2	11.2a	S46 *
	10	4	11.2a	T25 *
	10	10	11.2a	S53 *
16	3	3	11.2a	LS14 *
	3	6	11.2a	R35 *
	3	9	11.2a	R38 *
	4	4	11.2a	SR40 *
	4	6	11.2a	R36 *
	4	7	11.2a	S56, R37
	4	7	11.2a	LS12
	4	8	11.2a	LS15 *
	4	9	11.2a	R39, LS13
	5	5	11.6a	S1.11
	7	7	11.6a	LS58
	8	3	11.2a	S54
	8	6	11.2a	S55,* SR41 *
	8	7	11.2a	S57
	8	8	11.2a	SR43 *
	8	10	11.2a	SR44 *
	9	9	11.6a	R151
	9	9	11.6a	LS87

* Arranged to allow the elimination of variation in two directions; see section 13.4a

TABLE 9.2a PARTIALLY BALANCED INCOMPLETE BLOCK DESIGNS (*Continued*)

t	k	r	Reference	Design number
17	4	8	11.2a	C5 *
	8	8	11.2a	C6 *
	9	9	11.2a	C7 *
18	3	6	11.2a	SR45 *
	3	9	11.2a	R40
	3	9	11.2a	R41 *
	4	8	11.2a	S62 *
	6	4	11.2a	S59
	6	5	11.2a	S60
	6	6	11.2a	SR46 *
	6	9	11.2a	SR49 *
	6	10	11.2a	S65 *
	7	7	11.6a	R140
	8	8	11.2a	S64 *
	9	5	11.2a	S61
	9	6	11.2a	SR47 *
	9	8	11.2a	SR48
	9	10	11.2a	SR50
	10	10	11.2a	S67 *
	10	10	11.6a	R154
20	3	9	11.2a	R42 *
	4	5	11.2a	SR52
	4	9	11.2a	S73, R43
	4	10	11.2a	SR56 *
	5	4	11.2a	SR51
	5	8	11.2a	SR54
	6	9	11.2a	S74 *
	8	4	11.2a	S69 *
	8	6	11.2a	S70 *
	8	8	11.2a	S72 *
	10	3	11.2a	S68
	10	6	11.2a	S71,* SR53 *
	10	8	11.2a	SR55 *
	10	9	11.2a	S76
	10	10	11.2a	SR57 *
21	3	5	11.6a	T19
	3	7	11.2a	SR59
	3	9	11.2a	R44 *
	6	2	11.2a	T31 *
	6	6	11.2a	S78 *
	7	6	11.2a	SR58
	7	9	11.2a	SR60
	8	8	11.6a	R146
	9	3	11.2a	S77 *
	9	9	11.2a	S80 *
22	4	10	11.2a	S82 *
	10	5	11.2a	S81 *
	10	10	11.2a	S83 *
24	3	8	11.2a	SR61
	3	9	11.2a	R47 *
	3	10	11.2a	R48
	4	7	11.2a	R46
	4	10	11.2a	R49 *
	5	5	11.2a	R45 *
	6	7	11.2a	S85
	6	8	11.2a	SR62 *
	8	5	11.2a	S84
	8	9	11.2a	SR63
	8	10	11.2a	S86 *
	9	9	11.6a	R152
25	3	6	11.2a	LS16 *
	5	5	11.2a	SR64 *
	5	7	11.2a	R51
	5	8	11.2a	R52
	10	4	11.2a	S87 *
26	4	8	11.2a	R53 *
	6	3	11.2a	S1.4 *
	6	6	11.2a	S90,* S1.5 *
	8	4	11.2a	S89 *
	8	8	11.2a	S91 *
27	3	5	11.2a	S1.7
	3	9	11.2a	SR66 *
	4	8	11.2a	R54 *
	6	8	11.2a	S93 *
	9	4	11.2a	S92
	9	9	11.2a	SR67 *
	10	10	11.6a	R155
28	3	6	11.6a	T21
	4	7	11.2a	SR68
	4	8	11.2a	R55 *
	4	10	11.2a	R56 *
	7	2	11.2a	T32
	7	8	11.2a	SR69
	8	6	11.2a	S95 *
29	7	7	11.2a	C8 *
	8	8	11.2a	C9 *
30	3	10	11.2a	SR73
	4	10	11.2a	R57 *
	6	5	11.2a	SR70
	6	7	11.2a	S97
	6	9	11.2a	S98
	6	10	11.2a	SR72 *
	9	9	11.2a	S99 *
	10	5	11.2a	S96 *
	10	9	11.2a	SR71
	10	10	11.2a	S100 *
32	4	8	11.2a	SR74 *
	8	5	11.2a	S101
	8	7	11.2a	S102
	8	8	11.2a	SR75 *
33	6	10	11.2a	S104 *
	7	7	11.2a	R58 *
35	5	7	11.2a	SR76
	5	10	11.2a	R59 *
	7	3	11.2a	S1.9
	10	6	11.2a	S105 *
36	3	7	11.6a	T22
	4	9	11.2a	SR78
	5	5	11.2a	LS17 *
	8	2	11.2a	T33 *
	8	8	11.2a	S106 *
	9	7	11.2a	T34
	9	8	11.2a	SR77
37	3	9	11.2a	C10 *
38	6	9	11.2a	S107 *
39	5	10	11.2a	R60 *
	9	6	11.2a	S108 *
40	4	4	11.2a	S1.12 *
	5	8	11.2a	SR79
	5	9	11.2a	R61
	8	9	11.2a	S110
	10	7	11.2a	S109
42	6	7	11.2a	SR80
	6	10	11.2a	S112
	10	5	11.2a	S111 *
	10	10	11.2a	S113 *
44	8	10	11.2a	S114 *
45	3	8	11.6a	T23
	5	3	11.2a	S1.14
	5	9	11.2a	SR81
	5	10	11.2a	R62
	9	2	11.2a	T35
	9	7	11.2a	S115
	10	8	11.2a	S116 *
48	6	8	11.2a	SR82 *
	7	7	11.2a	R63 *
49	3	6	11.6a	LS19
	6	6	11.2a	LS18

* Arranged to allow the elimination of variation in two directions; see section 13.4a.

TABLE 9.2a PARTIALLY BALANCED INCOMPLETE BLOCK DESIGNS (*Continued*)

t	k	r	Reference	Design number
49	7	7	11.2a	SR83 *
	7	9	11.2a	R64
	7	10	11.2a	R65
50	8	4	11.2a	S1.17 *
	8	8	11.2a	S118 *
	10	6	11.2a	S117
	10	9	11.2a	S119
54	6	9	11.2a	SR84
55	3	9	11.6a	T24
	10	2	11.2a	T36 *
	10	10	11.2a	S120 *
56	7	8	11.2a	SR86
	8	7	11.2a	SR85

t	k	r	Reference	Design number
56	8	9	11.2a	S121
57	9	3	11.2a	S1.18 *
	9	9	11.2a	S122,* S1.20 *
63	7	9	11.2a	SR87
	8	8	11.2a	R66 *
	9	4	11.2a	S1.21
	9	10	11.2a	S123
64	7	7	11.2a	LS19 *
	8	8	11.2a	SR88 *
	8	10	11.2a	R67
70	10	3	11.2a	S1.22
	10	6	11.2a	S1.23 *

t	k	r	Reference	Design number
72	8	9	11.2a	SR90
	9	8	11.2a	SR89
80	9	9	11.2a	R68
81	9	9	11.2a	SR91 *
82	10	5	11.2a	S1.25 *
	10	10	11.2a	S124 *
85	5	5	11.2a	S1.26
100	9	9	11.2a	LS20 *
156	6	6	11.5a	

* Arranged to allow the elimination of variation in two directions; see section 13.4a.

REFERENCES

9.1 YATES, F. A new method of arranging variety trials involving a large number of varieties. *Jour. Agr. Sci.* 26, 424–455, 1936.

9.2 YATES, F. The recovery of inter-block information in variety trials arranged in three-dimensional lattices. *Ann. Eugen.* 9, 136–156, 1939.

9.3 YATES, F. The recovery of inter-block information in balanced incomplete block designs. *Ann. Eugen.* 10, 317–325, 1940.

9.4 COCHRAN, W. G. The analysis of lattice and triple lattice experiments in corn varietal tests. II. Mathematical theory. *Iowa Agr. Exp. Sta. Res. Bull.* 281, 1940.

9.5 RAO, C. R. General methods of analysis for incomplete block designs. *Jour. Amer. Stat. Assoc.* 42, 541–561, 1947.

9.6 COCHRAN, W. G. An examination of the accuracy of lattice and lattice square experiments on corn. *Iowa Agr. Exp. Sta. Res. Bull.* 289, 1941.

9.7 BLISS, C. I., and DEARBORN, R. B. The efficiency of lattice squares in corn selection tests in New England and Pennsylvania. *Proc. Amer. Soc. Hort. Sci.* 41, 324–342, 1942.

9.8 WELLHAUSEN, E. J. The accuracy of incomplete block designs in varietal trials in West Virginia. *Jour. Amer. Soc. Agron.* 35, 66–76, 1943.

9.1a BOSE, R. C., and NAIR, K. R. Partially balanced incomplete block designs. *Sankhya* 4, 337–372, 1939.

9.2a BOSE, R. C., CLATWORTHY, W. H., and SHRIKHANDE, S. S. Tables of partially balanced designs with two associate classes. *North Carolina Agr. Exp. Sta. Tech. Bull.* 107, 1954.

9.3a CLATWORTHY, W. H. Partially balanced incomplete block designs with two associate classes and two treatments per block. *Nat. Bur. Standards Jour. Res.* 54, 174–189, 1955.

9.4a KEMPTHORNE, O. *The design and analysis of experiments.* John Wiley and Sons, New York, 1952.

ADDITIONAL READING

NAIR, K. R. The recovery of inter-block information in incomplete block designs. *Sankhya* 6, 383–390, 1944.

TOCHER, K. D. Design and analysis of block experiments. *Jour. Roy. Stat. Soc. B*, 14, 45–100, 1952.

CHAPTER 10

Lattice Designs

10.1 Balanced Lattices

10.11 Description. The number of treatments must be an exact square. The size of block is the square root of this number. Incomplete blocks are combined in groups to form separate replications. The special feature of the balanced lattice, as distinguished from other lattices, is that every pair of treatments occurs once in the same incomplete block. Consequently, all pairs of treatments are compared with the same degree of precision.

The numbers of replications are rather severely restricted, as well as the numbers of treatments. The useful plans are indexed below. Balanced lattice designs cannot be constructed for 36 treatments, and none has been found for 100 or 144 treatments.

BALANCED LATTICES CONTAINED IN PLANS 10.1–10.6

Number of treatments	9	16	25	49	64	81
Units per block	3	4	5	7	8	9
Replicates	4	5	6	8	9	10

10.12 Statistical Analysis. The analysis is comparatively easy. It will be illustrated by an experiment on the effects of 9 feeding treatments on the growth rates of pigs, conducted by the North Carolina Agricultural Experiment Station. The analysis has been described in more detail by Comstock *et al.* (10.1), and the nutritional information obtained from the results by Peterson *et al.* (10.2).

For pigs of a given breed, previous experience indicated that a considerable part of the variance in growth rate between animals can be ascribed to the litter. Hence the experiment was planned so that litter differences would not contribute to the intra-block error. The pigs were divided into sets of 3 litter-mates. Two sets of 3 were assigned to each block. Within a block, each treatment received one member of each set. Thus the experimental unit was composed of 2 pigs each feeding in the same pen. The plan and growth rates are given in table 10.1; the plan has been rearranged so that it follows the same pattern as plan 10.1.

TABLE 10.1 Gains in weight (pounds per day) for a total of 2 pigs

Block	Rep. I			Totals	Block	Rep. II			Totals
(1)	(1)	(2)	(3)		(4)	(1)	(4)	(7)	
	2.20	1.84	2.18	6.22		1.19	1.20	1.15	3.54
(2)	(4)	(5)	(6)		(5)	(2)	(5)	(8)	
	2.05	0.85	1.86	4.76		2.26	1.07	1.45	4.78
(3)	(7)	(8)	(9)		(6)	(3)	(6)	(9)	
	0.73	1.60	1.76	4.09		2.12	2.03	1.63	5.78
				15.07					14.10
	Rep. III					Rep. IV			
(7)	(1)	(5)	(9)		(10)	(1)	(6)	(8)	
	1.81	1.16	1.11	4.08		1.77	1.57	1.43	4.77
(8)	(2)	(6)	(7)		(11)	(2)	(4)	(9)	
	1.76	2.16	1.80	5.72		1.50	1.60	1.42	4.52
(9)	(3)	(4)	(8)		(12)	(3)	(5)	(7)	
	1.71	1.57	1.13	4.41		2.04	0.93	1.78	4.75
				14.21					14.04

Treatment totals and adjustment factors

	T	B_t	$W = (3T - 4B_t + G)$	Adjusted total $T + \mu W$	Mean per unit
1	6.97	18.61	+3.89	7.21	1.80
2	7.36	21.24	−5.46	7.02	1.76
3	8.05	21.16	−3.07	7.86	1.96
4	6.42	17.23	+7.76	6.91	1.73
5	4.01	18.37	−4.03	3.76	0.94
6	7.62	21.03	−3.84	7.38	1.84
7	5.46	18.10	+1.40	5.55	1.39
8	5.61	18.05	+2.05	5.74	1.44
9	5.92	18.47	+1.30	6.00	1.50
	G = 57.42	172.26	0.00	57.43	

The steps in the analysis are as follows. The algebraic formulae refer to a $k \times k$ lattice in blocks of k units, with $r = (k + 1)$ replicates.

1. Calculate the block totals, the replication totals, the grand total G, and the treatment totals T, shown under the plan.

2. For each treatment, calculate the total B_t for all blocks in which

the treatment appears. For treatment 4, this is

$$B_t = 4.76 + 3.54 + 4.41 + 4.52 = 17.23$$

As a check, the total of the B_t values should be k times the total of the T values.

3. Compute the quantities

$$W = kT - (k + 1)B_t + G$$

whose sum should be exactly zero.

4. The analysis of variance is now obtained. The total s.s. and the sums of squares for replications and treatments are found in the usual way. The sum of squares for blocks within replications, adjusted for treatment effects, is

$$\frac{\sum W^2}{k^3(k+1)} = \frac{(3.89)^2 + (5.46)^2 + \cdots + (1.30)^2}{108} = 1.4206$$

5. Calculate the adjustment factor

$$\mu = \frac{(E_b - E_e)}{k^2 E_b} = \frac{0.1776 - 0.0773}{9 \times 0.1776} = 0.0628$$

where E_b and E_e are the blocks and intra-block m.s., respectively. The adjusted treatment total is $(T + \mu W)$ as shown in table 10.1. To avoid confusion, the adjusted means are shown *per unit* (total of 2 pigs) although a mean per pig would be more natural. If E_b is less than E_e, μ is taken as zero, and no adjustments are applied to the treatment totals.

TABLE 10.2 Analysis of variance for total growth rate of 2 pigs

	d.f.			
	General	$k = 3$	s.s.	m.s.
Replications	k	3	0.0774	
Treatments	$(k^2 - 1)$	8	3.2261	
Blocks (adj.)	$(k^2 - 1)$	8	1.4206	0.1776
Intra-block error	$(k - 1)(k^2 - 1)$	16	1.2368	0.0773
Total	$(k^3 + k^2 - 1)$	35	5.9609	

6. For t-tests, calculate the effective error m.s.

$$E_e' = E_e(1 + k\mu) = 0.0773(1 + 3 \times 0.0628) = 0.0919$$

The purpose of the adjustment factor is to increase E_e so as to take account of sampling errors in the block correction values μW. The ordinary rules for the calculation of t-tests may now be applied to E_e'. Thus the variance of the difference between two adjusted treatment totals is $2rE_e'$, while that for the difference between two adjusted treatment means is $2E_e'/r$, or 0.0460. The standard error of this difference is 0.214. If means per pig were taken, the standard error would be 0.107.*

7. The treatments m.s. in table 10.2 cannot be tested against the intra-block error m.s., since the former contains some block effects. For an approximate F-test, calculate the sum of squares of deviations of the adjusted treatment totals, which comes to 12.6771. This is divided by r or 4 to bring to a single-unit basis, and by $(k^2 - 1)$ or 8 to obtain the mean square, 0.3962. This is tested against the effective error m.s., 0.0919. The F-ratio, 0.3962/0.0919, or 4.31, has 8 and 16 d.f.

8. In order to estimate the precision relative to randomized blocks, pool the mean squares for blocks (adjusted) and the intra-block error. The result is 0.1107, with 24 d.f., and is an unbiased estimate of the error variance that would have been present if the experiment had been arranged in randomized blocks. This figure is compared with the effective error m.s., 0.0919. The relative precision is 0.1107/0.0919, or 120%. This means that 4 replications of the balanced lattice appear to have been slightly less accurate than 5 replicates of randomized blocks.

These designs may be used in factorial experiments where it is desired to avoid any sacrifice of replication on the interactions relative to that on the main effects. Some of the combinations of levels that may be tested in balanced lattices are: 3×3, 8×2, 4×4, $4 \times 2 \times 2$, 2^4, 5×5, 7×7, 8×8, 4^3, 2^6, 9×9, 3^4. In analyzing a factorial experiment, first carry through the procedure described in this section. The sum of squares for deviations of the adjusted treatment totals is then divided into the sums of squares for main effects and interactions in the usual way; a divisor r must be introduced in order to convert these sums of squares to a single-unit basis. The resulting mean squares are tested against the effective error m.s., $E_e(1 + k\mu)$, which has $(k - 1)(k^2 - 1)$, or in this case, 16 d.f.

10.12a Effects of Errors in the Weights. The fact that the recovery of inter-block information is based upon *estimated* weights has two principal consequences. There is a loss of precision in the sense that the estimates are less precise than they would be if the weights were known exactly. The magnitude of this loss has been investigated by Yates (10.12), Cochran (10.5), and Kempthorne (10.1a). Their results suggest the conclusion that the loss is unimportant if there are at least 10 d.f. for estimating the mean square for blocks.

A greater source of disturbance, more difficult to investigate, is that the formula for the standard error of the difference between two ad-

* If μ is taken as zero, use E_e as the effective error m.s., rather than the pooled m.s. for E_b and E_e. This practice is recommended for future designs also.

justed treatment means in an underestimate. The bias arises from two sources—from the assumption that true weights were used and from the substitution of estimated for true weights when the formula is computed in practice. For the 5 × 5 simple lattices, Meier (10.2a) has computed the bias in the average variance of the difference between two treatment means. His investigation also throws light on the question whether we ought to use E_e as the effective error mean square when μ is taken as zero, as recommended in the footnote to the previous page, or the pooled mean square for E_b and E_e in this situation. Results for both estimates, V_e and V_p, say, are shown below for a range of values of w/w'.

Bias of variance estimates expressed as percent errors

w/w'	1	2	3	4	5	6	7	8	9
V_e	+11.4	−5.0	−7.4	−7.3	−6.8	−6.1	−5.5	−5.0	−4.6
V_p	−6.2	−9.4	−9.0	−8.0	−7.1	−6.3	−5.6	−5.0	−4.6

Use of E_e when μ is zero produces a smaller negative bias over most of the range of w/w' but results in a positive bias when w/w' is near 1, i.e., when there is little or no block variance. Although overall there is little to choose between the two estimates, V_e appears less subject to bias on the average.

With V_e, the results suggest an average underestimation of about 6% in the variance, or 3% in the standard error. For 3 × 3 and 4 × 4 lattices, we may speculate that the standard error may be too low by as much as 10%. For 6 × 6 and larger lattices, the underestimation should be negligible. It would be desirable to develop a correction term in the standard error formula which removes most of the bias, but further work on this remains to be done.

10.13 Missing Data. Since lattice designs are often used in large experiments, involving perhaps several hundred observations, it is not easy to ensure that all the observations are accurately made. Even with careful management of the experiment, there is always a chance that mistakes or accidents will affect a few of the observations. Consequently, missing data tend to be more common with lattices than with small experiments.

Methods for the analysis of the results of lattice experiments with incomplete data have been developed by Cornish (10.3). As might be expected, the computations are lengthy. They become simpler in two special cases. The first arises when the incomplete blocks are ineffective, so that the analysis reduces to a "randomized blocks" analysis. In this event values are substituted for the missing observations by the formula which applies to randomized blocks. The second case occurs at the other extreme when the variation among blocks is so large that the inter-block information is negligible. Here the correct procedure is to insert values for the missing observations by minimizing the intra-block error m.s.

Since the general procedure must reduce to these two special procedures in the appropriate cases, it might be anticipated, as Cornish's solution shows, that both the "randomized blocks" and the "intra-block" estimates for the missing observations are required. Thus we have to find two estimates for each missing value. Similarly, two analyses of variance are necessary, one to obtain the correct value for the intra-block error m.s. and one for obtaining the mean square among blocks.

In an attempt to reduce the amount of arithmetic, some investigation has been made of the consequences of using a single estimate and a single analysis of variance. For this purpose the "intra-block" estimate seems the better of the two. It gives the correct intra-block error m.s.; the block m.s. is in general slightly too high. It provides an excellent approximation when the blocks are effective and is at its worst when blocks are ineffective; that is, when a "randomized blocks" analysis should have been used. In the latter case it still gives unbiased estimates of the treatment means, but they are not as accurate as the estimates obtained by the use of the "randomized blocks" formula. The chief defect of the "randomized blocks" formula is that when block variation is large it tends to give an overestimate of the intra-block error m.s., and sometimes the bias is substantial. For this reason the "randomized blocks" estimate is considered more hazardous for general use, despite its greater simplicity.

The procedure that is given for the incomplete block designs in this and succeeding chapters is to insert values for the missing observations by means of the "intra-block" formula. Thereafter the analysis is conducted in the usual way for complete data, except that in the analysis of variance 1 d.f. is subtracted from the intra-block error for each missing observation. This method, or the more accurate Cornish method, should be used whenever it is intended to recover inter-block information. If many observations are missing, or if it is evident on inspection that blocks are relatively ineffective, the experimenter may decide at the start to analyze the data as a randomized block experiment. In this case substitutes for the missing values should of course be obtained by the formula for randomized blocks (section 4.25).

The "intra-block" formula for the estimate of a missing observation in a $k \times k$ balanced lattice with $(k + 1)$ replications is as follows.

$$x = \frac{k^2T + k(k+1)B - R + G - kT_b - kB_t}{k(k-1)^2} \tag{10.1}$$

In this formula, T, B, and R denote as usual the totals for the treatment, block, and replication which contain the missing observation, while G is the grand total. Also

T_b = total (over all replications) of all treatments that appear in the block which has the missing observation.

B_t = total of all blocks in which the treatment with the missing observation appears.

Example. Suppose that the observation 2.20 for treatment 1 in replicate 1 of table 10.1 had been missing. Form the block, replicate, treatment, and grand totals, just as in the ordinary analysis. It is helpful to insert an x for the missing observation and to include this x in all totals where it should appear. When the value of x has been found, it can then be inserted in all the appropriate places and the data are ready for computing the analysis of variance. The list of treatment totals appears as shown below.

	T
1	$4.77 + x$
2	7.36
3	8.05
4	6.42
5	4.01
6	7.62
7	5.46
8	5.61
9	5.92
$G =$	$55.22 + x$

The quantities needed are

$$T = 4.77, \qquad B = 4.02, \qquad R = 12.87, \qquad G = 55.22$$

$$T_b = 4.77 + 7.36 + 8.05 = 20.18$$

$$B_t = 4.02 + 3.54 + 4.08 + 4.77 = 16.41$$

Hence

$$x = \frac{(9)(4.77) + (12)(4.02) - 12.87 + 55.22 - (3)(20.18) - (3)(16.41)}{12}$$

$$= 1.98$$

The analysis of variance is now computed in the usual way, except that the total degrees of freedom are reduced from 35 to 34 and the intra-block error degrees of freedom from 16 to 15. The same adjustment factor μ may be used (there is actually a slight change which can be ignored).

When several observations are missing, the values are inserted by the method of successive approximation (illustrated in section 4.25). The exact formulae for t-tests are complicated. The following approximate rule assigns an effective number of replicates to each of two treatments A and B whose means are being compared. In any replicate of the experiment, A is credited with 0 replication if it is absent; with 0 replication if A is present but B occurs *in the same block* and is absent; and with 1 replication otherwise. The same rule is applied to B. For instance, in a t-test of the difference between treatments (1) and (3) in table 10.1, treatment (1) is credited with 3 replications and treatment (3) with 3 replications (it loses 1 replication because in the first replicate of the experiment treatment (1) is in the same block and is missing). On the other hand, in comparing treatment (1) with treatment (4), the effective numbers of replications are 3 and 4, respectively.

10.2 Partially Balanced Lattices

10.21 Simple Lattices (Two Replicates). For an experiment with 2 replicates, use the first 2 replicates of the appropriate set shown below (plans 10.1–10.6). Notice that designs are available for 36, 100, and 144 treatments (plans 10.7–10.9) as well as for those numbers of treatments for which balanced designs are given. Plans for 121 and 169 treatments can be taken from plans 12.7 and 12.8.

The asymmetry of the designs is apparent from the plans; thus, with 9 treatments (plan 10.1) the first treatment appears in the same block as treatments (2), (3), (4), and (7), but not in the same block as any of the remaining treatments.

For 9 and 16 treatments, simple lattices are unlikely to be more accurate than randomized blocks unless the variation among incomplete blocks is great compared with that within incomplete blocks. Further, the numbers of degrees of freedom for estimating the error are only 4 and 9, as against 9 and 16, respectively, for randomized blocks.

10.22 Triple Lattices (Three Replicates). The first 3 replicates of plans 10.1–10.9, 12.7, and 12.8 are used. Designs may be obtained for all squares from 9 to 169. With 9 treatments the precaution mentioned in section 10.21 applies also to the triple lattice.

10.23 Four Replicates. These designs may be obtained either (i) by duplicating the simple lattice or (ii) by using a quadruple lattice, i.e., the first 4 replicates of the plans. With 36 or 100 treatments, only the first method can be used, since no quadruple lattice exists. The second procedure is slightly preferable, because the resulting design comes closer to symmetry, but the statistical analysis requires more time.

10.24 Five Replicates. The first 5 replicates from the balanced set are taken. Designs are not included for 9, 36, 100, or 144 treatments.

10.25 Higher Numbers of Replicates. Balanced designs should be used for the following numbers of treatments and replicates: 25, 6; 49, 8; 64, 9; and 81, 10. For 9 treatments in 8 or 12 replicates, 16 treatments in 10 replicates, and 25 treatments in 12 replicates, the plan for the balanced design should be repeated.

In other cases the following recommendations are made.

Six replicates. Use the triple lattice *twice.*

Eight replicates. Use the simple lattice *four times* or the quadruple lattice *twice.*

Nine replicates. Use the triple lattice *three times.*

Ten replicates. Use the simple lattice *five times* or the quintuple lattice *twice.*

The designs recommended are not always as fully balanced as they might be made. For instance with 6 replicates of 16 treatments the balanced design (5 replicates) plus 1 extra replicate gives a more nearly symmetrical arrangement than the triple lattice used twice. Since separate computing instructions would be required for such plans, we have preferred to adhere to a more uniform system. For this reason no account is given of designs with 7 replicates, though their statistical analysis presents no great difficulty to the reader who has mastered the principles.

10.26 Arrangement of Experimental Material. In the arrangement of a group of experimental material so as to apply one of those designs, the most important criterion is that units within the same incomplete block be *homogeneous.* In field trials, for example, if the plots are oblong the usual procedure is to have the incomplete blocks as nearly square as practicable, the plots extending the whole length of one side of the block. Uniformity trial investigations have shown that this layout gives on the average the most homogeneous block. In cases where there is detailed knowledge of the experimental site, some other method of grouping may be superior. Efforts should be made to keep the experimental technique uniform for all units in the same block. Changes in technique that are necessary should be made when changing from one block to another.

When the blocks have been formed, there is a secondary advantage in forming replications so that blocks within the same replicate are as similar to one another as the material permits, since this increases the precision of inter-block comparisons. If the full statistical *analysis is carried out,* it is more important to have the incomplete blocks homogeneous than to have the replications homogeneous. Thus, when the grouping of blocks into compact replications is troublesome, this criterion may be ignored without much loss of precision.

On the other hand, it is desirable to have homogeneous replications if the data may subsequently be analyzed as a randomized block design. As we have pointed out, the randomized block analysis may be needed in fields of work where certain experimental units or whole treatments are likely to be destroyed during the course of the experiment (e.g., in pasture plots subject to winterkill).

10.27 Randomization. The randomization consists of three steps.

1. Randomize the blocks separately and independently within each replication.
2. Randomize the treatments separately and independently within each block.
3. Allot the treatments to the treatment numbers at random.

For methods of randomization, see chapter 15.

Steps 1 and 2 give each treatment an equal chance of being allotted to any experimental unit. These steps correspond to the allotment of treatments to units at random in an ordinary randomized block design.

The function of step 3 is to decide by random choice which groups of treatments will form the blocks of the design. If differences among blocks are large, the error variance per plot for the mean of a group of treatments which lie in the same block may be considerably higher than the average error variance. This additional randomization ensures that the average error variance may be used, in nearly all cases, for comparisons among groups of treatments. For further discussion, see reference (10.4).

When a plan is repeated in order to obtain extra replications, a separate randomization must be made for every replicate.

10.28 Statistical Analysis. An account of the theory, with worked examples of the simple and triple lattice, is given in reference (10.5); a systematic presentation of the shortest computational methods, with worked examples of the simple, quadruple, and balanced lattices in

(10.6); and a detailed description of the methods for carrying out the computations on I.B.M. (Hollerith) punched-card machines in (10.7). The computational methods differ slightly from one reference to another, but all are basically the same.

Table 10.3 shows the plan and yields for a 5 × 5 simple lattice experiment on soybeans, the treatments being 25 varieties produced in the

TABLE 10.3 YIELDS OF A 5 × 5 SIMPLE LATTICE EXPERIMENT ON SOYBEANS

(Yields are in bushels per acre, minus 30 bu.)

Rep. I					B	C	μC
(1) 6	(2) 7	(3) 5	(4) 8	(5) 6	32	+ 61	+ 9.5
(6) 16	(7) 12	(8) 12	(9) 13	(10) 8	61	− 8	− 1.3
(11) 17	(12) 7	(13) 7	(14) 9	(15) 14	54	+ 48	+ 7.5
(16) 18	(17) 16	(18) 13	(19) 13	(20) 14	74	− 15	− 2.3
(21) 14	(22) 15	(23) 11	(24) 14	(25) 14	68	+ 17	+ 2.7
					289	+103	+16.1
Rep. II							
(1) 24	(6) 13	(11) 24	(16) 11	(21) 8	80	− 9	− 1.4
(2) 21	(7) 11	(12) 14	(17) 11	(22) 23	80	− 23	− 3.6
(3) 16	(8) 4	(13) 12	(18) 12	(23) 12	56	− 8	− 1.3
(4) 17	(9) 10	(14) 30	(19) 9	(24) 23	89	− 32	− 5.0
(5) 15	(10) 15	(15) 22	(20) 16	(25) 19	87	− 31	− 4.8
					392	−103	−16.1

Treatment totals (unadj.)					μC
(1) 30	(2) 28	(3) 21	(4) 25	(5) 21	+9.5
(6) 29	(7) 23	(8) 16	(9) 23	(10) 23	−1.3
(11) 41	(12) 21	(13) 19	(14) 39	(15) 36	+7.5
(16) 29	(17) 27	(18) 25	(19) 22	(20) 30	−2.3
(21) 22	(22) 38	(23) 23	(24) 37	(25) 33	+2.7
μC −1.4	−3.6	−1.3	−5.0	−4.8	0.0

Treatment totals (adj.)				
(1) 38.1	(2) 33.9	(3) 29.2	(4) 29.5	(5) 25.7
(6) 26.3	(7) 18.1	(8) 13.4	(9) 16.7	(10) 16.9
(11) 47.1	(12) 24.9	(13) 25.2	(14) 41.5	(15) 38.7
(16) 25.3	(17) 21.1	(18) 21.4	(19) 14.7	(20) 22.9
(21) 23.3	(22) 37.1	(23) 24.4	(24) 34.7	(25) 30.9

soybean breeding program of the North Carolina Experiment Station. The experiment actually contained 4 replications, obtained by duplicating the plan for the simple lattice. The first 2 replications will serve to illustrate the analysis for simple, triple, and quadruple lattices, which all follow the same general pattern. In section 10.29 the data from the whole experiment will be analyzed in order to indicate the procedure when the basic plan is repeated.

1. Find the block totals B, the replication totals, the treatment totals, and the grand total.

2. For each block calculate

$$C = \text{total (over all replicates) of all treatments in the block} - rB$$

For example, for block 1 in replication 1,

$$C = 30 + 28 + 21 + 25 + 21 - (2)(32) = +61$$

Find the replicate totals R_c of the C values. These replicate totals should add to zero.

3. The analysis of variance is as follows.

	d.f.		s.s.	m.s.
Replications	$(r-1)$	1	212.18	
Treatments (unadj.)	(k^2-1)	24	559.28	
Blocks within replications (adj.)	$r(k-1)$	8	501.84	$62.73E_b$
Intra-block error	$(k-1)(rk-k-1)$	16	218.48	$13.66E_e$
Total	(rk^2-1)	49	1491.78	

All sums of squares are found in the usual way except that for blocks (adjusted), which is given by

$$\frac{\sum C^2}{kr(r-1)} - \frac{\sum R_c^2}{k^2r(r-1)} = \frac{(61)^2 + (8)^2 + \cdots + (31)^2}{10} - \frac{(103)^2 + (103)^2}{50}$$

$$= 501.84$$

where R_c denotes a replication total of the C's.

4. The weighting factor used to obtain the adjusted treatment *totals* is

$$\mu = \frac{(E_b - E_e)}{k(r-1)E_b} = \frac{(62.73 - 13.66)}{(5)(62.73)} = 0.1564$$

where E_b and E_e are, respectively, the mean squares for blocks and intra-block error. If E_b is less than E_e, the factor is taken as zero and no adjustments are made for block effects, the experiment being analyzed as if in randomized blocks.

Each C value is now multiplied by μ to obtain the block corrections μC. These should sum to zero, apart from rounding errors. Each treatment *total* is adjusted by applying the appropriate correction for every block in which the treatment appears. For example, the adjusted total for treatment 4 is

$$25 + 9.5 - 5.0 = 29.5$$

In the case of the *simple* lattice it pays to write the μC values around the table of unadjusted treatment totals, putting the values for replication 1 in the right-hand column and those for replication 2 in the bottom row. Then the adjustments to any treatment total are at the end of the row and the bottom of the column in which the treatment lies.

5. The error variance of the difference between two treatment means is slightly smaller for treatments that appear in the same block than for those that do not. The formulae are:

Two treatments in the same block:

$$\frac{2E_e}{r}[1 + (r - 1)\mu] = \frac{(2)(13.66)}{2}[1 + 0.1564] \qquad = 15.80$$

Two treatments not in same block:

$$\frac{2E_e}{r}[1 + r\mu] = \frac{(2)(13.66)}{2}[1 + (2)(0.1564)] \qquad = 17.93$$

$$\text{Average: } \frac{2E_e}{r}\left[1 + \frac{rk\mu}{(k+1)}\right] = \frac{(2)(13.66)}{2}\left[1 + \frac{(2)(5)(0.1564)}{6}\right] = 17.22$$

The corresponding standard errors are 3.97, 4.23, and 4.15, respectively. Except with small designs it is sufficient to use the average value, 4.15, for all t-tests between pairs of treatments.

6. The analysis of variance does not supply an F-test of the adjusted treatment totals. A test of the unadjusted treatment totals can be obtained by analyzing the data as if the experiment were in randomized blocks. The error m.s. is the pooled mean square for blocks and intra-block error, as shown below.

	d.f.		s.s.	m.s.
Treatments (unadj.)	$(k^2 - 1)$	24	559.28	23.30
Error	$(k^2 - 1)(r - 1)$	24	720.32	30.01

This test is not fully sensitive, since it is based on unadjusted treatment totals, but it will be sufficient in cases where an F-test is of secondary interest or where differences among incomplete blocks are small.

An F-test of the adjusted treatment totals requires a little more calculation. This is most easily done by changing the unadjusted treatments s.s. (which is already available) so that it may be tested against the intra-block error m.s. Calculate B_u, the *unadjusted* sum of squares for blocks within replications, which comes to 350.00. If B_a is the adjusted sum of squares for blocks within replications (501.84), we subtract the following quantity from the treatments s.s.

$$k(r-1)\mu\left\{\left[\frac{r}{(r-1)(1+k\mu)}\right]B_u - B_a\right\}$$

$$= (5)(1)(0.1564)\left\{\left[\frac{2}{1+(5)(0.1564)}\right](350.00) - (501.84)\right\} = -85.30$$

In this case the subtraction term is negative, which is rather unusual. We add 85.30 to the unadjusted treatments s.s. The F-test is then completed by the following analysis.

	d.f.	s.s.	m.s.
Treatments	24	644.58	26.86
Intra-block error	16	218.48	13.66

The F-ratio, 1.97, lies between the 10 and the 5% levels. The F value differs considerably from that obtained by the randomized blocks test, as may happen when the adjustment for incomplete block effects results in a substantial increase in precision.

7. The error m.s. in the randomized blocks analysis was found to be 30.01. To estimate the gain in accuracy over randomized blocks, we compare this figure with the effective error variance

$$E_e\left[1+\frac{rk\mu}{(k+1)}\right] = (13.66)\left[1+\frac{(2)(5)(0.1564)}{6}\right] = 17.22$$

The relative accuracy is 30.01/17.22, or 174%.

10.29 Statistical Analysis for Repetitions of the Designs. Suppose that the basic design contains n replicates (where, e.g., $n = 2$ for a simple lattice) and that this is repeated p times, so that the total number of replications is $r = np$. For illustration, we present a joint analysis of the data in tables 10.3 and 10.4 (pp. 406 and 412), which show 4 replications of the 5×5 simple lattice. Replication III is a repetition of replication I, and IV of II. The field results have been rearranged so that treatments follow the same order within corresponding replications. In terms of our notation, $n = 2$, $p = 2$, and $r = 4$.

1. Calculate the block totals, the replication totals, the treatment totals (shown in table 10.4), and the grand total.

2. Arrange the block totals in supplementary tables as on the right of table 10.4. In these tables, blocks that contain the same set of k treatments are placed in the same row. Each table has k rows and p columns, while the number of tables is n. Form the row and column totals of each

table. For each group of similar blocks, compute the quantities

C = total (over all replicates) of the treatments appearing in the group $- n$(total of the group of blocks)

Thus for the first block in replications II and IV, which contains treatments (1), (6), (11), (16), and (21),

$$C = 59 + 51 + 80 + 63 + 58 - 2(157) = -3$$

Find the replication totals R_c of the C values; these should add to zero.

3. The analysis of variance is as follows.

	d.f.		s.s.	m.s.
Replications	$(r - 1)$	3	226.19	
Treatments (unadj.)	$(k^2 - 1)$	24	791.24	
Blocks within replications (adj.)	$r(k - 1)$	16	786.00	$49.12E_b$
Component (a)	$n(p - 1)(k - 1)$	8	164.72	
Component (b)	$n(k - 1)$	8	621.28	
Intra-block error	$(k - 1)(rk - k - 1)$	56	761.56	$13.60E_e$
Total	$(rk^2 - 1)$	99	2564.99	

The blocks s.s. contains two components, both of which are obtained from the supplementary table of block totals.

Component (a) is a new component, which arises only when the design has been repeated. It is composed of the differences between the totals of blocks that contain the same set of treatments. Since each supplementary table has k rows and p columns, it may be analyzed as follows.

	d.f.
Rows	$(k - 1)$
Columns	$(p - 1)$
Rows × columns	$(k - 1)(p - 1)$
Total	$(kp - 1)$

The sum of squares for component (a) is the sum of the rows × columns interactions over all n tables. It may be obtained from the following calculations.

$$\text{Total: } \frac{(32)^2 + (61)^2 + \cdots + (75)^2 + (84)^2}{k = 5} - \frac{(650)^2 + (749)^2}{pk^2 = 50} = 602.18$$

$$\text{Rows: } \frac{(104)^2 + (142)^2 + \cdots + (171)^2}{pk = 10} - \frac{(650)^2 + (749)^2}{50} = 309.28$$

$$\text{Columns: } \frac{(289)^2 + (361)^2 + (392)^2 + (357)^2}{k^2 = 25} - \frac{(650)^2 + (749)^2}{50} = 128.18$$

Component (a) s.s. $= 602.18 - 309.28 - 128.18 = 164.72$

The above is a general method that covers all cases. With a simple lattice and $p = 2$, it is quicker to calculate the differences between the totals of similar blocks. The sum of squares of deviations of these differences from their replication means, divided by $2k$, gives the sum of squares for component (a).

Component (b) is the component that is present even when there are no repetitions.

$$\text{Component } (b) \text{ s.s.} = \frac{\sum C^2}{kr(n-1)} - \frac{\sum R_c^{\,2}}{k^2r(n-1)}$$

$$= \frac{(63)^2 + (29)^2 + \cdots + (48)^2}{20} - \frac{(99)^2 + (99)^2}{100}$$

$$= 621.28$$

The sum of squares for blocks within replications (adjusted) is the pooled sum of squares for the two components. All other terms in the analysis of variance are computed in the usual way.

4. If E_b is the pooled mean square for blocks, and E_e that for error, the weighting factor is

$$\mu = \frac{p(E_b - E_e)}{k[(r-p)E_b + (p-1)E_e]} = \frac{(2)(35.52)}{(5)[(2)(49.12) + 13.60]} = 0.1270$$

where it will be recalled that p is the number of repetitions of the basic design, in this case 2. As usual, no adjustments are made if E_b is less than E_e.

The block corrections μC are obtained and entered in table 10.4; they should add to zero except for rounding errors. Each treatment total is adjusted by applying a correction for each group of blocks in which the treatment appears. For instance, treatment (13) appears in the third block in replications I and III and in the third block in replications II and IV. Its adjusted total is

$$44 + 8.8 - 0.1 = 52.7$$

5. The error variances for the difference between two treatment means are:

Two treatments in the same block:

$$\frac{2E_e}{r}[1 + (n-1)\mu] = \frac{(2)(13.60)}{4}[1 + 0.1270] = 7.70$$

Two treatments not in same block:

$$\frac{2E_e}{r}[1 + n\mu] = \frac{(2)(13.60)}{4}[1 + (2)(0.1270)] = 8.56$$

$$\text{Average: } \frac{2E_e}{r}\left[1 + \frac{nk\mu}{(k+1)}\right] = \frac{(2)(13.60)}{4}\left[1 + \frac{(2)(5)(0.1270)}{6}\right] = 8.28$$

TABLE 10.4 A 5 × 5 SIMPLE LATTICE (REPS. III AND IV)

Rep. III					Block totals					
						I	III	Sum	C	μC
(1) 13	(2) 26	(3) 9	(4) 13	(5) 11	72	32	72	104	+63	+ 8.0
(6) 15	(7) 18	(8) 22	(9) 11	(10) 15	81	61	81	142	−29	− 3.7
(11) 19	(12) 10	(13) 10	(14) 10	(15) 16	65	54	65	119	+69	+ 8.8
(16) 21	(17) 16	(18) 17	(19) 4	(20) 17	75	74	75	149	−34	− 4.3
(21) 15	(22) 12	(23) 13	(24) 20	(25) 8	68	68	68	136	+30	+ 3.8
					361	289	361	650	+99	+12.6

Rep. IV						II	IV	Sum	C	μC
(1) 16	(6) 7	(11) 20	(16) 13	(21) 21	77	80	77	157	− 3	− 0.4
(2) 15	(7) 10	(12) 11	(17) 7	(22) 14	57	80	57	137	+ 2	+ 0.3
(3) 7	(8) 11	(13) 15	(18) 15	(23) 16	64	56	64	120	− 1	− 0.1
(4) 19	(9) 14	(14) 20	(19) 6	(24) 16	75	89	75	164	−49	− 6.2
(5) 17	(10) 18	(15) 20	(20) 15	(25) 14	84	87	84	171	−48	− 6.1
					357	392	357	749	−99	−12.5

Treatment totals (4 reps.)

(1) 59	(2) 69	(3) 37	(4) 57	(5) 49	+8.0
(6) 51	(7) 51	(8) 49	(9) 48	(10) 56	−3.7
(11) 80	(12) 42	(13) 44	(14) 69	(15) 72	+8.8
(16) 63	(17) 50	(18) 57	(19) 32	(20) 62	−4.3
(21) 58	(22) 64	(23) 52	(24) 73	(25) 55	+3.8
−0.4	+0.3	−0.1	−6.2	−6.1	

Adjusted treatment totals

(1) 66.6	(2) 77.3	(3) 44.9	(4) 58.8	(5) 50.9
(6) 46.9	(7) 47.6	(8) 45.2	(9) 38.1	(10) 46.2
(11) 88.4	(12) 51.1	(13) 52.7	(14) 71.6	(15) 74.7
(16) 58.3	(17) 46.0	(18) 52.6	(19) 21.5	(20) 51.6
(21) 61.4	(22) 68.1	(23) 55.7	(24) 70.6	(25) 52.7

6. An approximate F-test of the adjusted treatment totals is obtained by the method described in the previous section, step 6. In this case the quantity that must be subtracted from the treatments s.s. is

$$k(n-1)\mu\left[\frac{n}{(n-1)(1+k\mu)}B_u - B_a\right]$$

where B_u is the unadjusted and B_a the adjusted sum of squares for component (b) of the blocks. Both quantities have already been computed: B_a, 621.28, appears in the analysis of variance, while B_u, 309.28, is the rows s.s. used in computing component (a).

7. A comparison with randomized blocks is made as in the previous section, step 7. The effective error variance of the lattice is taken as

$$E_e\left[1 + \frac{nk\mu}{(k+1)}\right]$$

10.210 Missing Data. If the missing observations are numerous, an analysis by the method for randomized blocks is recommended. If it is desired to carry out the full analysis, missing values are estimated by the formula which minimizes the intra-block error s.s., though as explained in section 10.13 this method is only approximate.

Experiments with no repetition of the basic design.

$$x = \frac{(r-1)k^2T - rR + G - rkC + kC'}{(r-1)(k-1)(rk-k-1)} \qquad (10.2)$$

In this formula T and R are the totals for the treatment and replicate that contain the missing value, G is the grand total, and C is the C value (defined in section 10.28) for the block which contains the missing observation. The only unfamiliar quantity is C':

C' = total of the C values for all blocks which contain the treatment that has the missing value

Example. For the experiment analyzed in section 10.28, suppose that 3 observations are missing: treatments (1) and (12) in replication I and

TABLE 10.5 SUMMARY TOTALS FOR A 5 × 5 SIMPLE LATTICE WITH THREE MISSING VALUES

Treatment totals

(1)	$24 + x$	(2)	28	(3)	21	(4)	25	(5)	21
(6)	29	(7)	23	(8)	16	(9)	23	(10)	23
(11)	41	(12)	$14 + y$	(13)	19	(14)	$9 + z$	(15)	36
(16)	29	(17)	27	(18)	25	(19)	22	(20)	30
(21)	22	(22)	38	(23)	23	(24)	37	(25)	33

Rep. I		Rep. II	
B	C	B	C
$26 + x$	$67 - x$	80	$-15 + x$
61	-8	80	$-30 + y$
$47 + y$	$25 - y + z$	56	-8
74	-15	$59 + z$	$-2 - z$
68	17	87	-31
$R_1 = 276 + x + y$	$86 - x - y + z$	$R_2 = 362 + z$	$-86 + x + y - z$

$$G = R_1 + R_2 = 638 + x + y + z$$

treatment (14) in replication II. Denote the estimates of the 3 missing values by x, y, and z, respectively; these will be obtained by the method of successive approximation.

It is best to start by finding all block, replicate, and treatment totals and all C values, as in steps 1 and 2 of the computing instructions for this design. Wherever a missing observation is involved, include the appropriate x, y, or z in its place. The summary totals are shown in table 10.5. Note that the total of the C values over the whole experiment is identically zero, so that a good check is available on this part of the calculations.

We must now find first approximations for y and z. In the replications in which they are present, their values are 14 and 9, respectively. However, replication II appears to have higher values than replication I, and it is probably worth while to make an adjustment for replication effect. If we ignore the missing values, the means per plot are about 12 for replication I and 15 for replication II. Since y is missing in replication I, we subtract 3 from its value in replication II, giving 11. For z we add 3, giving 12 as the first approximation. Using $y = 11$, $z = 12$, we now solve for x from formula (10.2). For $r = 2$, $k = 5$ the formula becomes

$$x = \frac{25T - 2R + G - 10C + 5C'}{16}$$

Put $y = 11$, $z = 12$. From table 10.5 we find for x

$$T = 24; \quad R = 276 + 11 = 287; \quad G = 638 + 11 + 12 = 661$$

$$C = 67; \quad C' = 67 - 15 = 52$$

$$x = \frac{(25)(24) - (2)(287) + 661 - (10)(67) + (5)(52)}{16} = \frac{277}{16} = 17$$

Put $x = 17$, $z = 12$. For y we have

$$T = 14; \quad R = 276 + 17 + 293; \quad G = 638 + 17 + 12 = 667$$

$$C = 25 + 12 = 37; \quad C' = 25 + 12 - 30 = 7$$

$$y = \frac{(25)(14) - (2)(293) + 667 - (10)(37) + (5)(7)}{16} = \frac{96}{16} = 6$$

Put $x = 17$, $y = 6$. For z we have

$$T = 9; \quad R = 362; \quad G = 638 + 17 + 6 = 661; \quad C = -2$$

$$C' = -2 + 25 - 6 = 17$$

$$z = \frac{(25)(9) - (2)(362) + 661 - (10)(-2) + (5)(17)}{16} = \frac{267}{16} = 17$$

This completes the first round. A second round leads to $x = 18, y = 5, z = 17$, and there is obviously no need for further calculation. The rest of the analysis proceeds as usual except that 3 d.f. are omitted from the total s.s. and the intra-block error s.s. For approximate t-tests, see section 10.13.

Experiments with repetitions of the basic design. If the basic design has n replicates and these are repeated p times to give $r = np$ replications, the formula becomes

$$x = \frac{(n-1)k^2T + (n-1)rR + G - nkC + kC' - n^2R'}{(n-1)(k-1)(rk-k-1)} \qquad (10.3)$$

All symbols have the same meaning as in (10.2) except that C and C' are now derived from the totals of groups of similar blocks, just as in the statistical analysis for this case in section 10.29. The new quantity R' is defined as

R' = total of all replications that are similar to the replication containing the missing value

Note that this total *includes* the replication with the missing value.

Example. In the 5×5 simple lattice with 4 replications, we have $n = 2$, $r = 4$. If treatment (1) is missing in the first replication, the reader may verify that

$$T = 53; \quad R = 283; \quad G = 1393; \quad C = 69; \quad C' = 69 - 9 = 60$$

$$R' = 283 + 361 = 644$$

This gives

$$x = \frac{(25)(53) + (4)(283) + 1393 - (10)(69) + (5)(60) - (4)(644)}{56}$$

$$= \frac{844}{56} = 16$$

10.3 Rectangular Lattices

10.31 Description. These designs were developed recently by Harshbarger (10.8) for $k(k+1)$ treatments in blocks of k units. They form a useful addition to the square lattices described in previous sections, since the allowable numbers of treatments, 12, 20, 30, 42, 56, 72, etc., fall about midway between the allowable numbers for square lattices. The statistical analysis is quite similar to that for simple and triple lattices, though it takes more time because the block adjustments are not so simple as with square lattices. The new designs are less symmetrical than the square lattices, in the sense that there is a greater variation in the accuracy with which two treatment means are compared. It will be recalled that only two standard errors are required for t-tests with the simple and triple lattice—one for two treatments that appear in the same block and one for two treatments that do not. The simple rectangular lattice requires four standard errors, while the triple rectangular lattice requires seven. For practical purposes it appears that these can nearly always be reduced to two.

There are several ways of constructing the designs. One (suggested by G. S. Watson) is by means of a latin square with $(k + 1)$ rows and columns, in which every letter in the leading diagonal is different. When writing down the square we attach a number to all letters except those in the leading diagonal, as illustrated below for a 4×4 square.

A	*B*1	*C*2	*D*3
*D*4	*C*	*B*5	*A*6
*B*7	*A*8	*D*	*C*9
*C*10	*D*11	*A*12	*B*

In the first replication we place in a block all numbers that lie in the same row of the latin square, in the second replication all numbers that lie in the same column, and in the third all numbers that have the same latin letter.

Block	Rep. I				Rep. II				Rep. III		
1	1	2	3	1	1	8	11	1	1	5	7
2	4	5	6	2	2	5	12	2	2	9	10
3	7	8	9	3	3	6	9	3	3	4	11
4	10	11	12	4	4	7	10	4	6	8	12

The important feature of this arrangement is that no two treatments are in the same block more than once. The use of a latin square with different letters down the leading diagonal ensures that in the third replication the three numbers associated with any letter are all in different rows and columns, and hence have not previously occurred together in a block.

Plans 10.10–10.16 were constructed by this method. By using the first 2 replications from any plan we obtain a rectangular lattice in 2 replications, which by means of repetitions can be used for an experiment in 4, 6, 8, etc., replications. By using all 3 replications of the plan we have designs for 3, 6, 9, etc., replications.

Alternatively (as pointed out by S. S. Shrikhande), we may construct the designs from a balanced lattice with $(k + 1)^2$ treatments. If the first replication is omitted, and if all treatments that appear in any one selected block in the first replicate are omitted in subsequent replications, it is easy to verify that we generate a design for $k(k + 1)$ treatments in which every block is of size k and in which no two treatments appear together more than once in the same block. This method gives any number of replicates up to k, though of course the method fails when no balanced lattice exists, as with 36 treatments. The discussion in this section is limited to simple and triple rectangular lattices. The properties and method of analysis for the designs with k replicates, called "near balance" rectangular lattices by Harshbarger, are described in (10.3a).

The method of randomizing is the same as for lattices (section 10.27).

10.32 Statistical Analysis. In each plan, the three basic replications are described as the X, Y, and Z replications. Further, if we take any block in one replication and examine another replication, we find that there is one and only one block in the other replication that has no treatments in common with the chosen block. These two blocks will be called *partners*. Since it is important to be able to distinguish partners in the statistical analysis, each block is denoted in the plan by two symbols, e.g., $Y2$, one to mark the replication and one to indicate the partners. All partners carry the same number: thus the partners of $Y2$ are $X2$ and $Z2$. It is advisable to write the block labels in the notebook in which the original results are recorded.

Example. The example is a triple rectangular lattice for 12 treatments in blocks of 3 ($k = 3$). Artificial data were assembled by taking true treatment effects as shown in table 10.6, and adding to them true block effects as given with each block in table 10.7. Thus the observation 7

TABLE 10.6 Treatment effects in artificial data for a 3×4 triple rectangular lattice

Treatment	True effect	True treatment total
1	14	56
2	7	35
3	2	20
4	0	14
5	3	23
6	11	47
7	9	41
8	16	62
9	8	38
10	1	17
11	5	29
12	6	32

for treatment (10) in block $X4$ of table 10.7 is obtained by adding the true treatment effect, 1, to the true block effect, 6. Since the average of the true block effects comes out to be 56/12, or 14/3, the true treatment *totals*, over the 3 replicates, will be $(3\tau + 14)$, where τ is the true treatment effect. These figures are shown in the column at the right in table 10.6.

TABLE 10.7 PLAN AND OBSERVATIONS FOR A 3×4 RECTANGULAR LATTICE

(Treatment numbers are enclosed in parentheses)

Block symbol	True block effect		Rep. X		Totals B	C_X	Adjustment factor
$X4$	6	(10) 7	(12) 12	(11) 11	30	−12	−0.6
$X1$	2	(2) 9	(3) 4	(1) 16	29	20	+3.4
$X3$	7	(7) 16	(9) 15	(8) 23	54	−5	−1.6
$X2$	0	(4) 0	(5) 3	(6) 11	14	30	+5.4
					127	33	
			Rep. Y		B	C_Y	
$Y4$	9	(3) 11	(6) 20	(9) 17	48	−32	−4.6
$Y2$	3	(1) 17	(11) 8	(8) 19	44	10	+1.4
$Y3$	3	(12) 9	(2) 10	(5) 6	25	10	+1.4
$Y1$	5	(10) 6	(4) 5	(7) 14	25	0	−0.6
					142	−12	
			Rep. Z		B	C_Z	
$Z1$	4	(8) 20	(6) 15	(12) 10	45	4	+0.2
$Z2$	8	(9) 16	(10) 9	(2) 15	40	−16	−3.8
$Z3$	1	(11) 6	(3) 3	(4) 1	10	19	+3.2
$Z4$	8	(5) 11	(1) 22	(7) 17	50	−28	−3.8
					145	−21	

Treatment totals

	1	2	3	4	5	6	7	8	9	10	11	12
Unadj.	55	34	18	6	20	46	47	62	48	22	25	31
Adj.	56	35	20	14	23	47	41	62	38	17	29	32

No intra-block error has been introduced into the data, which therefore provide two checks on the method of analysis. First, in the analysis of variance the intra-block error s.s. should be found to be identically zero. Second, the adjusted treatment totals as found from the analysis should be *exactly* equal to the true treatment totals.

The computing instructions given below apply to either the simple or the triple rectangular lattice without repetitions. The changes required when there are repetitions are given later.

1. Find the block totals, B, the replication totals, the treatment totals, and the grand total.

2. For each block calculate

$$C = \text{total (over all replicates) of all treatments in the block} - rB$$

Thus for block $Y4$,

$$C = 18 + 46 + 48 - (3)(48) = -32$$

As usual, the total of the C's is zero.

3. Arrange the C values in a supplementary table so that partners appear in the same row. The row totals of this table, S, give the sums of the C values for each set of partners. The column totals give values denoted by R_C.

TABLE 10.8 SUPPLEMENTARY TABLE OF C VALUES

Block symbol	C_X	C_Y	C_Z	S	λC_X	λC_Y	λC_Z	μS
1	20	0	4	24	4.0	0.0	0.8	0.6
2	30	10	−16	24	6.0	2.0	−3.2	0.6
3	−5	10	19	24	−1.0	2.0	3.8	0.6
4	−12	−32	−28	−72	−2.4	−6.4	−5.6	−1.8
Totals (R_C)	33	−12	−21	0	6.6	−2.4	−4.2	0.0

4. The analysis of variance is now obtained. All sums of squares are found in the usual way, except that for blocks, adjusted for treatments, which is

$$\frac{\sum C^2}{r(rk - k - 1)} - \frac{\sum R_C{}^2}{r(k + 1)(rk - k - 1)} - \frac{\sum S^2}{r(r - 1)(k + 1)(rk - k - 1)}$$

$$= \frac{(20)^2 + (30)^2 + \cdots + (28)^2}{15} - \frac{(33)^2 + (12)^2 + (21)^2}{60} - \frac{(24)^2 + \cdots + (72)^2}{120}$$

$$= 274.0 - 27.9 - 57.6 = 188.5 \qquad (10.4)$$

TABLE 10.9 ANALYSIS OF VARIANCE

Source of variation	d.f.		s.s.	m.s.
Replications	$(r - 1)$	2	15.5	
Treatments	$(k^2 + k - 1)$	11	1067.0	
Blocks	rk	9	188.5	$20.9E_b$
Intra-block error	$(r - 1)(k^2 - 1) - k$	13	0.0	$0.0E_e$
Total	$(rk^2 + rk - 1)$	35	1271.0	

As anticipated, the intra-block error s.s. in table 10.9 is zero.

5. We now calculate the weighting factors used to obtain the adjusted treatment totals. Although a general formula can be given, it is simpler to present these separately for simple and triple designs. Two weighting factors are necessary.*

Simple rectangular lattice

$$\lambda = \frac{r(E_b - E_e)}{r(k-1)E_b + (rk - 2k + r)E_e} \tag{10.5}$$

$$\mu = \frac{\lambda r(E_b - E_e)}{r(k+1)E_b + (rk - 2k - r)E_e} = \frac{\lambda^2}{1 + 2\lambda} \tag{10.6}$$

Triple rectangular lattice

$$\lambda = \frac{r(E_b - E_e)}{r(2k-1)E_b + (rk - 3k + r)E_e} \tag{10.7}$$

$$\mu = \frac{\lambda r(E_b - E_e)}{2r(k+1)E_b + (rk - 3k - 2r)E_e} = \frac{\lambda^2}{1 + 3\lambda} \tag{10.8}$$

In this example E_e is zero and we have

$$\lambda = \frac{3}{(3)(5)} = 0.2; \qquad \mu = \frac{(0.2)(3)}{(2)(3)(4)} = 0.025$$

6. Complete table 10.8 by adding the columns λC_X, λC_Y, λC_Z, and μS. The adjustment for any block is

$$\lambda C - \mu S$$

where S is taken from the row in which C lies in table 10.8. Thus for block $X1$ the adjustment is

$$+4.0 - 0.6 = +3.4$$

These adjustments are recorded on the plan in table 10.7. Note that the order in which blocks appear is different in tables 10.7 and 10.8; care must be taken to see that each adjustment is given to the appropriate block.

7. Finally, each treatment *total* is adjusted by adding the adjustments for every block in which the treatment appears. For instance, the adjusted total for treatment (4) is

$$6 + 5.4 - 0.6 + 3.2 = 14$$

* This method of estimating the weights is slightly different from that given by Harshbarger. In the interests of simplicity we have used only the pooled mean square for blocks and the intra-block error m.s. for estimating λ and μ.

Every adjusted treatment total will be found to be equal to the true treatment total as given in table 10.6.

Statistical analysis for repetitions of the designs. When the basic design with n replications is repeated p times to give $r = np$ replications, the changes required in the steps of the analysis are noted below.

1. No change.

2*a*. For any block there are $(p - 1)$ other blocks that have exactly the same set of treatments. Arrange the totals for such similar blocks in two-way tables, as exemplified below. There are n such tables.

Block no.	X Blocks Repetition 1 2 ⋯ p	Total ($\Sigma\, \beta$)
1		
2		
.		
.		
.		
$(k + 1)$		
Total		

2*b*. For each group of similar blocks calculate

$$C = \text{total (over all replicates) of all treatments in the group} - n(\textstyle\sum \beta)$$

3. No change.

4*a*. The blocks s.s. now has two components. For component (a) obtain the interaction s.s. for each of the tables in step 2*a*, and add these sums of squares. Since each block total contains k observations, a divisor k is required for the analysis of variance.

4*b*. The sum of squares for component b is

$$\frac{\sum C^2}{r(nk - k - 1)} - \frac{\sum R_C^2}{r(k + 1)(nk - k - 1)} - \frac{\sum S^2}{r(n - 1)(k + 1)(nk - k - 1)}$$

The separation of degrees of freedom in the analysis of variance is shown below.

	d.f.	m.s.
Replications	$(r - 1)$	
Treatments	$(k^2 + k - 1)$	
Blocks	rk	E_b
Component (a)	$(r - n)k$	
Component (b)	nk	
Intra-block error	$(r - 1)(k^2 - 1) - k$	E_e
Total	$(rk^2 + rk - 1)$	

5. The weighting factors in formulae (10.5)–(10.8) are unchanged. Note that E_b is the pooled mean square for blocks.

6. Unchanged.

7. Unchanged.

As mentioned previously, in order to have t-tests for every pair of treatments we would require to present 4 standard error formulae for simple rectangular lattices and 7 for triple rectangular lattices. With very little inaccuracy these can be reduced to 2, one for the case where the two treatments appear in a block and one for the case where this is not so. For the difference between two adjusted *means* the estimated error variances are:

Simple rectangular lattice

Two treatments in the same block: $\frac{2E_e}{r}(1 + \lambda - \mu)$

Two treatments not in same block: $\frac{2E_e}{r}(1 + 2\lambda - \mu)$

Average: * $$\frac{2E_e}{r}\left[\frac{2(k-1)}{k^2+k-1}(1+\lambda-\mu) + \frac{k^2-k+1}{k^2+k-1}(1+2\lambda-\mu)\right]$$

$$= \frac{2E_e}{r}\left[1 + \frac{2k^2\lambda - (k^2+k-1)\mu}{(k^2+k-1)}\right]$$

Triple rectangular lattice

Two treatments in the same block: $\frac{2E_e}{r}(1 + 2\lambda - \mu)$

Two treatments not in same block: $\frac{2E_e}{r}\left(1 + 3\lambda - \frac{3}{2}\mu\right)$

Average: * $$\frac{2E_e}{r}\left[\frac{3(k-1)}{k^2+k-1}(1+2\lambda-\mu) + \frac{k^2-2k+2}{k^2+k-1}\left(1+3\lambda-\frac{3}{2}\mu\right)\right]$$

In the example, where $\lambda = 0.2$, $\mu = 0.025$, the factors multiplying the usual term $2E_e/r$ are 1.375 for two treatments in the same block and 1.562 for two treatments not in the same block. The factor in the average variance is 1.460.

10.4 Cubic Lattices

10.41 Description. These designs were produced by Yates for plant-breeding work in which selections are to be made from an unusually large number of varieties. The number of treatments must be an exact cube.

* Owing to the fact that only two standard errors were used, this average is not quite equal to the average variance taken over all possible pairs, though it is very close to that value.

The most useful range comprizes 27, 64, 125, 216, 343, 512, 729, and 1000 treatments. The size of block is the cube root of the number of treatments, i.e., 3, 4, 5, 6, 7, 8, 9, and 10, respectively. Thus cubic lattices can accommodate a large number of treatments in a small size of incomplete block. The designs have been used, for example, in an experiment with 729 strains of *ponderosa* pine seedlings (10.9) and an experiment with 729 soybean varieties (10.10). The number of replicates must be 3 or some multiple of 3.

Since the plans are easy to construct and since they occupy a considerable amount of space for the higher numbers of treatments, they are not reproduced here. To obtain a plan, the k^3 treatments are numbered by means of a three-digit code in which each digit takes all values from 1 to k. For 27 treatments, the code is as follows:

T *	Code	T	Code	T	Code
1	111	4	121	7	131
2	211	5	221	8	231
3	311	6	321	9	331
10	112	13	122	16	132
11	212	14	222	17	232
12	312	15	322	18	332
19	113	22	123	25	133
20	213	23	223	26	233
21	313	24	323	27	333

* Treatment number.

The same principle applies with a larger number of treatments. For the first k treatments, the last two digits are fixed at (11) while the first digit runs from 1 to k. The next k treatments are coded by fixing the last two digits at (21) while the first digit again runs from 1 to k, and so on in a systematic manner until the final k treatments are reached, for which the last two digits have the fixed values (kk).

Within each of the 3 replications, the k^3 treatments are grouped into k^2 blocks, each of size k. In the first replication, the rule for this grouping is to keep the last two digits constant within a block, allowing the first digit to take all values from 1 to k. Thus, in the example above, the 9 groups of treatments constitute the 9 blocks, block 1 containing the treatments (111), (211), and (311).

To form blocks in the second replication, we keep the first and last digits fixed within any block and give the second digit all values from 1 to k. With 27 treatments, the first block therefore contains (111), (121), and (131), the second block (211), (221), and (231), and the last block (313), (323), and (333). In the third replication, the first and second digits are constant within each block.

The composition of the blocks in the second and third replications is shown in table 10.10.

TABLE 10.10 SECOND AND THIRD REPLICATES OF A CUBIC LATTICE WITH 27 TREATMENTS

Rep. II

Block	(1)		(2)		(3)		(4)		(5)	
	Code	T *	Code	T	Code	T	Code	T	Code	T
	111	1	211	2	311	3	112	10	212	11
	121	4	221	5	321	6	122	13	222	14
	131	7	231	8	331	9	132	16	232	17

	(6)		(7)		(8)		(9)	
	Code	T	Code	T	Code	T	Code	T
	312	12	113	19	213	20	313	21
	322	15	123	22	223	23	323	24
	332	18	133	25	233	26	333	27

Rep. III

Block	(1)		(2)		(3)		(4)		(5)	
	Code	T	Code	T	Code	T	Code	T	Code	T
	111	1	211	2	311	3	121	4	221	5
	112	10	212	11	312	12	122	13	222	14
	113	19	213	20	313	21	123	22	223	23

	(6)		(7)		(8)		(9)	
	Code	T	Code	T	Code	T	Code	T
	321	6	131	7	231	8	331	9
	322	15	132	16	232	17	332	18
	323	24	133	25	233	26	333	27

* Treatment number.

The original treatment numbers are shown as well as the code numbers. In an experiment it is simplest to record only the code numbers, which are required in order to follow the computing instructions.

Cubic lattices may be expected to be most useful when the number of treatments exceeds 100. The designs for $k = 3$, 4, 5, and 6 lie in the range which is also covered by lattice designs. We do not know of any investigations of the relative accuracy of lattice and cubic lattice designs in these cases. With highly variable experimental material, the use of a smaller block might give the cubic lattice some advantage; on the other hand the statistical analysis is more laborious owing to the greater number of block adjustments to be calculated. Like the lattice designs, cubic lattices cannot be appreciably less accurate than randomized blocks which occupy the same set of replications, and can be analyzed by the method for randomized blocks.

Some cubic lattices can also be arranged in quasi-latin squares (10.11) so as to allow the elimination of two sources of error variation.

10.42 Arrangement of Experimental Material. The most important rule is to have units in the same incomplete block as homogeneous as the experimental material permits; in field trials the blocks should be compact in shape. If practicable, blocks within the same replicate should also be similar to each other, so as to increase the accuracy of inter-block comparisons and to obtain more precise results if the analysis is subsequently carried out by the method for randomized blocks.

10.43 Randomization. The steps are:

1. Randomize the order of the blocks independently within each replication.
2. Randomize the positions of the treatment code numbers independently within each block.
3. Assign treatments to code numbers at random.

10.44 Statistical Analysis. The following reference should be consulted.

(10.12) YATES, F. The recovery of inter-block information in variety trials arranged in three-dimensional lattices. *Ann. Eugen.* 9, 136–156, 1939.

General formulae are given with a numerical example of the analysis for an experiment with 64 treatments. In following this example, the reader should note that the experiment was not arranged in separate replications; the changes introduced for this reason are pointed out by Yates (pp. 147–148). The method of analysis when 6 or 9 replicates are used is also indicated (p. 144). Reference (10.13) describes the analysis of an experiment with 125 treatments.

10.45 Error Variances. The formulae for the error variance of the difference between two treatment *means* are summarized below. In the analysis as described by Yates (10.12), the symbols

$$\lambda = \frac{w - w'}{w + 2w'}, \quad \mu = \frac{w - w'}{2w + w'}$$

play a prominent part. It is convenient to write $\lambda' = \lambda/k^2$, $\mu' = \mu/k^2$. The error variances for the different types of treatment comparisons are as follows:

Comparison	Variance of difference between means
$t_{211} - t_{111}$	$\frac{2E_e}{r}[1 + 2\lambda' + 2(k-1)\mu']$
$t_{122} - t_{111}$	$\frac{2E_e}{r}[1 + 4\lambda' + (3k-4)\mu']$
$t_{222} - t_{111}$	$\frac{2E_e}{r}[1 + 6\lambda' + 3(k-2)\mu']$
Average	$\frac{2E_e}{r}\left[1 + \frac{k^2}{(k^2+k+1)}\{6\lambda' + 3(k-1)\mu'\}\right]$

Here E_e is the intra-block error m.s. and r the number of replicates (3, 6, or 9). The formulae, though written in a slightly different form, are identical with those given by Yates. Our k corresponds to Yates's p, and our r to his n.

REFERENCES

10.1 Comstock, R. E., Peterson, W. J., and Stewart, H. A. An application of the balanced lattice design in a feeding trial with swine. *Jour. Animal Sci.* 7, 320–331, 1948.

10.2 Peterson, W. J., *et al.* Cystine and vitamins of the B-complex as supplements to raw soybeans in pig rations. *Jour. Animal Sci.* 7, 341–350, 1948.

10.3 Cornish, E. A. The recovery of inter-block information in quasi-factorial designs with incomplete data. 1. Square, triple, and cubic lattices. *Australian Coun. Sci. Ind. Res. Bull.* 158, 1943.

10.4 Cochran, W. G. Some additional lattice square designs. *Iowa Agr. Exp. Sta. Res. Bull.* 318, 1943.

10.5 Cox, G. M., Eckhardt, R. C., and Cochran, W. G. The analysis of lattice and triple lattice experiments in corn varietal tests. *Iowa Agr. Exp. Sta. Res. Bull.* 281, 1940.

10.6 Goulden, C. H. A uniform method of analysis for square lattice experiments. *Sci. Agr.* 25, 115–136, 1944.

10.7 Homeyer, P. G., Clem, M. A., and Federer, W. T. Punched card and calculating machine methods for analyzing lattice experiments including lattice squares and the cubic lattice. *Iowa Agr. Exp. Sta. Res. Bull.* 347, 1947.

10.8 Harshbarger, B. Rectangular lattices. *Va. Agr. Exp. Sta. Mem.* 1, 1947.

10.9 Day, B. B., and Austin, L. A three-dimensional lattice design for studies in forest genetics. *Jour. Agr. Res.* 59, 101–120, 1939.

10.10 Iowa Agric. Exp. Sta., *Project* 719. Data as yet unpublished.

10.11 Yates, F. A further note on the arrangement of variety trials: quasi-latin squares. *Ann. Eugen.* 7, 319–332, 1937.

10.12 Yates, F. The recovery of inter-block information in variety trials arranged in three-dimensional lattices. *Ann. Eugen.* 9, 136–156, 1939.

10.13 Phipps, I. F., *et al.* The analysis of cubic lattice designs in varietal trials. *Australian Coun. Sci. Ind. Res. Bull.* 176, 1944.

10.1a KEMPTHORNE, O. *The design and analysis of experiments.* Section 23.6. John Wiley & Sons, New York, 1952.

10.2a MEIER, P. *Weighted means and lattice designs.* Ph.D. thesis, Princeton University, 1951.

10.3a HARSHBARGER, B., and DAVIS, L. L. Latinized rectangular lattices. *Biometrics* 8, 73–84, 1952.

ADDITIONAL READING

BANCROFT, T. A., and SMITH, A. L. Efficiency of the simple lattice design relative to randomized complete blocks design in cotton variety and strain testing. *Jour. Amer. Soc. Agron.* 41, 157–160, 1949.

BOYCE, S. W. The efficiency of lattice designs. *New Zealand Jour. Sci. and Tech.* 270–275, 1945.

BOYCE, S. W. The analysis of lattice trials with incomplete data. *New Zealand Jour. Sci. and Tech.* 276–280, 1945.

DAWSON, C. D. R. An example of the quasi-factorial design applied to a corn breeding experiment. *Ann. Eugen.* 9, 157–173, 1939.

GRUNDY, P. M. The estimation of error in rectangular lattices. *Biometrics* 6, 25–33, 1950.

HARSHBARGER, B. Triple rectangular lattices. *Biometrics* 5, 1–13, 1949.

HEALY, M. J. R. The analysis of lattice designs when a variety is missing. *Emp. Jour. Exp. Agri.* 20, 220–226, 1952.

MEIER, P. Analysis of simple lattice designs with unequal sets of replications. *Jour. Am. Stat. Assoc.* 49, 786–813, 1954.

NAIR, K. R. Analysis of partially balanced incomplete block designs illustrated on the simple square and rectangular lattices. *Biometrics* 8, 122–155, 1952.

ROBINSON, H. F., and WATSON, G. S. Analysis of simple and triple rectangular lattice designs. *North Carolina Agr. Exp. Sta. Tech. Bull.* 88, 1949.

PLANS

Plan 10.1 3 × 3 balanced lattice

$t = 9, k = 3, r = 4, b = 12, \lambda = 1$*

Block	Rep. I			Block	Rep. II			Block	Rep. III			Block	Rep IV		
(1)	1	2	3	(4)	1	4	7	(7)	1	5	9	(10)	1	8	6
(2)	4	5	6	(5)	2	5	8	(8)	7	2	6	(11)	4	2	9
(3)	7	8	9	(6)	3	6	9	(9)	4	8	3	(12)	7	5	3

Plan 10.2 4 × 4 balanced lattice

$t = 16, k = 4, r = 5, b = 20, \lambda = 1$

Block	Rep. I				Block	Rep. II				Block	Rep. III			
(1)	1	2	3	4	(5)	1	5	9	13	(9)	1	6	11	16
(2)	5	6	7	8	(6)	2	6	10	14	(10)	5	2	15	12
(3)	9	10	11	12	(7)	3	7	11	15	(11)	9	14	3	8
(4)	13	14	15	16	(8)	4	8	12	16	(12)	13	10	7	4

Block	Rep. IV				Block	Rep. V			
(13)	1	14	7	12	(17)	1	10	15	8
(14)	13	2	11	8	(18)	9	2	7	16
(15)	5	10	3	16	(19)	13	6	3	12
(16)	9	6	15	4	(20)	5	14	11	4

Plan 10.3 5 × 5 balanced lattice

$t = 25, k = 5, r = 6, b = 30, \lambda = 1$

Block	Rep. I					Block	Rep. II					Block	Rep. III				
(1)	1	2	3	4	5	(6)	1	6	11	16	21	(11)	1	7	13	19	25
(2)	6	7	8	9	10	(7)	2	7	12	17	22	(12)	21	2	8	14	20
(3)	11	12	13	14	15	(8)	3	8	13	18	23	(13)	16	22	3	9	15
(4)	16	17	18	19	20	(9)	4	9	14	19	24	(14)	11	17	23	4	10
(5)	21	22	23	24	25	(10)	5	10	15	20	25	(15)	6	12	18	24	5

Block	Rep. IV					Block	Rep. V					Block	Rep. VI				
(16)	1	12	23	9	20	(21)	1	17	8	24	15	(26)	1	22	18	14	10
(17)	16	2	13	24	10	(22)	11	2	18	9	25	(27)	6	2	23	19	15
(18)	6	17	3	14	25	(23)	21	12	3	19	10	(28)	11	7	3	24	20
(19)	21	7	18	4	15	(24)	6	22	13	4	20	(29)	16	12	8	4	25
(20)	11	22	8	19	5	(25)	16	7	23	14	5	(30)	21	17	13	9	5

* The symbol λ denotes the number of times that two treatments appear in the same block.

Plan 10.4

7 × 7 balanced lattice

$t = 49, k = 7, r = 8, b = 56, \lambda = 1$

Block	Rep. I						
(1)	1	2	3	4	5	6	7
(2)	8	9	10	11	12	13	14
(3)	15	16	17	18	19	20	21
(4)	22	23	24	25	26	27	28
(5)	29	30	31	32	33	34	35
(6)	36	37	38	39	40	41	42
(7)	43	44	45	46	47	48	49

Block	Rep. II						
(8)	1	8	15	22	29	36	43
(9)	2	9	16	23	30	37	44
(10)	3	10	17	24	31	38	45
(11)	4	11	18	25	32	39	46
(12)	5	12	19	26	33	40	47
(13)	6	13	20	27	34	41	48
(14)	7	14	21	28	35	42	49

Block	Rep. III						
(15)	1	9	17	25	33	41	49
(16)	43	2	10	18	26	34	42
(17)	36	44	3	11	19	27	35
(18)	29	37	45	4	12	20	28
(19)	22	30	38	46	5	13	21
(20)	15	23	31	39	47	6	14
(21)	8	16	24	32	40	48	7

Block	Rep. IV						
(22)	1	37	24	11	47	34	21
(23)	15	2	38	25	12	48	35
(24)	29	16	3	39	26	13	49
(25)	43	30	17	4	40	27	14
(26)	8	44	31	18	5	41	28
(27)	22	9	45	32	19	6	42
(28)	36	23	10	46	33	20	7

Block	Rep. V						
(29)	1	30	10	39	19	48	28
(30)	22	2	31	11	40	20	49
(31)	43	23	3	32	12	41	21
(32)	15	44	24	4	33	13	42
(33)	36	16	45	25	5	34	14
(34)	8	37	17	46	26	6	35
(35)	29	9	38	18	47	27	7

Block	Rep. VI						
(36)	1	23	45	18	40	13	35
(37)	29	2	24	46	19	41	14
(38)	8	30	3	25	47	20	42
(39)	36	9	31	4	26	48	21
(40)	15	37	10	32	5	27	49
(41)	43	16	38	11	33	6	28
(42)	22	44	17	39	12	34	7

Block	Rep. VII						
(43)	1	16	31	46	12	27	42
(44)	36	2	17	32	47	13	28
(45)	22	37	3	18	33	48	14
(46)	8	23	38	4	19	34	49
(47)	43	9	24	39	5	20	35
(48)	29	44	10	25	40	6	21
(49)	15	30	45	11	26	41	7

Block	Rep. VIII						
(50)	1	44	38	32	26	20	14
(51)	8	2	45	39	33	27	21
(52)	15	9	3	46	40	34	28
(53)	22	16	10	4	47	41	35
(54)	29	23	17	11	5	48	42
(55)	36	30	24	18	12	6	49
(56)	43	37	31	25	19	13	7

Plan 10.5 8 × 8 balanced lattice

$t = 64, k = 8, r = 9, b = 72, \lambda = 1$

Block	Rep. I							
(1)	1	2	3	4	5	6	7	8
(2)	9	10	11	12	13	14	15	16
(3)	17	18	19	20	21	22	23	24
(4)	25	26	27	28	29	30	31	32
(5)	33	34	35	36	37	38	39	40
(6)	41	42	43	44	45	46	47	48
(7)	49	50	51	52	53	54	55	56
(8)	57	58	59	60	61	62	63	64

Block	Rep. II							
(9)	1	9	17	25	33	41	49	57
(10)	2	10	18	26	34	42	50	58
(11)	3	11	19	27	35	43	51	59
(12)	4	12	20	28	36	44	52	60
(13)	5	13	21	29	37	45	53	61
(14)	6	14	22	30	38	46	54	62
(15)	7	15	23	31	39	47	55	63
(16)	8	16	24	32	40	48	56	64

Block	Rep. III							
(17)	1	10	19	28	37	46	55	64
(18)	9	2	51	44	61	30	23	40
(19)	17	50	3	36	29	62	15	48
(20)	25	42	35	4	21	14	63	56
(21)	33	58	27	20	5	54	47	16
(22)	41	26	59	12	53	6	39	24
(23)	49	18	11	60	45	38	7	32
(24)	57	34	43	52	13	22	31	8

Block	Rep. IV							
(25)	1	18	27	44	13	62	39	56
(26)	17	2	35	60	53	46	31	16
(27)	25	34	3	12	45	54	23	64
(28)	41	58	11	4	29	22	55	40
(29)	9	50	43	28	5	38	63	24
(30)	57	42	51	20	37	6	15	32
(31)	33	26	19	52	61	14	7	48
(32)	49	10	59	36	21	30	47	8

Block	Rep. V							
(33)	1	26	43	60	21	54	15	40
(34)	25	2	11	52	37	62	47	24
(35)	41	10	3	20	61	38	31	56
(36)	57	50	19	4	45	30	39	16
(37)	17	34	59	44	5	14	55	32
(38)	49	58	35	28	13	6	23	48
(39)	9	42	27	36	53	22	7	64
(40)	33	18	51	12	29	46	63	8

Block	Rep. VI							
(41)	1	34	11	20	53	30	63	48
(42)	33	2	59	28	45	22	15	56
(43)	9	58	3	52	21	46	39	32
(44)	17	26	51	4	13	38	47	64
(45)	49	42	19	12	5	62	31	40
(46)	25	18	43	36	61	6	55	16
(47)	57	10	35	44	29	54	7	24
(48)	41	50	27	60	37	14	23	8

Plan 10.5 (Continued) 8 × 8 balanced lattice

	Rep. VII							
(49)	1	42	59	52	29	38	23	16
(50)	41	2	19	36	13	54	63	32
(51)	57	18	3	28	53	14	47	40
(52)	49	34	27	4	61	46	15	24
(53)	25	10	51	60	5	22	39	48
(54)	33	50	11	44	21	6	31	64
(55)	17	58	43	12	37	30	7	56
(56)	9	26	35	20	45	62	55	8

	Rep. VIII							
(57)	1	50	35	12	61	22	47	32
(58)	49	2	43	20	29	14	39	64
(59)	33	42	3	60	13	30	55	24
(60)	9	18	59	4	37	54	31	48
(61)	57	26	11	36	5	46	23	56
(62)	17	10	27	52	45	6	63	40
(63)	41	34	51	28	21	62	7	16
(64)	25	58	19	44	53	38	15	8

	Rep. IX							
(65)	1	58	51	36	45	14	31	24
(66)	57	2	27	12	21	38	55	48
(67)	49	26	3	44	37	22	63	16
(68)	33	10	43	4	53	62	23	32
(69)	41	18	35	52	5	30	15	64
(70)	9	34	19	60	29	6	47	56
(71)	25	50	59	20	13	46	7	40
(72)	17	42	11	28	61	54	39	8

Plan 10.6 9 × 9 balanced lattice

$t = 81, k = 9, r = 10, b = 90, \lambda = 1$

Block	Rep. I								
(1)	1	2	3	4	5	6	7	8	9
(2)	10	11	12	13	14	15	16	17	18
(3)	19	20	21	22	23	24	25	26	27
(4)	28	29	30	31	32	33	34	35	36
(5)	37	38	39	40	41	42	43	44	45
(6)	46	47	48	49	50	51	52	53	54
(7)	55	56	57	58	59	60	61	62	63
(8)	64	65	66	67	68	69	70	71	72
(9)	73	74	75	76	77	78	79	80	81

	Rep. II								
(10)	1	10	19	28	37	46	55	64	73
(11)	2	11	20	29	38	47	56	65	74
(12)	3	12	21	30	39	48	57	66	75
(13)	4	13	22	31	40	49	58	67	76
(14)	5	14	23	32	41	50	59	68	77
(15)	6	15	24	33	42	51	60	69	78
(16)	7	16	25	34	43	52	61	70	79
(17)	8	17	26	35	44	53	62	71	80
(18)	9	18	27	36	45	54	63	72	81

	Rep. III								
(19)	1	20	12	58	77	69	34	53	45
(20)	10	2	21	67	59	78	43	35	54
(21)	19	11	3	76	68	60	52	44	36
(22)	28	47	39	4	23	15	61	80	72
(23)	37	29	48	13	5	24	70	62	81
(24)	46	38	30	22	14	6	79	71	63
(25)	55	74	66	31	50	42	7	26	18
(26)	64	56	75	40	32	51	16	8	27
(27)	73	65	57	49	41	33	25	17	9

	Rep. IV								
(28)	1	11	21	31	41	51	61	71	81
(29)	19	2	12	49	32	42	79	62	72
(30)	10	20	3	40	50	33	70	80	63
(31)	55	65	75	4	14	24	34	44	54
(32)	73	56	66	22	5	15	52	35	45
(33)	64	74	57	13	23	6	43	53	36
(34)	28	38	48	58	68	78	7	17	27
(35)	46	29	39	76	59	69	25	8	18
(36)	37	47	30	67	77	60	16	26	9

Plan 10.6 (Continued) 9 × 9 balanced lattice

Rep. V									
(37)	1	29	57	22	50	78	16	44	72
(38)	55	2	30	76	23	51	70	17	45
(39)	28	56	3	49	77	24	43	71	18
(40)	10	38	66	4	32	60	25	53	81
(41)	64	11	39	58	5	33	79	26	54
(42)	37	65	12	31	59	6	52	80	27
(43)	19	47	75	13	41	69	7	35	63
(44)	73	20	48	67	14	42	61	8	36
(45)	46	74	21	40	68	15	34	62	9

Rep. VI									
(46)	1	56	30	13	68	42	25	80	54
(47)	28	2	57	40	14	69	52	26	81
(48)	55	29	3	67	41	15	79	53	27
(49)	19	74	48	4	59	33	16	71	45
(50)	46	20	75	31	5	60	43	17	72
(51)	73	47	21	58	32	6	70	44	18
(52)	10	65	39	22	77	51	7	62	36
(53)	37	11	66	49	23	78	34	8	63
(54)	64	38	12	76	50	24	61	35	9

Rep. VII									
(55)	1	47	66	76	14	33	43	62	27
(56)	64	2	48	31	77	15	25	44	63
(57)	46	65	3	13	32	78	61	26	45
(58)	37	56	21	4	50	69	79	17	36
(59)	19	38	57	67	5	51	34	80	18
(60)	55	20	39	49	68	6	16	35	81
(61)	73	11	30	40	59	24	7	53	72
(62)	28	74	12	22	41	60	70	8	54
(63)	10	29	75	58	23	42	52	71	9

Rep. VIII									
(64)	1	74	39	67	32	24	52	17	63
(65)	37	2	75	22	68	33	61	53	18
(66)	73	38	3	31	23	69	16	62	54
(67)	46	11	57	4	77	42	70	35	27
(68)	55	47	12	40	5	78	25	71	36
(69)	10	56	48	76	41	6	34	26	72
(70)	64	29	21	49	14	60	7	80	45
(71)	19	65	30	58	50	15	43	8	81
(72)	28	20	66	13	59	51	79	44	9

Rep. IX									
(73)	1	65	48	40	23	60	79	35	18
(74)	46	2	66	58	41	24	16	80	36
(75)	64	47	3	22	59	42	34	17	81
(76)	73	29	12	4	68	51	43	26	63
(77)	10	74	30	49	5	69	61	44	27
(78)	28	11	75	67	50	6	25	62	45
(79)	37	20	57	76	32	15	7	71	54
(80)	55	38	21	13	77	33	52	8	72
(81)	19	56	39	31	14	78	70	53	9

Rep. X									
(82)	1	38	75	49	59	15	70	26	36
(83)	73	2	39	13	50	60	34	71	27
(84)	37	74	3	58	14	51	25	35	72
(85)	64	20	30	4	41	78	52	62	18
(86)	28	65	21	76	5	42	16	53	63
(87)	19	29	66	40	77	6	61	17	54
(88)	46	56	12	67	23	33	7	44	81
(89)	10	47	57	31	68	24	79	8	45
(90)	55	11	48	22	32	69	43	80	9

Plan 10.7 6 × 6 triple lattice

Block	Rep. I					
(1)	1	2	3	4	5	6
(2)	7	8	9	10	11	12
(3)	13	14	15	16	17	18
(4)	19	20	21	22	23	24
(5)	25	26	27	28	29	30
(6)	31	32	33	34	35	36

Block	Rep. II					
(7)	1	7	13	19	25	31
(8)	2	8	14	20	26	32
(9)	3	9	15	21	27	33
(10)	4	10	16	22	28	34
(11)	5	11	17	23	29	35
(12)	6	12	18	24	30	36

Block	Rep. III					
(13)	1	8	15	22	29	36
(14)	31	2	9	16	23	30
(15)	25	32	3	10	17	24
(16)	19	26	33	4	11	18
(17)	13	20	27	34	5	12
(18)	7	14	21	28	35	6

Plan 10.8 10 × 10 triple lattice

Block	Rep. I									
(1)	1	2	3	4	5	6	7	8	9	10
(2)	11	12	13	14	15	16	17	18	19	20
(3)	21	22	23	24	25	26	27	28	29	30
(4)	31	32	33	34	35	36	37	38	39	40
(5)	41	42	43	44	45	46	47	48	49	50
(6)	51	52	53	54	55	56	57	58	59	60
(7)	61	62	63	64	65	66	67	68	69	70
(8)	71	72	73	74	75	76	77	78	79	80
(9)	81	82	83	84	85	86	87	88	89	90
(10)	91	92	93	94	95	96	97	98	99	100

Block	Rep. II									
(11)	1	11	21	31	41	51	61	71	81	91
(12)	2	12	22	32	42	52	62	72	82	92
(13)	3	13	23	33	43	53	63	73	83	93
(14)	4	14	24	34	44	54	64	74	84	94
(15)	5	15	25	35	45	55	65	75	85	95
(16)	6	16	26	36	46	56	66	76	86	96
(17)	7	17	27	37	47	57	67	77	87	97
(18)	8	18	28	38	48	58	68	78	88	98
(19)	9	19	29	39	49	59	69	79	89	99
(20)	10	20	30	40	50	60	70	80	90	100

Block	Rep. III									
(21)	1	12	23	34	45	56	67	78	89	100
(22)	91	2	13	24	35	46	57	68	79	90
(23)	81	92	3	14	25	36	47	58	69	80
(24)	71	82	93	4	15	26	37	48	59	70
(25)	61	72	83	94	5	16	27	38	49	60
(26)	51	62	73	84	95	6	17	28	39	50
(27)	41	52	63	74	85	96	7	18	29	40
(28)	31	42	53	64	75	86	97	8	19	30
(29)	21	32	43	54	65	76	87	98	9	20
(30)	11	22	33	44	55	66	77	88	99	10

Plan 10.9 12 × 12 quadruple lattice

Block	Rep. I											
(1)	1	2	3	4	5	6	7	8	9	10	11	12
(2)	13	14	15	16	17	18	19	20	21	22	23	24
(3)	25	26	27	28	29	30	31	32	33	34	35	36
(4)	37	38	39	40	41	42	43	44	45	46	47	48
(5)	49	50	51	52	53	54	55	56	57	58	59	60
(6)	61	62	63	64	65	66	67	68	69	70	71	72
(7)	73	74	75	76	77	78	79	80	81	82	83	84
(8)	85	86	87	88	89	90	91	92	93	94	95	96
(9)	97	98	99	100	101	102	103	104	105	106	107	108
(10)	109	110	111	112	113	114	115	116	117	118	119	120
(11)	121	122	123	124	125	126	127	128	129	130	131	132
(12)	133	134	135	136	137	138	139	140	141	142	143	144

Block	Rep. II											
(13)	1	13	25	37	49	61	73	85	97	109	121	133
(14)	2	14	26	38	50	62	74	86	98	110	122	134
(15)	3	15	27	39	51	63	75	87	99	111	123	135
(16)	4	16	28	40	52	64	76	88	100	112	124	136
(17)	5	17	29	41	53	65	77	89	101	113	125	137
(18)	6	18	30	42	54	66	78	90	102	114	126	138
(19)	7	19	31	43	55	67	79	91	103	115	127	139
(20)	8	20	32	44	56	68	80	92	104	116	128	140
(21)	9	21	33	45	57	69	81	93	105	117	129	141
(22)	10	22	34	46	58	70	82	94	106	118	130	142
(23)	11	23	35	47	59	71	83	95	107	119	131	143
(24)	12	24	36	48	60	72	84	96	108	120	132	144

Block	Rep. III											
(25)	1	14	27	40	57	70	83	96	101	114	127	140
(26)	2	13	28	39	58	69	84	95	102	113	128	139
(27)	3	16	25	38	59	72	81	94	103	116	125	138
(28)	4	15	26	37	60	71	82	93	104	115	126	137
(29)	5	18	31	44	49	62	75	88	105	118	131	144
(30)	6	17	32	43	50	61	76	87	106	117	132	143
(31)	7	20	29	42	51	64	73	86	107	120	129	142
(32)	8	19	30	41	52	63	74	85	108	119	130	141
(33)	9	22	35	48	53	66	79	92	97	110	123	136
(34)	10	21	36	47	54	65	80	91	98	109	124	135
(35)	11	24	33	46	55	68	77	90	99	112	121	134
(36)	12	23	34	45	56	67	78	89	100	111	122	133

Plan 10.9 (Continued) 12 × 12 quadruple lattice

Rep. IV

(37)	1	24	30	43	53	64	82	95	105	116	122	135
(38)	2	23	29	44	54	63	81	96	106	115	121	136
(39)	3	22	32	41	55	62	84	93	107	114	124	133
(40)	4	21	31	42	56	61	83	94	108	113	123	134
(41)	5	16	34	47	57	68	74	87	97	120	126	139
(42)	6	15	33	48	58	67	73	88	98	119	125	140
(43)	7	14	36	45	59	66	76	85	99	118	128	137
(44)	8	13	35	46	60	65	75	86	100	117	127	138
(45)	9	20	26	39	49	72	78	91	101	112	130	143
(46)	10	19	25	40	50	71	77	92	102	111	129	144
(47)	11	18	28	37	51	70	80	89	103	110	132	141
(48)	12	17	27	38	52	69	79	90	104	109	131	142

Plan 10.10 3 × 4 rectangular lattice

Block	Rep. *X*			Block	Rep. *Y*			Block	Rep. *Z*		
*X*1	1	2	3	*Y*1	4	7	10	*Z*1	6	8	12
*X*2	4	5	6	*Y*2	1	8	11	*Z*2	2	9	10
*X*3	7	8	9	*Y*3	2	5	12	*Z*3	3	4	11
*X*4	10	11	12	*Y*4	3	6	9	*Z*4	1	5	7

Plan 10.11 4 × 5 rectangular lattice

Block	Rep. *X*				Block	Rep. *Y*				Block	Rep. *Z*			
*X*1	1	2	3	4	*Y*1	5	9	13	17	*Z*1	8	11	15	18
*X*2	5	6	7	8	*Y*2	1	10	14	18	*Z*2	2	9	16	20
*X*3	9	10	11	12	*Y*3	2	6	15	19	*Z*3	4	7	14	17
*X*4	13	14	15	16	*Y*4	3	7	11	20	*Z*4	1	5	12	19
*X*5	17	18	19	20	*Y*5	4	8	12	16	*Z*5	3	6	10	13

Plan 10.12 5 × 6 rectangular lattice

Block	Rep. *X*					Block	Rep. *Y*					Block	Rep. *Z*				
*X*1	1	2	3	4	5	*Y*1	6	11	16	21	26	*Z*1	7	13	19	25	27
*X*2	6	7	8	9	10	*Y*2	1	12	17	22	27	*Z*2	5	14	16	23	29
*X*3	11	12	13	14	15	*Y*3	2	7	18	23	28	*Z*3	1	8	20	21	30
*X*4	16	17	18	19	20	*Y*4	3	8	13	24	29	*Z*4	2	9	15	22	26
*X*5	21	22	23	24	25	*Y*5	4	9	14	19	30	*Z*5	3	10	11	17	28
*X*6	26	27	28	29	30	*Y*6	5	10	15	20	25	*Z*6	4	6	12	18	24

Plan 10.13 6 × 7 rectangular lattice

Block	Rep. X						Block	Rep. Y					
X1	1	2	3	4	5	6	Y1	7	13	19	25	31	37
X2	7	8	9	10	11	12	Y2	1	14	20	26	32	38
X3	13	14	15	16	17	18	Y3	2	8	21	27	33	39
X4	19	20	21	22	23	24	Y4	3	9	15	28	34	40
X5	25	26	27	28	29	30	Y5	4	10	16	22	35	41
X6	31	32	33	34	35	36	Y6	5	11	17	23	29	42
X7	37	38	39	40	41	42	Y7	6	12	18	24	30	36

Block	Rep. Z					
Z1	12	17	22	28	33	38
Z2	2	13	24	29	35	40
Z3	4	9	20	25	36	42
Z4	6	11	16	27	32	37
Z5	1	7	18	23	34	39
Z6	3	8	14	19	30	41
Z7	5	10	15	21	26	31

Plan 10.14 7 × 8 rectangular lattice

Block	Rep. X							Block	Rep. Y						
X1	1	2	3	4	5	6	7	Y1	8	15	22	29	36	43	50
X2	8	9	10	11	12	13	14	Y2	1	16	23	30	37	44	51
X3	15	16	17	18	19	20	21	Y3	2	9	24	31	38	45	52
X4	22	23	24	25	26	27	28	Y4	3	10	17	32	39	46	53
X5	29	30	31	32	33	34	35	Y5	4	11	18	25	40	47	54
X6	36	37	38	39	40	41	42	Y6	5	12	19	26	33	48	55
X7	43	44	45	46	47	48	49	Y7	6	13	20	27	34	41	56
X8	50	51	52	53	54	55	56	Y8	7	14	21	28	35	42	49

Block	Rep. Z						
Z1	9	17	25	33	41	49	51
Z2	7	18	26	34	38	43	53
Z3	1	10	27	35	36	47	55
Z4	2	11	19	29	42	44	56
Z5	3	12	20	28	37	45	50
Z6	4	13	21	22	30	46	52
Z7	5	14	15	23	31	39	54
Z8	6	8	16	24	32	40	48

Plan 10.15 8 × 9 rectangular lattice

Block	Rep. X							
X1	1	2	3	4	5	6	7	8
X2	9	10	11	12	13	14	15	16
X3	17	18	19	20	21	22	23	24
X4	25	26	27	28	29	30	31	32
X5	33	34	35	36	37	38	39	40
X6	41	42	43	44	45	46	47	48
X7	49	50	51	52	53	54	55	56
X8	57	58	59	60	61	62	63	64
X9	65	66	67	68	69	70	71	72

	Rep. Y							
Y1	9	17	25	33	41	49	57	65
Y2	1	18	26	34	42	50	58	66
Y3	2	10	27	35	43	51	59	67
Y4	3	11	19	36	44	52	60	68
Y5	4	12	20	28	45	53	61	69
Y6	5	13	21	29	37	54	62	70
Y7	6	14	22	30	38	46	63	71
Y8	7	15	23	31	39	47	55	72
Y9	8	16	24	32	40	48	56	64

	Rep. Z							
Z1	16	23	30	37	45	52	59	66
Z2	2	17	32	39	46	54	61	68
Z3	4	11	26	33	48	55	63	70
Z4	6	13	20	35	42	49	64	72
Z5	8	15	22	29	44	51	58	65
Z6	1	9	24	31	38	53	60	67
Z7	3	10	18	25	40	47	62	69
Z8	5	12	19	27	34	41	56	71
Z9	7	14	21	28	36	43	50	57

Plan 10.16 9 × 10 rectangular lattice

Block	Rep. X								
X1	1	2	3	4	5	6	7	8	9
X2	10	11	12	13	14	15	16	17	18
X3	19	20	21	22	23	24	25	26	27
X4	28	29	30	31	32	33	34	35	36
X5	37	38	39	40	41	42	43	44	45
X6	46	47	48	49	50	51	52	53	54
X7	55	56	57	58	59	60	61	62	63
X8	64	65	66	67	68	69	70	71	72
X9	73	74	75	76	77	78	79	80	81
X10	82	83	84	85	86	87	88	89	90

	Rep. Y								
Y1	10	19	28	37	46	55	64	73	82
Y2	1	20	29	38	47	56	65	74	83
Y3	2	11	30	39	48	57	66	75	84
Y4	3	12	21	40	49	58	67	76	85
Y5	4	13	22	31	50	59	68	77	86
Y6	5	14	23	32	41	60	69	78	87
Y7	6	15	24	33	42	51	70	79	88
Y8	7	16	25	34	43	52	61	80	89
Y9	8	17	26	35	44	53	62	71	90
Y10	9	18	27	36	45	54	63	72	81

	Rep. Z								
Z1	11	21	31	41	51	61	71	81	83
Z2	9	22	32	42	52	62	64	76	84
Z3	1	12	33	43	53	63	68	73	87
Z4	2	13	23	44	54	55	65	79	89
Z5	3	14	24	34	46	56	72	75	90
Z6	4	15	25	35	45	57	67	74	82
Z7	5	16	26	36	37	47	66	77	85
Z8	6	17	27	28	38	48	58	78	86
Z9	7	18	19	29	39	49	59	69	88
Z10	8	10	20	30	40	50	60	70	80

CHAPTER 11

Balanced and Partially Balanced Incomplete Block Designs

11.1 Balanced Incomplete Blocks

The balanced lattices described in the last chapter are a particular group of a general class of designs known as balanced incomplete blocks. All these designs have the property that any pair of treatments appears together equally often within some block. Thus in plan 11.1 every pair of treatments appears together once in the same block, in plan 11.4 twice, and in plan 11.5 four times. This property insures that the same standard error may be used for comparing every pair of treatments; it also facilitates the statistical analysis, since any treatment total is adjusted in a single operation for all the blocks in which the treatment appears.

The construction of the designs presents interesting mathematical problems, (11.1), (11.2), (11.3). Although a design can be found for any number of treatments, t, and any size of block, k, most of these are of no interest for our purpose, since they require too many replications. The plans in this chapter have been restricted to those in which the number of replicates does not exceed 10.

The balanced lattices are the particular set of designs for which the relation $t = k^2$ holds. They also have the property that the blocks can be grouped in separate replications. The majority of the balanced incomplete block arrangements do not possess this property, which is possible only when t is a multiple of k. A few designs can be sorted into groups which comprize two or more replicates each.

In most plans, the block contains six or fewer units. Accordingly, the designs are adapted for experiments in which the appropriate size of block is small. They have been applied in greenhouse pot experiments, where the block is restricted to the width of the bench; in experiments on plant virus diseases, where the block consists of a small number of leaves on each plant; in experimental cookery where there is a limited number of stoves, in tests of mosquito repellents, reference (11.4), where the block consists of the two exposed arms of a subject; and in nutritional experiments where each child constitutes a block, 3 different foods being given during the 3 terms of the school year.

Examples from field experimental work occur in blueberry fertilizer trials where, owing to the uneven growth of different bushes, only a small number of them can be grouped into a homogeneous block. A similar consideration may limit the size of block in experiments on the control of fruit pests, where the tree constitutes the experimental unit.

A few types of factorial design—the 5×2, 5×3, 7×3, and 7×4—can be arranged in balanced incomplete blocks with the result that main effects and interactions are confounded to the same extent.

11.1a Balanced Incomplete Blocks in Taste and Preference Testing. These designs have found fruitful applications in experiments in which individuals are asked to make a comparative rating of different objects that are presented to them. Examples are the rating of different ways of preparing a food as to palatability, different colors in which some article is made as to acceptability, and different occupations as to their social status. In most of these experiments the individual finds it increasingly difficult to reach a satisfactory rating as the number of objects presented to him becomes larger. The optimum number of items that can be rated varies with the type of experiment, but it is often less than the total number that are to be compared in the experiment.

One solution to this difficulty is to arrange the objects in a balanced incomplete block design. For instance, if there are 10 objects to be rated and it has been found that an individual should not be asked to compare more than 5 items at a time, the design for 10 treatments in blocks of size 5 (plan 11.17) is appropriate. Each individual constitutes an incomplete block. Sometimes the individual rates only one block, sometimes several blocks, with intervening rest periods.

These applications have produced a number of interesting problems in statistical analysis. Bradley (11.8a) gives an introductory review. The individual may merely rate the objects in order of preference (e.g., 1, 2, 3), so that the analysis must be based on ranked data. Durbin (11.9a) develops the analogue of the F-test for ranked data, and Bradley and Terry (11.10a, 11.11a) create a mathematical model which relates the ranked data to a continuous scale of preferences.

When the observer is asked to rate on a *fixed* scale, e.g., 1, 2, 3, $\cdots$ 10, it is often found that the rating assigned to an object depends on the other objects with which it is being compared in the incomplete block, being, for instance, lower if the other objects are "good" than if they are "poor." This suggests the presence of a type of intra-block correlation that is not provided for in the standard analysis of variance of these designs. Calvin (11.12a) has extended the mathematical model and the analysis so as to take account of this correlation. He shows that this analysis becomes simpler if the design has an additional property of symmetry—namely, that every trio of treatments appears equally often within the same block—and gives a catalogue of doubly balanced designs of this type.

11.2 Comparisons with Other Designs

Much experimental material is of the type exemplified above, in that the natural grouping for use as a block accommodates only a small number of units. Where the number of treatments exceeds the number of units in a block, there are three alternatives to the use of a balanced incomplete block design. (1) A common treatment is placed in each block, so as to provide a basis for comparison between treatments which are in different blocks. (2) The treatments are divided into groups such that the number of treatments per group is equal to the number of units per block. The groups are compared in a split-plot arrangement (chapter 7). (3) Complete replicates are assembled by the grouping of units from more than one block, in order that a randomized block design may be used. The merits of the first two alternatives were discussed in section 9.5.

The relative efficiency of randomized blocks and balanced incomplete blocks has been discussed by Yates (11.5). Balanced incomplete block designs which are arranged in replications cannot be appreciably less accurate than randomized blocks. With balanced incomplete blocks which cannot be arranged in replications, a comparison with randomized blocks is not easy except by the use of uniformity data. Certain conclusions can be drawn from the nature of the experimental material. If there is some association among the members of different blocks, it may be possible to form replicates which are only slightly less homogeneous than the blocks. For example, suppose that in an animal experiment the litter is the natural block. If the responses of the animals are influenced by their genetic constitution and if replications are constructed from closely related litters, the experimental error may be only slightly greater within a replication than within a block. In this event randomized blocks may be the more accurate layout.

Sometimes the investigator has no information about any associations between the groups which constitute the blocks; he knows merely that the experimental responses are likely to differ from group to group. In these circumstances Yates (11.5) has shown that incomplete blocks (with the recovery of inter-block information) are more accurate on the average except when the blocks are ineffective. There is the additional advantage that the block differences can be measured as a guide to the design of future experiments, whereas with replications composed of units from several blocks it would be difficult to estimate the block effects.

It is worth noting the *efficiency factor* (E) of any balanced incomplete block design whose use is under consideration. This factor, which is shown as a decimal fraction in table **11.3 (p. 469)** and in the plans, is a lower limit to the efficiency of balanced incomplete blocks relative to randomized blocks. With 37 treatments in blocks of 9 units, for instance, the efficiency factor is 91%, so that the loss of efficiency could not exceed 9%. The loss equals 9% only in the rather unlikely situation in which the replications of 37 units are as uniform as the blocks of 9 units, yet the variation among blocks is large compared to that within blocks. With any reasonable prospect that the blocks are more homogeneous than the replicates, the incomplete block design is to be preferred in this case. On the other hand, with 7 treatments in blocks of 2 units, the efficiency factor is only 58%. In this case incomplete blocks will be more accurate only if there is a substantial reduction in error variation because of the reduction in block size.

11.3 Arrangement of Experimental Material

Units within the same incomplete block should be as homogeneous as possible. For those designs which can also be grouped in replications, it is advisable to place similar blocks, so far as these can be discerned, within the same replicate.

Certain of these designs can be rearranged to form an incomplete latin square (chapter 13) in which variation associated with two types of grouping can be eliminated from the experimental errors. The experimenter should consider whether advantage can be taken of this double control so as to reduce the error.

11.4 Randomization

The steps are:

1. Rearrange the blocks at random. (If the design is arranged in complete replications, the blocks are randomized only within each replication and the replications are kept separate.)

2. Randomize the positions of the treatment numbers within each block.
3. Assign treatments at random to the numbers in the plan.

11.5 Statistical Analysis

In order to present the shortest computational methods, it is advisable to divide the plans into five types, which are shown in the index at the end of this chapter.

11.51 Type I. Designs Arranged in Replications. We will analyze an example with 6 treatments in blocks of 2 (plan 11.3). The experiment was conducted by Dr. Pauline Paul at Iowa State College (11.6). Its object was to compare the effects of length of cold storage on the tenderness and flavor of beef roasts. Six periods of storage (0, 1, 2, 4, 9, and 18 days) were tested: these are denoted by treatment symbols 1, 2, 3, ···, 6, respectively.

Thirty roasts from the round of an animal were used. Four muscles each provided 6 roasts, while 3 muscles each provided 2 roasts. The roasts on any muscle group themselves naturally into pairs, since to each roast on the left side of an animal there corresponds another roast on the right side. From previous experience it was believed that the 2 roasts in any pair would give closely similar results. Variation among different pairs from the same muscle was expected to be somewhat larger, and variation among muscles to be still larger.

These opinions prompted the use of a design in blocks of 2, each block comprizing the left and right roasts in a pair. When grouping the blocks into replications, it was natural to put roasts from the same muscle into the same replicate. With the first 4 muscles a separate replicate could be made from each muscle. The remaining replication consisted of the 3 smaller muscles.

The plan and the scores for tenderness are given in table 11.1. Scoring was done by 4 judges, each marking on a scale from 0 to 10. The scores shown are their totals (out of 40), a high score indicating very tender beef.

The following symbols are used: $t = 6$ = the number of treatments; $k = 2$ = number of units per block; $r = 5$ = number of replications; $b = 15$ = number of blocks.

1. Find the block totals, B, the replicate totals, the grand total, G, and the treatment totals, T, placed below the plan.

TABLE 11.1 SCORES FOR TENDERNESS OF BEEF

Rep. I		B	Rep. II		B	Rep. III		B
(1) 7	(2) 17	24	(1) 17	(3) 27	44	(1) 10	(4) 25	35
(3) 26	(4) 25	51	(2) 23	(5) 27	50	(2) 26	(6) 37	63
(5) 33	(6) 29	62	(4) 29	(6) 30	59	(3) 24	(5) 26	50
		137			153			148

Rep. IV		B	Rep. V		B
(1) 25	(5) 40	65	(1) 11	(6) 27	38
(2) 25	(4) 34	59	(2) 24	(3) 21	45
(3) 34	(6) 32	66	(4) 26	(5) 32	58
		190			141

Storage time (days)	Treatment no.	T	B_t	Q	W	$T + \mu W$	Adj. means
0	1	70	206	−66	19	71.8	14.4
1	2	115	241	−11	24	117.3	23.5
2	3	132	256	8	17	133.6	26.7
4	4	139	262	16	15	140.4	28.1
9	5	158	285	31	−24	155.7	31.1
18	6	155	288	22	−51	150.2	30.0
		G = 769	1538	0	0	769.0	

2. For each treatment, calculate the following quantities:

B_t = total of all blocks in which the treatment appears

$$Q = kT - B_t = 2T - B_t$$

$$W = (t - k)T - (t - 1)B_t + (k - 1)G = 4T - 5B_t + G$$

The total of the Q's and the W's should each be zero.

3. The analysis of variance is as follows:

	d.f.		s.s.	m.s.
Replications	$(r - 1)$	4	298.5	
Treatments (unadj.)	$(t - 1)$	5	1059.8	
Blocks within reps. (adj.)	$(b - r)$	10	213.4	$21.34E_b$
Intra-block error	$(tr - t - b + 1)$	10	77.3	$7.73E_e$
Total	$(tr - 1)$	29	1649.0	

The only sum of squares requiring special instructions is that for blocks within replications, adjusted for treatment effects. Since we do not know any simple way of finding this directly, it is obtained by a roundabout method. Calculate in the usual way the *unadjusted* sum of squares for blocks within replications: this comes to 753.0. Then compute the sum of squares for treatments, adjusted for blocks, which is

$$\frac{(t-1)\sum Q^2}{ktr(k-1)} = \frac{5}{(2)(6)(5)(1)} \cdot [(66)^2 + (11)^2 + \cdots + (22)^2] = 520.2$$

The adjusted sum of squares for blocks is then given by the following combination of sums of squares

Blocks (unadj.)	+	treatments (adj.)	−	treatments (unadj.)	
753.0	+	520.2	−	1059.8	= 213.4

4. Calculate the weighting factor

$$\mu = \frac{r(E_b - E_e)}{rt(k-1)E_b + k(b-r-t+1)E_e}$$

$$= \frac{5(21.34 - 7.73)}{(30)(21.34) + (10)(7.73)} = 0.09484$$

The adjusted treatment totals are

$$T + \mu W$$

and the means per unit are found on division by r.

5. The effective error variance per unit is estimated as

$$E_e[1 + (t-k)\mu] = 7.73[1 + (4)(0.0948)] = 10.66$$

We derive t-tests from this figure by the ordinary rules. For instance, the standard error of the difference between two adjusted totals is $\sqrt{2r(10.66)}$, or 10.32; for the difference between two adjusted *means* the standard error is $\sqrt{(2)(10.66)/r}$, or 2.06.

6. If an F-test is desired, calculate the sum of squares of the adjusted treatment totals. This figure, 4718.0, is divided by $r(t-1)$, or in this case 25, in order to find the mean square on a unit basis. The mean square, 188.7, is compared with the effective error m.s., 10.66, which is assigned $(tr - t - b + 1)$, or 10, d.f. The F-ratio, 188.7/10.66, is highly significant. The test is not exact, because it ignores the effects of errors in the weighting factor μ.

In an experiment of this type, interest centers on the shape of the response curve rather than on comparisons between pairs of treatments.

The adjusted means indicate the common "diminishing returns" curve. An exponential of the form $y = A - Be^{-ct}$, where t is the time of storage, fits the data satisfactorily. A parabolic regression on t, which is easier to compute, will be found to be inadequate. The conclusion is that storage up to about a week increases tenderness.

7. The gain in precision over randomized blocks is estimated as follows. By pooling the sums of squares for blocks within replications and intra-block error, we obtain an estimate of the error m.s. that would have applied with randomized blocks. This figure, 14.54, is compared with the effective error m.s., 10.66. The relative precision is estimated as 14.54/10.66, or 136%. Since the randomized block design has 20 error d.f. as against 10 for the design actually used, this estimate should be reduced to 126% by the adjustment described in section 2.31.

11.52 Type II. Designs Arranged in Groups of Replications. In our first edition, only 2 plans (11.6 and 11.13) were of this type. Since then, Rupp (11.7a) has shown that 8 additional plans can be arranged in groups of replications. The analysis is similar to that in the previous section. Two changes should be noted. If c is the number of groups, the d.f. in the analysis of variance subdivide as shown below.

	d.f.	m.s.
Groups	$(c - 1)$	
Treatments (unadj.)	$(t - 1)$	
Blocks within groups (adj.)	$(b - c)$	E_b
Intra-block error	$(tr - t - b + 1)$	E_e
Total	$(tr - 1)$	

The second change is that the weighting factor μ becomes

$$\mu = \frac{(b - c)(E_b - E_e)}{t(k - 1)(b - c)E_b + (t - k)(b - c - t + 1)E_e}$$

11.53 Type III. Designs Not Arranged in Replications or Groups of Replications. Here again the method of section 11.51 may be followed except that no replication effects appear. The analysis of variance now reads as follows.

	d.f.	m.s.
Treatments (unadj.)	$(t - 1)$	
Blocks (adj.)	$(b - 1)$	E_b
Intra-block error	$(tr - t - b + 1)$	E_e
Total	$(tr - 1)$	

As before, the sum of squares for blocks (adjusted) is found by a subtraction process. The weighting factor becomes

$$\mu = \frac{(b-1)(E_b - E_e)}{t(k-1)(b-1)E_b + (t-k)(b-t)E_e}$$

Otherwise the method of section 11.51 applies. Of course, the gain in accuracy relative to randomized blocks cannot be estimated. A worked example with $t = 9$, $k = 4$, has been given by Yates (11.5, 11.7). His method of calculating the adjusted sum of squares for blocks is slightly different from ours, since he divides this sum of squares into two components so that the calculation becomes partially self-checking. Results should be identical by either method.

11.54 Type IV. Experiments with $t = b$. These designs, which cannot be arranged in replications, are covered by the computing instructions given in the previous section. However, the analysis simplifies slightly when $t = b$, because the sum of squares for blocks (adjusted) can be found directly. The data in table 11.2 are from an experiment on corn hybrids with $t = b = 13$, $r = k = 4$. For field experiments a design that can be laid out in separate replications is usually preferred. In this experiment, conducted in 1943 by the North Carolina Agricultural Experiment Station, it was desired to compare 10 hybrids, all potentially suitable for commercial use, at a number of places throughout the state, and at each place to test in addition 3 standard varieties that were adapted to that place. With 13 treatments no design that can be put into replications is available, and since 4 was a convenient number of replications, the design with $t = 13$, $r = k = 4$ was chosen. Randomized blocks might have been equally good. Incidentally, the plan used is different from that given in plan 11.22, a different method of construction having been employed. Both plans have the same structural properties.

1. Find the block totals, B, the treatment totals, T, and for each treatment find the total B_t of all blocks in which the treatment appears. Time is saved if T and B_t are found simultaneously. The B_t values should add to k times the grand total G.

2. For each treatment find

$$W = (t-k)T - (t-1)B_t + (k-1)G$$

$$= 9T - 12B_t + 3G \quad \text{(in this example)}$$

The W's should add to zero.

TABLE 11.2 PLAN AND YIELDS OF CORN (POUNDS PER PLOT)

Block									Block totals
1	(3)	25.3	(6)	19.9	(9)	29.0	(11)	24.6	98.8
2	(3)	23.0	(4)	19.8	(8)	33.3	(12)	22.7	98.8
3	(10)	16.2	(11)	19.3	(12)	31.7	(13)	26.6	93.8
4	(2)	27.3	(5)	27.0	(8)	35.6	(11)	17.4	107.3
5	(7)	23.4	(8)	30.5	(9)	30.8	(10)	32.4	117.1
6	(4)	30.6	(5)	32.4	(6)	27.2	(10)	32.8	123.0
7	(1)	34.7	(5)	31.1	(9)	25.7	(12)	30.5	122.0
8	(3)	34.4	(5)	32.4	(7)	33.3	(13)	36.9	137.0
9	(1)	38.2	(2)	32.9	(3)	37.3	(10)	31.3	139.7
10	(2)	28.7	(4)	30.7	(9)	26.9	(13)	35.3	121.6
11	(1)	36.6	(4)	31.1	(7)	31.1	(11)	28.4	127.2
12	(1)	31.8	(6)	33.7	(8)	27.8	(13)	41.1	134.4
13	(2)	30.3	(6)	31.5	(7)	39.3	(12)	26.7	127.8
									1548.5

No.	T	B_t	W	$T + \mu W$
1	141.3	523.3	−362.4	136.7
2	119.2	496.4	−238.5	116.2
3	120.0	474.3	33.9	120.4
4	112.2	470.6	8.1	112.3
5	122.9	489.3	−120.0	121.4
6	112.3	484.0	−151.8	110.4
7	127.1	509.1	−319.8	123.0
8	127.2	457.6	299.1	131.0
9	112.4	459.5	143.1	114.2
10	112.7	473.6	− 23.4	112.4
11	89.7	427.1	327.6	93.9
12	111.6	442.4	341.1	115.9
13	139.9	486.8	63.0	140.7
	1548.5	6194.0	0	1548.5

3. The analysis of variance is:

		d.f.	s.s.	m.s.
Treatments (unadj.)	$(t-1)$	12	542.67	
Blocks (adj.)	$(b-1)$	12	475.27	$39.61E_b$
Intra-block error	$(tr-2t+1)$	27	538.21	$19.93E_e$
Total	$(tr-1)$	51	1556.15	

The sum of squares for blocks (adjusted) is found directly as the sum of squares of the W's, divided by $tr(t-k)(k-1)$, or in this case $(4)(13)(9)(3) = 1404$.

4. The weighting factor μ simplifies to

$$\mu = \frac{E_b - E_e}{t(k-1)E_b} = \frac{19.68}{(13)(3)(39.61)} = 0.0127$$

The adjusted treatment totals are the quantities $(T + \mu W)$.

5. As before, the effective error m.s. per unit is taken as

$$E_e[1 + (t - k)\mu] = 19.93[1 + (9)(0.0127)] = 22.2$$

Other features of the analysis are the same as in previous sections. Yates (11.5) presents an example with $t = 21$, $k = 5$.

11.55 Type V. Small Experiments. In certain of the smaller designs, the numbers of degrees of freedom in the mean squares for blocks and intra-block error are rather small. Consequently, the estimates of the relative weights assigned to inter- and intra-block comparisons are poor. In these cases, unless the whole plan has been repeated to secure extra replication, it is best to use the original method of analysis that Yates developed (11.8). This method adjusts the treatment means for differences between the blocks, but makes no use of inter-block information. The analysis is a little simpler than when inter-block information is used.

For purposes of illustration, this analysis will be applied to the example in the previous section, with $t = b = 13$, $r = k = 4$, though in practice inter-block information should be utilized in an experiment of this size. The data are presented in table 11.2.

1. Find the block totals, B, and the treatment totals, T. For each treatment find the total B_t over all blocks in which the treatment appears.

2. For each treatment compute

$$Q = kT - B_t = 4T - B_t$$

$$t' = m + \frac{t-1}{tr(k-1)}Q = m + \frac{12}{(13)(4)(3)}Q = m + 0.0769Q$$

where $m = 1548.5/52 = 29.78$ is the mean of the whole experiment. The t' values are the adjusted treatment means. The Q values should add to zero.

3. The analysis of variance is shown below.

	d.f.		s.s.	m.s.
Blocks (unadj.)	$(b-1)$	12	689.38	
Treatments (adj.)	$(t-1)$	12	328.55	27.38
Intra-block error	$(tr - t - b + 1)$	27	538.22	$19.93E_e$
Total	$(tr-1)$	51	1556.15	

Note that the blocks sum of squares is *unadjusted.* It is found in the usual way as the sum of squares of deviations of the B's, divided by k.

This rule applies even when the experiment is arranged in replications or groups of replications, though if the experimenter wishes, he may divide the blocks sum of squares into that due to replications (or groups) and that due to blocks within replications.

The adjusted sum of squares for treatments is

$$\frac{t-1}{rtk(k-1)} \sum Q^2 = \frac{12}{(4)(13)(4)(3)} [(41.9)^2 + (19.6)^2 + \cdots + (72.8)^2]$$
$$= 328.55$$

As usual, the intra-block error s.s. is found by subtraction.

4. For t-tests, the effective error variance is

$$E_e \left[1 + \frac{t-k}{t(k-1)}\right] = E_e \frac{k(t-1)}{t(k-1)} = \frac{(19.93)(12)(4)}{(13)(3)} = 24.5$$

For the variance of the difference between two adjusted treatment means, we multiply this quantity by $2/r$.

5. The analysis above provides an exact F-test of the adjusted treatment means. The F-ratio is 27.38/19.93, with 12 and 27 d.f.

11.56 Missing Data. For reasons given in section 10.13, it is recommended that the analysis of incomplete data be carried out by inserting for the missing observations values which minimize the intra-block error s.s. In our notation the formula, first given by Cornish (11.9), is

$$x = \frac{tr(k-1)B + k(t-1)Q - (t-1)Q'}{(k-1)[tr(k-1) - k(t-1)]}$$

where B is the total of the block containing the missing observation, Q is the Q value ($= kT - B_t$) for the treatment that contains the missing value, and Q' is the sum of the Q values for all treatments that are in the block with the missing value.

Example 1. In the experiment on roast beef in section 11.51, suppose that treatment (5) in replication IV is absent. The first step is to find the block totals, the treatment totals, the B_t and the Q values as in the ordinary analysis. An x should be inserted for the missing observation. The data read as follows.

Treatment number	T	B_t	$Q = kT - B_t$
1	70	$166 + x$	$-26 - x$
2	115	241	-11
3	132	256	8
4	139	262	16
5	$118 + x$	$245 + x$	$-9 + x$
6	155	288	22
Totals	$729 + x$	$1458 + 2x$	0

Since the block with the missing value contains treatments (1) and (5)

$$Q' = -26 - 9 = -35$$

Hence, since B is 25,

$$x = \frac{30B + 10Q - 5Q'}{20} = \frac{6B + 2Q - Q'}{4}$$

$$= \frac{(6)(25) + (2)(-9) - (-35)}{4} = 42$$

On substituting this value in all places where an x occurs in the table above, we are ready to proceed with the analysis of variance.

Example 2. For an example with 2 missing observations, suppose that treatments (8) and (13) are missing from block 12 in the corn experiment in section 11.54. Denote the values to be inserted by x and y, respectively. To save space, the preliminary data are presented only for the treatments that occur in the same block as the missing observations; these are treatments (1), (6), (8), and (13).

Treatment number	T	B_t	$Q = 4T - B_t$
1	141.3	$454.4 + x + y$	$110.8 - x - y$
6	112.3	$415.1 + x + y$	$34.1 - x - y$
8	$99.4 + x$	$388.7 + x + y$	$8.9 + 3x - y$
13	$98.8 + y$	$417.9 + x + y$	$-22.7 - x + 3y$
			$131.1 = Q'$
		$B = 65.5 + x + y$	

In the 3 replications in which it appears, the average for treatment (13) is 33: this will be chosen as the first approximation to y. Then for x we have

$$B = 65.5 + 33.0 = 98.5; \qquad Q = 8.9 - 33 = -24.1; \qquad Q' = 131.1$$

Since $t = 13$, $k = r = 4$,

$$x = \frac{156B + 48Q - 12Q'}{324} = \frac{13B + 4Q - Q'}{27}$$

$$= \frac{(13)(98.5) + (4)(-24.1) - (131.1)}{27} = 39.0$$

We now substitute $x = 39.0$ in the preliminary data and obtain a second approximation to y. For this we have

$$B = 65.5 + 39.0 = 104.5; \qquad Q = -22.7 - 39.0 = -61.7$$

$$Q' = 131.1$$

Note that Q', which does not involve x or y, remains fixed throughout the calculation. The second approximation to y is

$$y = \frac{(13)(104.5) + (4)(-61.7) - (131.1)}{27} = 36.3$$

The next 2 approximations will be found to be $x = 40.1$, $y = 36.7$, which may be used as the values to be inserted.

The rest of the analysis proceeds as usual; the intra-block error degrees of freedom are reduced by 1 for each missing observation. In the calculation of standard errors for t-tests of the adjusted treatment means, the method given in section 10.13 should be followed.

11.57a Statistical Analysis for Repetitions of the Designs. With some of the smaller plans, the whole plan may be repeated p times when extra replication is wanted. In order to continue to use as many of the formulae already given as possible, it is advisable to let r denote the *total* number of replications and b the *total* number of blocks. If the basic design as given in the plans has r' replications and b' blocks, then $r = pr'$ and $b = pb'$.

Type I. Designs arranged in replications. With these definitions of r and b, all the steps and formulae given in section 11.51 remain the same. As before the weighting factor is

$$\mu = \frac{r(E_b - E_e)}{rt(k-1)E_b + k(b - r - t + 1)E_e}$$

Formulae will not be presented for the designs arranged in groups of replications, *Type II*, since these designs will seldom be duplicated.

Type III. Designs not arranged in replications or groups of replications. It is assumed that the repetitions will be kept apart during the course of the experiment. The $(b - 1)$ d.f. between blocks then subdivide into $(p - 1)$ d.f. between repetitions and $(b\text{-}p)$ d.f. between blocks within repetitions. In the analysis of variance, the d.f. are allocated as follows:

	d.f.
Repetitions	$(p - 1)$
Treatments	$(t - 1)$
Blocks within repetitions	$(b - p)$
Intra-block error	$(tr - t - b + 1)$
Total	$(tr - 1)$

The analysis follows the pattern described in section 11.53. Calculate the total sum of squares and the sums of squares for repetitions, treatments (unadjusted), and blocks within repetitions (unadjusted) in the usual way. The sum of squares for treatments (adjusted) is

$$\frac{(t-1)\sum Q^2}{ktr(k-1)}$$

as given in section 11.51. The sum of squares for blocks within repetitions (adjusted) is obtained by the indirect route as

Blocks within repetitions (unadj.)

+ treatments (adj.) − treatments (unadj.)

The weighting factor becomes

$$\mu = \frac{(b-p)(E_b - E_e)}{t(k-1)(b-p)E_b + (t-k)(b-t-p+1)E_e}$$

where E_b is the mean square for blocks within repetitions (adjusted). No further changes are needed.

Type IV. Experiments with $t = b$. It is suggested that the same method as for Type III be used in these experiments.

Type V. Small experiments. The procedure described in section 11.55 is still valid, remembering that r and b are the *total* numbers of replications and blocks, respectively.

11.6a Partially Balanced Incomplete Block Designs

11.61a Description. As mentioned previously, Bose and Nair (11.1a) showed that an incomplete block design which lacks full balance must possess certain properties of symmetry if the statistical analysis is not to be too cumbersome. These properties will be illustrated by the 3 × 3 triple lattice, which is a partially balanced design in the sense of Bose and Nair.

TABLE 11.1a A 3 × 3 TRIPLE LATTICE

Block	Rep. I	Block	Rep. II	Block	Rep. III
(1)	1 2 3	(4)	1 4 7	(7)	1 5 9
(2)	4 5 6	(5)	2 5 8	(8)	7 2 6
(3)	7 8 9	(6)	3 6 9	(9)	4 8 3

First, there are some properties that hold in all the designs which we have considered:

(1) Every block contains k units. Every treatment occurs r times and no treatment appears more than once in a block.

Consider treatment 1. The other treatments that appear in the same block as treatment 1 are treatments 2 and 3 (block (1)), treatments 4 and 7 (block (4)), and treatments 5 and 9 (block (7)). There are two other treatments (6 and 8) which do not appear in the same block as treatment 1. Similarly, if we start with any other treatment, we find six treatments that are in the same block with it, and two that are not. Pairs of treatments that occur in the same block are called *first associates;* pairs that do not are called *second associates*, and the design is said to have two associate classes.

This requirement may be put in general terms as follows (for a design with two associate classes):

(2) Every pair of treatments occurs together in either λ_1 or λ_2 blocks. Pairs that occur together in λ_1 blocks are called first associates; pairs that occur together in λ_2 blocks are called second associates.

For the triple lattice, $\lambda_1 = 1$ and $\lambda_2 = 0$, as we have seen. In general, λ_1 and λ_2 can be any two whole numbers that are different.

We must now dig a little deeper. Take a pair of treatments that are first associates, say 1 and 2, and consider the relationships of the other treatments to these two. Treatments 3, 4, 5, 7, and 9 are in the same block as 1, while 6 and 8 are not. For treatment 2, the other treatments that are in the same block are 3, 5, 6, 7, and 8, while 4 and 9 do not appear in the same block. These relationships may be presented in a 2×2 table.

TABLE 11.2a RELATIONSHIPS OF THE OTHER TREATMENTS TO 1 AND 2

		Relation to 1	
		1st associate	2nd associate
Relation to 2	1st associate	3, 5, 7	6, 8
	2nd associate	4, 9	None

Treatments 3, 5, and 7 are first associates of both 1 and 2; treatments 4 and 9 occur in the same block with 1 but not with 2, and hence are first associates of 1 and second associates of 2, and so on.

The numbers of other treatments falling into the four cells in table 11.2a are, respectively,

$$\begin{matrix} 3 & 2 \\ 2 & 0 \end{matrix}$$

The third basic property of a partially balanced design is that these four numbers must remain the same, no matter which pair of first associates we start with. The reader may verify, for instance, that the same set of four numbers is obtained when starting with treatments 3 and 9.

If this process of classification is repeated, starting with a pair of treatments like 1 and 6 that are *second* associates, we find a different set of four numbers, namely

$$\begin{matrix} 6 & 0 \\ 0 & 1 \end{matrix}$$

Once again, however, these same four numbers will emerge for any pair of second associates with which we start. In technical language, this property is expressed as follows:

(3) Given any two treatments that are ith associates, the number of treatments that are jth associates of the first treatment and kth associates of the second treatment is the same no matter which pair of ith associates we start with. This number is denoted by the symbol $p_{jk}^{\,i}$. Further, $p_{jk}^{\,i} = p_{kj}^{\,i}$.

Any design satisfying properties (1), (2), and (3) above is a partially balanced incomplete block (*pbib*) design with two associate classes. The most intricate property is, of course, (3). This property makes the least squares analysis of the results tractable, and guarantees that only two standard errors are required for testing differences between pairs of treatments. In practice, these standard errors are often so close together that a single error suffices.

The user of *pbib* designs need not master these properties. They do, however, indicate the difficulties that may arise when "home-made" designs are used without an investigation of the relationships among the treatments and blocks. The rectangular lattices (section 10.3) are not partially balanced in this sense, with the result that 4 different standard errors are necessary in the simple rectangular lattice and 7 in the triple rectangular lattice, although usually we can dispense with most of these in practice.

11.62a Index to Partially Balanced Incomplete Block Designs. Although the number of designs is too great to enable us to present the individual plans, table 9.2a (p. 302) shows the number of treatments, size of block (k), and number of replications (r) for each plan that is available, and contains a reference to the source of the plan. This table deals with designs in blocks of sizes from 3 to 10. For designs for which $k = 2$ (two treatments per block) see reference (11.3a).

In the following sections we present an example of the statistical analysis of a *pbib* design, in order that the reader may appraise the degree of complexity. The intra-block analysis is little more involved than that for a fully balanced design. The analysis with recovery of inter-block information takes considerably longer. Thus the designs are especially suitable for experimental situations in which the grouping into blocks is highly effective, so that inter-block information is unimportant and may be neglected.

11.63a Example of the Intra-block Analysis. This experiment has 15 treatments arranged in blocks of size 4, with 4 replications. The data are the pounds of seed cotton per plot in a uniformity trial on which dummy treatments have been superimposed. The plan and yields are shown in table 11.3a.

TABLE 11.3a PLAN AND YIELDS OF COTTON (POUNDS PER PLOT)

Block					Block totals
1	(15) 2.4	(9) 2.5	(1) 2.6	(13) 2.0	9.5
2	(5) 2.7	(7) 2.8	(8) 2.4	(1) 2.7	10.6
3	(10) 2.6	(1) 2.8	(14) 2.4	(2) 2.2	10.0
4	(15) 3.4	(11) 3.1	(2) 2.1	(3) 2.3	10.9
5	(6) 4.1	(15) 3.3	(4) 3.3	(7) 2.9	13.6
6	(12) 3.4	(4) 3.2	(3) 2.8	(1) 3.0	12.4
7	(12) 3.2	(14) 2.5	(15) 2.4	(8) 2.6	10.7
8	(6) 2.3	(3) 2.3	(14) 2.4	(5) 2.7	9.7
9	(5) 2.8	(4) 2.8	(2) 2.6	(13) 2.5	10.7
10	(10) 2.5	(12) 2.7	(13) 2.8	(6) 2.6	10.6
11	(9) 2.6	(7) 2.6	(10) 2.3	(3) 2.4	9.9
12	(8) 2.7	(6) 2.7	(2) 2.5	(9) 2.6	10.5
13	(5) 3.0	(9) 3.6	(11) 3.2	(12) 3.2	13.0
14	(7) 3.0	(13) 2.8	(14) 2.4	(11) 2.5	10.7
15	(10) 2.4	(4) 2.5	(8) 3.2	(11) 3.1	11.2
					$G = 164.0$

This design is of the type known as "group divisible." The key to the design is the following association scheme among the treatment numbers.

1	6	11
2	7	12
3	8	13
4	9	14
5	10	15

Every treatment appears once in a block with all other treatments except the two other treatments in the same row of the association scheme. For instance, every treatment except 6 and 11 occurs somewhere in the same block as treatment 1. If we regard treatments in the same row as first associates, we have $\lambda_1 = 0$, and $\lambda_2 = 1$.

This example is also worked in the monograph (11.2a) from which the design was taken. The analysis given here follows the same method, but utilizes totals instead of means until the final step.* In this design,

$$t = 15; \quad k = 4; \quad r = 4; \quad b = 15$$

1. Find the block totals B and the grand total G and insert them in the plan (table 11.3a).

2. For each treatment, find the treatment total T and the total B_t of all blocks in which the treatment occurs. For treatment 1,

$$T = 2.6 + 2.7 + 2.8 + 3.0 = 11.1$$

$$B_t = 9.5 + 10.6 + 10.0 + 12.4 = 42.5$$

These quantities can be found simultaneously on a desk computing machine by placing the treatment yields on the left of the keyboard and the corresponding block totals on the right. As a check, the sum of the T's is G and the sum of the B_t's is kG. The T's and B_t's are written in the first two columns of the working table 11.4a.

TABLE 11.4a COMPUTATIONS FOR THE INTRA-BLOCK ANALYSIS OF A *pbib* DESIGN

Treat-ment	(1) T	(2) B_t	(3) $Q'' = kT - B_t$ $= 4T - B_t$	(4) G''	(5) $180\hat{t}$	(6) Adjusted mean
1	11.1	42.5	1.9	6.1	22.4	2.86
2	9.4	42.1	−4.5	−0.8	−66.7	2.36
3	9.8	42.9	−3.7	−4.2	−51.3	2.45
4	11.8	47.9	−0.7	−0.7	−9.8	2.68
5	11.2	44.0	0.8	−0.4	12.4	2.80
6	11.7	44.4	2.4	6.1	29.9	2.90
7	11.3	44.8	0.4	−0.8	6.8	2.77
8	10.9	43.0	0.6	−4.2	13.2	2.81
9	11.3	42.9	2.3	−0.7	35.2	2.93
10	9.8	41.7	−2.5	−0.4	−37.1	2.53
11	11.9	45.8	1.8	6.1	20.9	2.85
12	12.5	46.7	3.3	−0.8	50.3	3.01
13	10.1	41.5	−1.1	−4.2	−12.3	2.66
14	9.7	41.1	−2.3	−0.7	−33.8	2.55
15	11.5	44.7	1.3	−0.4	19.9	2.84
Totals	164.0	656.0	0.0	0.0	0.0	

* Our results differ slightly owing to an error in the total sum of squares in (11.2a).

3. From these two columns, form a third column of the values

$$Q'' = kT - B_t = 4T - B_t$$

The Q'' values should sum exactly to zero. (In previous designs we have used Q for this quantity instead of Q''; however, the monograph (11.2a) uses Q for our Q''/k, and the change in our notation has been made to avoid confusion.)

4. We are now ready to compute the treatment means $\hat{t}$, adjusted for block effects. The formula is

$$rk(k-1)\hat{t} = (k - c_2)Q'' + (c_1 - c_2)S_1(Q'') \quad (1)$$

In this formula the quantities c_1 and c_2 are functions of the structural constants of the design, i.e., of the λ's and the $p_{jk}^{\ i}$ of section 11.61a. The numerical values of c_1 and c_2 are given with the plan for each design contained in references (11.2a) and (11.3a). For this design

$$c_1 = 0; \qquad c_2 = \tfrac{4}{15}$$

The quantity $S_1(Q'')$ denotes the sum of the Q'', taken over the first associates of the treatment in question. Thus in finding the adjusted mean of treatment 1 we add the Q'' values for treatments 6 and 11, which are the first associates of treatment 1.

Formula (1) for $\hat{t}$ holds for any *pbib* design with two associate classes. At this point, however, it is usually possible to shorten the calculation slightly by rewriting the formula, the particular method of rewriting depending on the type of design. For this design, let

G'' = Sum of Q'' over all treatments in the same row in the association scheme

For treatment 1, G'' is the sum of the Q'' over treatments 1, 6, and 11. Then it follows that

$$G'' = S_1(Q'') + Q''$$

Substituting for $S_1(Q'')$ in terms of G'' in the equation for $\hat{t}$, we have

$$\begin{aligned} rk(k-1)\hat{t} &= (k - c_2 - c_1 + c_2)Q'' + (c_1 - c_2)G'' \\ &= (k - c_1)Q'' + (c_1 - c_2)G'' \quad (2) \end{aligned}$$

The advantage of introducing the G'' is that there are only five values of G'' to be computed, one for each row in the association scheme. These values are entered in column (4) of table 11.4a. For treatments 1, 6, and 11,

$$G'' = 1.9 + 2.4 + 1.8 = +6.1$$

5. Finally, substituting numerical values for the constants in equation (2) for $\hat{t}$,

$$(4)(4)(3)\hat{t} = 4Q'' - \tfrac{4}{15}G''$$

This equation can be written as

$$180\hat{t} = 15Q'' - G''$$

The values of $180\hat{t}$ appear in column (5) of table 11.4*a*. They should add to zero. To obtain the adjusted treatment means, multiply the values in column (5) by 0.005556, the reciprocal of 180, and add the general mean, 164.0/60, or 2.733. For treatment 1 this gives

$$(0.005556)(22.4) + 2.733 = 2.86$$

6. In the analysis of variance, the total sum of squares and the sum of squares for blocks are found in the usual way. The general formula for the adjusted treatments sum of squares is

$$\frac{1}{k} \sum \hat{t}Q'' = \frac{1}{4} \sum \hat{t}Q''$$

where the sum is taken over all treatments. This may be written

$$\frac{1}{720} \sum (180\hat{t})Q'' = \frac{\text{Sum of products of columns (5) and (3)}}{720}$$

TABLE 11.5a INTRA-BLOCK ANALYSIS OF VARIANCE

	d.f.		s.s.	m.s.
Blocks (unadj.)	$b - 1$	14	4.9233	
Treatments (adj.)	$t - 1$	14	1.5641	0.1117
Error	$tr - t - b + 1$	31	2.6859	$0.0866 = E_e$
Total	$tr - 1$	59	9.1733	

For two treatments that are first associates (like 1 and 6), the error variance of the difference between their adjusted means is

$$\frac{2E_e}{r} \frac{(k - c_1)}{(k - 1)} = \frac{2(0.0866)}{4} \cdot \frac{(4 - 0)}{(4 - 1)}$$
$$= 0.0577$$

For second associates, the error variance of the difference between the adjusted means is

$$\frac{2E_e}{r} \frac{(k - c_2)}{(k - 1)} = \frac{2(0.0866)}{4} \cdot \frac{(4 - \frac{4}{15})}{(4 - 1)}$$
$$= 0.0539$$

The two variances are so close together that an average variance can safely be used for all comparisons between pairs. Since any treatment has 2 first associates and 12 second associates, an appropriate average is

$$\frac{2(0.0577) + 12(0.0539)}{14} = 0.0544$$

The standard error, $\sqrt{0.0544} = 0.233$, with 31 d.f., is used for t-tests among the adjusted treatment means. In this example, which has dummy treatments, t-tests are of no interest. From table 11.5a, the F-ratio is $0.1117/0.0866 = 1.29$, with 14 and 31 d.f.

11.64a Analysis with Recovery of Inter-block Information. All the computations already made in section 11.63a are needed, except the adjusted treatment means in column (6) of table 11.4a. We continue as follows:

7. Find the unadjusted treatments sum of squares in the usual way. This comes to 3.1383. From this value and the analysis of variance in table 11.5a we obtain the adjusted blocks sum of squares by the relation

$$\begin{aligned}\text{Blocks (adj.)} &= \text{blocks (unadj.)} + \text{treatments (adj.)} \\ &\qquad - \text{treatments (unadj.)} \\ &= 4.9233 + 1.5641 - 3.1383 = 3.3491\end{aligned}$$

The mean square for blocks (adjusted) is

$$E_b = \frac{3.3491}{14} = 0.2392$$

The intra-block error mean square (from table 11.5a) is

$$E_e = 0.0866$$

If E_b is less than or equal to E_e, proceed no further and use the unadjusted treatment means as the final estimates.

8. Various weighting coefficients are now computed.

$$w = \frac{1}{E_e} = 11.55$$

$$w' = \frac{t(r-1)}{k(b-1)E_b - (t-k)E_e} = \frac{45}{(56)(0.2392) - (11)(0.0866)} = 3.62$$

$$W = \frac{w'}{w - w'} = \frac{3.62}{11.55 - 3.62} = 0.46$$

We also require coefficients d_1 and d_2 which take the place of the c_1 and c_2 in the intra-block analysis. These d's depend on W, on the c's and on other structural constants Δ, H, λ_1 and λ_2. For this design

$$c_1 = 0; \quad c_2 = \tfrac{4}{15}; \quad \Delta = \tfrac{45}{4}; \quad H = \tfrac{27}{4}; \quad \lambda_1 = 0; \quad \lambda_2 = 1$$

$$d_1 = \frac{c_1\Delta + r\lambda_1 W}{\Delta + rHW + r^2W^2} = 0, \text{ since } c_1 \text{ and } \lambda_1 \text{ are both zero}$$

$$d_2 = \frac{c_2\Delta + r\lambda_2 W}{\Delta + rHW + r^2W^2} = \frac{(\frac{4}{15})(\frac{45}{4}) + 4(0.46)}{(\frac{45}{4}) + 4(\frac{27}{4})(0.46) + 16(0.46)^2}$$

$$= \frac{4.84}{27.06} = 0.1789$$

9. To replace the Q'' we compute values P''.

$$P'' = w'B_t + wQ'' - \frac{w'kG}{t}$$

where G is the grand total, 164.0. Numerically,

$$P'' = 3.62B_t + 11.55Q'' - 158.315$$

The P'' are shown in column (1) of table 11.6a. For treatment 1,

TABLE 11.6a Additional computations for recovery of inter-block information

Treatment	(1) P''	(2) G'''	(3) Adjusted mean
1	17.480	75.884	2.82
2	−57.888	−0.553	2.37
3	−45.752	−62.267	2.47
4	6.998	−5.552	2.78
5	10.205	−7.517	2.80
6	30.133	75.884	2.90
7	8.481	−0.553	2.79
8	4.275	−62.267	2.78
9	23.548	−5.552	2.88
10	−36.236	−7.517	2.51
11	28.271	75.884	2.89
12	48.854	−0.553	3.04
13	−20.790	−62.267	2.62
14	−36.098	−5.552	2.51
15	18.514	−7.517	2.85
	−0.005	−0.015	

$$P'' = (3.62)(42.5) + (11.55)(1.9) - 158.315$$
$$= 17.480$$

The P'' add to zero, apart from rounding errors.

10. The adjusted treatment means $\hat{t}'$ are given by the equation

$$kr\{w' + w(k-1)\}\hat{t}' = (k - d_2)P'' + (d_1 - d_2)S_1(P'')$$

where S_1 denotes summation over the first associates of the treatment in question. This is the general formula for all *pbib* designs with two associate classes.

For this design the formula can be rewritten, as in step 4 of the intra-block analysis,

$$kr\{w' + w(k-1)\}\hat{t}' = (k - d_1)P'' + (d_1 - d_2)G'''$$

where the G''' for any treatment is the sum of the P'' over the row in which the treatment appears in the association scheme at the beginning of section 11.63a. The G''' values are placed in column (2) of table 11.6a. For treatment 1

$$G''' = 17.480 + 30.133 + 28.271 = 75.884$$

Finally, the adjusted treatment means are

$$M + \frac{(k - d_2)}{kr\{w' + w(k-1)\}}P'' + \frac{(d_1 - d_2)}{kr\{w' + w(k-1)\}}G'''$$

The constant M is the general mean and is added so that the adjusted means will average to the general mean of the experiment. Numerically we have

$$\text{Adjusted mean} = 2.733 + 0.00624P'' - 0.00029G'''$$

These values appear in column (3) of table 11.6a.

11. The estimated variance of the difference between the adjusted means of two treatments that are first associates, like 1 and 6, is

$$\frac{2(k - d_1)}{r\{w' + w(k-1)\}} = 0.0523$$

For two treatments that are second associates, the corresponding formula is

$$\frac{2(k - d_2)}{r\{w' + w(k-1)\}} = 0.0499$$

As in the intra-block analysis, an average variance of the difference between two adjusted treatment means is computed as

$$\frac{2(0.0523) + 12(0.0499)}{14} = 0.0502$$

The method of analysis, as illustrated by this example, is essentially the same for all *pbib* designs with two associate classes, except for slight simplifications that can be made at the point at which we introduced the G'''.

In some of the designs the blocks can be combined to form replications (as in the lattices) or groups of replications. No change is required in the intra-block analysis, although, if desired, the $(b - 1)$ d.f. for blocks may be partitioned into $(r' - 1)$ d.f. for replication groups and $(b - r')$ d.f. for blocks within replication groups. For the recovery of inter-block information, this partition is necessary. The sum of squares for blocks within replication groups (adjusted) is computed as:

blocks within rep. groups (unadj.)

+ treatments (adj.) − treatments (unadj.)

and has $(b - r')$ d.f. If E_b is the corresponding mean square, then

$$w' = \frac{k(b - r') - (t - k)}{k(b - r')E_b - (t - k)E_e}$$

The remainder of the analysis proceeds as in the example.

In a substantial number of the designs, replications can be formed at right angles to the blocks, permitting a two-way elimination of heterogeneity as in the latin square. These designs are discussed in section 13.4a.

11.7a Chain Block Designs

11.71a Description. These designs were developed by Youden and Connor (11.4a) for experimental situations in which

(i) The size of block is limited, and the number of treatments to be tested considerably exceeds this size,

(ii) Within the blocks, comparisons are of such high precision that only one or two replications are needed.

The designs have been used in precise experimentation in the physical sciences. They are unlike any of the other designs in this book in that some treatments are replicated only once while others are replicated twice. Although their utility is likely to be restricted by condition (ii) above, they are flexible and easy to construct, and adapt themselves to situations in which even a partially balanced design, if one exists, demands too many replications.

The example in table 11.7a, for 17 treatments in blocks of 6 or 5 units, illustrates the principle of the design.

TABLE 11.7a A CHAIN BLOCK DESIGN FOR 17 TREATMENTS IN BLOCKS OF 6 OR 5

Block 1	2	3	4	5
A_1	C_2	E_3	G_4	I_5
B_1	D_2	F_3	H_4	J_5
C_1	E_2	G_3	I_4	A_5
D_1	F_2	H_3	J_4	B_5
k_1	l_2	m_3	n_4	o_5
p_1	q_2			

The letters denote the treatments. Capital letters refer to treatments that are replicated twice; small letters to treatments that are replicated once. The suffixes indicate the blocks.

Treatments C and D appear in both blocks 1 and 2, and form the link between these blocks. Treatments E and F form a similar link between blocks 2 and 3, treatments G and H between blocks 3 and 4, and so on until we reach treatments A and B, which complete the chain by linking blocks 5 and 1. This linking is the basic property of the design. Every block contains one group of treatments linking it to the preceding block in the chain, and one group linking it to the succeeding block.

Now as to flexibility. The basic groups like (A, B) may contain 1, 2, 3 or more letters, i.e., have 1, 2, 3 or more treatments. This number should, however, be the same for all groups. Any number of additional treatments that are replicated once, like k, l, m, may be included, subject only to the limitations imposed by the size of the block. In table 11.7a, seven treatments are replicated once. Note that in the first two blocks, all 6 units are occupied by treatments, whereas in the last three blocks only 5 units are utilized. It is advisable to distribute the extra treatments so that block sizes are as equal as possible.

In the selection of a design for a given number of treatments and units per block, the main question to decide is how many of the treatments are to be replicated twice. Then use the largest size of basic group that will give these treatments double replication and will fit into the blocks. Suppose that we have blocks of size 7, and that 9 treatments are to be replicated twice and 3 replicated once. The basic group cannot exceed 3 treatments, since there must be two basic groups in each block. A moment's trial shows that a plan can be constructed with basic groups of size 3, as shown in table 11.8a.

TABLE 11.8a CHAIN BLOCK DESIGN FOR 12 TREATMENTS IN BLOCKS OF 7

Block 1	2	3
A_1	D_2	G_3
B_1	E_2	H_3
C_1	F_2	I_3
D_1	G_2	A_3
E_1	H_2	B_3
F_1	I_2	C_3
j_1	k_2	l_3

Since the number of error d.f. is sometimes small, it is well to check this before adopting a design. This number is

$$b(n_2 - 1) + 1$$

where b is the number of blocks and n_2 the number of treatments in a basic group. The designs in tables 11.7a and 11.8a have 6 and 7 error d.f., respectively.

When treatments that are replicated once are present, the designs are not partially balanced in the sense of Bose and Nair, and several standard errors must be computed if t-tests are wanted. This is the chief disadvantage of the designs. In allocating treatments to blocks, it is worth noting that the standard error of the difference between two treatment means is smallest for pairs of treatments in the same block. Next come pairs of treatments in blocks that are neighbors in the chain of blocks; next come pairs in blocks 2 units apart (by the shortest path), and so on. Of course, treatments replicated twice are more precisely determined than those replicated once.

11.72a Statistical Analysis of Chain Block Designs. The statistical analysis is not difficult. The easiest method is first to estimate the block effects. Each treatment mean is then adjusted for the effects of the block or blocks in which it lies. In following the instructions it is helpful to think of the blocks as arranged in a circular chain, each block having two neighbors in the chain.

For each block, calculate the quantity

D' = (Total of that part of the block containing treatments that are replicated twice) − (total of these same treatments in the neighboring blocks)

Since it is important that this instruction be clearly grasped, two examples are given. For block 1 in table 11.7a

$$D' = (A_1 + B_1 + C_1 + D_1) - (A_5 + B_5 + C_2 + D_2)$$

For block 2 in table 11.8a

$$D' = (D_2 + E_2 + F_2 + G_2 + H_2 + I_2) - (D_1 + E_1 + F_1 + G_3 + H_3 + I_3)$$

The quantities D' add to zero over the blocks.

The formula for the estimated block mean $\hat{b}$ varies with the number of blocks. Table 11.9a presents these formulae for experiments with from

TABLE 11.9a FORMULAE FOR THE ESTIMATED BLOCK MEAN IN CHAIN BLOCK DESIGNS

No. of blocks	Estimated block mean $\hat{b}_0$
3	$D_0'/3n_2$
4	$(4D_0' + D_1' + D_{-1}')/8n_2$
5	$(3D_0' + D_1' + D_{-1}')/5n_2$
6	$(8D_0' + 3D_1' + 3D_{-1}' - D_3')/12n_2$

3 to 6 blocks. The suffix 0 denotes the block in question; suffixes 1 and -1 refer to the two neighboring blocks. As before, n_2 is the number of treatments in a basic group. In the last formula, D_3' is the D' for the block most distant in the chain from the block in question; e.g., when computing the effect of block 2, D_3' refers to block 5.

If $\hat{b}_i$ is the estimated mean of block i in the *original* block numbering, the estimated mean for a treatment that occurs in blocks i and j is

$$\hat{t} = \frac{(T - \hat{b}_i - \hat{b}_j)}{2}$$

where T is the total of the two units which have this treatment. For a treatment occurring only in block i

$$\hat{t} = T - \hat{b}_i$$

In the analysis of variance, the total sum of squares and the sum of squares for blocks are computed in the standard way. Note, however, that if blocks contain differing numbers of units, we require a weighted sum of squares

$$\sum \frac{B_i^2}{k_i} - \frac{G^2}{N}$$

where k_i is the number of units in the ith block and B_i is the block total.

Instead of finding the adjusted sum of squares for treatments, we compute the error sum of squares directly, obtaining that for treatments by subtraction. The error sum of squares is found as follows. For each group of treatments that is replicated twice, form differences between the two replicates of each treatment, as illustrated below.

		Diff.
A_1	A_3	d_1
B_1	B_3	d_2
C_1	C_3	d_3
Total		d

Then

$$\frac{1}{2}\left\{\sum_{i=1}^{n_2} d_i^2 - \frac{d^2}{n_2}\right\}$$

contributes $(n_2 - 1)$, or in this case 2, d.f. to the error sum of squares. Compute and add these terms over all b treatment groups, obtaining $b(n_2 - 1)$ d.f. One additional d.f. comes from the difference between the total (over all blocks) of the *upper* groups in the block and the corresponding total of the *lower* groups. The square of this difference is divided by $2bn_2$. In table 11.8a the sum of squares for this single d.f. is

$$\frac{\left\{\begin{matrix}(A_1 + B_1 + C_1 + D_2 + E_2 + F_2 + G_3 + H_3 + I_3 \\ - D_1 - E_1 - F_1 - G_2 - H_2 - I_2 - A_3 - B_3 - C_3)\end{matrix}\right\}^2}{18}$$

The partition of d.f. is as follows.

	d.f.
Blocks	$b - 1$
Treatments	$t - 1$
Error	$N - b - t + 1 = b(n_2 - 1) + 1$
Total	$N - 1$

For t-tests, the variance of the difference between two adjusted treatment means depends on the number of replications which the treatments have, and on the shortest distance between the *groups* in which they occur. The formulae may be written:

Both treatments with 2 reps.: $V(\text{diff.}) = \left\{1 + \frac{f_{22}}{n_2 b}\right\} s^2$

Both treatments with 1 rep.: $V(\text{diff.}) = 2\left\{1 + \frac{f_{11}}{n_2 b}\right\} s^2$

One with 1 rep.; one with 2: $V(\text{diff.}) = \frac{3}{2}\left\{1 + \frac{f_{12}}{n_2 b}\right\} s^2$

where s^2 is the error mean square. Table 11.10a gives the factors f for most practical cases.

TABLE 11.10a Factors for obtaining variances in chain block designs

Treatments	f_{22}	f_{11}	f_{12}
In same group	0	0	$(b-1)/3$
1 group apart	$(b-2)$	$(b-1)$	$(b-1)/3$
2 groups apart	$(3b-8)$	$(2b-4)$	$(5b-9)/3$
3 groups apart	$(5b-18)$	$(3b-9)$	$(9b-25)/3$

In finding the shortest distance apart, number the groups as follows.

Block	1	2	3	$\cdots$	b
	G_b	G_1	G_2		G_{b-1}
	G_1	G_2	G_3		G_b
	g_1	g_2	g_3		g_b

where g_i stands for the groups of treatments that are replicated once. In table 11.7a, for instance, A and I are one group apart, since A is in G_5 and I in G_4. Similarly A (in G_5) and q (in g_2) are 2 groups apart while C (in G_1) and k (in g_1) are considered in the same group.

REFERENCES

11.1 Bose, R. C. On the construction of balanced incomplete block designs. *Ann. Eugen.* 9, 353–400, 1939.

11.2 Cox, G. M. Enumeration and construction of balanced incomplete block configurations. *Ann. Math. Stat.* 11, 72–85, 1940.

11.3 Fisher, R. A. An examination of the different possible solutions of a problem in incomplete blocks. *Ann. Eugen.* 10, 52–75, 1940.

11.4 Wadley, F. M. Incomplete-block design adapted to paired tests of mosquito repellents. *Biom. Bull.* 2, 30–31, 1946.

11.5 Yates, F. The recovery of inter-block information in balanced incomplete block designs. *Ann. Eugen.* 10, 317–325, 1940.

11.6 Paul, Pauline C. Ph.D. Thesis, Iowa State College, 1943.

11.7 Fisher, R. A., and Yates, F. *Statistical tables for biological, agricultural and medical research.* Oliver and Boyd, Edinburgh, 4th ed., 1953.

11.8 Yates, F. Incomplete randomized blocks. *Ann. Eugen.* 7, 121–140, 1936.

11.9 Cornish, E. A. The estimation of missing values in incomplete randomized block experiments. *Ann. Eugen.* 10, 112–118, 1940.

11.1a Bose, R. C., and Nair, K. R. Partially balanced incomplete block designs. *Sankhya* 4, 337–372, 1939.

11.2a Bose, R. C., Clatworthy, W. H., and Shrikhande, S. S. Tables of partially balanced designs with two associate classes. *North Carolina Agr. Exp. Sta. Tech. Bull.* 107, 1954.

11.3a Clatworthy, W. H. Partially balanced incomplete block designs with two associate classes and two treatments per block. *Nat. Bur. Standards Jour. Res.* 54, 174–189, 1955.

11.4a Youden, W. J., and Connor, W. S. The chain block design. *Biometrics* 9, 127–140, 1953.

11.5a CLATWORTHY, W. H. A geometrical configuration which is a partially balanced design. *Proc. Amer. Math. Soc.* 5, 47–55, 1954.

11.6a CLATWORTHY, W. H. On partially balanced incomplete block designs with two associate classes. *Nat. Bur. Standards Applied Math. Series* 47, 1956.

11.7a RUPP, M. K. The complete solutions of the balanced incomplete block designs with ten or fewer replications. *Nat. Bur. Standards unpublished report*, 1952.

11.8a BRADLEY, R. A. Some statistical methods in taste testing and quality evaluation. *Biometrics* 9, 22–38, 1953.

11.9a DURBIN, J. Incomplete blocks in ranking experiments. *Brit. Jour. Psych.* 4, 85–90, 1951.

11.10a BRADLEY, R. A., and TERRY, M. E. The rank analysis of incomplete block designs. I. *Biometrika* 39, 324–345, 1952.

11.11a BRADLEY, R. A. Incomplete block rank analysis: on the appropriateness of the model for a method of paired comparisons. *Biometrics* 10, 374–390, 1954.

11.12a CALVIN, L. D. Doubly balanced incomplete block designs for experiments in which the treatment effects are correlated. *Biometrics* 10, 61–88, 1954.

ADDITIONAL READING

CORNISH, E. A. Factorial treatments in incomplete randomized blocks. *Jour. Australian Inst. Agr. Sci.* 4, 199–203, 1938.

HOPKINS, J. W. Incomplete block rank analysis: some taste test results. *Biometrics* 10, 391–399, 1954.

ZELEN, M. Analysis for some partially balanced incomplete block designs having a missing block. *Biometrics* 10, 273–281, 1954.

TABLE 11.3 INDEX TO PLANS

t	k	r	b	λ †	E	Plan	Type
4	2	3	6	1	.67	11.1	V
	3	3	4	2	.89	*	V
5	2	4	10	1	.62	11.2	V
	3	6	10	3	.83	11.1a	V
	4	4	5	3	.94	*	V
6	2	5	15	1	.60	11.3	I
	3	5	10	2	.80	11.4	III
	3	10	20	4	.80	11.5	I
	4	10	15	6	.90	11.6	II
	5	5	6	4	.96	*	V
7	2	6	21	1	.58	11.2a	II
	3	3	7	1	.78	11.7	V
	4	4	7	2	.88	11.8	V
	6	6	7	5	.97	*	V
8	2	7	28	1	.57	11.9	I
	4	7	14	3	.86	11.10	I
	7	7	8	6	.98	*	V
9	2	8	36	1	.56	11.3a	II
	4	8	18	3	.84	11.11	II

TABLE 11.3 INDEX TO PLANS (*Continued*)

t	k	r	b	λ †	E	Plan	Type
9	5	10	18	5	.90	11.12	II
	6	8	12	5	.94	11.13	II
	8	8	9	7	.98	*	IV
10	2	9	45	1	.56	11.14	I
	3	9	30	2	.74	11.15	II
	4	6	15	2	.83	11.16	III
	5	9	18	4	.89	11.17	III
	6	9	15	5	.93	11.18	III
	9	9	10	8	.99	*	IV
11	2	10	55	1	.55	11.4a	II
	5	5	11	2	.88	11.19	IV
	6	6	11	3	.92	11.20	IV
	10	10	11	9	.99	*	IV
13	3	6	26	1	.72	11.21	II
	4	4	13	1	.81	11.22	IV
	9	9	13	6	.96	11.23	IV
15	3	7	35	1	.71	11.24	I
	7	7	15	3	.92	11.25	IV
	8	8	15	4	.94	11.26	IV
16	6	6	16	2	.89	11.27	IV
	6	9	24	3	.89	11.28	II
	10	10	16	6	.96	11.29	IV
19	3	9	57	1	.70	11.30	II
	9	9	19	4	.94	11.31	IV
	10	10	19	5	.95	11.32	IV
21	3	10	70	1	.70	11.33	I
	5	5	21	1	.84	11.34	IV
	7	10	30	3	.90	11.35	III
25	4	8	50	1	.78	11.36	II
	9	9	25	3	.93	11.37	IV
28	4	9	63	1	.78	11.38	I
	7	9	36	2	.89	11.39	III
31	6	6	31	1	.86	11.40	IV
	10	10	31	3	.93	11.41	IV
37	9	9	37	2	.91	11.42	IV
41	5	10	82	1	.82	11.43	II
57	8	8	57	1	.89	11.44	IV
73	9	9	73	1	.90	11.45	IV
91	10	10	91	1	.91	11.46	IV

† Number of times that two treatments appear together in the same block.

* These plans are constructed by forming all possible combinations of the t numbers in groups of size k. The number of blocks b serves as a check that no group has been missed.

PLANS

Plan 11.1 $t = 4, k = 2, r = 3, b = 6, \lambda = 1, E = .67$, Type V

Block	Rep. I		Block	Rep. II		Block	Rep. III	
(1)	1	2	(3)	1	3	(5)	1	4
(2)	3	4	(4)	2	4	(6)	2	3

Plan 11.2 $t = 5, k = 2, r = 4, b = 10, \lambda = 1, E = .62$, Type V

Block	Reps. I and II		Block	Reps. III and IV	
(1)	1	2	(6)	1	4
(2)	3	4	(7)	2	3
(3)	2	5	(8)	3	5
(4)	1	3	(9)	1	5
(5)	4	5	(10)	2	4

Plan 11.1a $t = 5, k = 3, r = 6, b = 10, \lambda = 3, E = .83$, Type V

Block	Reps. I, II, and III			Block	Reps. IV, V, and VI		
(1)	1	2	3	(6)	1	2	4
(2)	1	2	5	(7)	1	3	4
(3)	1	4	5	(8)	1	3	5
(4)	2	3	4	(9)	2	3	5
(5)	3	4	5	(10)	2	4	5

Plan 11.3 $t = 6, k = 2, r = 5, b = 15, \lambda = 1, E = .60$, Type I

Block	Rep. I		Block	Rep. II		Block	Rep. III		Block	Rep. IV		Block	Rep. V	
(1)	1	2	(4)	1	3	(7)	1	4	(10)	1	5	(13)	1	6
(2)	3	4	(5)	2	5	(8)	2	6	(11)	2	4	(14)	2	3
(3)	5	6	(6)	4	6	(9)	3	5	(12)	3	6	(15)	4	5

Plan 11.4 $t = 6, k = 3, r = 5, b = 10, \lambda = 2, E = .80$, Type III

Block				Block			
(1)	1	2	5	(6)	2	3	4
(2)	1	2	6	(7)	2	3	5
(3)	1	3	4	(8)	2	4	6
(4)	1	3	6	(9)	3	5	6
(5)	1	4	5	(10)	4	5	6

Plan 11.5 $t = 6, k = 3, r = 10, b = 20, \lambda = 4, E = .80$, Type I

Block	Rep. I	Block	Rep. II	Block	Rep. III	Block	Rep. IV
(1)	1 2 3	(3)	1 2 4	(5)	1 2 5	(7)	1 2 6
(2)	4 5 6	(4)	3 5 6	(6)	3 4 6	(8)	3 4 5

Block	Rep. V	Block	Rep. VI	Block	Rep. VII	Block	Rep. VIII
(9)	1 3 4	(11)	1 3 5	(13)	1 3 6	(15)	1 4 5
(10)	2 5 6	(12)	2 4 6	(14)	2 4 5	(16)	2 3 6

Block	Rep. IX	Block	Rep. X
(17)	1 4 6	(19)	1 5 6
(18)	2 3 5	(20)	2 3 4

Plan 11.6 $t = 6, k = 4, r = 10, b = 15, \lambda = 6, E = .90$, Type II

Block	Reps. I and II	Block	Reps. III and IV	Block	Reps. V and VI
(1)	1 2 3 4	(4)	1 2 3 5	(7)	1 2 3 6
(2)	1 4 5 6	(5)	1 2 4 6	(8)	1 3 4 5
(3)	2 3 5 6	(6)	3 4 5 6	(9)	2 4 5 6

Block	Reps. VII and VIII	Block	Reps. IX and X
(10)	1 2 4 5	(13)	1 2 5 6
(11)	1 3 5 6	(14)	1 3 4 6
(12)	2 3 4 6	(15)	2 3 4 5

Plan 11.2a $t = 7, k = 2, r = 6, b = 21, \lambda = 1, E = .58$, Type II

Block	Reps. I and II	Block	Reps. III and IV	Block	Reps. V and VI
(1)	1 2	(8)	1 3	(15)	1 4
(2)	2 6	(9)	2 4	(16)	2 3
(3)	3 4	(10)	3 5	(17)	3 6
(4)	4 7	(11)	4 6	(18)	4 5
(5)	1 5	(12)	5 7	(19)	2 5
(6)	5 6	(13)	1 6	(20)	6 7
(7)	3 7	(14)	2 7	(21)	1 7

Plan 11.7 $t = 7, k = 3, r = 3, b = 7, \lambda = 1, E = .78$, Type V

Block		Block		Block		Block	
(1)	1 2 4	(3)	3 4 6	(5)	5 6 1	(7)	7 1 3
(2)	2 3 5	(4)	4 5 7	(6)	6 7 2		

Plan 11.8 $t = 7, k = 4, r = 4, b = 7, \lambda = 2, E = .88$, Type V

Block		Block		Block	
(1)	3 5 6 7	(4)	1 2 3 6	(7)	2 4 5 6
(2)	1 4 6 7	(5)	2 3 4 7		
(3)	1 2 5 7	(6)	1 3 4 5		

Plan 11.9 $t = 8, k = 2, r = 7, b = 28, \lambda = 1, E = .57$, Type I

Block	Rep. I		Rep. II		Rep. III		Rep. IV
(1)	1 2	(5)	1 3	(9)	1 4	(13)	1 5
(2)	3 4	(6)	2 8	(10)	2 7	(14)	2 3
(3)	5 6	(7)	4 5	(11)	3 6	(15)	4 7
(4)	7 8	(8)	6 7	(12)	5 8	(16)	6 8

	Rep. V		Rep. VI		Rep. VII
(17)	1 6	(21)	1 7	(25)	1 8
(18)	2 4	(22)	2 6	(26)	2 5
(19)	3 8	(23)	3 5	(27)	3 7
(20)	5 7	(24)	4 8	(28)	4 6

Plan 11.10 $t = 8, k = 4, r = 7, b = 14, \lambda = 3, E = .86$, Type I

Block	Rep. I		Rep. II		Rep. III		Rep. IV
(1)	1 2 3 4	(3)	1 2 7 8	(5)	1 3 6 8	(7)	1 4 6 7
(2)	5 6 7 8	(4)	3 4 5 6	(6)	2 4 5 7	(8)	2 3 5 8

	Rep. V		Rep. VI		Rep. VII
(9)	1 2 5 6	(11)	1 3 5 7	(13)	1 4 5 8
(10)	3 4 7 8	(12)	2 4 6 8	(14)	2 3 6 7

Plan 11.3a $t = 9, k = 2, r = 8, b = 36, \lambda = 1, E = .56$, Type II

Block	Reps. I and II		Reps. III and IV		Reps. V and VI		Reps. VII and VIII
(1)	1 2	(10)	1 3	(19)	1 4	(28)	1 5
(2)	2 8	(11)	2 5	(20)	2 6	(29)	2 4
(3)	3 4	(12)	3 6	(21)	2 3	(30)	3 8
(4)	4 7	(13)	4 9	(22)	4 5	(31)	4 6
(5)	5 6	(14)	5 8	(23)	5 7	(32)	3 5
(6)	1 6	(15)	6 7	(24)	6 8	(33)	6 9
(7)	3 7	(16)	1 7	(25)	7 9	(34)	2 7
(8)	8 9	(17)	4 8	(26)	1 8	(35)	7 8
(9)	5 9	(18)	2 9	(27)	3 9	(36)	1 9

Plan 11.11 $t = 9, k = 4, r = 8, b = 18, \lambda = 3, E = .84$, Type II

Block	Reps. I, II, III, and IV		Reps. V, VI, VII, and VIII
(1)	1 4 6 7	(10)	1 2 5 7
(2)	2 6 8 9	(11)	2 3 5 6
(3)	1 3 8 9	(12)	3 4 7 9
(4)	1 2 3 4	(13)	1 2 4 9
(5)	1 5 7 8	(14)	1 5 6 9
(6)	4 5 6 9	(15)	1 3 6 8
(7)	2 3 6 7	(16)	4 6 7 8
(8)	2 4 5 8	(17)	3 4 5 8
(9)	3 5 7 9	(18)	2 7 8 9

Plan 11.12 $t = 9, k = 5, r = 10, b = 18, \lambda = 5, E = .90$, Type II

Block	Reps. I, II, III, IV, and V		Reps. VI, VII, VIII, IX, and X
(1)	1 2 3 7 8	(10)	1 2 3 5 9
(2)	1 2 4 6 8	(11)	1 2 5 6 8
(3)	2 3 5 8 9	(12)	1 3 4 5 6
(4)	2 3 4 6 9	(13)	2 3 4 7 8
(5)	1 3 4 5 7	(14)	2 4 5 7 9
(6)	2 4 5 6 7	(15)	3 5 6 7 8
(7)	1 3 6 7 9	(16)	1 4 7 8 9
(8)	1 4 5 8 9	(17)	3 4 6 8 9
(9)	5 6 7 8 9	(18)	1 2 6 7 9

Plan 11.13 $t = 9, k = 6, r = 8, b = 12, \lambda = 5, E = .94$, Type II

Block	Reps. I and II		Reps. III and IV
(1)	1 2 4 5 7 8	(4)	1 2 5 6 7 9
(2)	2 3 5 6 8 9	(5)	1 3 4 5 8 9
(3)	1 3 4 6 7 9	(6)	2 3 4 6 7 8
	Reps. V and VI		**Reps. VII and VIII**
(7)	1 3 5 6 7 8	(10)	4 5 6 7 8 9
(8)	1 2 4 6 8 9	(11)	1 2 3 4 5 6
(9)	2 3 4 5 7 9	(12)	1 2 3 7 8 9

Plan 11.14 $t = 10, k = 2, r = 9, b = 45, \lambda = 1, E = .56$, Type I

Block	Rep. I		Rep. II		Rep. III		Rep. IV		Rep. V
(1)	1 2	(6)	1 3	(11)	1 4	(16)	1 5	(21)	1 6
(2)	3 4	(7)	2 7	(12)	2 10	(17)	2 8	(22)	2 9
(3)	5 6	(8)	4 8	(13)	3 7	(18)	3 10	(23)	3 8
(4)	7 8	(9)	5 9	(14)	5 8	(19)	4 9	(24)	4 10
(5)	9 10	(10)	6 10	(15)	6 9	(20)	6 7	(25)	5 7

	Rep. VI		Rep. VII		Rep. VIII		Rep. IX
(26)	1 7	(31)	1 8	(36)	1 9	(41)	1 10
(27)	2 6	(32)	2 3	(37)	2 4	(42)	2 5
(28)	3 9	(33)	4 6	(38)	3 5	(43)	3 6
(29)	4 5	(34)	5 10	(39)	6 8	(44)	4 7
(30)	8 10	(35)	7 9	(40)	7 10	(45)	8 9

Plan 11.15 $t = 10, k = 3, r = 9, b = 30, \lambda = 2, E = .74$, Type II

Block	Reps. I, II, and III		Reps. IV, V, and VI		Reps. VII, VIII, and IX
(1)	1 2 3	(11)	1 2 4	(21)	1 3 5
(2)	2 5 8	(12)	2 3 6	(22)	2 6 7
(3)	3 4 7	(13)	3 4 8	(23)	3 8 9
(4)	1 4 6	(14)	4 5 9	(24)	2 4 10
(5)	5 7 8	(15)	1 5 7	(25)	3 5 6
(6)	4 6 9	(16)	6 8 9	(26)	1 6 8
(7)	1 7 9	(17)	3 7 10	(27)	2 7 9
(8)	2 8 10	(18)	1 8 10	(28)	4 7 8
(9)	3 9 10	(19)	2 5 9	(29)	1 9 10
(10)	5 6 10	(20)	6 7 10	(30)	4 5 10

Plan 11.16 $t = 10, k = 4, r = 6, b = 15, \lambda = 2, E = .83$, Type III

Block					
(1)	1 2 3 4	(6)	1 6 8 10	(11)	3 5 9 10
(2)	1 2 5 6	(7)	2 3 6 9	(12)	3 6 7 10
(3)	1 3 7 8	(8)	2 4 7 10	(13)	3 4 5 8
(4)	1 4 9 10	(9)	2 5 8 10	(14)	4 5 6 7
(5)	1 5 7 9	(10)	2 7 8 9	(15)	4 6 8 9

Plan 11.17 $t = 10, k = 5, r = 9, b = 18, \lambda = 4, E = .89$, Type III

Block					
(1)	1 2 3 4 5	(7)	1 4 5 6 10	(13)	2 5 6 8 10
(2)	1 2 3 6 7	(8)	1 4 8 9 10	(14)	2 6 7 9 10
(3)	1 2 4 6 9	(9)	1 5 7 9 10	(15)	3 4 6 7 10
(4)	1 2 5 7 8	(10)	2 3 4 8 10	(16)	3 4 5 7 9
(5)	1 3 6 8 9	(11)	2 3 5 9 10	(17)	3 5 6 8 9
(6)	1 3 7 8 10	(12)	2 4 7 8 9	(18)	4 5 6 7 8

Plan 11.18 $t = 10, k = 6, r = 9, b = 15, \lambda = 5, E = .93$, Type III

Block							Block							Block						
(1)	1	2	4	5	8	9	(6)	2	3	4	6	8	10	(11)	1	4	5	7	8	10
(2)	5	6	7	8	9	10	(7)	1	2	6	7	9	10	(12)	1	2	3	5	7	10
(3)	2	4	5	6	9	10	(8)	1	3	5	6	8	9	(13)	2	3	5	6	7	8
(4)	1	2	4	6	7	8	(9)	1	2	3	8	9	10	(14)	1	3	4	5	6	10
(5)	3	4	7	8	9	10	(10)	2	3	4	5	7	9	(15)	1	3	4	6	7	9

Plan 11.4a $t = 11, k = 2, r = 10, b = 55, \lambda = 1, E = .55$, Type II

Block	Reps. I and II			Reps. III and IV			Reps. V and VI			Reps. VII and VIII			Reps. IX and X	
(1)	1	2	(12)	1	3	(23)	1	4	(34)	1	5	(45)	1	6
(2)	2	11	(13)	2	6	(24)	2	3	(35)	2	9	(46)	2	5
(3)	3	10	(14)	3	5	(25)	3	7	(36)	3	6	(47)	3	4
(4)	4	5	(15)	4	10	(26)	4	6	(37)	2	4	(48)	4	7
(5)	5	6	(16)	5	9	(27)	5	10	(38)	5	7	(49)	5	8
(6)	6	7	(17)	6	8	(28)	6	9	(39)	6	10	(50)	6	11
(7)	1	7	(18)	2	7	(29)	7	11	(40)	7	8	(51)	7	10
(8)	3	8	(19)	1	8	(30)	2	8	(41)	4	8	(52)	8	9
(9)	4	9	(20)	7	9	(31)	1	9	(42)	9	11	(53)	3	9
(10)	9	10	(21)	10	11	(32)	8	10	(43)	1	10	(54)	2	10
(11)	8	11	(22)	4	11	(33)	5	11	(44)	3	11	(55)	1	11

Plan 11.19 $t = 11, k = 5, r = 5, b = 11, \lambda = 2, E = .88$, Type IV

Block						Block					
(1)	1	2	3	5	8	(7)	7	8	9	11	3
(2)	2	3	4	6	9	(8)	8	9	10	1	4
(3)	3	4	5	7	10	(9)	9	10	11	2	5
(4)	4	5	6	8	11	(10)	10	11	1	3	6
(5)	5	6	7	9	1	(11)	11	1	2	4	7
(6)	6	7	8	10	2						

Plan 11.20 $t = 11, k = 6, r = 6, b = 11, \lambda = 3, E = .92$, Type IV

Block							Block						
(1)	4	6	7	9	10	11	(7)	1	2	4	5	6	10
(2)	1	5	7	8	10	11	(8)	2	3	5	6	7	11
(3)	1	2	6	8	9	11	(9)	1	3	4	6	7	8
(4)	1	2	3	7	9	10	(10)	2	4	5	7	8	9
(5)	2	3	4	8	10	11	(11)	3	5	6	8	9	10
(6)	1	3	4	5	9	11							

Plan 11.21 $t = 13, k = 3, r = 6, b = 26, \lambda = 1, E = .72$, Type II

Block	Reps. I, II, and III			Block	Reps. IV, V, and VI		
(1)	1	3	9	(14)	2	5	6
(2)	2	4	10	(15)	3	6	7
(3)	3	5	11	(16)	4	7	8
(4)	4	6	12	(17)	5	8	9
(5)	5	7	13	(18)	6	9	10
(6)	1	6	8	(19)	7	10	11
(7)	2	7	9	(20)	8	11	12
(8)	3	8	10	(21)	9	12	13
(9)	4	9	11	(22)	1	10	13
(10)	5	10	12	(23)	1	2	11
(11)	6	11	13	(24)	2	3	12
(12)	1	7	12	(25)	3	4	13
(13)	2	8	13	(26)	1	4	5

Plan 11.22 $t = 13, k = 4, r = 4, b = 13, \lambda = 1, E = .81$, Type IV

Block					Block					Block				
(1)	1	2	4	10	(6)	6	7	9	2	(11)	11	12	1	7
(2)	2	3	5	11	(7)	7	8	10	3	(12)	12	13	2	8
(3)	3	4	6	12	(8)	8	9	11	4	(13)	13	1	3	9
(4)	4	5	7	13	(9)	9	10	12	5					
(5)	5	6	8	1	(10)	10	11	13	6					

Plan 11.23 $t = 13, k = 9, r = 9, b = 13, \lambda = 6, E = .96$, Type IV

Block									
(1)	3	5	6	7	8	9	11	12	13
(2)	1	4	6	7	8	9	10	12	13
(3)	1	2	5	7	8	9	10	11	13
(4)	1	2	3	6	8	9	10	11	12
(5)	2	3	4	7	9	10	11	12	13
(6)	1	3	4	5	8	10	11	12	13
(7)	1	2	4	5	6	9	11	12	13
(8)	1	2	3	5	6	7	10	12	13
(9)	1	2	3	4	6	7	8	11	13
(10)	1	2	3	4	5	7	8	9	12
(11)	2	3	4	5	6	8	9	10	13
(12)	1	3	4	5	6	7	9	10	11
(13)	2	4	5	6	7	8	10	11	12

Plan 11.24 $t = 15, k = 3, r = 7, b = 35, \lambda = 1, E = .71$, Type I

Block	Rep. I			Block	Rep. II			Block	Rep. III			Block	Rep. IV		
(1)	1	2	3	(6)	1	4	5	(11)	1	6	7	(16)	1	8	9
(2)	4	8	12	(7)	2	8	10	(12)	2	9	11	(17)	2	13	15
(3)	5	10	15	(8)	3	13	14	(13)	3	12	15	(18)	3	4	7
(4)	6	11	13	(9)	6	9	15	(14)	4	10	14	(19)	5	11	14
(5)	7	9	14	(10)	7	11	12	(15)	5	8	13	(20)	6	10	12

Block	Rep. V			Block	Rep. VI			Block	Rep. VII		
(21)	1	10	11	(26)	1	12	13	(31)	1	14	15
(22)	2	12	14	(27)	2	5	7	(32)	2	4	6
(23)	3	5	6	(28)	3	9	10	(33)	3	8	11
(24)	4	9	13	(29)	4	11	15	(34)	5	9	12
(25)	7	8	15	(30)	6	8	14	(35)	7	10	13

Plan 11.25 $t = 15, k = 7, r = 7, b = 15, \lambda = 3, E = .92$, Type IV

See incomplete latin squares Plan 13.7; randomize units in blocks ignoring replications.

Plan 11.26 $t = 15, k = 8, r = 8, b = 15, \lambda = 4, E = .94$, Type IV

See incomplete latin squares Plan 13.8; randomize units in blocks ignoring replications.

Plan 11.27 $t = 16, k = 6, r = 6, b = 16, \lambda = 2, E = .89$, Type IV

See incomplete latin squares Plan 13.9; randomize units in blocks ignoring replications.

Plan 11.28 $t = 16, k = 6, r = 9, b = 24, \lambda = 3, E = .89$, Type II

Block	Reps. I, II, and III						Block	Reps. IV, V, and VI						Block	Reps. VII, VIII, and IX					
(1)	1	2	5	6	11	12	(9)	1	3	6	8	13	15	(17)	1	4	5	8	10	11
(2)	3	4	7	8	9	10	(10)	2	4	5	7	14	16	(18)	2	3	6	7	9	12
(3)	5	6	9	10	13	14	(11)	5	7	9	11	13	15	(19)	5	8	9	12	13	16
(4)	7	8	11	12	15	16	(12)	6	8	10	12	14	16	(20)	1	4	6	7	13	16
(5)	1	2	9	10	15	16	(13)	2	4	6	8	9	11	(21)	1	4	9	12	14	15
(6)	3	4	11	12	13	14	(14)	1	3	5	7	10	12	(22)	6	7	10	11	14	15
(7)	1	2	7	8	13	14	(15)	2	4	10	12	13	15	(23)	2	3	10	11	13	16
(8)	3	4	5	6	15	16	(16)	1	3	9	11	14	16	(24)	2	3	5	8	14	15

Plan 11.29 $t = 16, k = 10, r = 10, b = 16, \lambda = 6, E = .96$, Type IV

See incomplete latin squares Plan 13.10; randomize units in blocks ignoring replications.

Plan 11.30 $t = 19, k = 3, r = 9, b = 57, \lambda = 1, E = .70$, Type II

See extended incomplete latin squares Plan 13.15a; randomize units in blocks ignoring replications.

Plan 11.31 $t = 19, k = 9, r = 9, b = 19, \lambda = 4, E = .94$, Type IV

See incomplete latin squares Plan 13.11; randomize units in blocks ignoring replications.

Plan 11.32 $t = 19, k = 10, r = 10, b = 19, \lambda = 5, E = .95$, Type IV

See incomplete latin squares Plan 13.12; randomize units in blocks ignoring replications.

Plan 11.33 $t = 21, k = 3, r = 10, b = 70, \lambda = 1, E = .70$, Type I

Block	Rep. I		Rep. II		Rep. III		Rep. IV		Rep. V
(1)	1 2 3	(8)	1 4 15	(15)	1 5 17	(22)	1 6 9	(29)	1 7 21
(2)	4 5 6	(9)	2 5 11	(16)	2 4 14	(23)	2 7 16	(30)	2 13 17
(3)	7 8 9	(10)	3 9 16	(17)	3 7 11	(24)	3 8 21	(31)	3 10 18
(4)	10 11 12	(11)	6 17 20	(18)	6 10 19	(25)	4 17 19	(32)	4 8 11
(5)	13 14 15	(12)	7 12 19	(19)	8 16 20	(26)	5 10 13	(33)	5 16 19
(6)	16 17 18	(13)	8 13 18	(20)	9 15 18	(27)	11 15 20	(34)	6 12 15
(7)	19 20 21	(14)	10 14 21	(21)	12 13 21	(28)	12 14 18	(35)	9 14 20

	Rep. VI		Rep. VII		Rep. VIII		Rep. IX		Rep. X
(36)	1 8 10	(43)	1 11 18	(50)	1 12 20	(57)	1 13 19	(64)	1 14 16
(37)	2 18 19	(44)	2 10 20	(51)	2 6 8	(58)	2 9 12	(65)	2 15 21
(38)	3 15 17	(45)	3 5 12	(52)	3 14 19	(59)	3 4 20	(66)	3 6 13
(39)	4 12 16	(46)	4 9 13	(53)	4 18 21	(60)	5 8 14	(67)	4 7 10
(40)	5 9 21	(47)	6 16 21	(54)	5 7 15	(61)	6 7 18	(68)	5 18 20
(41)	6 11 14	(48)	7 14 17	(55)	9 10 17	(62)	10 15 16	(69)	8 12 17
(42)	7 13 20	(49)	8 15 19	(56)	11 13 16	(63)	11 17 21	(70)	9 11 19

Plan 11.34 $t = 21, k = 5, r = 5, b = 21, \lambda = 1, E = .84$, Type IV

See incomplete latin squares Plan 13.13; randomize units in blocks ignoring replications.

Plan 11.35 $t = 21, k = 7, r = 10, b = 30, \lambda = 3, E = .90$, Type III

Block							
(1)	2	5	10	11	17	19	20
(2)	3	6	11	12	18	20	21
(3)	4	7	12	13	19	21	15
(4)	5	1	13	14	20	15	16
(5)	6	2	14	8	21	16	17
(6)	7	3	8	9	15	17	18
(7)	1	4	9	10	16	18	19
(8)	3	4	8	13	17	19	20
(9)	4	5	9	14	18	20	21
(10)	5	6	10	8	19	21	15
(11)	6	7	11	9	20	15	16
(12)	7	1	12	10	21	16	17
(13)	1	2	13	11	15	17	18
(14)	2	3	14	12	16	18	19
(15)	1	6	9	12	17	19	20
(16)	2	7	10	13	18	20	21
(17)	3	1	11	14	19	21	15
(18)	4	2	12	8	20	15	16
(19)	5	3	13	9	21	16	17
(20)	6	4	14	10	15	17	18
(21)	7	5	8	11	16	18	19
(22)	1	2	4	8	9	11	21
(23)	2	3	5	9	10	12	15
(24)	3	4	6	10	11	13	16
(25)	4	5	7	11	12	14	17
(26)	5	6	1	12	13	8	18
(27)	6	7	2	13	14	9	19
(28)	7	1	3	14	8	10	20
(29)	1	2	3	4	5	6	7
(30)	8	9	10	11	12	13	14

Plan 11.36 $t = 25, k = 4, r = 8, b = 50, \lambda = 1, E = .78$, Type II

See extended incomplete latin squares Plan 13.16a; randomize units in blocks ignoring replications.

Plan 11.37 $t = 25, k = 9, r = 9, b = 25, \lambda = 3, E = .93$, Type IV

See incomplete latin squares, Plan 13.1a; randomize units in blocks ignoring replications.

Plan 11.38 $t = 28, k = 4, r = 9, b = 63, \lambda = 1, E = .78$, Type I

Block	Rep. I				Block	Rep. II				Block	Rep. III			
(1)	28	1	10	19	(8)	28	2	11	20	(15)	28	3	12	21
(2)	2	9	13	16	(9)	3	1	14	17	(16)	4	2	15	18
(3)	3	8	11	18	(10)	4	9	12	10	(17)	5	1	13	11
(4)	4	7	23	24	(11)	5	8	24	25	(18)	6	9	25	26
(5)	5	6	20	27	(12)	6	7	21	19	(19)	7	8	22	20
(6)	12	17	22	25	(13)	13	18	23	26	(20)	14	10	24	27
(7)	14	15	21	26	(14)	15	16	22	27	(21)	16	17	23	19

Block	Rep. IV				Block	Rep. V				Block	Rep. VI			
(22)	28	4	13	22	(29)	28	5	14	23	(36)	28	6	15	24
(23)	5	3	16	10	(30)	6	4	17	11	(37)	7	5	18	12
(24)	6	2	14	12	(31)	7	3	15	13	(38)	8	4	16	14
(25)	7	1	26	27	(32)	8	2	27	19	(39)	9	3	19	20
(26)	8	9	23	21	(33)	9	1	24	22	(40)	1	2	25	23
(27)	15	11	25	19	(34)	16	12	26	20	(41)	17	13	27	21
(28)	17	18	24	20	(35)	18	10	25	21	(42)	10	11	26	22

Block	Rep. VII				Block	Rep. VIII				Block	Rep. IX			
(43)	28	7	16	25	(50)	28	8	17	26	(57)	28	9	18	27
(44)	8	6	10	13	(51)	9	7	11	14	(58)	1	8	12	15
(45)	9	5	17	15	(52)	1	6	18	16	(59)	2	7	10	17
(46)	1	4	20	21	(53)	2	5	21	22	(60)	3	6	22	23
(47)	2	3	26	24	(54)	3	4	27	25	(61)	4	5	19	26
(48)	18	14	19	22	(55)	10	15	20	23	(62)	11	16	21	24
(49)	11	12	27	23	(56)	12	13	19	24	(63)	13	14	20	25

Plan 11.39 $t = 28, k = 7, r = 9, b = 36, \lambda = 2, E = .89$, Type III

Block								Block							
(1)	4	7	8	9	14	23	28	(19)	4	8	11	17	19	21	25
(2)	1	5	9	10	11	15	24	(20)	1	13	14	18	23	25	26
(3)	6	8	13	15	16	18	21	(21)	2	4	5	6	16	22	23
(4)	7	12	13	17	22	24	25	(22)	3	4	10	11	12	14	18
(5)	4	10	16	17	20	26	27	(23)	1	9	14	16	17	19	22
(6)	2	11	18	19	22	26	28	(24)	1	2	4	13	20	24	28
(7)	1	3	6	12	19	23	27	(25)	3	5	8	17	23	24	26
(8)	2	3	5	14	20	21	25	(26)	5	6	7	10	19	25	28
(9)	1	2	8	10	12	16	25	(27)	1	6	7	8	11	20	26
(10)	2	3	6	9	11	13	17	(28)	9	10	13	19	20	21	23
(11)	4	5	12	13	15	19	26	(29)	2	8	14	15	19	24	27
(12)	3	7	16	18	19	20	24	(30)	3	9	15	16	25	26	28
(13)	6	10	14	21	22	24	26	(31)	5	8	9	12	18	20	22
(14)	11	15	20	22	23	25	27	(32)	11	12	16	21	23	24	28
(15)	1	5	17	18	21	27	28	(33)	1	3	4	7	15	21	22
(16)	2	7	9	12	21	26	27	(34)	5	7	11	13	14	16	27
(17)	3	8	10	13	22	27	28	(35)	4	6	9	18	24	25	27
(18)	6	12	14	15	17	20	28	(36)	2	7	10	15	17	18	23

Plan 11.40 $t = 31, k = 6, r = 6, b = 31, \lambda = 1, E = .86$, Type IV

See incomplete latin squares Plan 13.14; randomize units in blocks ignoring replications.

Plan 11.41 $t = 31, k = 10, r = 10, b = 31, \lambda = 3, E = .93$, Type IV

See incomplete latin squares Plan 13.2a; randomize units in blocks ignoring replications.

Plan 11.42 $t = 37, k = 9, r = 9, b = 37, \lambda = 2, E = .91$, Type IV

See incomplete latin squares Plan 13.15; randomize units in blocks ignoring replications.

Plan 11.43 $t = 41, k = 5, r = 10, b = 82, \lambda = 1, E = .82$, Type III

See extended latin squares Plan 13.17a; randomize units in blocks ignoring replications.

Plan 11.44 $t = 57, k = 8, r = 8, b = 57, \lambda = 1, E = .89$, Type IV

See incomplete latin squares Plan 13.3a; randomize units in blocks ignoring replications.

Plan 11.45 $t = 73, k = 9, r = 9, b = 73, \lambda = 1, E = .90$, Type IV

See incomplete latin squares Plan 13.4a; randomize units in blocks ignoring replications.

Plan 11.46 $t = 91, k = 10, r = 10, b = 91, \lambda = 1, E = .91$, Type IV

See incomplete latin squares Plan 13.5a; randomize units in blocks ignoring replications.

CHAPTER 12

Lattice Squares

12.1 Description

12.11 Balanced Lattice Squares. The number of treatments must be an exact square. Within each replicate, the k^2 treatments are arranged on the plan in a $k \times k$ square. The method of grouping into rows and columns, which varies in successive replications, is such that the treatment means can be adjusted for differences among the rows and columns of each square. Thus, in addition to the elimination of differences among replicates from the experimental errors, the design permits a "double control" within each replicate, similar to that obtained in a latin square.

The principle which governs the grouping into rows and columns may be seen from the plans 12.1–12.8. With 9 treatments, for instance, treatment 1 occurs in the same row as treatments 2, 3, 6, and 8, and in the same column as treatments 4, 5, 7, and 9. Thus every other treatment appears with treatment 1 either in the same row or in the same column. More generally, any pair of treatments occurs once in the same row or in the same column.

This property holds for all plans having an odd number of treatments, i.e., for 9, 25, 49, 81, 121, and 169 treatments. The number of replicates, for k^2 treatments, is $(k + 1)/2$. The standard error of the difference is not exactly the same for all pairs of treatments, though the variation in accuracy is small. Separate formulae are given for the two standard errors for cases in which they may be needed.

A design with twice the basic number of replicates (e.g., 25 treatments in 6 replicates) is obtained by repeating the plan, *the rows being interchanged with the columns.* In designs so produced, every pair of treatments occurs together once in the same row *and* once in the same column. On account of this property, the standard error of the difference is the same for all pairs of treatments, whether the row and column differences are large or small.

Within the most useful range, the only even numbers of treatments which provide lattice square designs are 16 and 64. These designs have, respectively, 5 and 9, i.e. $(k + 1)$ replicates. Every pair of treatments occurs once both in the same row and in the same column.

The available selection of designs (up to 12 replicates) is summarized in table 12.1. By analogy with balanced lattices, the designs may be called *balanced* lattice squares.

TABLE 12.1 Available numbers of treatments and replications for balanced lattice squares

Number of treatments	9	16	25	49	64	81	121	169
Number of replicates	4, 8	5, 10	3, 6	4, 8	9	5, 10	6, 12	7

For 16 treatments in 10 replicates, the basic plan is used twice. Although the basic plan (plan 12.1) requires only 2 replicates, the design for 9 treatments is not recommended with less than 4 replicates. Even in this case there are only 8 d.f. for estimating row and column variances and only 8 d.f. for error. Little would be gained over randomized blocks unless the extra controls were highly effective. A design with 9 treatments and 8 replicates is found by repeating the design for 9 treatments and 4 replicates.

Certain factorial experiments can be arranged in *balanced* lattice squares; e.g., 3×3, 8×2, 4×4, $4 \times 2 \times 2$, 2^4, 5×5, 7×7, 8×8, 4^3, 2^6, 9×9, 3^4. All main effects and interactions are confounded to the same extent with rows and columns.

12.12 Partially Balanced Squares. For a balanced design, the available numbers of replications are rather severely restricted. Although designs are possible for other numbers of replicates, they lack the symmetry of the designs described above, with the consequence that the statistical analysis usually becomes more complicated. It happens, however, that if the number of replications is less than that required for a balanced design, the statistical analysis follows the same procedure as for balanced designs, apart from minor changes in some formulae. In experimentation where double grouping has proved effective but where the replications are not sufficient for a balanced lattice square, the additional designs shown in table 12.2 may be useful.

TABLE 12.2 Other lattice square designs

Number of treatments	49	64	81	121	169
Number of replicates	3	3, 4	3, 4	3, 4, 5	3, 4, 5, 6

When the number of treatments is odd, the plans for these designs are obtained simply by taking the desired number of replicates from the plans for the balanced designs. Thus, for 49 treatments in threefold replication, we use squares I to III from plan 12.4. With 64 treatments, squares I, III, and V are used for 3 replicates, and squares I, III, V, and VII for 4 replicates.

Partially balanced lattice squares are analogous to the triple, quadruple, and quintuple lattices.

12.13 Arrangement of Experimental Material. The method of arranging the experimental material is similar to that for an ordinary latin square. With k^2 treatments, the units are arranged in $k \times k$ squares so that the rows and columns of each square correspond to the two types of variation whose effects we wish to eliminate from the errors. In field experiments, the plots of each replicate are usually laid out in square formation, in which case row and column differences represent fertility variations in two directions at right angles to each other. If the width of a greenhouse bench accommodates 5 pots, 25 treatments can be laid out in replicates which consist of 5 rows of 5 pots each.

12.14 Randomization. Within each replicate the rows and columns of the basic plan should be permuted separately at random before applying the treatments. It is also advisable to assign the treatments at random to the treatment numbers in the plan.

12.2 Statistical Analysis

12.21 Designs with $(k + 1)/2$, or Fewer, Replications. The same instructions apply to the partially balanced designs and to the balanced designs with $(k + 1)/2$ replications. Table 12.3 shows the plan and yields for a lattice square with 25 corn hybrids in 3 replications, conducted by the North Carolina Experiment Station in 1942. Each plot contained 4 rows, with 10 hills per row.

The computations proceed as follows. The number of treatments is k^2, and the number of replications r.

1. Find the row, column, treatment, replication, and grand totals. The treatment totals are placed in a square under the plan.

2. For each row, calculate the L value, where

L = total (from all replications) of all treatments included in the row $- \, r$ times row total

e.g.,

$$L_1 = 89.8 + 93.3 + 82.1 + 88.9 + 91.1 - 3(159.1) = -32.1$$

As a check, the total of the L values over a replication equals

$$(\text{Grand total}) - r(\text{replication total})$$

Thus,

$$-76.2 = 2189.4 - 3(755.2)$$

Similarly for each column we obtain M values.

M = total (from all replications) of all treatments included in the column $- \, r$ times column total

TABLE 12.3 YIELDS (POUNDS CORN PER PLOT) FOR A 5 × 5 LATTICE SQUARE IN 3 REPLICATIONS

Treatment numbers are shown in parentheses

	Square I				Total	L	δ
(18) 33.3	(9) 30.7	(11) 35.4	(2) 30.1	(25) 29.6	159.1	−32.1	−2.58
(24) 24.6	(15) 30.8	(17) 28.8	(8) 34.8	(1) 32.5	151.5	−10.5	−0.84
(12) 28.5	(3) 24.0	(10) 28.4	(21) 25.0	(19) 35.1	141.0	+22.4	+1.80
(6) 26.7	(22) 27.2	(4) 25.6	(20) 25.0	(13) 29.4	133.9	+11.0	+0.88
(5) 40.1	(16) 35.7	(23) 30.1	(14) 30.3	(7) 33.5	169.7	−67.0	−5.39
Total 153.2	148.4	148.3	145.2	160.1	755.2		−6.13
M −16.0	+3.8	−12.8	−10.0	−41.2		−76.2	
ϵ − 0.83	+0.20	− 0.66	− 0.52	− 2.13	−3.94		
	Square II						
(20) 30.9	(17) 33.3	(19) 38.8	(16) 27.7	(18) 34.4	165.1	− 45.9	− 3.69
(15) 37.2	(12) 31.2	(14) 27.9	(11) 27.3	(13) 21.6	145.2	− 5.8	− 0.47
(25) 32.7	(22) 43.0	(24) 28.5	(21) 24.7	(23) 22.7	151.6	− 46.2	− 3.71
(5) 32.0	(2) 32.8	(4) 31.8	(1) 28.7	(3) 32.3	157.6	− 33.8	− 2.72
(10) 39.8	(7) 37.3	(9) 31.9	(6) 34.0	(8) 34.3	177.3	− 69.3	− 5.57
Total 172.6	177.6	158.9	142.4	145.3	796.8		−16.16
M −62.2	−85.5	−30.1	−12.6	−10.6		−201.0	
ϵ − 3.22	− 4.43	− 1.56	− 0.65	− 0.55	−10.41		
	Square III						
(19) 28.7	(15) 26.3	(23) 21.7	(6) 21.9	(2) 26.0	124.6	+ 69.1	+ 5.56
(11) 19.4	(7) 17.3	(20) 16.9	(3) 22.6	(24) 24.2	100.4	+ 98.0	+ 7.88
(22) 18.3	(18) 22.1	(1) 17.5	(14) 25.0	(10) 26.9	109.8	+105.9	+ 8.51
(5) 30.2	(21) 27.5	(9) 30.7	(17) 28.1	(13) 27.6	144.1	+ 9.3	+ 0.75
(8) 34.4	(4) 32.8	(12) 31.9	(25) 28.8	(16) 30.6	158.5	− 5.1	− 0.41
Total 131.0	126.0	118.7	126.4	135.3	637.4		+22.29
M +86.0	+61.6	+54.8	+46.8	+28.0		+277.2	
ϵ + 4.45	+ 3.19	+ 2.84	+ 2.42	+ 1.45	+14.35		
				Grand total	2189.4	0	0

Treatment totals

(1) 78.7	(2) 88.9	(3) 78.9	(4) 90.2	(5) 102.3	
(6) 82.6	(7) 88.1	(8) 103.5	(9) 93.3	(10) 95.1	
(11) 82.1	(12) 91.6	(13) 78.6	(14) 83.2	(15) 94.3	
(16) 94.0	(17) 90.2	(18) 89.8	(19) 102.6	(20) 72.8	
(21) 77.2	(22) 88.5	(23) 74.5	(24) 77.3	(25) 91.1	2189.4

Adjusted treatment totals

(1) 83.7	(2) 85.7	(3) 87.9	(4) 88.9	(5) 95.3
(6) 84.4	(7) 81.6	(8) 100.1	(9) 87.4	(10) 97.4
(11) 90.1	(12) 90.1	(13) 78.5	(14) 86.2	(15) 98.7
(16) 85.5	(17) 83.8	(18) 93.8	(19) 107.0	(20) 77.0
(21) 78.1	(22) 94.4	(23) 72.6	(24) 79.7	(25) 81.5

3. We now compute the analysis of variance. The total s.s. and the sums of squares for replications and treatments are found in the usual way. If the symbol L_r denotes a replication total of the L's, the sum of squares for rows within replications, adjusted for treatments, is

$$\frac{\sum L^2}{kr(r-1)} - \frac{\sum L_r^2}{k^2r(r-1)}$$

$$= \frac{(32.1)^2 + (10.5)^2 + \cdots + (5.1)^2}{30} - \frac{(76.2)^2 + (201.0)^2 + (277.2)^2}{150}$$

$$= 1405.95 - 820.32 = 585.63$$

The sum of squares for columns, eliminating treatments, is

$$\frac{(16.0)^2 + (3.8)^2 + \cdots + (28.0)^2}{30} - \frac{(76.2)^2 + (201.0)^2 + (277.2)^2}{150}$$

$$= 1058.53 - 820.32 = 238.21$$

The complete analysis of variance is shown in table 12.4.

TABLE 12.4 ANALYSIS OF VARIANCE FOR A 5×5 LATTICE SQUARE

	d.f.		s.s.	m.s.
Replications	$(r-1)$	2	546.88	273.44
Treatments	(k^2-1)	24	611.09	25.46
Rows (adj.)	$r(k-1)$	12	585.63	$48.80E_r$
Columns (adj.)	$r(k-1)$	12	238.21	$19.85E_c$
Error	$(k-1)(rk-r-k-1)$	24	229.79	$9.57E_e$
Total	(rk^2-1)	74	2211.60	

4. This step leads to the adjusted treatment means. Let

$$E_r = \text{mean square for rows}$$

$$E_c = \text{mean square for columns}$$

$$E_e = \text{mean square for error}$$

$$\lambda' = \frac{(E_r - E_e)}{k(r-1)E_r} = \frac{48.80 - 9.57}{(5)(2)(48.80)} = 0.0804$$

$$\mu' = \frac{(E_c - E_e)}{k(r-1)E_c} = \frac{19.85 - 9.57}{(5)(2)(19.85)} = 0.0518$$

If E_r or E_c is less than E_e, then λ' or μ', as the case may be, is taken as 0.

5. Multiply the L's by λ' to give δ values, and the M's by μ' to give ϵ values, as presented in table 12.3.

6. The adjusted total for any treatment is secured by adding to the unadjusted total the δ and ϵ values for each row and column in which the treatment appears. For treatment (1), we have

$$78.7 + (-0.84) + (-2.13) + (-2.72) + (-0.65) + (8.51) + (2.84) = 83.7$$

To obtain the treatment means (not shown) we divide by the number of replicates.

7. The *average variance* of the difference between two adjusted treatment means is

$$\frac{2E_e}{r}\left[1 + \frac{rk}{(k+1)}(\lambda' + \mu')\right] = \frac{2(9.57)}{3}\left[1 + \frac{(3)(5)}{6}(0.1322)\right]$$

$$= 8.49; \quad \text{s.e.} = 2.91$$

Except perhaps with the 5 × 5 and smaller squares, it is sufficiently accurate to use this figure for comparisons between any pair of treatments. More accurately, for two treatments that appear in the same row, the variance of the difference is

$$\frac{2E_e}{r}[1 + (r-1)\lambda' + r\mu'] = 8.40; \quad \text{s.e.} = 2.90$$

For two treatments in the same column.

$$\frac{2E_e}{r}[1 + r\lambda' + (r-1)\mu'] = 8.58; \quad \text{s.e.} = 2.93$$

Clearly the average error is good enough in this experiment. For the partially balanced designs one more formula is needed, since some pairs of treatments do not occur together either in a row or in a column. The variance of the difference is

$$\frac{2E_e}{r}[1 + r(\lambda' + \mu')]$$

8. As is typical of these designs, the analysis does not provide an exact F-test. The treatments m.s. in the analysis (table 12.4) cannot be compared with the error m.s., since the former has not been adjusted for row and column effects. For an approximate F-test, compute the sum of squares of deviations of the adjusted treatment totals in table 12.3. This comes to 1606.05. Division by 3, since there are 3 replicates, gives

535.35, with a mean square of 22.31. This may be compared with the effective error m.s., which is

$$E_e\left[1+\frac{rk}{k+1}(\lambda'+\mu')\right] = 9.57\left[1+\frac{15}{6}(0.1322)\right] = 12.73$$

The F-ratio is 22.31/12.73, or 1.75, with 24 and 24 d.f.

9. The gain in precision relative to randomized blocks is estimated as follows. If the experiment were analyzed as randomized blocks, rows and columns would be amalgamated with error, giving

	d.f.	s.s.	m.s.
Replications	2	546.88	
Treatments	24	611.09	
Error	48	1053.63	21.95

This error, 21.95, is compared with the effective error m.s., 12.73. The relative information is estimated as 21.95/12.73, or 172%. Thus 3 replicates of the lattice square appear about as precise as 5 with randomized blocks.

Often a comparison with a lattice design will be of more interest. This experiment could have been planned as a triple lattice, and in that event the rows would probably have been chosen as blocks. The intra-block error would be derived from the pooled sum of squares for columns and error in table 12.4. The mean square would be 13.00, with 36 d.f. Consequently, for the triple lattice we have $E_b = 48.80$ (m.s. for rows), $E_e = 13.00$. The effective error m.s. (section 10.28) is

$$E_e\left[1+\frac{rk\mu}{k+1}\right] = E_e\left[1+\frac{r(E_b-E_e)}{(k+1)(r-1)E_b}\right]$$

$$= 13.00\left[1+\frac{3(48.80-13.00)}{(6)(2)(48.80)}\right] = 15.38$$

The relative accuracy of the lattice square to the triple lattice is estimated as 15.38/12.73, or 121%.

An account of the theory and a worked example are given by Yates (12.1) for balanced squares and by Cochran (12.2) for partially balanced squares.

12.22 Designs with $(k+1)$ Replications. In this case the analysis is slightly different, since it is possible to apply a single adjustment for all the rows in which a treatment lies, instead of making a separate adjust-

TABLE 12.5 PERCENTAGES OF SQUARES ATTACKED BY BOLL WEEVILS FOR A 4×4 LATTICE SQUARE IN $(k + 1)$ REPLICATIONS

Boll weevil infestation

Square I

									Row totals
	(10)	9.0	(12)	20.3	(9)	17.7	(11)	26.3	73.3
	(2)	4.7	(4)	9.0	(1)	7.3	(3)	8.3	29.3
	(14)	9.0	(16)	6.7	(13)	11.7	(15)	4.3	31.7
	(6)	4.0	(8)	5.0	(5)	5.7	(7)	14.3	29.0
Column totals		26.7		41.0		42.4		53.2	163.3

Square II

	(5)	19.0	(12)	8.7	(15)	13.0	(2)	15.7	56.4
	(10)	12.0	(7)	6.0	(4)	15.3	(13)	12.0	45.3
	(16)	12.7	(1)	6.3	(6)	1.7	(11)	13.0	33.7
	(3)	3.7	(14)	3.7	(9)	8.0	(8)	13.3	28.7
Column totals		47.4		24.7		38.0		54.0	164.1

Square III

	(10)	17.0	(15)	7.0	(8)	10.3	(1)	1.3	35.6
	(9)	11.3	(16)	12.3	(7)	3.0	(2)	5.3	31.9
	(12)	12.3	(13)	8.7	(6)	8.0	(3)	9.3	38.3
	(11)	30.3	(14)	22.3	(5)	11.0	(4)	12.7	76.3
Column totals		70.9		50.3		32.3		28.6	182.1

Square IV

	(16)	5.0	(12)	10.3	(8)	5.7	(4)	12.7	33.7
	(11)	2.7	(15)	6.7	(3)	10.3	(7)	5.7	25.4
	(1)	1.0	(5)	10.3	(9)	11.3	(13)	11.7	34.3
	(6)	11.0	(2)	19.0	(14)	20.7	(10)	29.7	80.4
Column totals		19.7		46.3		48.0		59.8	173.8

Square V

	(3)	2.0	(16)	5.0	(5)	4.0	(10)	13.7	24.7
	(6)	9.3	(9)	1.7	(4)	6.3	(15)	12.3	29.6
	(12)	16.7	(7)	4.3	(14)	18.7	(1)	8.7	48.4
	(13)	16.7	(2)	30.0	(11)	25.7	(8)	14.0	86.4
Column totals		44.7		41.0		54.7		48.7	189.1
									872.4

ment for every row. On the whole, however, the analysis is more complex than with $(k + 1)/2$ replications.

The example is a 4×4 lattice square conducted by the U. S. Bureau of Entomology and Plant Quarantine at Florence, S. C., and described by Wadley (12.3). The treatments were 16 arsenical insecticides applied to cotton with a hand dusting machine. Plots were 10 rows wide and 70 feet long, being about 1/18 acre. To allow for border effects, records were taken only from the 4 center rows. The data in table 12.5 are the percentages of squares * showing attack by boll weevils. These figures were obtained by examining 100 squares per plot, 25 from each of the 4 center rows. Such counts were made at intervals during the summer: the data are averages from 3 counts made in August.

The simplest computational routine, devised by Yates (12.1), will be presented here, though it is not too well adapted for making clear the meaning of the various steps. Yates's paper should be consulted for an account of the theory.

TABLE 12.6 ORIGINAL AND ADJUSTED TREATMENT TOTALS

	Treatment totals, T	R_t	C_t	D	L'	J	K	M'	Adj. treatment totals	Adj. means
1	24.6	181.3	164.1	17.2	64.3	81.5	133.1	150.3	32.24	6.45
2	74.7	284.4	196.6	87.8	−250.8	−163.0	100.4	188.2	68.41	13.68
3	33.6	146.4	221.9	−75.5	274.8	199.3	− 27.2	−102.7	43.64	8.73
4	56.0	214.2	222.1	− 7.9	25.4	17.5	− 6.2	− 14.1	56.79	11.36
5	50.0	220.7	223.1	− 2.4	− 31.1	− 33.5	− 40.7	− 43.1	47.20	9.44
6	34.0	211.0	161.4	49.6	− 46.6	3.0	151.8	201.4	37.89	7.58
7	33.3	180.0	211.0	−31.0	105.6	74.6	− 18.4	− 49.4	36.85	7.37
8	48.3	213.4	224.0	−10.6	− 1.4	− 12.0	− 43.8	− 54.4	46.58	9.32
9	50.0	197.8	240.3	−42.5	83.4	40.9	− 86.6	−129.1	50.07	10.01
10	81.4	259.3	253.5	5.8	− 98.5	− 92.7	− 75.3	− 69.5	74.57	14.91
11	98.0	295.1	252.5	42.6	−211.1	−168.5	− 40.7	1.9	87.95	17.59
12	68.3	250.1	227.6	22.5	−104.9	− 82.4	− 14.9	7.6	63.51	12.70
13	60.8	236.0	251.2	−15.2	− 64.4	− 79.6	−125.2	−140.4	53.45	10.69
14	74.4	265.5	204.4	61.1	−157.5	− 96.4	86.9	148.0	71.36	14.27
15	43.3	178.7	236.5	−57.8	152.1	94.3	− 79.1	−136.9	46.42	9.28
16	41.7	155.7	199.4	−43.7	260.7	217.0	85.9	42.2	55.46	11.09
Totals	872.4	3489.6	3489.6	0	0	0	0	0	872.39	174.47

1. Find the row, column, replication, and grand totals, all shown in table 12.5.

2. Find the treatment totals T (table 12.6). For each treatment find

$$R_t = \text{total of all rows in which the treatment appears}$$
$$C_t = \text{total of all columns in which the treatment appears}$$

* A "square" is the name given to the young flower bud.

It may save time to find all 3 quantities, T, R_t, and C_t, by simultaneous addition, since this means that the positions of a treatment in the different replicates are found only once. This can be done if the machine carriage is moved so as to accommodate 3 sets of running totals. The sum of the R_t values and the sum of the C_t values should each equal k times the sum of the T values.

Thereafter the successive columns in table 12.6 are filled out as follows.

$$D = R_t - C_t$$

$$L' = kT - (k+1)R_t + G$$

$$J = D + L'$$

$$K = J + (k-1)D$$

$$M' = D + K$$

Note that the total of each of these quantities is zero.

3. Compute the analysis of variance. The only sums of squares requiring special instructions are those for rows and columns. After adjustment for treatment effects, it happens unfortunately that the sums of squares for rows and columns are not mutually orthogonal. By a well-known result in the analysis of variance, their combined sum of squares may be expressed in either of two ways:

Rows (adj. for treatments) + columns (adj. for treatments and rows) = columns (adj. for treatments) + rows (adj. for treatments and columns)

Both expressions are computed.

Sum of squares for rows adjusted for treatments

$$= \frac{S(L')^2}{k^3(k+1)} = \frac{(64.3)^2 + (250.8)^2 + \cdots + (260.7)^2}{320} = 1093.02$$

Sum of squares for rows adjusted for treatments and columns

$$= \frac{S(J^2)}{k^3(k-1)} = \frac{(81.5)^2 + (163.0)^2 + \cdots + (217.0)^2}{192} = 1026.76$$

Sum of squares for columns adjusted for treatments $= \dfrac{S(M')^2}{k^3(k+1)} = 625.85$

Sum of squares for columns adjusted for treatments and rows

$$= \frac{S(K^2)}{k^3(k-1)} = 559.59$$

Note that $1093.02 + 559.59 = 1026.76 + 625.85$.

TABLE 12.7 ANALYSIS OF VARIANCE

	d.f.		s.s.	m.s.
Replications	k	4	31.56	7.89
Treatments	$k^2 - 1$	15	1244.20	82.95
Rows (adj. for treatments)	$k^2 - 1$	15	1093.02	
Rows (adj. for tr. and col.)	$k^2 - 1$	15	1026.76	$68.45E_r$
Columns (adj. for tr.)	$k^2 - 1$	15	625.85	
Columns (adj. for tr. and rows)	$k^2 - 1$	15	559.59	$37.31E_c$
Error	$(k^2 - 1)(k - 2)$	30	680.17	$22.67E_e$
Total	$k^3 + k^2 - 1$	79	3608.54	

In finding the error s.s. by subtraction, note that we subtract the combined effect of rows and columns *only once*.

4. This leads to the adjusted treatment totals. Compute

$$q = k^2E_rE_c - E_e^2 = (16)(68.45)(37.31) - (22.67)^2 = 40{,}348$$

$$Q = (k - 1)q = 121{,}044$$

$$\lambda' = \frac{(E_r - E_e)(kE_c - E_e)}{Q} = \frac{(45.78)(126.57)}{121{,}044} = 0.04787$$

$$\mu' = \frac{(E_c - E_e)(kE_r - E_e)}{Q} = \frac{(14.64)(251.13)}{121{,}044} = 0.03037$$

The adjusted treatment totals are

$$T + \lambda'L' + \mu'M'$$

and are inserted on the right of the M' column in table 12.6. For the means, divide by r.

5. The error variance of the difference between two adjusted means is

$$\frac{2E_e}{r}[1 + k(\lambda' + \mu')] = \frac{2(22.67)}{5}[1 + 4(0.07824)] = 11.9$$

To obtain an approximate F-test of the treatments, compute the sum of squares of deviations of the adjusted treatment totals, and divide by $(k + 1)(k^2 - 1)$ to give the mean square on a single unit basis. This is tested against the effective error m.s., $E_e[1 + k(\lambda' + \mu')]$.

The gain in precision relative to a randomized blocks or lattice design is estimated by the procedure for the lattice square with $(k + 1)/2$ replications.

12.23 Designs with $2(k + 1)$ Replications. This case is unlikely to be of much practical interest except possibly with the 3×3 design in 8 replicates and the 4×4 in 10. The analysis is similar to that for $(k + 1)$ replications; the changes will be noted very briefly.

The partition of degrees of freedom is as follows.

	d.f.
Replications	$(2k + 1)$
Treatments	$(k^2 - 1)$
Rows	
Component (*a*)	$(k^2 - 1)$
Component (*b*)	$(k^2 - 1)$
Columns	
Component (*a*)	$(k^2 - 1)$
Component (*b*)	$(k^2 - 1)$
Error	$(2k - 3)(k^2 - 1)$
Total	$2k^2(k + 1) - 1$

Corresponding to any row, there is another row with the same set of treatments. Find the differences D_r between the totals of corresponding rows. The sum of squares of deviations of the D_r from their replication means, divided by $2k$, gives component (*a*) of the rows s.s. Component (*b*) corresponds to the component that was obtained with $(k + 1)$ replicates. As before, it has two forms, an L' and a J form. Both are found in the same way as with $(k + 1)$ replications, except that the divisors must be doubled. The sums of squares for columns are derived similarly.

The formulae for the adjustment factors are a little more complicated. Let E_r be the mean square found by pooling component (*a*) and the J component of the rows s.s., with an analogous definition for E_c. Calculate

$$W_i = \frac{1}{E_e}; \quad W_r = \frac{(2k - 1)}{2kE_r - E_e}; \quad W_c = \frac{(2k - 1)}{2kE_c - E_e}$$

$$\lambda' = \frac{(W_i - W_r)}{k[W_r + W_c + (k - 1)W_i]}; \quad \mu' = \frac{W_i - W_c}{k[W_r + W_c - (k - 1)W_i]}$$

The remainder of the analysis proceeds without change.

12.24 Missing Data. The procedure for analyzing lattice squares when certain observations are missing has been worked out by Cornish (12.4). As in the case of lattices (section 10.13), the method requires two different estimates of each missing observation, one being that found by minimizing the intra-row-and-column error, the other being that appropriate to the case where the data are analyzed as a randomized block experiment. In order to reduce the amount of computation, we present a cruder method which should be satisfactory if only a small fraction of the total observations is missing.

In this approach we estimate each missing observation by the first method mentioned above, that is, by minimizing the intra-row-and-column error s.s. Thereafter the analysis proceeds as usual: 1 d.f. is deleted from the total and intra-row-and-column error s.s. for each missing value. Special rules for t-tests are given later.

The following totals are used in the formulae for inserting estimates in place of the missing values. R, C, T, and P are the totals of the row, column, treatment, and replication, respectively, that contain the missing observation, while G is the grand total. Further,

S_x = total of all *other* rows and columns in which the treatment with the missing value appears (note that this sum does not include the row and column which contain the missing value)

T_x = total (from all replicates) of all *other* treatments which appear in the row or column that contains the missing value

$Z = kR_1 - P_1 + kC_2 - P_2$

With $(k + 1)$ replicates, there is one other row with exactly the same treatments as C. Let R_1 be its total and P_1 the total of the replicate containing R_1, with similar definitions for C_2 and P_2. With $(k + 1)/2$ replicates, Z is not used.

Experiment with $(k + 1)/2$ replications

$$x = \frac{k(k-1)(R+C) - (k+3)P + 2k(k-2)T + 6G - 2kS_x - 2kT_x}{(k-1)^2(k-3)}$$

Experiment with $(k + 1)$ replications

$$x = \frac{k(k-1)(R+C) - (k+1)P + k(k-2)T + 3G - kS_x - kT_x + Z}{(k-1)^2(k-2)}$$

Experiment with less than $(k + 1)/2$ replications. Because of the lack of symmetry in this design, an additional symbol is needed.

U_x = total (from all replicates) of all treatments that appear in the same row or column as the treatment that has the missing value.

The distinction between U_x and T_x should be realized. T_x is a total over only those other treatments that are in the actual row or column that contains the missing value, while U_x is a total over all other treatments that appear somewhere in the experiment in the same row or column as the treatment that has the missing value. Thus T_x is a part of U_x. It may help to note that T_x is a total over $2(k - 1)$ treatments, or $2r(k-1)$ observations, where r is the number of replications, while U_x is a total over $2r(k-1)$ treatments, or $2r^2(k-1)$ observations.

$$x = \frac{\begin{Bmatrix} kr(r-1)(R+C) - r(r+1)P + k(k-2)(r-1)T \\ + (r+1)G - krS_x - krT_x + kU_x \end{Bmatrix}}{(k-1)(r-1)(kr-k-r-1)}$$

Example. In the corn experiment in table 12.3, suppose that treatment (18) in replication I is missing. Here $k = 5$, $r = (k+1)/2 = 3$. Omitting the observation 33.3 for this treatment, we find

$$R = 125.8; \quad C = 119.9; \quad R + C = 245.7; \quad P = 721.9$$

$$T = 56.5; \quad G = 2156.1$$

S_x is the total of the rows and columns in which treatment (18) appears in squares II and III. Thus

$$S_x = 165.1 + 145.3 + 109.8 + 126.0 = 546.2$$

Finally, T_x is the total for treatments (2), (9), (11), and (25), which appear in the row that has the missing value, plus that for treatments (5), (6), (12), and (24) which appear in the column with the missing value. These totals are to be taken from the table of treatment totals. This gives $T_x = 709.2$. Then

$$x = \frac{\begin{Bmatrix} (20)(245.7) - (8)(721.9) + (30)(56.5) \\ + (6)(2156.1) - (10)(546.2) - (10)(709.2) \end{Bmatrix}}{32} = 38.0$$

The exact formulae for t-tests are complicated, and it appears that any good approximate rule would also be rather complicated. The following rule is suggested, though it somewhat underestimates the standard errors for the smaller squares and also lacks the simplicity that might be desired. For a comparison between the means of two treatments A and B, we assign to each an effective amount of replication which depends on the number and situation of the missing values. When scoring A we examine each replicate in turn and assign a score by the following rules.

Case	Score to A
I. A missing	0
II. A present, and B in the same row or column as A	
i. B missing	0
ii. B present	1
III. A present, and B not in the same row or column as A	
i. Other values missing in both row and column	⅓
ii. Other values missing in row *or* column but not in both	⅔
iii. No values missing in row or column	1

The effective replication for A is of course its total score over all replications in the experiment. B is scored similarly. Suppose that in the example the mean of treatment (18) were being compared with that of treatment (9), which occurs in the same row in square I. The effective replication for treatment (18) is 2, since case I arises in square I. The effective replication for treatment (9) is also 2, since case II(i) arises in square I.

REFERENCES

12.1 YATES, F. Lattice squares. *Jour. Agr. Sci.* 30, 672–687, 1940.

12.2 COCHRAN, W. G. Some additional lattice square designs. *Iowa Agr. Exp. Sta. Res. Bull.* 318, 731–748, 1943.

12.3 WADLEY, F. M. Incomplete block experimental designs in insect population problems. *Jour. Econ. Ent.* 38, 651–654, 1946.

12.4 CORNISH, E. A. The recovery of inter-block information in quasi-factorial designs with incomplete data. II. Lattice squares. *Australian Coun. Sci. Ind. Res. Bull.* 175, 1944.

ADDITIONAL READING

KEMPTHORNE, O. Recent developments in the design of field experiments. IV. Lattice squares with split-plots. *Jour. Agr. Sci.* 37, 156–162, 1947.

PLANS

Plan 12.1 3×3 balanced lattice square

$t = 9, k = 3, r = 2$, rows = 6, columns = 6, $\lambda = 1$*

Square I

1	2	3
4	5	6
7	8	9

Square II

1	6	8
9	2	4
5	7	3

Plan 12.2 4×4 balanced lattice square

$t = 16, k = 4, r = 5$, rows = 20, columns = 20, $\lambda = 2$

Square I

1	5	9	13
2	6	10	14
3	7	11	15
4	8	12	16

Square II

1	2	3	4
6	5	8	7
11	12	9	10
16	15	14	13

Square III

1	11	16	6
12	2	5	15
14	8	3	9
7	13	10	4

Square IV

1	7	12	14
8	2	13	11
10	16	3	5
15	9	6	4

Square V

1	10	15	8
9	2	7	16
13	6	3	12
5	14	11	4

* Number of times that two treatments appear in the same row or column.

Plan 12.3 **5 × 5 balanced lattice square**

$t = 25,\ k = 5,\ r = 3$, rows = 15, columns = 15, $\lambda = 1$

Square I				
1	2	3	4	5
6	7	8	9	10
11	12	13	14	15
16	17	18	19	20
21	22	23	24	25

Square II				
1	10	14	18	22
23	2	6	15	19
20	24	3	7	11
12	16	25	4	8
9	13	17	21	5

Square III				
1	8	15	17	24
25	2	9	11	18
19	21	3	10	12
13	20	22	4	6
7	14	16	23	5

Plan 12.4 **7 × 7 balanced lattice square**

$t = 49,\ k = 7,\ r = 4$, rows = 28, columns = 28, $\lambda = 1$

Square I						
1	2	3	4	5	6	7
8	9	10	11	12	13	14
15	16	17	18	19	20	21
22	23	24	25	26	27	28
29	30	31	32	33	34	35
36	37	38	39	40	41	42
43	44	45	46	47	48	49

Square II						
1	38	26	14	44	32	20
21	2	39	27	8	45	33
34	15	3	40	28	9	46
47	35	16	4	41	22	10
11	48	29	17	5	42	23
24	12	49	30	18	6	36
37	25	13	43	31	19	7

Square III						
1	19	30	48	10	28	39
40	2	20	31	49	11	22
23	41	3	21	32	43	12
13	24	42	4	15	33	44
45	14	25	36	5	16	34
35	46	8	26	37	6	17
[illegible]	29	47	9	27	38	7

Square IV						
1	42	27	12	46	31	16
17	2	36	28	13	47	32
33	18	3	37	22	14	48
49	34	19	4	38	23	8
9	43	35	20	5	39	24
25	10	44	29	21	6	40
41	26	11	45	30	15	7

Plan 12.5 **8 × 8 balanced lattice square**

$t = 64, k = 8, r = 9$, rows = 72, columns = 72, $\lambda = 2$

Square I

1	9	17	25	33	41	49	57
2	10	18	26	34	42	50	58
3	11	19	27	35	43	51	59
4	12	20	28	36	44	52	60
5	13	21	29	37	45	53	61
6	14	22	30	38	46	54	62
7	15	23	31	39	47	55	63
8	16	24	32	40	48	56	64

Square II

1	10	19	28	37	46	55	64
9	2	51	44	61	30	23	40
17	50	3	36	29	62	15	48
25	42	35	4	21	14	63	56
33	58	27	20	5	54	47	16
41	26	59	12	53	6	39	24
49	18	11	60	45	38	7	32
57	34	43	52	13	22	31	8

Square III

1	44	62	56	27	39	18	13
46	2	17	35	16	53	60	31
64	23	3	25	54	12	45	34
55	40	29	4	58	41	11	22
28	9	50	63	5	24	38	43
37	51	15	42	20	6	32	57
19	61	48	14	33	26	7	52
10	30	36	21	47	59	49	8

Square IV

1	60	54	40	43	15	26	21
62	2	25	11	24	37	52	47
56	31	3	41	38	20	61	10
39	16	45	4	50	57	19	30
44	17	34	55	5	32	14	59
13	35	23	58	28	6	48	49
27	53	64	22	9	42	7	36
18	46	12	29	63	51	33	8

Square V

1	11	20	30	34	48	53	63
15	2	56	45	59	28	22	33
21	52	3	39	32	58	9	46
26	47	38	4	17	13	64	51
40	62	31	19	5	49	42	12
43	25	61	16	55	6	36	18
54	24	10	57	44	35	7	29
60	37	41	50	14	23	27	8

Square VI

1	59	52	38	42	16	29	23
63	2	32	13	19	36	54	41
53	28	3	47	40	18	57	14
34	15	46	4	49	61	24	27
48	22	39	51	5	25	10	60
11	33	21	64	31	6	44	50
30	56	58	17	12	43	7	37
20	45	9	26	62	55	35	8

Square VII

1	32	47	61	22	50	12	35
29	2	14	49	39	64	43	20
42	13	3	24	60	33	30	55
59	54	18	4	48	31	37	9
23	36	57	46	5	11	56	26
52	63	40	27	10	6	17	45
16	41	28	34	51	21	7	62
38	19	53	15	25	44	58	8

Square VIII

1	14	24	31	36	45	51	58
12	2	55	48	57	27	21	38
22	49	3	37	26	63	16	44
32	43	33	4	23	10	62	53
35	64	30	18	5	52	41	15
47	29	60	9	56	6	34	19
50	20	13	59	46	40	7	25
61	39	42	54	11	17	28	8

Square IX

1	2	3	4	5	6	7	8
14	12	16	10	15	9	13	11
24	21	22	23	18	19	20	17
31	27	26	32	30	29	25	28
36	38	37	33	35	34	40	39
45	48	44	43	41	47	46	42
51	55	49	53	52	56	50	54
58	57	63	62	64	60	59	61

Plan 12.6 9 × 9 balanced lattice square

$t = 81, k = 9, r = 5$, rows = 45, columns = 45, $\lambda = 1$

Square I

1	2	3	4	5	6	7	8	9
10	11	12	13	14	15	16	17	18
19	20	21	22	23	24	25	26	27
28	29	30	31	32	33	34	35	36
37	38	39	40	41	42	43	44	45
46	47	48	49	50	51	52	53	54
55	56	57	58	59	60	61	62	63
64	65	66	67	68	69	70	71	72
73	74	75	76	77	78	79	80	81

Square II

1	12	20	34	45	53	58	69	77
21	2	10	54	35	43	78	59	67
11	19	3	44	52	36	68	76	60
61	72	80	4	15	23	28	39	47
81	62	70	24	5	13	48	29	37
71	79	63	14	22	6	38	46	30
31	42	50	55	66	74	7	18	26
51	32	40	75	56	64	27	8	16
41	49	33	65	73	57	17	25	9

Square III

1	57	29	16	72	44	22	78	50
30	2	55	45	17	70	51	23	76
56	28	3	71	43	18	77	49	24
25	81	53	4	60	32	10	66	38
54	26	79	33	5	58	39	11	64
80	52	27	59	31	6	65	37	12
13	69	41	19	75	47	7	63	35
42	14	67	48	20	73	36	8	61
68	40	15	74	46	21	62	34	9

Square IV

1	33	62	27	47	76	14	43	66
63	2	31	77	25	48	64	15	44
32	61	3	46	78	26	45	65	13
17	37	69	4	36	56	21	50	79
67	18	38	57	5	34	80	19	51
39	68	16	35	55	6	49	81	20
24	53	73	11	40	72	7	30	59
74	22	54	70	12	41	60	8	28
52	75	23	42	71	10	29	58	9

Square V

1	60	35	18	65	40	23	79	48
36	2	58	41	16	66	46	24	80
59	34	3	64	42	17	81	47	22
26	73	51	4	63	29	12	68	43
49	27	74	30	5	61	44	10	69
75	50	25	62	28	6	67	45	11
15	71	37	20	76	54	7	57	32
38	13	72	52	21	77	33	8	55
70	39	14	78	53	19	56	31	9

Plan 12.7 **11 × 11 balanced lattice square**

$t = 121$, $k = 11$, $r = 6$, rows = 66, columns = 66, $\lambda = 1$

Square I

1	2	3	4	5	6	7	8	9	10	11
12	13	14	15	16	17	18	19	20	21	22
23	24	25	26	27	28	29	30	31	32	33
34	35	36	37	38	39	40	41	42	43	44
45	46	47	48	49	50	51	52	53	54	55
56	57	58	59	60	61	62	63	64	65	66
67	68	69	70	71	72	73	74	75	76	77
78	79	80	81	82	83	84	85	86	87	88
89	90	91	92	93	94	95	96	97	98	99
100	101	102	103	104	105	106	107	108	109	110
111	112	113	114	115	116	117	118	119	120	121

Square II

1	102	82	62	42	22	112	92	72	52	32
33	2	103	83	63	43	12	113	93	73	53
54	23	3	104	84	64	44	13	114	94	74
75	55	24	4	105	85	65	34	14	115	95
96	76	45	25	5	106	86	66	35	15	116
117	97	77	46	26	6	107	87	56	36	16
17	118	98	67	47	27	7	108	88	57	37
38	18	119	99	68	48	28	8	109	78	58
59	39	19	120	89	69	49	29	9	110	79
80	60	40	20	121	90	70	50	30	10	100
101	81	61	41	21	111	91	71	51	31	11

Square III

1	115	108	90	83	76	58	51	44	26	19
20	2	116	109	91	84	77	59	52	34	27
28	21	3	117	110	92	85	67	60	53	35
36	29	22	4	118	100	93	86	68	61	54
55	37	30	12	5	119	101	94	87	69	62
63	45	38	31	13	6	120	102	95	88	70
71	64	46	39	32	14	7	121	103	96	78
79	72	65	47	40	33	15	8	111	104	97
98	80	73	66	48	41	23	16	9	112	105
106	99	81	74	56	49	42	24	17	10	113
114	107	89	82	75	57	50	43	25	18	11

Plan 12.7 (Continued) **11 × 11 balanced lattice square**

Square IV

1	40	68	107	14	53	81	120	27	66	94
95	2	41	69	108	15	54	82	121	28	56
57	96	3	42	70	109	16	55	83	111	29
30	58	97	4	43	71	110	17	45	84	112
113	31	59	98	5	44	72	100	18	46	85
86	114	32	60	99	6	34	73	101	19	47
48	87	115	33	61	89	7	35	74	102	20
21	49	88	116	23	62	90	8	36	75	103
104	22	50	78	117	24	63	91	9	37	76
77	105	12	51	79	118	25	64	92	10	38
39	67	106	13	52	80	119	26	65	93	11

Square V

1	119	105	91	88	74	60	46	43	29	15
16	2	120	106	92	78	75	61	47	44	30
31	17	3	121	107	93	79	76	62	48	34
35	32	18	4	111	108	94	80	77	63	49
50	36	33	19	5	112	109	95	81	67	64
65	51	37	23	20	6	113	110	96	82	68
69	66	52	38	24	21	7	114	100	97	83
84	70	56	53	39	25	22	8	115	101	98
99	85	71	57	54	40	26	12	9	116	102
103	89	86	72	58	55	41	27	13	10	117
118	104	90	87	73	59	45	42	28	14	11

Square VI

1	110	87	64	41	18	116	93	70	47	24
25	2	100	88	65	42	19	117	94	71	48
49	26	3	101	78	66	43	20	118	95	72
73	50	27	4	102	79	56	44	21	119	96
97	74	51	28	5	103	80	57	34	22	120
121	98	75	52	29	6	104	81	58	35	12
13	111	99	76	53	30	7	105	82	59	36
37	14	112	89	77	54	31	8	106	83	60
61	38	15	113	90	67	55	32	9	107	84
85	62	39	16	114	91	68	45	33	10	108
109	86	63	40	17	115	92	69	46	23	11

Plan 12.8 **13 × 13 balanced lattice square**

$t = 169$, $k = 13$, $r = 7$, rows $= 91$, columns $= 91$, $\lambda = 1$

Square I

1	2	3	4	5	6	7	8	9	10	11	12	13
14	15	16	17	18	19	20	21	22	23	24	25	26
27	28	29	30	31	32	33	34	35	36	37	38	39
40	41	42	43	44	45	46	47	48	49	50	51	52
53	54	55	56	57	58	59	60	61	62	63	64	65
66	67	68	69	70	71	72	73	74	75	76	77	78
79	80	81	82	83	84	85	86	87	88	89	90	91
92	93	94	95	96	97	98	99	100	101	102	103	104
105	106	107	108	109	110	111	112	113	114	115	116	117
118	119	120	121	122	123	124	125	126	127	128	129	130
131	132	133	134	135	136	137	138	139	140	141	142	143
144	145	146	147	148	149	150	151	152	153	154	155	156
157	158	159	160	161	162	163	164	165	166	167	168	169

Square II

1	26	38	50	62	74	86	98	110	122	134	146	158
159	2	14	39	51	63	75	87	99	111	123	135	147
148	160	3	15	27	52	64	76	88	100	112	124	136
137	149	161	4	16	28	40	65	77	89	101	113	125
126	138	150	162	5	17	29	41	53	78	90	102	114
115	127	139	151	163	6	18	30	42	54	66	91	103
104	116	128	140	152	164	7	19	31	43	55	67	79
80	92	117	129	141	153	165	8	20	32	44	56	68
69	81	93	105	130	142	154	166	9	21	33	45	57
58	70	82	94	106	118	143	155	167	10	22	34	46
47	59	71	83	95	107	119	131	156	168	11	23	35
36	48	60	72	84	96	108	120	132	144	169	12	24
25	37	49	61	73	85	97	109	121	133	145	157	13

Plan 12.8 (Continued) **13 × 13 balanced lattice square**

Square III

1	24	34	44	54	77	87	97	107	130	140	150	160
161	2	25	35	45	55	78	88	98	108	118	141	151
152	162	3	26	36	46	56	66	89	99	109	119	142
143	153	163	4	14	37	47	57	67	90	100	110	120
121	131	154	164	5	15	38	48	58	68	91	101	111
112	122	132	155	165	6	16	39	49	59	69	79	102
103	113	123	133	156	166	7	17	27	50	60	70	80
81	104	114	124	134	144	167	8	18	28	51	61	71
72	82	92	115	125	135	145	168	9	19	29	52	62
63	73	83	93	116	126	136	146	169	10	20	30	40
41	64	74	84	94	117	127	137	147	157	11	21	31
32	42	65	75	85	95	105	128	138	148	158	12	22
23	33	43	53	76	86	96	106	129	139	149	159	13

Square IV

1	22	30	51	59	67	88	96	117	125	133	154	162
163	2	23	31	52	60	68	89	97	105	126	134	155
156	164	3	24	32	40	61	69	90	98	106	127	135
136	144	165	4	25	33	41	62	70	91	99	107	128
129	137	145	166	5	26	34	42	63	71	79	100	108
109	130	138	146	167	6	14	35	43	64	72	80	101
102	110	118	139	147	168	7	15	36	44	65	73	81
82	103	111	119	140	148	169	8	16	37	45	53	74
75	83	104	112	120	141	149	157	9	17	38	46	54
55	76	84	92	113	121	142	150	158	10	18	39	47
48	56	77	85	93	114	122	143	151	159	11	19	27
28	49	57	78	86	94	115	123	131	152	160	12	20
21	29	50	58	66	87	95	116	124	132	153	161	13

Plan 12.8 (Continued) 13 × 13 balanced lattice square

Square V

1	20	39	45	64	70	89	95	114	120	139	145	164
165	2	21	27	46	65	71	90	96	115	121	140	146
147	166	3	22	28	47	53	72	91	97	116	122	141
142	148	167	4	23	29	48	54	73	79	98	117	123
124	143	149	168	5	24	30	49	55	74	80	99	105
106	125	131	150	169	6	25	31	50	56	75	81	100
101	107	126	132	151	157	7	26	32	51	57	76	82
83	102	108	127	133	152	158	8	14	33	52	58	77
78	84	103	109	128	134	153	159	9	15	34	40	59
60	66	85	104	110	129	135	154	160	10	16	35	41
42	61	67	86	92	111	130	136	155	161	11	17	36
37	43	62	68	87	93	112	118	137	156	162	12	18
19	38	44	63	69	88	94	113	119	138	144	163	13

Square VI

1	18	35	52	56	73	90	94	111	128	132	149	166
167	2	19	36	40	57	74	91	95	112	129	133	150
151	168	3	20	37	41	58	75	79	96	113	130	134
135	152	169	4	21	38	42	59	76	80	97	114	118
119	136	153	157	5	22	39	43	60	77	81	98	115
116	120	137	154	158	6	23	27	44	61	78	82	99
100	117	121	138	155	159	7	24	28	45	62	66	83
84	101	105	122	139	156	160	8	25	29	46	63	67
68	85	102	106	123	140	144	161	9	26	30	47	64
65	69	86	103	107	124	141	145	162	10	14	31	48
49	53	70	87	104	108	125	142	146	163	11	15	32
33	50	54	71	88	92	109	126	143	147	164	12	16
17	34	51	55	72	89	93	110	127	131	148	165	13

Plan 12.8 (Continued) 13 × 13 balanced lattice square

Square VII

1	16	31	46	61	76	91	93	108	123	138	153	168
169	2	17	32	47	62	77	79	94	109	124	139	154
155	157	3	18	33	48	63	78	80	95	110	125	140
141	156	158	4	19	34	49	64	66	81	96	111	126
127	142	144	159	5	20	35	50	65	67	82	97	112
113	128	143	145	160	6	21	36	51	53	68	83	98
99	114	129	131	146	161	7	22	37	52	54	69	84
85	100	115	130	132	147	162	8	23	38	40	55	70
71	86	101	116	118	133	148	163	9	24	39	41	56
57	72	87	102	117	119	134	149	164	10	25	27	42
43	58	73	88	103	105	120	135	150	165	11	26	28
29	44	59	74	89	104	106	121	136	151	166	12	14
15	30	45	60	75	90	92	107	122	137	152	167	13

CHAPTER 13

Incomplete Latin Squares

13.1 Description

13.11 Youden Squares. These designs, which are constructed by a rearrangement of certain of the balanced incomplete blocks, possess the characteristic "double control" of the latin square, without the restriction that the number of replicates must equal the number of treatments. A latin square with 13 treatments is rarely used, because it necessitates 13 replications. There are, however, incomplete latin squares for 13 treatments in either 4 or 9 replicates. Every treatment occurs once in a column (replication), and every pair of treatments appears together an equal number of times in the same block. Column differences are eliminated automatically from the treatment comparisons, while block differences may be removed by adjusting the treatment mean yields in the same way as with balanced incomplete blocks. Most of the designs were developed by Youden (13.1, 13.2), whose name is commonly associated with them; previously Yates (13.3) had drawn attention to the group of designs in which the number of replicates is one less than the number of treatments.

Plans are presented for three series of designs. The first are the incomplete latin squares themselves, formed by omitting certain columns in an ordinary latin square. For extra replication, the basic design may be repeated. A second series is constructed by adding an incomplete latin square to a complete latin square. This combination is sometimes useful when the number of treatments is small. Recently, a third series, called extended Youden squares (13.1a), has been developed by putting together two or more parts of latin squares in such a way that the whole design has the balanced incomplete blocks property, although the individual parts do not. Table 13.2 at the end of this chapter gives an index to the designs for numbers of treatments up to 91.

Some further possibilities are discussed briefly without giving detailed plans. These include a latin square with a row and a column added (section 13.34a), chain block designs with two-way elimination of heterogeneity (section 13.35a), and partially balanced designs (section 13.4a).

An interesting application of the incomplete latin square for the control of plant variability was made by Youden (13.1) in greenhouse experiments on tobacco-mosaic virus. The experimental unit was a single leaf, and the data consisted of the number of lesions produced per leaf by rubbing the leaf with a solution which contained the virus. The numbers of lesions had been found to depend much more on inherent qualities of the plant than on the position of the plant on the greenhouse bench. Consequently, each block of the design was a single plant, so that the large differences in responsiveness which existed among plants did not contribute to the experimental errors. The columns were the positions, from top to bottom, of the five leaves which were used on each plant. That is, the first replication contained the top leaf of every plant. Since there was a fairly consistent gradient in responsiveness down each plant, this control also proved effective.

As in the example above, when laying out an incomplete latin square the general principle is to group the units so that differences among blocks and differences among columns represent the major sources of variation that are known or suspected.

13.12 Randomization. In plans 13.1 to 13.26 and plans 13.1a to 13.5a, the steps are (1) rearrange the blocks of the plan at random, and (2) rearrange the replications of the plan at random.

When a plan is repeated for additional replication, blocks and columns are randomized separately within each *repetition*. The repetitions may be kept separate.

In plans 13.6a to 13.17a, where the plan consists of two or more incomplete latin squares, randomize the blocks and columns separately within each incomplete square.

13.2 Statistical Analysis

In experiments with more than 10 treatments, the numbers of degrees of freedom for blocks are large enough to allow the use of inter-block information. The appropriate analysis is described in the next section. For small experiments, where inter-block information should be ignored, the analysis is given in section 13.22. The recommended method of analysis for any experiment is indicated both in table 13.2 and in the plans by the "type."

13.21 Type I. Analysis with Recovery of Inter-block Information. The steps are as follows (t = number of treatments, k = number of units per block = r).

1. Calculate the treatment totals T, the column totals, the block totals, and the grand total G.

2. For each treatment, calculate the total B_t of all the blocks which contain the treatment. Place these quantities in a column next to the column of treatment totals. Then form a third column of the quantities:

$$W = (t - k)T - (t - 1)B_t + (k - 1)G$$

The W's should sum to zero.

3. The separation of degrees of freedom in the analysis of variance is as follows.

	d.f.	m.s.
Columns (replications)	$(k - 1)$	
Blocks (adjusted)	$(t - 1)$	E_b
Treatments (unadjusted)	$(t - 1)$	
Error	$(k - 2)(t - 1)$	E_e
Total	$(tk - 1)$	

The total s.s. and the sums of squares for columns and treatments are found by the usual methods. The sum of squares for blocks (adjusted for treatments) is the sum of squares of the W's, divided by $kt(t - k)(k - 1)$. The error s.s. is obtained by subtraction.

4. Calculate the factor

$$\mu = \frac{(E_b - E_e)}{t(k - 1)E_b}$$

The adjusted *total* for any treatment is

$$Y = T + \mu W$$

Should E_b be less than E_e, μ is taken as zero and no adjustments are made to the treatment totals. The adjusted treatment *means* are obtained on dividing each Y by r. The estimated error variance of the difference between two adjusted treatment means is

$$\frac{2E_e}{r}[1 + (t - k)\mu]$$

13.21a Type V. Analysis with Recovery of Inter-block Information. These are the new designs formed by putting several parts of squares together so that the whole design is balanced. The analysis proceeds as in section 13.21 except for the changes noted below.

3. The separation of degrees of freedom in the analysis of variance is as follows:

	d.f.	m.s.
Columns (replications)	$(r-1)$	
Blocks (adjusted)	$(b-1)$	E_b
Treatments (unadjusted)	$(t-1)$	
Error	$(t-1)(r-1)-(b-1)$	E_e
Total	$(tr-1)$	

The sum of squares for blocks (adjusted) is obtained by a roundabout method.

$$\text{Blocks (adj.)} = \text{Blocks (unadj.)} + \text{Treatments (adj.)} - \text{Treatments (unadj.)}$$

The sums of squares for blocks (unadj.) and treatments (unadj.) are computed in the usual way. To find the sum of squares for treatments (adj.), compute for each treatment the quantity

$$Q = kT - B_t$$

Then

$$\text{Treatments (adj.)} = \frac{1}{tk\lambda} \sum Q^2$$

4. The weighting factor is

$$\mu = \frac{(b-1)(E_b - E_e)}{t(k-1)(b-1)E_b + (t-k)(b-t)E_e}$$

This completes the necessary changes. Note: if it is desired to analyze any type V design *without* recovery of inter-block information, follow the method in section 13.22, except that the partition of degrees of freedom is as given in this section.

13.22 Type II. Analysis without Recovery of Inter-block Information. The analysis proceeds as follows:

1. Calculate the column totals, the block totals, the treatment totals T, and the grand total G.

2. For each treatment, obtain the quantity

$$Q = kT - B_t$$

where B_t is the total of all the blocks in which the treatment appears. The quantities Q should sum to zero.

3. The degrees of freedom in the analysis of variance are partitioned as in section 13.21. (For plans 13.6a and 13.7a, use the partition of de-

grees of freedom given in section 13.21a.) In this case the blocks s.s. is calculated without eliminating treatment effects, while the treatments s.s. is adjusted for block differences. The total s.s. and the sums of squares for blocks are computed by the standard procedure. The treatments s.s. (adjusted) is the sum of the squares of the Q's, divided by $tk\lambda$. The error s.s. is found by subtraction.

4. In order to obtain any treatment mean, adjusted for block differences, divide the corresponding Q by $t\lambda$ and add to the quotient the mean for the whole experiment.

The estimated error variance of the difference between two adjusted treatment means is

$$\frac{2E_e}{r} \cdot \frac{k(t-1)}{t(k-1)}$$

Numerical examples of this analysis for $t = 7$, $r = 3$, and $t = 21$, $r = 5$ are given in reference (13.1), and an example for $t = 6$, $r = 5$ in (13.3).

13.23 Type I*a*. Repetitions of Type I.

Let n be the number of replications in the basic plan, which is used p times in an experiment, so that the total number of replications $r = np$.

The analysis presented here may be used with repetitions of *any* of the type I and II designs, since even with small numbers of treatments the repetition provides sufficient degrees of freedom for estimating the inter-block variation. If, however, the efficiency factor exceeds 95%, it is scarcely worth while to utilize inter-block information.

The changes necessary in the procedure of section 13.21 are outlined below, the numbers referring to the steps in that section.

1. Unchanged. Corresponding to any block, there are $(p - 1)$ other blocks which contain the same set of treatments. The block totals should be arranged in a table (table A, say) with t rows and p columns.
2. Unchanged. Note that the calculation of the quantities B_t is expedited by the use of the row totals of table A.
3. The analysis of variance is as follows.

	d.f.	m.s.
Columns (replications)	$(kp - 1)$	
Blocks		
Component (*a*)	$(p-1)(t-1)$	
Component (*b*)	$(t - 1)$	
Block total	$p(t - 1)$	E_b
Treatments	$(t - 1)$	
Error	$(t-1)(pk - p - 1)$	E_e
Total	$(pkt - 1)$	

In this analysis, the $(p - 1)$ degrees of freedom among repetitions have been ascribed to the columns, so that the blocks s.s. is actually a sum of squares among blocks *within repetitions*.

Component (*a*) of the sum of squares for blocks consists of comparisons among blocks which contain the same set of treatments and is the interaction s.s. for table *A*.

Component (*b*) is the sum of squares of the *W*'s, with a divisor $pkt(t - k)(k - 1)$.

4. Unchanged except that

$$\mu = \frac{p(E_b - E_e)}{pt(k - 1)E_b - (t - k)(p - 1)E_e}$$

The estimated error variance of the difference between two adjusted treatment means is

$$\frac{2E_e}{r}[1 + (t - n)\mu]$$

13.24 Type II*a*. Repetitions of Type II. As we have pointed out, in designs where the efficiency factor is high, inter-block information can be neglected. The changes required from section 13.22 should present no difficulty. The degrees of freedom subdivide as in section 13.23. The $p(t - 1)$ degrees of freedom among blocks within repetitions are calculated in the usual way, each repetition furnishing $(t - 1)$ degrees of freedom. Notice that the divisors for the treatments s.s. and for the quantities Q must be increased by the factor p.

13.25 Missing Data. The formula for inserting an estimate in place of a missing observation is obtained by minimizing the intra-block error. This is the correct estimate for the smaller experiments where inter-block information is ignored. For reasons given in section 10.13, this estimate will also be used to provide an approximate solution when inter-block information is recovered. The formula is quite similar to that for balanced incomplete blocks, with the addition of an extra term involving the column (or replication) total.

Let C, B, and T be the totals of the column, block, and treatment that contain the missing value, and let G be the grand total. Further, as in the statistical analysis, let B_t be the total of all blocks in which the treatment with the missing value appears (there will, of course, be r such blocks). Finally,

T' = total (over all replicates) of all *other* treatments that appear in the block which has the missing value

B_t' = total of the B_t values for all *other* treatments that appear in the block with the missing value

The estimate x of the missing value is

$$x = \frac{\lambda[rC + tB + (t-1)T - G] - rT' - (r-1)B_t + B_t'}{r(r-1)(r-2)}$$

The symbol λ is given for each design in table 13.2. Care must be taken not to confuse the block totals with the column totals.

If the experiment contains p repetitions, the formula is

$$x = \frac{p[pk\lambda C + pt\lambda B + k(k-1)T - p\lambda R - kT' - (k-1)B_t + B_t']}{r(k-1)(pk - p - 1)}$$

where R is the total of the repetition in which the missing value occurs, and λ is as in table 13.2.

13.3 Other Designs for Small Numbers of Treatments

13.31 Description. When the number of treatments is small, it is sometimes useful to have a design of the "latin square" type in which the number of replicates exceeds the number of treatments. Some designs can be obtained by repetition of an ordinary latin square or of a Youden square. Additional plans can be constructed by adding a Youden square to a latin square.

Plans are available for most numbers of replicates up to 10. Actually, plans can be made for any number of replicates up to 10; those shown here have been selected for ease of analysis.

13.32 Type III. Statistical Analysis When $k = it - 1$. In this case the size of block, which equals the number of replicates, is 1 less than some multiple (i) of the number of treatments. Since these designs have not been discussed in the literature, an example is given from uniformity data. Table 13.1 shows the tomato yields in pounds of 28 single-row plots, Hartman and Stair's data (13.4), on which the design for 4 treatments in 7 replicates has been superimposed.

TABLE 13.1 Incomplete latin square for 4 treatments in 7 replicates

Treatment symbols and yields of tomatoes (pounds)

Block	Column I	II	III	IV	V	VI	VII	Total
1	2	2	4	4	3	3	1	
	50	72	83	82	76	89	74	526
2	1	3	3	1	4	2	4	
	40	59	71	91	59	73	52	445
3	4	1	2	3	1	4	2	
	43	57	58	98	54	71	51	432
4	3	4	1	2	2	1	3	
	48	54	74	97	75	75	54	477
Totals	181	242	286	368	264	308	231	1880

Block	T	$Q = kT + B'$	$Q \div 48$	Adjusted means (subtract 11.2)
1	465	3,781	78.8	67.6
2	476	3,777	78.7	67.5
3	495	3,897	81.2	70.0
4	444	3,585	74.7	63.5
Check totals or mean	1,880	15,040		67.2

Since the plots measured 6 feet × 24 feet, each column (replication) is compact, being 24 feet square. There are obviously substantial differences among replicates. The blocks, although very oblong, may also exhibit differences in yield, because each block is a separate row of plants.

The efficiency factors of these designs are all very high, so that interblock information is ignored in the analysis.

The steps in the analysis are as follows:

1. Calculate the column totals, the block totals, the treatment totals T, and the grand total G.

2. A property of these designs is that each treatment is replicated less in one block than in the other blocks. Thus, in the example, treatment (1) appears only once in the first block, but twice in all other blocks. Similarly, treatment (2) is deficient in block 2 and so on. For each treatment calculate the quantity

$$Q = kT + B'$$

where B' is the total for the block in which the treatment is deficient. For instance

$$Q_1 = 7 \times 465 + 526 = 3781$$

As a check, the quantities Q should sum to $(k + 1)G$.

3. All sums of squares in the analysis of variance are obtained in the usual way except that for treatments, which is given by the sum of squares of deviations of the Q's, divided by $k(k^2 - 1)$. In this case we have

$$\frac{(3781)^2 + (3777)^2 + (3897)^2 + (3585)^2 - \frac{1}{4}(15{,}040)^2}{336} = 149$$

The analysis of variance is shown below. The degrees of freedom subdivide as in section 13.21.

	d.f.	s.s.	m.s.
Columns (replications)	6	5387	897.8
Blocks	3	750	250.0
Treatments	3	149	50.0
Error	15	767	51.1
Total	27	7053	

The elimination of block differences has substantially reduced the error m.s.

4. To obtain the adjusted treatment mean yields, we first divide each Q by $(k^2 - 1)$, in this case 48. From the resulting quantities we subtract $G/tk(k - 1)$, or $1880/168 = 11.2$. As a further check, the mean of the adjusted treatment means should equal the mean yield of the whole experiment (67.2). The efficiency factor is $(k^2 - 1)/k^2$, and the estimated error variance of the difference between two adjusted treatment means is

$$\frac{2E_e r}{r^2 - 1}$$

13.33 Type IV. Statistical Analysis When $k = it + 1$. The analysis closely resembles that of section 13.32. For each treatment there is one block in which the treatment has extra replication. Thus, in plan 13.17 for $t = 3$, $r = 7$, treatment (1) appears 3 times in the first block but only twice in any other block. The only changes in the computing instructions of section 13.32 are (i) $Q = kT - B'$, where B' is the total of the block in which the treatment has extra replication, (ii) the Q's sum to $(k - 1)G$, and (iii) for the adjusted treatment means, divide Q by $(k^2 - 1)$, and *add* to the quotient the quantity $G/tk(k + 1)$. The change in the divisor of G should be noted.

The error variance of the difference between two adjusted means remains as at the end of section 13.32 since $k = r$ for these plans.

13.34a Other Modifications of the Latin Square. There are other modifications of the latin square for which the analysis is slightly more complicated. These designs may occasionally supply a need in situations where variations in two directions at right angles are encountered.

Latin square with a row and column added. Table 13.1a shows an example (not randomized) with 4 treatments. Treatments (1), (2), and (3) are replicated 6 times, but treatment (4) has to be replicated 7 times.

TABLE 13.1a A 4 × 4 LATIN SQUARE WITH A ROW AND COLUMN ADDED

1	2	3	4	1		d.f.
2	3	4	1	2	Rows	4
3	4	1	2	3	Columns	4
4	1	2	3	4	Treatments	3
1	2	3	4	4	Error	13
					Total	24

Latin square with a column added and a row omitted. In this design for 5 treatments in a 6 × 4 rectangular array, the first 4 of the treatments have 5 replicates each; however treatment (5) has only 4 replicates.

TABLE 13.2a A 5 × 5 LATIN SQUARE WITH A COLUMN ADDED AND A ROW OMITTED

1	2	3	4	5	1		d.f.
2	3	4	5	1	2	Rows	3
3	4	5	1	2	3	Columns	5
4	5	1	2	3	4	Treatments	4
						Error	11
						Total	23

Latin square with a row and column omitted. This plan accommodates 6 treatments in a 5 × 5 square array. One treatment, in this case (5),

TABLE 13.3a A 6 × 6 LATIN SQUARE WITH A ROW AND COLUMN OMITTED

1	2	3	4	5		d.f.
2	3	4	5	6	Rows	4
3	4	5	6	1	Columns	4
4	5	6	1	2	Treatments	5
5	6	1	2	3	Error	11
					Total	24

has one extra replicate. The statistical analysis for the first two designs above is given by Pearce (13.2a), who proposed them for certain situations in horticultural research. For example, the trees available for an experiment may lie in a 6 × 4 rectangular array in which there is a likelihood of gradients along both rows and columns. A new experiment may be started on a site which previously held an old experiment in randomized blocks. Since some residual effects of the old treatments may persist, a design of the latin square type suggests itself, with the old blocks as rows and the old treatments as columns. Yates (13.3a) describes the analysis for the designs illustrated in table 13.3a.

13.35a Generalized Chain Block Designs. Mandel (13.4a) has shown that a few of the chain block designs (section 11.7a) can be written so as to adjust for differences between rows and columns. Table 13.4a shows a design with this property for 8 treatments. If the columns are

TABLE 13.4a A GENERALIZED CHAIN BLOCK DESIGN

	Blocks (columns)			
Rows	1	2	3	4
1	*A*	*C*	*E*	*G*
2	*B*	*D*	*F*	*H*
3	*C*	*F*	*G*	*B*
4	*D*	*E*	*H*	*A*

regarded as blocks, this is a typical chain block design. Treatments (*C*, *D*) are the link between blocks 1 and 2, (*E*, *F*) between blocks 2 and 3, (*G*, *H*) between blocks 3 and 4, and (*A*, *B*) between blocks 4 and 1. Now examine the arrangement from the point of view of the rows. Treatments (*C*, *G*) form a link between rows 1 and 3, (*B*, *F*) between 3 and 2, (*D*, *H*) between 2 and 4, and finally (*A*, *E*) between 4 and 1. Thus the chain in the rows goes 1, 3, 2, 4, 1. The arrangement is a chain block design for both rows and columns, and treatment means may be adjusted to remove the effects of the rows and columns in which they lie.

This design was used to compare the wearing qualities of 8 tires. The columns represented the 4 positions which a tire can occupy on the car, and the rows were four different test runs by the same car. The design provides only 2 d.f. for error, but this test was part of a larger one involving 32 tires.

Mandel gives designs of this type for 12, 16, 18, 20, 24, and 30 treatments, with instructions for analysis.

13.4a Partially Balanced Designs

13.41a Description. Many of the partially balanced designs recently published by Bose *et al.* (13.5a) can be arranged so as to allow the elimination of variation in two directions at right angles. Table 13.5a shows this arrangement for the design with $t = 15$, $k = 4$, that was used in section 11.63a to illustrate the analysis of the results of a *pbib* experiment. The rows represent the incomplete blocks and the columns are replications.

TABLE 13.5a A *pbib* DESIGN IN INCOMPLETE LATIN SQUARE FORM

$t = 15, k = 4$

Block	Replications I	II	III	IV
(1)	1	3	4	12
(2)	2	4	5	13
(3)	3	5	6	14
(4)	4	6	7	15
(5)	5	7	8	1
(6)	6	8	9	2
(7)	7	9	10	3
(8)	8	10	11	4
(9)	9	11	12	5
(10)	10	12	13	6
(11)	11	13	14	7
(12)	12	14	15	8
(13)	13	15	1	9
(14)	14	1	2	10
(15)	15	2	3	11

Some of the *pbib* designs can be set out so that two or more columns form a replication or group of replications, as illustrated by the plan in table 13.6a for 8 treatments in incomplete blocks of size 6. This design

TABLE 13.6a A *pbib* DESIGN IN WHICH A PAIR OF COLUMNS FORMS A REPLICATION

$t = 8, k = 6$

Block	Replication I		II		III	
(1)	1	5	2	7	3	6
(2)	2	6	1	5	4	8
(3)	3	7	4	8	1	5
(4)	4	8	3	6	2	7

is of the "group divisible" type, with the following association scheme.

1	5
2	6
3	7
4	8

Treatments in the same row of the association scheme (like 4 and 8) occur together in three of the blocks; treatments not in the same row occur together in two blocks.

Finally, in some designs a group of rows (incomplete blocks) forms a replication, while at the same time a group of columns also forms a replication. Table 13.7a for $t = 12$, $k = 6$, is an example. This design

TABLE 13.7a A *pbib* DESIGN IN WHICH PAIRS OF ROWS AND PAIRS OF COLUMNS FORM REPLICATES

$t = 12, k = 6$

		Replication					
Reps.	Block	I		II		III	
I	(1)	1	2	5	6	9	10
	(2)	3	4	7	8	11	12
II	(3)	9	11	1	3	5	7
	(4)	10	12	2	4	6	8
III	(5)	5	8	9	12	1	4
	(6)	6	7	10	11	2	3

is also "group divisible," treatments in the same row of the association scheme appearing together in three blocks, and treatments in different rows appearing once in a block. The association scheme is

1	5	9
2	6	10
3	7	11
4	8	12

An index to the available *pbib* designs of the types illustrated above has been given in table 9.2a, p. 392. Randomization and statistical analysis are described by Bose (13.5a).

REFERENCES

13.1 Youden, W. J. Use of incomplete block replications in estimating tobacco-mosaic virus. *Contr. Boyce Thompson Inst.* 9, 41–48, 1937.

13.2 Youden, W. J. Experimental designs to increase accuracy of greenhouse studies. *Contr. Boyce Thompson Inst.* 11, 219–228, 1940.

13.3 YATES, F. Incomplete latin squares. *Jour. Agr. Sci.* 26, 301–315, 1936.
13.4 HARTMAN, J. D., and STAIR, E. C. Field plot technique studies with tomatoes. *Proc. Amer. Soc. Hort. Sci.* 41, 315–320, 1942.
13.1a RUPP, M. K. The complete solutions of the balanced incomplete block designs with ten or fewer replications. *Nat. Bur. Standards unpublished report.* 1952.
13.2a PEARCE, S. C. Some new designs of latin square type. *Jour. Roy. Stat. Soc. B*, 14, 101–106, 1952.
13.3a YATES, F. Incomplete latin squares. *Jour. Agr. Sci.* 26, 301–315, 1936.
13.4a MANDEL, J. Chain block designs with two-way elimination of heterogeneity. *Biometrics* 10, 251–272, 1954.
13.5a BOSE, R. C., CLATWORTHY, W. H., and SHRIKHANDE, S. S. Tables of partially balanced designs with two associate classes. *North Carolina Agr. Exp. Sta. Tech. Bull.* 107, 1954.

TABLE 13.2 INDEX TO PLANS

(Combines tables 13.1 and 13.2 of first edition)

t	*k*	*r*	*b*	λ	*E*	Plan	Type †
3	3	6	6	6	1.00	2LS	
	3	9	9	9	1.00	3LS	
	5	5	3	5	.96	13.16	III
	5	10	6	10	.96	13.16	IIa
	7	7	3	7	.98	13.17	IV
	8	8	3	8	.98	13.18	III
	10	10	3	10	.99	13.19	IV
4	3	3	4	2	.89	*	II
	3	6	8	4	.89	**	Ia
	3	9	12	6	.89	**	Ia
	4	8	8	8	1.00	2LS	
	5	5	4	5	.96	13.20	IV
	5	10	8	10	.96	13.20	IIa
	7	7	4	7	.98	13.21	III
	9	9	4	9	.99	13.22	IV
5	2	4	10	1	.67	13.6a	II
	3	6	10	3		13.7a	II
	4	4	5	3	.94	*	II
	4	8	10	6	.94	**	Ia
	5	10	10	10	1.00	2LS	
	6	6	5	6	.97	13.23	IV
	9	9	5	9	.99	13.24	III
6	5	5	6	4	.96	*	II
	5	10	12	8	.96	**	Ia
	7	7	6	7	.98	13.25	IV
7	2	6	21	1	.58	13.8a	V
	3	3	7	1	.78	13.1	II
	3	9	21	3	.78	13.1	Ia
	4	4	7	2	.88	13.2	II
	4	8	14	4	.88	13.2	Ia

TABLE 13.2 Index to plans (*Continued*)

t	k	r	b	λ	E	Plan	Type †
7	6	6	7	5	.97	*	II
	8	8	7	8	.98	13.26	IV
8	7	7	8	6	.98	*	II
9	2	8	36	1	.56	13.9a	V
	4	8	18	3	.84	13.10a	V
	5	10	18	5	.90	13.11a	V
	8	8	9	7	.98	*	II
10	3	9	30	2	.74	13.12a	V
	9	9	10	8	.99	*	II
11	2	10	55	1	.55	13.13a	V
	5	5	11	2	.88	13.3	I
	6	6	11	3	.92	13.4	I
	10	10	11	9	.99	*	II
13	3	6	26	1	.72	13.14a	V
	4	4	13	1	.81	13.5	I
	9	9	13	6	.95	13.6	I
15	7	7	15	3	.92	13.7	I
	8	8	15	4	.94	13.8	I
16	6	6	16	2	.89	13.9	I
	10	10	16	6	.96	13.10	I
19	3	9	57	1	.70	13.15a	V
	9	9	19	4	.94	13.11	I
	10	10	19	5	.95	13.12	I
21	5	5	21	1	.84	13.13	I
25	4	8	50	1	.78	13.16a	V
	9	9	25	3	.93	13.1a	I
31	6	6	31	1	.86	13.14	I
	10	10	31	3	.93	13.2a	I
37	9	9	37	2	.91	13.15	I
41	5	10	82	1	.82	13.17a	V
57	8	8	57	1	.89	13.3a	I
73	9	9	73	1	.90	13.4a	I
91	10	10	91	1	.91	13.5a	I

* Constructed from a $t \times t$ latin square by omission of the last column.

** By repetition of the plan for $r = t - 1$, which is constructed by taking a $t \times t$ latin square and omitting the last column.

† This refers to the method of analysis. For types I, Ia, II, IIa, and V, see section 13.2; for types III and IV, see section 13.3.

PLANS

Plan 13.1 $t = 7, k = 3, r = 3, b = 7, \lambda = 1, E = .78$, Type II

Block	Reps. I	II	III
(1)	7	1	3
(2)	1	2	4
(3)	2	3	5
(4)	3	4	6
(5)	4	5	7
(6)	5	6	1
(7)	6	7	2

Plan 13.2 $t = 7, k = 4, r = 4, b = 7, \lambda = 2, E = .88$, Type II

Block	Reps. I	II	III	IV
(1)	3	5	6	7
(2)	4	6	7	1
(3)	5	7	1	2
(4)	6	1	2	3
(5)	7	2	3	4
(6)	1	3	4	5
(7)	2	4	5	6

Plan 13.3 $t = 11, k = 5, r = 5, b = 11, \lambda = 2, E = .88$, Type I

Block	Reps. I	II	III	IV	V
(1)	1	2	3	4	5
(2)	7	1	6	10	3
(3)	9	8	1	6	2
(4)	11	9	7	1	4
(5)	10	11	5	8	1
(6)	8	7	2	3	11
(7)	2	6	4	11	10
(8)	6	3	11	5	9
(9)	3	4	10	9	8
(10)	5	10	9	2	7
(11)	4	5	8	7	6

Plan 13.4 $t = 11, k = 6, r = 6, b = 11, \lambda = 3, E = .92$, Type I

Block	Reps. I	II	III	IV	V	VI
(1)	6	7	8	9	10	11
(2)	5	8	4	11	2	9
(3)	4	5	7	3	11	10
(4)	3	10	2	6	5	8
(5)	2	3	9	7	4	6
(6)	1	6	10	4	9	5
(7)	9	1	3	5	8	7
(8)	8	2	1	10	7	4
(9)	7	11	5	1	6	2
(10)	11	4	6	8	1	3
(11)	10	9	11	2	3	1

Plan 13.5 $t = 13, k = 4, r = 4, b = 13, \lambda = 1, E = .81$, Type I

Block	Reps. I	II	III	IV
(1)	13	1	3	9
(2)	1	2	4	10
(3)	2	3	5	11
(4)	3	4	6	12
(5)	4	5	7	13
(6)	5	6	8	1
(7)	6	7	9	2
(8)	7	8	10	3
(9)	8	9	11	4
(10)	9	10	12	5
(11)	10	11	13	6
(12)	11	12	1	7
(13)	12	13	2	8

Plan 13.6 $t = 13, k = 9, r = 9, b = 13, \lambda = 6, E = .95$, Type I

	Reps.								
Block	I	II	III	IV	V	VI	VII	VIII	IX
(1)	2	5	6	7	9	10	11	12	13
(2)	3	6	7	8	10	11	12	13	1
(3)	4	7	8	9	11	12	13	1	2
(4)	5	8	9	10	12	13	1	2	3
(5)	6	9	10	11	13	1	2	3	4
(6)	7	10	11	12	1	2	3	4	5
(7)	8	11	12	13	2	3	4	5	6
(8)	9	12	13	1	3	4	5	6	7
(9)	10	13	1	2	4	5	6	7	8
(10)	11	1	2	3	5	6	7	8	9
(11)	12	2	3	4	6	7	8	9	10
(12)	13	3	4	5	7	8	9	10	11
(13)	1	4	5	6	8	9	10	11	12

Plan 13.7 $t = 15, k = 7, r = 7, b = 15, \lambda = 3, E = .92$, Type I

	Reps.						
Block	I	II	III	IV	V	VI	VII
(1)	13	8	12	6	7	1	9
(2)	5	14	10	7	12	2	8
(3)	15	12	11	5	8	3	6
(4)	12	11	6	9	2	4	14
(5)	4	5	8	1	14	9	15
(6)	11	9	7	2	13	15	5
(7)	1	2	3	4	5	6	7
(8)	2	3	1	13	15	14	12
(9)	8	6	4	15	10	13	2
(10)	10	4	5	11	1	12	13
(11)	9	13	14	10	6	5	3
(12)	14	7	13	3	4	8	11
(13)	7	15	9	12	3	10	4
(14)	3	1	2	8	9	11	10
(15)	6	10	15	14	11	7	1

Plan 13.8 $t = 15, k = 8, r = 8, b = 15, \lambda = 4, E = .94$, Type I

Reps.

Block	I	II	III	IV	V	VI	VII	VIII
(1)	11	4	2	5	10	3	14	15
(2)	4	1	3	15	13	11	6	9
(3)	9	2	14	4	7	1	10	13
(4)	15	3	1	10	8	13	7	5
(5)	7	13	10	12	11	2	3	6
(6)	6	10	12	1	4	14	8	3
(7)	12	9	13	14	15	10	11	8
(8)	10	11	9	7	6	8	5	4
(9)	5	14	7	3	1	9	12	11
(10)	8	15	6	2	3	7	9	14
(11)	1	7	11	8	2	4	15	12
(12)	2	6	15	9	5	12	1	10
(13)	13	8	5	11	14	6	2	1
(14)	14	5	4	6	12	15	13	7
(15)	3	12	8	13	9	5	4	2

Plan 13.9 $t = 16, k = 6, r = 6, b = 16, \lambda = 2, E = .89$, Type I

Reps.

Block	I	II	III	IV	V	VI
(1)	1	2	3	4	5	6
(2)	2	7	8	9	10	1
(3)	3	1	13	7	11	12
(4)	4	8	1	11	14	15
(5)	5	12	14	1	16	9
(6)	6	10	15	13	1	16
(7)	7	14	2	16	15	3
(8)	8	16	12	2	4	13
(9)	9	15	11	5	13	2
(10)	10	11	6	12	2	14
(11)	11	4	16	3	9	10
(12)	12	3	10	15	8	5
(13)	13	6	9	14	3	8
(14)	14	13	5	10	7	4
(15)	15	9	4	6	12	7
(16)	16	5	7	8	6	11

Plan 13.10 $t = 16, k = 10, r = 10, b = 16, \lambda = 6, E = .96$, Type I

Block	Reps. I	II	III	IV	V	VI	VII	VIII	IX	X
(1)	8	7	9	10	11	12	13	14	15	16
(2)	3	4	5	13	16	11	12	6	14	15
(3)	9	2	4	5	6	8	10	15	16	14
(4)	2	6	3	7	5	10	9	16	12	13
(5)	6	8	2	3	4	7	15	10	13	11
(6)	4	5	14	8	3	9	7	2	11	12
(7)	1	10	11	4	13	5	6	12	8	9
(8)	5	1	15	6	14	3	11	9	7	10
(9)	16	3	1	14	12	6	4	8	10	7
(10)	7	13	16	1	15	4	3	5	9	8
(11)	12	14	7	15	1	2	8	13	6	5
(12)	14	11	13	16	9	1	2	7	4	6
(13)	10	15	12	2	7	16	1	11	5	4
(14)	11	12	6	9	8	15	16	1	2	3
(15)	13	16	8	11	10	14	5	3	1	2
(16)	15	9	10	12	2	13	14	4	3	1

Plan 13.11 $t = 19, k = 9, r = 9, b = 19, \lambda = 4, E = .94$, Type I

Block	Reps. I	II	III	IV	V	VI	VII	VIII	IX
(1)	1	2	3	4	5	6	7	8	9
(2)	14	1	2	3	4	12	11	10	13
(3)	17	16	1	2	15	10	6	5	11
(4)	12	13	16	1	2	7	18	15	8
(5)	9	10	12	16	1	3	17	19	7
(6)	8	11	13	19	17	1	3	6	18
(7)	7	18	11	14	16	19	1	4	5
(8)	6	9	17	12	14	18	15	1	4
(9)	5	8	9	10	13	15	19	14	1
(10)	13	5	14	9	18	2	16	3	17
(11)	10	6	7	15	19	14	2	18	3
(12)	18	17	5	8	12	4	10	2	19
(13)	19	4	15	7	9	17	13	11	2
(14)	2	19	6	11	8	9	14	16	12
(15)	3	15	10	18	11	8	4	9	16
(16)	4	3	19	5	6	16	12	13	15
(17)	15	7	8	17	3	11	5	12	14
(18)	16	14	4	6	7	13	8	17	10
(19)	11	12	18	13	10	5	9	7	6

Plan 13.12 $t = 19, k = 10, r = 10, b = 19, \lambda = 5, E = .95$, **Type I**

Block	I	II	III	IV	V	VI	VII	VIII	IX	X
					Reps.					
(1)	10	11	12	13	14	15	16	17	18	19
(2)	19	5	6	7	8	9	15	16	17	18
(3)	18	19	3	4	7	8	9	12	13	14
(4)	14	17	19	11	4	5	6	9	10	3
(5)	13	2	15	14	18	4	5	6	11	8
(6)	12	14	2	10	15	7	4	5	9	16
(7)	15	13	17	2	12	6	8	10	3	9
(8)	16	10	11	19	2	3	7	13	8	5
(9)	11	16	18	17	6	2	3	4	12	7
(10)	1	8	10	15	11	19	12	7	4	6
(11)	17	1	8	9	16	12	13	11	5	4
(12)	6	15	1	16	9	14	11	3	7	13
(13)	5	6	14	1	3	10	18	8	16	12
(14)	4	7	5	3	1	13	17	18	15	10
(15)	7	12	13	5	19	1	2	14	6	17
(16)	8	9	7	18	10	17	1	2	14	11
(17)	9	4	16	6	13	18	10	1	19	2
(18)	3	18	9	12	5	11	19	15	2	1
(19)	2	3	4	8	17	16	14	19	1	15

Plan 13.13 $t = 21, k = 5, r = 5, b = 21, \lambda = 1, E = .84$, Type I

Block	Reps. I	II	III	IV	V
(1)	21	1	4	14	16
(2)	1	2	5	15	17
(3)	2	3	6	16	18
(4)	3	4	7	17	19
(5)	4	5	8	18	20
(6)	5	6	9	19	21
(7)	6	7	10	20	1
(8)	7	8	11	21	2
(9)	8	9	12	1	3
(10)	9	10	13	2	4
(11)	10	11	14	3	5
(12)	11	12	15	4	6
(13)	12	13	16	5	7
(14)	13	14	17	6	8
(15)	14	15	18	7	9
(16)	15	16	19	8	10
(17)	16	17	20	9	11
(18)	17	18	21	10	12
(19)	18	19	1	11	13
(20)	19	20	2	12	14
(21)	20	21	3	13	15

Plan 13.1a $t = 25, k = 9, r = 9, b = 25, \lambda = 3, E = .93$, Type I

Block	Reps. I	II	III	IV	V	VI	VII	VIII	IX
(1)	1	2	3	4	5	6	7	8	9
(2)	2	4	9	10	24	17	15	22	12
(3)	3	24	8	23	18	21	13	4	10
(4)	4	22	25	8	20	12	11	3	19
(5)	5	15	17	18	8	11	2	13	20
(6)	6	8	12	13	1	14	24	25	15
(7)	7	16	5	22	3	10	25	15	13
(8)	8	10	11	16	6	22	23	1	17
(9)	9	13	20	5	12	23	1	21	22
(10)	10	19	14	12	16	2	8	5	21
(11)	11	18	19	24	10	1	5	9	25
(12)	12	6	10	25	7	18	20	2	23
(13)	13	11	4	9	23	25	14	16	2
(14)	14	3	7	17	11	5	12	23	24
(15)	15	20	21	1	14	7	10	11	4
(16)	16	17	18	7	13	19	4	12	1
(17)	17	14	13	20	19	3	9	10	6
(18)	18	9	22	14	25	8	21	17	7
(19)	19	7	23	15	9	20	16	24	8
(20)	20	5	16	6	4	24	22	14	18
(21)	21	12	6	11	15	9	3	18	16
(22)	22	21	24	19	2	13	6	7	11
(23)	23	1	2	3	22	15	18	19	14
(24)	24	25	1	2	21	16	17	20	3
(25)	25	23	15	21	17	4	19	6	5

Plan 13.14 $t = 31, k = 6, r = 6, b = 31, \lambda = 1, E = .86$, Type I

	Reps.					
Block	I	II	III	IV	V	VI
(1)	31	1	3	8	12	18
(2)	1	2	4	9	13	19
(3)	2	3	5	10	14	20
(4)	3	4	6	11	15	21
(5)	4	5	7	12	16	22
(6)	5	6	8	13	17	23
(7)	6	7	9	14	18	24
(8)	7	8	10	15	19	25
(9)	8	9	11	16	20	26
(10)	9	10	12	17	21	27
(11)	10	11	13	18	22	28
(12)	11	12	14	19	23	29
(13)	12	13	15	20	24	30
(14)	13	14	16	21	25	31
(15)	14	15	17	22	26	1
(16)	15	16	18	23	27	2
(17)	16	17	19	24	28	3
(18)	17	18	20	25	29	4
(19)	18	19	21	26	30	5
(20)	19	20	22	27	31	6
(21)	20	21	23	28	1	7
(22)	21	22	24	29	2	8
(23)	22	23	25	30	3	9
(24)	23	24	26	31	4	10
(25)	24	25	27	1	5	11
(26)	25	26	28	2	6	12
(27)	26	27	29	3	7	13
(28)	27	28	30	4	8	14
(29)	28	29	31	5	9	15
(30)	29	30	1	6	10	16
(31)	30	31	2	7	11	17

Plan 13.2a $t = 31, k = 10, r = 10, b = 31, \lambda = 3, E = .93$, Type I

Block	Reps. I	II	III	IV	V	VI	VII	VIII	IX	X
(1)	1	2	4	8	9	11	15	16	18	28
(2)	2	3	12	9	10	17	16	19	5	22
(3)	3	4	20	10	17	13	6	18	11	23
(4)	4	5	7	11	12	21	18	14	19	24
(5)	5	6	1	12	13	8	19	20	15	25
(6)	6	7	13	16	14	9	20	21	2	26
(7)	7	1	15	14	8	10	21	17	3	27
(8)	8	11	17	25	16	23	29	7	26	5
(9)	9	12	24	29	27	18	1	26	17	6
(10)	10	13	18	19	29	25	2	27	28	7
(11)	11	14	22	26	19	20	3	28	29	1
(12)	12	8	27	23	20	29	4	22	21	2
(13)	13	9	29	28	21	15	5	23	24	3
(14)	14	10	25	22	15	16	24	29	6	4
(15)	15	24	26	5	2	27	11	10	30	20
(16)	16	25	6	30	3	28	12	11	27	21
(17)	17	26	28	7	30	22	13	12	4	15
(18)	18	27	23	1	5	30	14	13	22	16
(19)	19	28	30	2	6	14	8	24	23	17
(20)	20	22	8	3	7	24	9	30	25	18
(21)	21	23	10	4	1	26	30	25	9	19
(22)	22	21	11	17	24	1	25	2	31	13
(23)	23	15	3	18	25	2	26	31	12	14
(24)	24	16	19	31	26	3	27	4	13	8
(25)	25	17	14	27	31	4	28	5	20	9
(26)	26	18	5	21	28	31	22	6	8	10
(27)	27	19	31	15	22	6	23	9	7	11
(28)	28	20	16	24	23	7	31	1	10	12
(29)	29	30	2	6	4	5	7	3	1	31
(30)	30	31	9	13	11	12	10	8	14	29
(31)	31	29	21	20	18	19	17	15	16	30

Plan 13.15 $t = 37, k = 9, r = 9, b = 37, \lambda = 2, E = .91$, Type I

	Reps.								
Block	I	II	III	IV	V	VI	VII	VIII	IX
(1)	1	7	9	10	12	16	26	33	34
(2)	2	8	10	11	13	17	27	34	35
(3)	3	9	11	12	14	18	28	35	36
(4)	4	10	12	13	15	19	29	36	37
(5)	5	11	13	14	16	20	30	37	1
(6)	6	12	14	15	17	21	31	1	2
(7)	7	13	15	16	18	22	32	2	3
(8)	8	14	16	17	19	23	33	3	4
(9)	9	15	17	18	20	24	34	4	5
(10)	10	16	18	19	21	25	35	5	6
(11)	11	17	19	20	22	26	36	6	7
(12)	12	18	20	21	23	27	37	7	8
(13)	13	19	21	22	24	28	1	8	9
(14)	14	20	22	23	25	29	2	9	10
(15)	15	21	23	24	26	30	3	10	11
(16)	16	22	24	25	27	31	4	11	12
(17)	17	23	25	26	28	32	5	12	13
(18)	18	24	26	27	29	33	6	13	14
(19)	19	25	27	28	30	34	7	14	15
(20)	20	26	28	29	31	35	8	15	16
(21)	21	27	29	30	32	36	9	16	17
(22)	22	28	30	31	33	37	10	17	18
(23)	23	29	31	32	34	1	11	18	19
(24)	24	30	32	33	35	2	12	19	20
(25)	25	31	33	34	36	3	13	20	21
(26)	26	32	34	35	37	4	14	21	22
(27)	27	33	35	36	1	5	15	22	23
(28)	28	34	36	37	2	6	16	23	24
(29)	29	35	37	1	3	7	17	24	25
(30)	30	36	1	2	4	8	18	25	26
(31)	31	37	2	3	5	9	19	26	27
(32)	32	1	3	4	6	10	20	27	28
(33)	33	2	4	5	7	11	21	28	29
(34)	34	3	5	6	8	12	22	29	30
(35)	35	4	6	7	9	13	23	30	31
(36)	36	5	7	8	10	14	24	31	32
(37)	37	6	8	9	11	15	25	32	33

Plan 13.3a $t = 57, k = 8, r = 8, b = 57, \lambda = 1, E = .89$, Type I

Block	I	II	III	IV	V	VI	VII	VIII
	Reps.							
(1)	1	2	3	4	5	6	7	50
(2)	2	51	16	23	30	37	9	44
(3)	3	10	51	24	31	38	45	17
(4)	4	11	18	51	32	39	46	25
(5)	5	12	19	26	51	40	33	47
(6)	6	13	20	27	34	51	48	41
(7)	7	14	21	28	35	42	49	51
(8)	8	9	10	11	12	14	50	13
(9)	9	15	28	34	52	46	40	3
(10)	10	52	22	35	41	47	4	16
(11)	11	17	23	29	42	52	5	48
(12)	12	18	52	30	36	49	24	6
(13)	13	19	25	31	37	43	52	7
(14)	14	20	26	32	38	44	1	52
(15)	15	16	17	18	19	50	21	20
(16)	16	28	33	38	53	11	6	43
(17)	17	22	34	12	44	7	39	53
(18)	18	53	35	40	45	1	13	23
(19)	19	24	29	41	46	2	53	14
(20)	20	25	30	53	3	8	47	42
(21)	21	36	53	36	48	4	31	9
(22)	22	23	24	25	50	27	28	26
(23)	23	34	38	49	4	54	8	19
(24)	24	35	39	54	43	9	20	5
(25)	25	29	40	44	6	10	54	21
(26)	26	54	41	45	7	15	30	11
(27)	27	31	54	42	1	12	16	46
(28)	28	32	36	47	2	13	17	54
(29)	29	30	31	50	33	32	34	35
(30)	30	40	43	55	14	17	27	4
(31)	31	41	44	5	8	18	55	28
(32)	32	42	55	6	9	19	22	45
(33)	33	36	46	7	55	20	23	10
(34)	34	55	47	1	11	21	37	24
(35)	35	38	48	2	15	55	25	12
(36)	36	37	50	39	40	41	42	38
(07)	07	40	0	8	17	20	35	50
(38)	38	47	7	9	18	56	29	27
(39)	39	48	1	19	10	28	56	30
(40)	40	49	2	56	20	22	11	31
(41)	41	56	12	43	21	23	3	32
(42)	42	44	4	13	56	24	15	33
(43)	43	50	45	46	47	48	44	49
(44)	44	3	11	57	27	35	19	36
(45)	45	4	57	20	28	29	12	37
(46)	46	5	13	21	57	30	38	22
(47)	47	6	14	15	23	31	57	39
(48)	48	7	8	16	24	57	32	40
(49)	49	1	9	17	25	33	41	57
(50)	50	57	56	52	54	53	51	55
(51)	51	8	15	22	29	36	43	1
(52)	52	21	27	33	39	45	2	8
(53)	53	27	32	37	49	5	10	15
(54)	54	33	37	48	22	3	14	18
(55)	55	39	49	3	13	16	26	29
(56)	56	45	5	14	16	25	36	34
(57)	57	43	42	10	26	34	18	2

Plan 13.4a $t = 73, k = 9, r = 9, b = 73, \lambda = 1, E = .90$, Type I

Block	Reps. I	II	III	IV	V	VI	VII	VIII	IX
(1)	1	2	3	4	5	6	7	8	65
(2)	2	66	18	26	34	42	50	10	58
(3)	3	11	66	27	35	19	51	43	59
(4)	4	12	20	66	36	44	52	28	60
(5)	5	13	21	29	37	45	53	61	66
(6)	6	14	22	30	38	66	54	62	46
(7)	7	15	23	31	39	47	66	63	55
(8)	8	16	24	32	40	48	56	66	64
(9)	9	10	11	12	13	65	16	14	15
(10)	10	19	28	1	46	55	67	64	37
(11)	11	67	25	40	47	54	18	4	61
(12)	12	17	67	39	48	53	26	3	62
(13)	13	24	31	34	41	52	59	6	67
(14)	14	23	32	33	42	51	60	67	5
(15)	15	22	29	67	43	50	57	36	8
(16)	16	21	30	35	67	49	58	44	7
(17)	17	18	19	20	21	22	65	23	24
(18)	18	30	68	43	56	60	1	39	13
(19)	19	31	38	42	68	57	4	53	16
(20)	20	32	37	41	54	58	68	15	3
(21)	21	25	36	68	51	63	6	48	10
(22)	22	68	26	47	52	64	5	35	9
(23)	23	27	34	46	49	61	8	68	12
(24)	24	28	33	45	50	68	62	7	11
(25)	25	26	27	28	29	30	31	65	32
(26)	26	37	69	49	63	43	14	24	4
(27)	27	40	42	69	62	1	15	52	21
(28)	28	39	41	51	69	2	61	16	22
(29)	29	69	48	54	60	7	9	19	34
(30)	30	33	47	53	59	8	10	20	69
(31)	31	36	46	56	58	5	11	69	17
(32)	32	35	45	55	57	69	12	18	6
(33)	33	34	35	36	65	37	39	37	40
(34)	34	47	57	14	3	70	21	56	28
(35)	35	46	53	70	2	15	24	60	25
(36)	36	45	54	59	1	16	23	70	26
(37)	37	44	51	62	8	31	70	9	18
(38)	38	43	52	61	7	10	17	32	70
(39)	39	70	49	64	6	11	20	42	29
(40)	40	41	5	63	70	12	19	50	30
(41)	41	42	43	65	44	46	47	45	48
(42)	42	54	63	8	71	17	28	13	35
(43)	43	55	62	5	16	20	25	34	71
(44)	44	56	61	6	15	26	71	33	19
(45)	45	49	60	3	10	71	22	40	31
(46)	46	71	50	21	9	4	32	59	39
(47)	47	51	58	71	12	24	29	1	38
(48)	48	52	71	37	11	23	30	2	57
(49)	49	50	65	52	53	56	55	54	51
(50)	50	61	1	48	20	14	35	31	72
(51)	51	64	4	15	17	72	34	30	45
(52)	52	63	72	16	18	29	3	46	33
(53)	53	58	6	9	23	28	40	72	43
(54)	54	57	39	10	24	27	72	5	44
(55)	55	60	8	11	72	21	38	26	41
(56)	56	59	7	72	22	25	37	12	42
(57)	57	65	59	60	61	62	65	58	63
(58)	58	8	14	19	25	39	45	73	52
(59)	59	5	15	18	28	73	48	38	49
(60)	60	6	16	17	27	37	73	47	50
(61)	61	3	9	24	30	36	42	55	73
(62)	62	4	10	23	73	35	41	29	56
(63)	63	1	73	22	32	34	44	11	53
(64)	64	73	12	2	31	33	43	21	54
(65)	65	72	70	73	66	67	69	71	68
(66)	66	9	17	25	33	41	49	57	1
(67)	67	20	56	38	45	9	63	27	2
(68)	68	29	40	44	55	59	2	17	14
(69)	69	38	44	50	64	3	13	25	23
(70)	70	48	55	58	4	13	33	22	27
(71)	71	53	64	7	14	18	27	41	36
(72)	72	62	2	13	19	32	36	49	47
(73)	73	7	13	57	26	40	46	51	20

Plan 13.5a $t = 91, k = 10, r = 10, b = 91, \lambda = 1, E = .91$, Type I

Block	I	II	III	IV	V	VI	VII	VIII	IX	X
(1)	1	2	7	11	24	27	35	42	54	56
(2)	2	3	8	12	25	28	36	43	55	57
(3)	3	4	9	13	26	29	37	44	56	58
(4)	4	5	10	14	27	30	38	45	57	59
(5)	5	6	11	15	28	31	39	46	58	60
(6)	6	7	12	16	29	32	40	47	59	61
(7)	7	8	13	17	30	33	41	48	60	62
(8)	8	9	14	18	31	34	42	49	61	63
(9)	9	10	15	19	32	35	43	50	62	64
(10)	10	11	16	20	33	36	44	51	63	65
(11)	11	12	17	21	34	37	45	52	64	66
(12)	12	13	18	22	35	38	46	53	65	67
(13)	13	14	19	23	36	39	47	54	66	68
(14)	14	15	20	24	37	40	48	55	67	69
(15)	15	16	21	25	38	41	49	56	68	70
(16)	16	17	22	26	39	42	50	57	69	71
(17)	17	18	23	27	40	43	51	58	70	72
(18)	18	19	24	28	41	44	52	59	71	73
(19)	19	20	25	29	42	45	53	60	72	74
(20)	20	21	26	30	43	46	54	61	73	75
(21)	21	22	27	31	44	47	55	62	74	76
(22)	22	23	28	32	45	48	56	63	75	77
(23)	23	24	29	33	46	49	57	64	76	78
(24)	24	25	30	34	47	50	58	65	77	79
(25)	25	26	31	35	48	51	59	66	78	80
(26)	26	27	32	36	49	52	60	67	79	81
(27)	27	28	33	37	50	53	61	68	80	82
(28)	28	29	34	38	51	54	62	69	81	83
(29)	29	30	35	39	52	55	63	70	82	84
(30)	30	31	36	40	53	56	64	71	83	85
(31)	31	32	37	41	54	57	65	72	84	86
(32)	32	33	38	42	55	58	66	73	85	87
(33)	33	34	39	43	56	59	67	74	86	88
(34)	34	35	40	44	57	60	68	75	87	89
(35)	35	36	41	45	58	61	69	76	88	90
(36)	36	37	42	46	59	62	70	77	89	91
(37)	37	38	43	47	60	63	71	78	90	1
(38)	38	39	44	48	61	64	72	79	91	2
(39)	39	40	45	49	62	65	73	80	1	3
(40)	40	41	46	50	63	66	74	81	2	4
(41)	41	42	47	51	64	67	75	82	3	5
(42)	42	43	48	52	65	68	76	83	4	6
(43)	43	44	49	53	66	69	77	84	5	7
(44)	44	45	50	54	67	70	78	85	6	8
(45)	45	46	51	55	68	71	79	86	7	9
(46)	46	47	52	56	69	72	80	87	8	10

Block	I	II	III	IV	V	VI	VII	VIII	IX	X
(47)	47	48	53	57	70	73	81	88	9	11
(48)	48	49	54	58	71	74	82	89	10	12
(49)	49	50	55	59	72	75	83	90	11	13
(50)	50	51	56	60	73	76	84	91	12	14
(51)	51	52	57	61	74	77	85	1	13	15
(52)	52	53	58	62	75	78	86	2	14	16
(53)	53	54	59	63	76	79	87	3	15	17
(54)	54	55	60	64	77	80	88	4	16	18
(55)	55	56	61	65	78	81	89	5	17	19
(56)	56	57	62	66	79	82	90	6	18	20
(57)	57	58	63	67	80	83	91	7	19	21
(58)	58	59	64	68	81	84	1	8	20	22
(59)	59	60	65	69	82	85	2	9	21	23
(60)	60	61	66	70	83	86	3	10	22	24
(61)	61	62	67	71	84	87	4	11	23	25
(62)	62	63	68	72	85	88	5	12	24	26
(63)	63	64	69	73	86	89	6	13	25	27
(64)	64	65	70	74	87	90	7	14	26	28
(65)	65	66	71	75	88	91	8	15	27	29
(66)	66	67	72	76	89	1	9	16	28	30
(67)	67	68	73	77	90	2	10	17	29	31
(68)	68	69	74	78	91	3	11	18	30	32
(69)	69	70	75	79	1	4	12	19	31	33
(70)	70	71	76	80	2	5	13	20	32	34
(71)	71	72	77	81	3	6	14	21	33	35
(72)	72	73	78	82	4	7	15	22	34	36
(73)	73	74	79	83	5	8	16	23	35	37
(74)	74	75	80	84	6	9	17	24	36	38
(75)	75	76	81	85	7	10	18	25	37	39
(76)	76	77	82	86	8	11	19	26	38	40
(77)	77	78	83	87	9	12	20	27	39	41
(78)	78	79	84	88	10	13	21	28	40	42
(79)	79	80	85	89	11	14	22	29	41	43
(80)	80	81	86	90	12	15	23	30	42	44
(81)	81	82	87	91	13	16	24	31	43	45
(82)	82	83	88	1	14	17	25	32	44	46
(83)	83	84	89	2	15	18	26	33	45	47
(84)	84	85	90	3	16	19	27	34	46	48
(85)	85	86	91	4	17	20	28	35	47	49
(86)	86	87	1	5	18	21	29	36	48	50
(87)	87	88	2	6	19	22	30	37	49	51
(88)	88	89	3	7	20	23	31	38	50	52
(89)	89	90	4	8	21	24	32	39	51	53
(90)	90	91	5	9	22	25	33	40	52	54
(91)	91	1	6	10	23	26	34	41	53	55

Plan 13.16 $t = 3, k = 5, r = 5, b = 3, \lambda = 5, E = .96$, Type III

Reps.

Block	I	II	III	IV	V
(1)	1	2	3	2	3
(2)	2	3	1	1	2
(3)	3	1	2	3	1

Plan 13.17 $t = 3, k = 7, r = 7, b = 3, \lambda = 7, E = .98$, Type IV

Reps.

Block	I	II	III	IV	V	VI	VII
(1)	1	2	3	1	2	3	1
(2)	2	3	1	3	1	2	2
(3)	3	1	2	2	3	1	3

Plan 13.18 $t = 3, k = 8, r = 8, b = 3, \lambda = 8, E = .98$, Type III

Reps

Block	I	II	III	IV	V	VI	VII	VIII
(1)	1	2	3	1	2	3	1	2
(2)	2	3	1	3	1	2	2	3
(3)	3	1	2	2	3	1	3	1

Plan 13.19 $t = 3, k = 10, r = 10, b = 3, \lambda = 10, E = .99$, Type IV

Reps.

Block	I	II	III	IV	V	VI	VII	VIII	IX	X
(1)	1	2	3	1	2	3	1	2	3	1
(2)	2	3	1	3	1	2	2	3	1	3
(3)	3	1	2	2	3	1	3	1	2	2

Plan 13.20 $t = 4, k = 5, r = 5, b = 4, \lambda = 5, E = .96$, Type IV

Reps.

Block	I	II	III	IV	V
(1)	1	2	3	4	1
(2)	2	3	4	1	4
(3)	3	4	1	2	3
(4)	4	1	2	3	2

Plan 13.21 $t = 4, k = 7, r = 7, b = 4, \lambda = 7, E = .98$, Type III

Reps.

Block	I	II	III	IV	V	VI	VII
(1)	1	2	3	4	1	2	3
(2)	2	1	4	3	3	4	1
(3)	3	4	1	2	4	3	2
(4)	4	3	2	1	2	1	4

Plan 13.22 $t = 4, k = 9, r = 9, b = 4, \lambda = 9, E = .99$, Type IV

Reps.

Block	I	II	III	IV	V	VI	VII	VIII	IX
(1)	1	2	3	4	1	2	3	4	1
(2)	2	1	4	3	3	4	1	2	4
(3)	3	4	1	2	4	3	2	1	2
(4)	4	3	2	1	2	1	4	3	3

Plan 13.23 $t = 5, k = 6, r = 6, b = 5, \lambda = 6, E = .97$, Type IV

Reps.

Block	I	II	III	IV	V	VI
(1)	1	2	3	4	5	1
(2)	2	4	5	3	1	3
(3)	3	1	2	5	4	4
(4)	4	5	1	2	3	2
(5)	5	3	4	1	2	5

Plan 13.24 $t = 5, k = 9, r = 9, b = 5, \lambda = 9, E = .99$, Type III

Reps.

Block	I	II	III	IV	V	VI	VII	VIII	IX
(1)	1	2	3	4	5	1	2	3	4
(2)	2	3	4	5	1	4	5	1	2
(3)	3	4	5	1	2	2	3	4	5
(4)	4	5	1	2	3	5	1	2	3
(5)	5	1	2	3	4	3	4	5	1

Plan 13.25 $t = 6, k = 7, r = 7, b = 6, \lambda = 7, E = .98$, Type IV

Block	Reps. I	II	III	IV	V	VI	VII
(1)	1	2	3	4	5	6	1
(2)	2	3	6	1	4	5	3
(3)	3	6	2	5	1	4	5
(4)	4	5	1	2	6	3	4
(5)	5	1	4	6	3	2	6
(6)	6	4	5	3	2	1	2

Plan 13.26 $t = 7, k = 8, r = 8, b = 7, \lambda = 8, E = .98$, Type IV

Block	Reps. I	II	III	IV	V	VI	VII	VIII
(1)	1	2	3	4	5	6	7	1
(2)	2	5	1	7	6	4	3	3
(3)	3	6	7	2	4	1	5	7
(4)	4	7	5	6	3	2	1	6
(5)	5	4	2	3	1	7	6	5
(6)	6	3	4	1	7	5	2	4
(7)	7	1	6	5	2	3	4	2

EXTENDED INCOMPLETE LATIN SQUARES

Plan 13.6a $t = 5, k = 2, r = 4, b = 10, \lambda = 1, E = .67$, Type II

Block	Reps. I	II
(1)	1	2
(2)	2	5
(3)	3	4
(4)	4	1
(5)	5	3

	Reps. III	IV
(6)	1	3
(7)	2	4
(8)	3	2
(9)	4	5
(10)	5	1

Plan 13.7a $t = 5, k = 3, r = 6, b = 10, \lambda = 3, E = .83$, Type II

Block	Reps. I	II	III
(1)	1	2	3
(2)	2	1	5
(3)	3	4	2
(4)	4	5	1
(5)	5	3	4

	Reps. IV	V	VI
(6)	1	2	4
(7)	2	3	5
(8)	3	4	1
(9)	4	5	2
(10)	5	1	3

Plan 13.8a $t = 7, k = 2, r = 6, b = 21, \lambda = 1, E = .58$, Type V

Block	Reps. I	II	Block	Reps. III	IV	Block	Reps. V	VI
(1)	1	2	(8)	1	3	(15)	1	4
(2)	2	6	(9)	2	4	(16)	2	3
(3)	3	4	(10)	3	5	(17)	3	6
(4)	4	7	(11)	4	6	(18)	4	5
(5)	5	1	(12)	5	7	(19)	5	2
(6)	6	5	(13)	6	1	(20)	6	7
(7)	7	3	(14)	7	2	(21)	7	1

Plan 13.9a $t = 9, k = 2, r = 8, b = 36, \lambda = 1, E = .56$, Type V

Block	Reps. I	II	Block	Reps. III	IV	Block	Reps. V	VI	Block	Reps. VII	VIII
(1)	1	2	(10)	1	3	(19)	1	4	(28)	1	5
(2)	2	8	(11)	2	5	(20)	2	6	(29)	2	4
(3)	3	4	(12)	3	6	(21)	3	2	(30)	3	8
(4)	4	7	(13)	4	9	(22)	4	5	(31)	4	6
(5)	5	6	(14)	5	8	(23)	5	7	(32)	5	3
(6)	6	1	(15)	6	7	(24)	6	8	(33)	6	9
(7)	7	3	(16)	7	1	(25)	7	9	(34)	7	2
(8)	8	9	(17)	8	4	(26)	8	1	(35)	8	7
(9)	9	5	(18)	9	2	(27)	9	3	(36)	9	1

Plan 13.10a $t = 9, k = 4, r = 8, b = 18, \lambda = 3, E = .84$, Type V

Block	Reps. I	II	III	IV	Block	Reps. V	VI	VII	VIII
(1)	1	4	6	7	(10)	1	2	5	7
(2)	2	6	8	9	(11)	2	3	6	5
(3)	3	8	9	1	(12)	3	4	7	9
(4)	4	1	3	2	(13)	4	9	2	1
(5)	5	7	1	8	(14)	5	1	9	6
(6)	6	9	4	5	(15)	6	8	1	3
(7)	7	3	2	6	(16)	7	6	4	8
(8)	8	2	5	4	(17)	8	5	3	4
(9)	9	5	7	3	(18)	9	7	8	2

Plan 13.11a $t = 9, k = 5, r = 10, b = 18, \lambda = 5, E = .90$, Type V

Block	Reps. I	II	III	IV	V	Block	Reps. VI	VII	VIII	IX	X
(1)	1	2	3	7	8	(10)	1	2	3	5	9
(2)	2	6	8	4	1	(11)	2	6	5	1	8
(3)	3	8	5	9	2	(12)	3	5	1	4	6
(4)	4	3	9	2	6	(13)	4	3	2	8	7
(5)	5	1	7	3	4	(14)	5	7	9	2	4
(6)	6	4	2	5	7	(15)	6	8	7	3	5
(7)	7	9	1	6	3	(16)	7	4	8	9	1
(8)	8	5	4	1	9	(17)	8	9	4	6	3
(9)	9	7	6	8	5	(18)	9	1	6	7	2

Plan 13.12a $t = 10, k = 3, r = 9, b = 30, \lambda = 2, E = .74$, Type V

Block	Reps. I	II	III	Block	Reps. IV	V	VI	Block	Reps. VII	VIII	IX
(1)	1	2	3	(11)	1	2	4	(21)	1	3	5
(2)	2	5	8	(12)	2	3	6	(22)	2	7	6
(3)	3	7	4	(13)	3	4	8	(23)	3	8	9
(4)	4	1	6	(14)	4	9	5	(24)	4	2	10
(5)	5	8	7	(15)	5	7	1	(25)	5	6	3
(6)	6	4	9	(16)	6	8	9	(26)	6	1	8
(7)	7	9	1	(17)	7	10	3	(27)	7	9	2
(8)	8	10	2	(18)	8	1	10	(28)	8	4	7
(9)	9	3	10	(19)	9	5	2	(29)	9	10	1
(10)	10	6	5	(20)	10	6	7	(30)	10	5	4

Plan 13.13a $t = 11, k = 2, r = 10, b = 55, \lambda = 1, E = .55$, Type V

Block	Reps. I	II		III	IV		V	VI
(1)	1	2	(12)	1	3	(23)	1	4
(2)	2	11	(13)	2	6	(24)	2	3
(3)	3	10	(14)	3	5	(25)	3	7
(4)	4	5	(15)	4	10	(26)	4	6
(5)	5	6	(16)	5	9	(27)	5	10
(6)	6	7	(17)	6	8	(28)	6	9
(7)	7	1	(18)	7	2	(29)	7	11
(8)	8	3	(19)	8	1	(30)	8	2
(9)	9	4	(20)	9	7	(31)	9	1
(10)	10	9	(21)	10	11	(32)	10	8
(11)	11	8	(22)	11	4	(33)	11	5

	Reps. VII	VIII		Reps. IX	X
(34)	1	5	(45)	1	6
(35)	2	9	(46)	2	5
(36)	3	6	(47)	3	4
(37)	4	2	(48)	4	7
(38)	5	7	(49)	5	8
(39)	6	10	(50)	6	11
(40)	7	8	(51)	7	10
(41)	8	4	(52)	8	9
(42)	9	11	(53)	9	3
(43)	10	1	(54)	10	2
(44)	11	3	(55)	11	1

Plan 13.14a $t = 13, k = 3, r = 6, b = 26, \lambda = 1, E = .72$, Type V

	Reps.				Reps.		
Block	I	II	III		IV	V	VI
(1)	1	3	9	(14)	2	6	5
(2)	2	4	10	(15)	3	7	6
(3)	3	5	11	(16)	4	8	7
(4)	4	6	12	(17)	5	9	8
(5)	5	7	13	(18)	6	10	9
(6)	6	8	1	(19)	7	11	10
(7)	7	9	2	(20)	8	12	11
(8)	8	10	3	(21)	9	13	12
(9)	9	11	4	(22)	10	1	13
(10)	10	12	5	(23)	11	2	1
(11)	11	13	6	(24)	12	3	2
(12)	12	1	7	(25)	13	4	3
(13)	13	2	8	(26)	1	5	4

Plan 13.15a $t = 19, k = 3, r = 9, b = 57, \lambda = 1, E = .70$, Type V

	Reps.				Reps.				Reps.		
Block	I	II	III		IV	V	VI		VII	VIII	IX
(1)	1	7	11	(20)	2	3	14	(39)	4	6	9
(2)	2	8	12	(21)	3	4	15	(40)	5	7	10
(3)	3	9	13	(22)	4	5	16	(41)	6	8	11
(4)	4	10	14	(23)	5	6	17	(42)	7	9	12
(5)	5	11	15	(24)	6	7	18	(43)	8	10	13
(6)	6	12	16	(25)	7	8	19	(44)	9	11	14
(7)	7	13	17	(26)	8	9	1	(45)	10	12	15
(8)	8	14	18	(27)	9	10	2	(46)	11	13	16
(9)	9	15	19	(28)	10	11	3	(47)	12	14	17
(10)	10	16	1	(29)	11	12	4	(48)	13	15	18
(11)	11	17	2	(30)	12	13	5	(49)	14	16	19
(12)	12	18	3	(31)	13	14	6	(50)	15	17	1
(13)	13	19	4	(32)	14	15	7	(51)	16	18	2
(14)	14	1	5	(33)	15	16	8	(52)	17	19	3
(15)	15	2	6	(34)	16	17	9	(53)	18	1	4
(16)	16	3	7	(35)	17	18	10	(54)	19	2	5
(17)	17	4	8	(36)	18	19	11	(55)	1	3	6
(18)	18	5	9	(37)	19	1	12	(56)	2	4	7
(19)	19	6	10	(38)	1	2	13	(57)	3	5	8

Plan 13.16a $t = 25, k = 4, r = 8, b = 50, \lambda = 1, E = .78$, Type V

Block	Reps. I	II	III	IV	Block	Reps. V	VI	VII	VIII
(1)	1	2	6	25	(26)	1	3	11	19
(2)	2	3	7	21	(27)	2	4	12	20
(3)	3	4	8	22	(28)	3	5	13	16
(4)	4	5	9	23	(29)	4	1	14	17
(5)	5	1	10	24	(30)	5	2	15	18
(6)	6	7	11	5	(31)	6	8	16	24
(7)	7	8	12	1	(32)	7	9	17	25
(8)	8	9	13	2	(33)	8	10	18	21
(9)	9	10	14	3	(34)	9	6	19	22
(10)	10	6	15	4	(35)	10	7	20	23
(11)	11	12	16	10	(36)	11	13	21	4
(12)	12	13	17	6	(37)	12	14	22	5
(13)	13	14	18	7	(38)	13	15	23	1
(14)	14	15	19	8	(39)	14	11	24	2
(15)	15	11	20	9	(40)	15	12	25	3
(16)	16	17	21	15	(41)	16	18	1	9
(17)	17	18	22	11	(42)	17	19	2	10
(18)	18	19	23	12	(43)	18	20	3	6
(19)	19	20	24	13	(44)	19	16	4	7
(20)	20	16	25	14	(45)	20	17	5	8
(21)	21	22	1	20	(46)	21	23	6	14
(22)	22	23	2	16	(47)	22	24	7	15
(23)	23	24	3	17	(48)	23	25	8	11
(24)	24	25	4	18	(49)	24	21	9	12
(25)	25	21	5	19	(50)	25	22	10	13

Plan 13.17a $t = 41, k = 5, r = 10, b = 82, \lambda = 1, E = .82$, Type V

Block	Reps. I	II	III	IV	V	Block	Reps. VI	VII	VIII	IX	X
(1)	1	10	16	18	37	(42)	1	19	31	32	35
(2)	2	11	17	19	38	(43)	2	20	32	33	36
(3)	3	12	18	20	39	(44)	3	21	33	34	37
(4)	4	13	19	21	40	(45)	4	22	34	35	38
(5)	5	14	20	22	41	(46)	5	23	35	36	39
(6)	6	15	21	23	1	(47)	6	24	36	37	40
(7)	7	16	22	24	2	(48)	7	25	37	38	41
(8)	8	17	23	25	3	(49)	8	26	38	39	1
(9)	9	18	24	26	4	(50)	9	27	39	40	2
(10)	10	19	25	27	5	(51)	10	28	40	41	3
(11)	11	20	26	28	6	(52)	11	29	41	1	4
(12)	12	21	27	29	7	(53)	12	30	1	2	5
(13)	13	22	28	30	8	(54)	13	31	2	3	6
(14)	14	23	29	31	9	(55)	14	32	3	4	7
(15)	15	24	30	32	10	(56)	15	33	4	5	8
(16)	16	25	31	33	11	(57)	16	34	5	6	9
(17)	17	26	32	34	12	(58)	17	35	6	7	10
(18)	18	27	33	35	13	(59)	18	36	7	8	11
(19)	19	28	34	36	14	(60)	19	37	8	9	12
(20)	20	29	35	37	15	(61)	20	38	9	10	13
(21)	21	30	36	38	16	(62)	21	39	10	11	14
(22)	22	31	37	39	17	(63)	22	40	11	12	15
(23)	23	32	38	40	18	(64)	23	41	12	13	16
(24)	24	33	39	41	19	(65)	24	1	13	14	17
(25)	25	34	40	1	20	(66)	25	2	14	15	18
(26)	26	35	41	2	21	(67)	26	3	15	16	19
(27)	27	36	1	3	22	(68)	27	4	16	17	20
(28)	28	37	2	4	23	(69)	28	5	17	18	21
(29)	29	38	3	5	24	(70)	29	6	18	19	22
(30)	30	39	4	6	25	(71)	30	7	19	20	23
(31)	31	40	5	7	26	(72)	31	8	20	21	24
(32)	32	41	6	8	27	(73)	32	9	21	22	25
(33)	33	1	7	9	28	(74)	33	10	22	23	26
(34)	34	2	8	10	29	(75)	34	11	23	24	27
(35)	35	3	9	11	30	(76)	35	12	24	25	28
(36)	36	4	10	12	31	(77)	36	13	25	26	29
(37)	37	5	11	13	32	(78)	37	14	26	27	30
(38)	38	6	12	14	33	(79)	38	15	27	28	31
(39)	39	7	13	15	34	(80)	39	16	28	29	32
(40)	40	8	14	16	35	(81)	40	17	29	30	33
(41)	41	9	15	17	36	(82)	41	18	30	31	34

CHAPTER 14

Analysis of the Results of a Series of Experiments

14.1 Initial Steps in the Analysis

14.11 Introduction. In a program of research it is quite common to repeat the same experiment at a number of different places, on a number of different occasions. There may be several reasons for this. Sometimes the object of the research is to produce recommendations which are to apply to a population that is extensive either in space or in time or in both. Thus in agricultural field experimentation, many projects are undertaken in the hope that their results can be applied in practical farming. The conclusions drawn from such research, if they are to be of use, must be valid for at least several seasons in the future and over a reasonably large area of farm land. It has been found that the effectiveness of the common plant nutrients, of different varieties of a crop, and of different cultivation practices usually varies from field to field and, even more markedly, from season to season. A single experiment, however well conducted, supplies information about only one place and one season. Consequently such experiments are carried out at several different places in the area for which recommendations are wanted, and are repeated for a number of seasons.

In other cases we may be interested, not in making inferences about some specific population, but in studying the influence of external conditions on some measurement or on the responses to treatments. For example, is the vitamin A content of a vegetable affected by the climate in which it is grown? Do the relative durabilities of two types of road surface depend on weather or topography or traffic load? Repetition of such experiments in different places is necessary in order to have variations in the external factors that are under investigation.

A third example is supplied by collaborative experiments on biological assay by a group of laboratories. Here the objects may be to obtain information about the accuracy with which the potency of a drug can be estimated and to discover whether different laboratories reach the same conclusions about the relative potencies of different drugs. Bliss (14.1) gives an interesting account of a series of experiments of this type, where sixteen laboratories estimated the potency of preparations of digitalis by injection into cats.

The appropriate statistical analysis for the data from a series of experiments will of course vary with the object of the research. Nevertheless, the preliminary stages of the analysis tend to be the same in all cases. In this chapter an introductory account is given of these initial procedures, which present difficulties that are not always appreciated. The first step, which is to analyze and interpret the data from each individual experiment, will be assumed to have been completed.

When we begin to combine the data, the first point of interest is to examine whether the differences among treatments are the same in all experiments. This question is likely to be relevant whatever the purpose of the experiments. In experiments designed to lead to the recommendation of a "best" treatment for some operation, we wish to know whether there is a consistent superiority of certain treatments, or whether on the other hand we may have to consider recommending different treatments for different circumstances. In the other two types of experimentation mentioned above, the test would indicate whether the responses to treatments have varied with the external conditions of the experiment, or whether the different laboratories agreed in their estimates of relative potency.

Secondly, it is often, though not always, desirable to estimate and compare the average effects of treatments over the whole series of experiments.

14.12 Numerical Example. The data come from a group of 6 experiments on Irish potatoes conducted in two counties of North Carolina in 1945 and 1946. The results have been described by Nelson and Hawkins (14.2), and form part of a study of the responses to applications of superphosphate on soils of varying degree of fertility. In 1945, on each of the 6 sites, the amount of readily soluble phosphorus in the soil itself was estimated from soil samples by the modified Truog method. These amounts varied from 48 to 850 lb. P_2O_5 per acre. The treatments comprised 5 different levels of application, 0, 40, 80, 120, and 160 lb. P_2O_5 per acre.

These experiments are an example of the second type discussed above, in that their object was to find out the extent to which the responses to treatments were influenced by the condition of the soil. The soils were not intended to be a representative sample of the soils in the counties but were chosen so as to give a wide range in fertility.

Potatoes were grown on the plots in 1945 and 1946, the treatments being applied in both years, with in addition substantial applications of nitrogen and potash on all plots so that these nutrients would not be deficient. The data to be analyzed are the 1946 results. The mean yields and error m.s. per plot are shown in table 14.1. In five of the six experi-

TABLE 14.1 TREATMENT YIELDS (100 LB. PER ACRE) OF IRISH POTATOES, 1946

Pounds P_2O_5 applied per acre	Experiment number					
	1	2	3	4	5	6
	Amount of readily soluble phosphorus (pounds P_2O_5) in soil, 1945					
	48	310	410	710	790	850
0	114	142	180	170	130	170
40	206	176	188	171	140	178
80	231	201	207	188	150	187
120	237	205	208	185	152	188
160	252	217	217	189	156	189
s_i^2 *	106	83	158	43	204	52
d.f.	21	12	12	12	12	12
r †	4	5	5	5	5	5

* Error m.s. per plot.
† Number of replications.

ments the F-ratios for treatments against error were significant. One feature of the results in table 14.1 is that experiment 1, which has the lowest amount of available phosphorus in the soil according to the soil tests, actually gave the highest total yield in 1946. It should be remembered that the soil tests were made in 1945, and that some residual effects of the 1945 applications may have persisted. Whatever the reason, this soil was apparently very responsive to phosphorus in 1946; the plots without phosphorus do have the lowest yield of all places.

It is possible to compute from table 14.1 a combined analysis of variance for all six experiments. As will be seen later, there are limitations to the use of such an analysis. For the present we will ignore any complexities that may arise. Further, in order to discuss only the simplest case at first, we will omit experiment 1, which differs in size and structure from the other experiments. This omission is of course inappropriate from the agronomic point of view, but the data are being used to throw light on the general procedure in analysis rather than on the agronomic questions involved.

14.13 Preliminary Combined Analysis When all Experiments Have the Same Design. Formally, the analysis subdivides into the following components.

Places
Treatments
Treatments × places
Pooled experimental error

In addition, in a complete analysis there are terms representing the differences among rows and columns in the individual experiments. Since these are not relevant to the interpretation, they are not included.

The analysis can be computed from either treatment means or totals; the totals will be used here and are shown in table 14.2.

TABLE 14.2 TREATMENT TOTALS FOR EXPERIMENTS 2 TO 6 (IN 100 LB.)

Pounds P_2O_5	Experiment 2	3	4	5	6	Total
0	710	900	850	650	850	3,960
40	880	940	855	700	890	4,265
80	1,005	1,035	940	750	935	4,665
120	1,025	1,040	925	760	940	4,690
160	1,085	1,085	945	780	945	4,840
Total	4,705	5,000	4,515	3,640	4,560	22,420

The first 3 components are computed in the same way as for a single randomized blocks experiment, except that all sums of squares are divided by an extra 5 in order to reduce them to a single-plot basis. Thus we have

$$\text{Places:} \quad \frac{(4705)^2 + \cdots + (4560)^2}{25} - \frac{(22{,}420)^2}{125} = 41{,}367$$

$$\text{Treatments:} \quad \frac{(3960)^2 + \cdots + (4840)^2}{25} - \frac{(22{,}420)^2}{125} = 20{,}979$$

To obtain the treatments × places s.s. we calculate the total s.s. for table 14.2, which comes to 69,199. Then

$$\text{Treatments} \times \text{places: } 69{,}199 - 41{,}367 - 20{,}979 = 6853$$

Since all experiments have the same number of error degrees of freedom, the pooled error m.s. may be found as the simple average of the five s^2 values in table 14.1. This comes to 108, as given in table 14.3.

The null hypothesis that the treatment differences are the same at all places (i.e., that there are no treatment × place interactions) is tested by the F-ratio 428/108, or 3.96, with 16 and 60 d.f., respectively. Since the 5% level is 1.81, the ratio is definitely significant.

TABLE 14.3 COMBINED ANALYSIS OF VARIANCE (ON A SINGLE-PLOT BASIS)

Source of variation	d.f.	s.s.	m.s.
Places	4	41,367	
Treatments	4	20,979	5,245
Treatments × places	16	6,853	428
Pooled error	60	6,480	108

A test of the average responses to treatments taken over all five places is of minor interest in this example, because the places do not constitute a random sample from a population about which we wish to make inferences. For purposes of illustration, however, we will assume that the places *were* selected as a random sample of fields on which potatoes might be grown commercially.

In the F-test of the average effects of treatments there are two possible candidates for the denominator of F—the mean square for the treatments × places interactions (428) or the pooled error m.s. (108). Sometimes this competition does not arise, because our knowledge of the data and the F-test of the interactions both indicate that there is no reason to suppose real interactions to be present. In that event the mean squares for interactions and error may be pooled to form a single denominator for the F-test of treatments. But frequently the experimental conditions are such that we expect interactions to be present. This is so in the example, where previous work in a number of countries has shown a relation between the response to phosphorus and the amount found in the soil by a soil test. Consequently, even if the interactions m.s. had not proved significant, we might have been unwilling to assume that there were no interactions.

In discussing the two F-ratios it is helpful to examine the mathematical model on which the combined analysis is based. If x_{ij} is the observed mean for the jth treatment at the ith place, we postulate that

$$x_{ij} = \mu + \pi_i + \tau_j + \mu_{ij} + \bar{e}_{ij} \tag{14.1}$$

where π_i, τ_j represent the effects of the place and the treatment, respectively, μ_{ij} that of the treatment × place interaction, and $\bar{e}_{ij}$ that of the experimental error. $\bar{e}_{ij}$ is the average of the errors on the r plots that receive the treatment at that place.

With this model we can study the nature of the mean squares in the analysis of variance. If the experimental errors on individual plots have a variance σ_e^2, and if the interaction terms μ_{ij} may be considered to have a variance σ_μ^2, the average values of the mean squares work out as follows.

TABLE 14.4 Expected values of mean squares

Source of variation	Expected value of mean square
Treatments	$\sigma_e^2 + r\sigma_\mu^2 + \frac{rp}{(t-1)} \sum (\tau_j - \bar{\tau})^2$
Treatments × places	$\sigma_e^2 + r\sigma_\mu^2$
Pooled experimental error	σ_e^2

The symbols r, p, and t stand respectively for the numbers of replications, places, and treatments.

It will be noted that the treatments m.s. is influenced by three components—the experimental error variance, the variance of treatment × place interactions, and the variance among the true treatment means τ_j. The pooled experimental error takes account of only the first of these components. Hence the treatments m.s. may be statistically significant, as compared with the pooled error m.s., either because there are real differences among the τ's or because treatment × place interactions are present. This F-test is informative only when we are indifferent as to whether a significant treatments m.s. was due to real treatment differences or to interactions. This situation is rare in practice. In most cases, on the contrary, having established that interactions are present, or at least that it is not safe to assume them absent, we wish to know whether *in addition* there are consistent differences among the effects of treatments. As table 14.4 shows, the appropriate denominator of F for this test is the interactions m.s. This contains both the interaction and error components of variation in exactly the same way as they enter into the treatments m.s. The F-ratio, 5245/428, or 12.26, with 4 and 16 d.f., is significant at the 1% level.

The conclusions from this initial analysis are: (i) there are real differences in response that are consistent from place to place, and (ii) there are real variations in responsiveness from place to place. It need not be stressed that these statements do not constitute a competent summary of the results. They merely indicate what should be examined next. In the present case we would consider which treatments have proved consistently superior, and, if possible, why this is so. Similarly, we should investigate the nature of the interactions and try to find a rational explanation for them.

14.2 Criticisms of the Preliminary Analysis

14.21 Heterogeneity of the Interaction Variance. The previous analysis is open to several criticisms. Although these criticisms are not valid for all series of experiments, our experience is that it is well to be on one's guard against them. They deal essentially with the assumptions on which the combined analysis was based.

These assumptions have been presented in section 14.13. Specifically, we postulate that the model (14.1) holds. Further, the experimental errors e_{ij} of individual observations are assumed to be normally and independently distributed with the same variance σ_e^2. Finally, for the F-test of the treatments m.s. against the treatments $\times$ places interactions, we require also the assumptions that the interaction terms μ_{ij} are normally and independently distributed, with zero population means and the same variance σ_μ^2, and are independent of the e's.

The first criticism is that some components of the treatments $\times$ places s.s. may be much larger than others, or in mathematical terms that the "interaction" variance σ_μ^2 is not constant. This will happen if the effectiveness of some treatments varies greatly from place to place, while that of others varies less or not at all.

If the interaction m.s. is heterogeneous in this sense, the F-test of treatments against interactions is vitiated. The general effect is that the F value read from the tables is too low, i.e., that too many significant results are obtained. Some idea of the extent of the bias can be obtained in certain extreme cases. In the numerical example, if one comparison among the treatment means has a much larger interaction variance than any other comparisons, the F-ratio is distributed approximately as an F value with 1 and 4 d.f., respectively, instead of 4 and 16 d.f. The 5% significance level would be 7.71 instead of 3.01. More generally, if there are p places, F is distributed approximately with 1 and $(p - 1)$ d.f. This situation produces about the greatest distortion in F that is likely to arise, so that 7.71 could be regarded as an upper limit to the significance level of F.

The exact distribution of F can be worked out, but is not available in a form adapted for practical use. Even so, in many instances, uncertainty about the correct significance level for F does not preclude us from drawing conclusions from the test. For, in the example above, if F turned out to be 1.38 we would be confident that it is *not* significant, since the 5% level is at least 3.01. Similarly, there seems little doubt that the F value actually obtained in the experiments, 12.26, *is* significant, since it lies well above 7.71. On the other hand, we would be uncertain about significance if F happened to lie between 3 and 7.

A better method of coping with this difficulty is to divide the treatment s.s. into a set of orthogonal components that will supply all or most of the information of interest to us. The interactions s.s. is partitioned in the same way so as to isolate the interactions of each component with places. By Bartlett's test for homogeneity of variance, reference (14.3), we can then test whether σ_μ^2 is the same for all the components of the interactions s.s. If we decide that σ_μ^2 *can* be assumed constant, the difficulty vanishes. If σ_μ^2 is not constant, it is valid to test any component of the treatments s.s. against *its own* interaction with places. Apart from the extra computations, the only drawback to this procedure is that the degrees of freedom in the denominator of F are reduced.

These remarks will be illustrated by the example. In these experiments the principal point of interest is to see whether the average response to a given amount of phosphorus decreases when the amount presumed to be in the soil increases. It may also be worth while to examine whether the rate of decline in response with the higher levels of dressing changes with the nature of the soil. These questions can be considered conveniently by fitting at each place a parabolic regression of the yield on the amount of dressing. Although a parabolic regression would scarcely be regarded as the true form of the response curve, it appears to fit the data well.

Since the dressings increase by equal amounts, the orthogonal polynomials given by Fisher and Yates (14.4) are suitable. The calculations are shown below for experiment 3. The linear term is proportional to

		Multipliers for	
Amount of P_2O_5	Total yield	Linear term	Quadratic term
0	900	−2	2
40	940	−1	−1
80	1035	0	−2
120	1040	1	−1
160	1085	2	2
Sum of products		470	−80
Divisor for square		50	70

the average increase in yield per 40 lb. of P_2O_5: the quadratic term may be interpreted as measuring the rate of decline in response with increased dressings. The values found for the two terms are as follows.

	Experiment 2	3	4	5	6	Total
L	895	470	260	320	240	2185
Q	−325	−80	−70	−100	−110	−685

The contributions to the treatments s.s. are

$$\text{Linear: } \frac{(2185)^2}{250} = 19{,}097; \qquad \text{quadratic: } \frac{(685)^2}{350} = 1341$$

In the treatments × places s.s. we have

$$L \times \text{places:} \frac{(895)^2 + \cdots + (240)^2}{50} - 19{,}097 = 5893$$

$$Q \times \text{places:} \frac{(325)^2 + \cdots + (110)^2}{70} - 1341 = 645$$

In table 14.5 the three interaction mean squares are in the same order of size as the corresponding treatment mean squares. This result is typical, in that large effects tend to have large interactions. It is obvious on inspection that the components of the interaction m.s. cannot be regarded as homogeneous. Accordingly, we test the linear component of treatments against the mean square 1473 instead of against the complete treatments × places m.s. of 428 in table 14.3. This makes a substantial difference to the F-ratio, though it remains significant. The quadratic × places m.s. (161) is not significantly above the error m.s. (108), but it seems prudent to use it as the denominator in the F-test of the quadratic term in treatments. The remainder of the interactions shows no sign of significance, and could be pooled with error for an F-test of the deviations from regression.

TABLE 14.5 Subdivision of the analysis of variance

Source of variation	d.f.	s.s.	m.s.
Treatments			
Linear	1	19,097	19,097
Quadratic	1	1,341	1,341
Deviations from regression	2	541	270
Treatments × places			
Linear × places	4	5,893	1,473
Quadratic × places	4	645	161
Deviations × places	8	315	39
Pooled error	60	6,480	108

14.22 Heterogeneity of the Experimental Error Variances. A second criticism concerns the assumption that the experimental error variances σ_e^2 are the same in all experiments. In general this assumption will hold only if all experiments have been conducted in the same way, with the same amount of control over environmental conditions and with experimental material of the same variability. In experiments with crops or animals this degree of uniformity is seldom attainable, because the natural variability among pieces of land or among animals at one place differs from that at other places. Hence in this and many other types of cooperative experimentation we expect *a priori* that experimental error variances change from place to place.

When there is doubt about this point, Bartlett's test of homogeneity of variances can be applied to the error m.s., s^2, in the experiments. In the example (table 14.1) the s^2 values varied from 43 in experiment 4 to 204 in experiment 5, and the test of homogeneity shows that σ_e^2 cannot be assumed constant.

Variation in σ_e^2 invalidates the F-test of the interactions m.s. against the pooled error m.s. Although the effect on the significance level of F is not known exactly, it operates so that use of the tabular F produces too many significant results. As in the previous section, the extent of the distortion in F can be seen in extreme cases. If one experiment has a much higher error variance than any of the others, F will be distributed approximately as the tabular F with $(t - 1)$ and n' degrees of freedom, where t is the number of treatments and n' is the number of error degrees of freedom in the experiment with the high error variance. In the example, if one of the experiments with 12 error d.f. happened to be much less accurate than the other experiments, F would have 4 and 12 d.f. instead of 16 and 60 as used in the test in section 14.13. The 5% level would be 3.26 instead of 1.81. Since this case appears to be the most unfavorable that would occur, we may conclude that the true significance level of F lies somewhere between these values.

If the observed F in our data falls outside these limits, a definite conclusion can be reached from the test without further knowledge of the exact significance level of F. This happens in the example, where the observed F-ratio for the interactions m.s. is 3.96. However, even if the exact significance level were known, the test can be criticized because the F-ratio is no longer the most sensitive test criterion. Studies have shown that, with the amount of variation in σ_e^2 that appears typical of agricultural experimentation, this loss of sensitivity might be equivalent to discarding 10 to 20% of the data. An approximate test that avoids some of the loss in sensitivity will be presented in section 14.4.

Thus far we have been discussing how the F-test of the interactions is affected by heterogeneity in the error variances. The F-test of treatments may now be considered. If the interactions are regarded as negligible, so that the denominator of F is the pooled m.s. for interactions and error, the F-test of treatments is invalidated in about the same way as that of interactions. On the other hand, when interactions are present, and especially when they are large, the F-test of the treatments m.s. against the interactions m.s. is much less disturbed. The reason may be clearer if we consider the test of the linear component of treatments in table 14.5. The suggested denominator for this test was the linear $\times$ places component, 1473. This is nearly 14 times the pooled error m.s. It appears, therefore, that heterogeneity in the error variances can influence only a small part of the interaction m.s., so that its effect is, as it were, greatly reduced. These remarks apply with much less force to the quadratic $\times$ places m.s., to which, taking the data at their face value, the pooled error contributes more than half.

14.23 Summary. When all experiments are identical in structure, a combined analysis of variance can be computed relatively easily. Application of the ordinary tests of significance to this analysis is frequently open to question because of heterogeneity in the error and interaction variances. Despite this, it is advisable to draw preliminary conclusions as far as possible from this analysis, at least in the present state of our knowledge, because tests that are fully efficient and theoretically sound have not yet been discovered, and the approximate tests that have been devised to meet the criticisms are more laborious.

When the interactions are sizable, there is usually little difficulty in interpreting the combined analysis. For, if the F-ratio for interactions is large, we may be confident that it is statistically significant in spite of some uncertainty about the exact significance level of F, and the criticism that the F-test is not fully sensitive carries less weight if F establishes significance. Further, as we have seen, the F-test of the treatments m.s. against the interactions m.s. is little affected by inequality in the error variances when the interactions are large. The chief point to remember is that the interactions m.s. may itself be heterogeneous. Subdivision of the treatments and interactions m.s. according to the treatment comparisons that are of greatest importance is often useful.

Another situation that presents little difficulty occurs when the average differences among treatments are substantial, yet interactions are negligible. In this case it may be found that neither the F value for interactions nor that for treatments is close to the significance level. The cases that leave us in doubt are those where the F values are just slightly above the tabular significance levels.

14.3 Experiments of Unequal Size

14.31 Numerical Example. Provided that the same set of treatments appears in all experiments, a combined analysis can usually be constructed even when the experiments differ in size and structure. Some issues arise, however, that are not encountered when all experiments are identical in design. The problems will be illustrated by the inclusion of the first experiment in the previous example. This was arranged in randomized blocks with 4 replications, whereas all other experiments were in

5 × 5 latin squares*. The treatment totals and numbers of replications are shown in table 14.6.

TABLE 14.6 TREATMENT TOTALS (in 100 lb.)

Pounds P_2O_5	Experiment 1	2	3	4	5	6	Total	No. of plots
0	456	710	900	850	650	850	4,416	29
40	824	880	940	855	700	890	5,089	29
80	924	1,005	1,035	940	750	935	5,589	29
120	948	1,025	1,040	925	760	940	5,638	29
160	1,008	1,085	1,085	945	780	945	5,848	29
Total	4,160	4,705	5,000	4,515	3,640	4,560	26,580	145
No. of plots	20	25	25	25	25	25		

A combined analysis can be obtained by following the standard procedure for data based on unequal numbers. The square of any quantity is divided by the number of replications involved. The sums of squares are given below.

$$\text{Total: } \frac{(456)^2 + \cdots + (1008)^2}{4} + \frac{(710)^2 + \cdots + (945)^2}{5} - \frac{(26{,}580)^2}{145} = 131{,}925$$

$$\text{Places: } \frac{(4160)^2}{20} + \frac{(4705)^2 + \cdots + (4560)^2}{25} - \frac{(26{,}580)^2}{145} = 55{,}509$$

$$\text{Treatments: } \frac{(4416)^2 + \cdots + (5848)^2}{29} - \frac{(26{,}580)^2}{145} = 45{,}613$$

$$\text{Treatments} \times \text{places: } 131{,}925 - 55{,}509 - 45{,}613 = 30{,}803$$

The pooled error m.s. requires a little thought. If all experiments have the same true error variance σ_e^2, the best procedure is to weight each error m.s., s^2, by the number of degrees of freedom n_i. But if the experiments vary in accuracy, this weighted mean is a biased estimate of the component of error variance that enters into the interaction and treatments m.s. The correct component is $\sum r_i\sigma_i^2/\sum r_i$ instead of $\sum n_i\sigma_i^2/\sum n_i$. Consequently, unless we are confident that σ_e^2 is constant, it is best to weight the s_i^2 values by the numbers of replications. This gives

$$\bar{s}_e^2 = \frac{4(106) + 5(83 + 158 + \cdots + 52)}{29} = 108$$

* Since experiment 1 contained additional treatments (not discussed here), it provided 21 error d.f.

The F-test of the interactions m.s. against the error m.s. is carried out in the same way as with experiments of equal size, and is subject to the criticisms previously discussed. The F-test of the treatments m.s. against the interactions m.s. encounters a new difficulty that is due to

TABLE 14.7 PRELIMINARY ANALYSIS OF VARIANCE FOR EXPERIMENTS OF UNEQUAL SIZE

Source of variation	d.f.	s.s.	m.s.
Places	5	55,509	11,102
Treatments	4	45,613	11,403
Treatments × places	20	30,803	1,540
Pooled error m.s.	81		108

the unequal numbers of replications and is present even if all the assumptions required for the analysis of variance are satisfied. Under these assumptions, the expectations of the principal mean squares in table 14.7 are as given in table 14.8.

TABLE 14.8 EXPECTED VALUES OF MEAN SQUARES (WITH UNEQUAL NUMBERS OF REPLICATIONS)

Source of variation	Expected value of mean square
Treatments	$\sigma_e^2 + \bar{r}_1\sigma_\mu^2 + (\sum r_i)\dfrac{\sum(\tau_j - \bar{\tau})^2}{(t-1)}$
Treatments × places	$\sigma_e^2 + \bar{r}_2\sigma_\mu^2$
Pooled error	σ_e^2

where

$$\bar{r}_1 = \frac{\sum r_i^2}{\sum r_i} = \frac{141}{29} = 4.862$$

$$\bar{r}_2 = \frac{1}{(p-1)}\left(\sum r_i - \bar{r}_1\right) = \frac{1}{5}(29 - 4.862) = 4.828$$

The complicating factor is that the coefficient of σ_μ^2 in the expected treatments m.s. is not the same as that in the expected treatments × places m.s.; in fact, it is always larger. The difference is very small in this example, as it is whenever the experiments do not vary much in numbers of replications. In addition, and for the same reason, the interactions m.s. is not distributed as a multiple of chi-square; that is, it does not have the type of distribution required for the validity of the F-test. Both factors tend to make the F-test give too many significant results.

A crude method which at least prevents us from being led too far astray is to adjust the F-ratio so as to remove the upward bias in expectation. Let the expected values of the three mean squares be denoted by θ_t, θ_{tp}, and θ_e, respectively. Then, if the null hypothesis is true (all τ_j equal), it is seen from table 14.8 that the three expected values are connected by the equation

$$\theta_t + (k - 1)\theta_e = k\theta_{tp}$$

where $k = \bar{r}_1/\bar{r}_2$. In other words, the test that we seek is a test that this relation holds. This suggests that F might be computed as

$$F' = \frac{s_t^2 + (k - 1)\bar{s}_e^2}{ks_{tp}^2} = \frac{11{,}403 + 0.007(108)}{(1.007)(1540)} = 7.35$$

as compared with the original F-ratio of 11,403/1540, or 7.40. The significance level of F' will also be altered slightly from that in the table.

When interactions are likely to be large, and the chief interest centers in the test of the treatments m.s. against the interactions m.s., an alternative approach has much to commend it. This is to compute the analysis from the *unweighted* treatment means given in table 14.1, paying no attention to the numbers of replications. With this type of analysis the bias disappears, the expectations of the mean squares being shown in table 14.9. It will be noted that σ_e^2 and σ_μ^2 carry the same coefficients

TABLE 14.9 EXPECTED VALUES OF MEAN SQUARES IN AN UNWEIGHTED ANALYSIS

Source of variation	Expected value of mean square
Treatments	$\dfrac{\sigma_e^2}{\bar{r}_h} + \sigma_\mu^2 + p\dfrac{\sum (\tau_j - \bar{\tau})^2}{(t-1)}$
Treatments × places	$\dfrac{\sigma_e^2}{\bar{r}_h} + \sigma_\mu^2$

in both expectations. The divisor $\bar{r}_h$ is the harmonic mean of the numbers of replications, given by $\bar{r}_h = p/\sum (1/r_i)$. Moreover, if σ_μ^2 is much larger than $\sigma_e^2/\bar{r}_h$, the interactions m.s. tends to be distributed as a multiple of chi-square, so that the conditions for the F-test are closely approximated.

It is suggested that the pooled error m.s. for insertion in an analysis of this type be calculated from the formula

$$\bar{s}_e^2 = \frac{1}{p}\left(\frac{s_1^2}{r_1} + \frac{s_2^2}{r_2} + \cdots + \frac{s_p^2}{r_p}\right)$$

By an extension of the results in table 14.9 to the case where the experiments have different error variances, this quantity is found to be an unbiased estimate of the error component that enters into the mean squares for treatments and treatments × places. For the data in table 14.1 we have

$$\bar{s}_e^2 = \frac{1}{6}\left(\frac{106}{4} + \frac{83 + 158 + \cdots + 52}{5}\right) = 22.4$$

TABLE 14.10 UNWEIGHTED ANALYSIS OF VARIANCE OF TREATMENT MEANS

Source of variation	d.f.	s.s.	m.s.
Places	5	11,692	2,338
Treatments	4	10,554	2,638
Treatments × places	20	7,158	358
Pooled error	81		22.4

Table 14.10, being in terms of treatment means, is in different units from table 14.7 (p. 557). As a conversion factor, the harmonic mean $\bar{r}_h$, which in this case is 4.8, may be used to multiply the mean squares above for comparison with those in table 14.7, though this conversion is not needed for our purpose.

In a choice between a weighted and an unweighted analysis, the following are the relevant considerations. The statements below are strictly true only if all experiments have the same error variance per observation (or plot), though they should remain substantially true with a moderate variation in σ_i^2. The weighted analysis is superior for the F-test of the interactions. It gives a more powerful test, and the F-ratio approximates the tabular distribution of F more closely. It is also superior for the F-test of treatments if interactions are non-existent or small, since in this case the bias in the test is negligible. The unweighted analysis is preferable for the F-test of treatments when the interactions are not negligible. As it happens, the results in the example are not too well in accord with these statements. For the interactions test, the F-ratio is 14.26 with the weighted analysis and 15.98 with the unweighted analysis. The explanation of the higher value with the unweighted analysis is probably that experiment 1, whose responses differed from those in the other experiments, had the smallest number of replications. The values of F in the test of treatments were practically identical, being 7.35 (after adjustment) in the weighted analysis and 7.37 in the unweighted analysis.

14.32 The Combined Analysis for a Series of Lattice Experiments. In the discussion of lattice experiments it was pointed out that if inter-block information is recovered, the F-test of treatments in an individual experiment is not exact, because the relative weights attached to inter- and intra-block estimates are subject to sampling errors. For the same reason any combined analysis must also be approximative rather than exact.

The designs need not be identical at all places. With 25 treatments, for example, some experiments might be simple lattices in 4 replications, while others are lattice squares in 3 replications. Also, inter-block information need not be recovered in all analyses; at some places we might have used a randomized blocks analysis.

If the experiments have different numbers of replications, the previous section has indicated that there is a question whether to use a weighted or an unweighted analysis. The former is preferable when the chief purpose is to test the interactions with places, and when such interactions are negligible; the latter when interactions are sizable and the principal interest is in a test of the average effects of treatments. With lattice experiments as used in agriculture, we are often interested in both interactions and average effects. But as interactions are seldom absent, the unweighted analysis is frequently advisable, and has the virtue of being slightly simpler.

The weighted analysis will be described first. First form a two-way "treatments × places" table similar to table 14.6. The entries in the table are the *adjusted* treatment totals, except in experiments analyzed by randomized blocks, where unadjusted totals are used. This table is analyzed into components for

Places
Treatments
Treatments × places

If the numbers of replications r_i differ, the weighted analysis follows the same procedure as in table 14.6 (p. 556). The square of every marginal treatment total is divided by $\sum r_i$ in order to reduce the analysis of variance to a single-observation basis. The square of a place total, summed over all treatments, is divided by tr_i, where t is the number of treatments.

The only additional component is the pooled error m.s. In any experiment where inter-block information has been recovered, we use as the estimate of the error variance per observation (or plot) the *effective* error m.s. E_i'. This quantity is obtained by adjusting the intra-block m.s. upwards so as to allow for sampling errors in the block corrections, and its calculation is explained along with the analysis for each type of lattice. For the triple lattice, for example, E_i' is $E_e\left[1 + \frac{rk\mu}{k+1}\right]$. If there are any experiments in which the intra-block analysis has been carried out, E_i' is given the value which it takes when E_b becomes very large relative to E_i. For experiments analyzed by randomized blocks, E_i' is the ordinary error m.s.

On the assumption that the true effective variances are unlikely to be the same in all experiments, the pooled error m.s. is calculated as $\sum r_i E_i'/\sum r_i$. When r_i is constant, this reduces to the unweighted mean of the E_i'. Tests of significance are made as in the numerical example, and are subject to the same general criticisms. Since the larger lattice designs provide ample degrees of freedom for error, the pooled error may be relatively well determined even if some experiments are much more accurate than others, so that there is less uncertainty about the true significance levels of the F-ratios.

For an unweighted analysis with experiments of unequal size, construct a two-way table of adjusted treatment *means*. This is analyzed by the simple standard procedure. The comparable pooled error m.s. is taken as

$$\bar{s}_e^2 = \frac{1}{p}\left(\frac{E_1'}{r_1} + \frac{E_2'}{r_2} + \cdots + \frac{E_p'}{r_p}\right)$$

With balanced designs, the method given above is a natural extension of the technique used to obtain an approximate F-test of treatments in an individual experiment. With partially balanced designs the method is more crude than that used in single experiments, where a special supplementary calculation was made for the F-test. This calculation can be extended to apply to a combined analysis, but it is doubtful whether the elaboration is worth while.

14.4 A Test of the Treatments × Places Interactions

In this section we give an approximate test of the treatments × places interactions for series of experiments where there is considerable variation in the experimental error variances. It was suggested that the F-test, despite its imperfections, will often serve our purpose. The present test differs from the F-test in that it gives less weight to experiments which have a high error variance, and consequently may be expected to be more sensitive. It may be useful in cases where there is doubt about the verdict given by the F-test and in cases where the most efficient analysis is desired.

As before, let x_{ij} be the mean of the jth treatment, and σ_i^2 be the true error variance per observation in the ith experiment. We assume that all experiments have the same design, and let s_i^2 be the error m.s. per observation in the ith experiment, based on n degrees of freedom.

For *known* values of σ_i^2 the theory of least squares indicates that the mean x_{ij} should receive a weight $W_i = r/\sigma_i^2$. In a weighted analysis of variance of the treatment means, the interaction s.s. is known to be distributed as chi-square with $(p - 1)(t - 1)$ degrees of freedom. This sum of squares appears to be the best test criterion available, unless we possess specific information about the type of interaction that may exist, in which event a more specialized criterion would be constructed.

In default of knowledge of the σ_i^2, the natural step is to consider a weighted analysis with weights $w_i = r/s_i^2$. That is, we weight each mean inversely as its *estimated* variance, since we do not know its true variance. This is satisfactory, provided that the s_i^2 are good estimates of the σ_i^2. Upon examination it appears that the s_i^2 should be based on at least 15 d.f. To illustrate the calculations, the test will be applied to the data from the last 5 experiments, though the number of degrees of freedom, 12, is slightly too low to use the test with full confidence.

TABLE 14.11 Treatment mean yields (100 lb. per acre) for a weighted analysis of variance

Pounds P_2O_5	Experiment 2	3	4	5	6	Weighted total $\sum w_i x_{ij} = T_j$
0	142	180	170	130	170	53.570
40	176	188	171	140	178	57.000
80	201	207	188	150	187	62.194
120	205	208	185	152	188	62.264
160	217	217	189	156	189	63.932
w_i	0.060	0.032	0.116	0.025	0.096	0.329 = W
Total (P_i)	941	1000	903	728	912	
w_iP_i	56.460	32.000	104.748	18.200	87.552	298.960 = G
s.s. (S_i)	180,655	200,946	163,431	106,440	166,618	

The arrangement of the data in table 14.11 should be noted. The treatment means and s_i^2 values come from table 14.1 (p. 547). The weights for individual entries in the table are placed in the row immediately below the entries; for example, $w_1 = 5/83 = 0.060$. The total of the weights is denoted by W.

1. Form the column totals P_i and the products w_iP_i. Form the weighted row totals. The corner total, $G = 298.960$, supplies a check on both the column and row totals.

2. The items in the analysis of variance are obtained as follows.

Correction term:

$$C = \frac{G^2}{tW} = \frac{(298.960)^2}{(5)(0.329)} = 54,332.57$$

Total: First compute and record in the table the sum of squares S_i (uncorrected) of the entries in each column. The total s.s. is then found as

$$\sum (w_i S_i) - C = (0.060)(180{,}655) + \cdots + (0.096)(166{,}618) - C = 551.33$$

Places: A column (place) total has variance $t\sigma_i^2/r$, and hence receives a weight w_i/t. The sum of squares is

$$\tfrac{1}{5}\sum (w_i P_i^2) - C = 230.09$$

This is found readily by using the products $w_i P_i$.

Treatments: Each treatment total has estimated weight W. The sum of squares is

$$\frac{\sum T_j^2}{W} - C = \frac{(53.570)^2 + \cdots + (63.932)^2}{0.329} - C = 229.57$$

Treatments × places: Since this is the only component in which we are currently interested, it is unfortunate that it must be found by means of the others. As usual, it is given by

$$551.33 - 230.09 - 229.57 = 91.67 = I$$

The sum of squares I does not follow a chi-square distribution, being inflated by errors in the weights. It can be reduced to a quantity that is distributed approximately as chi-square.* We take

$$\chi^2 = \frac{(n-4)(n-2)}{n(n+t-3)} I, \quad \text{with} \; \frac{(p-1)(t-1)(n-4)}{(n+t-3)} \; \text{d.f.}$$

In this case

$$\chi^2 = \frac{(8)(10)}{(12)(14)}(91.67) = 43.65, \quad \text{with} \; \frac{(16)(8)}{(14)} = 9.14 \; \text{d.f.}$$

In this approximation the degrees of freedom ascribed to chi-square are not integral. Significance levels are obtained by linear interpolation in the tables of chi-square. Since the 1% levels of chi-square are 21.67 and 23.21 for 9 and 10 d.f., respectively, I is obviously significant.

The test can also be applied to any component of the interactions s.s. As an example, we will test the interaction of the quadratic component of the regression on amount of phosphorus. This was previously tested by the F-test in table 14.5, and found non-significant. The values given

* This test is an extension and modification of a test previously given by Cochran (14.6).

for the quadratic components in section 14.21 were derived from treatment *totals;* to obtain corresponding values from means, we divide by 5.

	Experiment					
	2	3	4	5	6	Total
Q_i	−65	−16	−14	−20	−22	
w_iQ_i	− 3.900	− 0.512	− 1.624	− 0.500	− 2.112	−8.648

Each quantity has variance $14\,\sigma_i^2/r$, and receives a weight $w_i/14$. The sum of squares is

$$\frac{1}{14}\left(\sum w_iQ_i^2 - \frac{(\sum w_iQ_i)^2}{W}\right) = \frac{1}{14}(340.89 - 227.32) = 8.11 = I_q$$

The quantity I_q is the sum of squares for the interaction of a single treatment comparison with all p places, and would normally have $(p - 1)$ degrees of freedom. To present the general formula for conversion to chi-square, suppose that we have computed the sum of squares for the interaction of t_1 treatment comparisons with p_1 of the places, so that there would normally be $t_1(p_1 - 1)$ degrees of freedom. Then

$$\chi^2 = \frac{(n-4)(n-2)}{n(n+t_1-2)} I_q, \quad \text{with} \frac{t_1(p_1-1)(n-4)}{(n+t_1-2)} \text{ d.f.}$$

This formula is in accord with that used for the whole $(t - 1)(p - 1)$ degrees of freedom in the interactions. Since the t treatments provide $(t - 1)$ independent treatment comparisons, we would take $t_1 = (t - 1)$ in applying the formula to the complete interactions s.s. For $t_1 = 1$, $p = 5$,

$$\chi^2 = \frac{(8)(10)}{(12)(11)}(8.11) = 4.92, \quad \text{with} \frac{(4)(8)}{(11)} = 2.91 \text{ d.f.}$$

The tables show that this value has a probability between 0.2 and 0.1, since by interpolation the 20% level for 2.91 d.f. is 4.51, the 10% level 6.10. The earlier F-test gave a probability slightly over 0.2, so that the two tests are in close agreement.

If the experiments differ in size, the weight w_i becomes r_i/s_i^2. The numbers of error degrees of freedom will also vary. This necessitates a more elaborate conversion formula: as a rough approximation the average number of error degrees of freedom per experiment is used in place of n.

The test is not recommended for low values of n, because the s_i^2 are relatively poor estimates of the σ_i^2, and frequently one or two experiments receive such high weights that they dominate the analysis. Appropriate methods are presented by Yates and Cochran (14.5) and Cochran (14.6). Somewhat more exact tests of a weighted sum of squares of deviations, as in the test of the quadratic components, have been worked out by James (14.1a) and Welch (14.2a).

14.5 Repetitions in Both Space and Time

As has been mentioned, agricultural field experiments are often repeated both at a number of places and for a number of years. We will give only an introduction to the simplest case, in which experiments have been carried out at each of p places for y years. At any place a new site is chosen each year for the experiment, and a new randomization employed, so that the data from successive years may be regarded as independent. The experimental arrangement need not be uniform throughout the whole series of experiments: in fact, a change in design between seasons is not uncommon.

The mathematical representation of the mean of the jth treatment at the ith place in the kth year is now more lengthy. In addition to the symbols previously used for place and treatment effects, γ_k will denote the effect of the year. Interactions are denoted by multiple symbols: thus $(\pi\tau)_{ij}$ is the contribution of the place $\times$ treatment interaction in this experiment. With this notation

$$x_{ijk} = \mu + \pi_i + \tau_j + \gamma_k + (\pi\tau)_{ij} + (\pi\gamma)_{ik} + (\tau\gamma)_{jk} + (\pi\tau\gamma)_{ijk} + \bar{e}_{ijk}$$

Note that there are treatment $\times$ place, treatment $\times$ year, and treatment $\times$ place $\times$ year interactions.

In experimental programs of this type, it is usually hoped that the places and years constitute a representative sample of the population of places and years to which the results will be applied. There are obvious practical difficulties in choosing places and years that can be confidently asserted to be such a representative sample, and sometimes little effort is made to ensure that this will be so. The hard fact is that any statistical inferences drawn from an analysis of the data will apply only to the population (if one exists) of which the experiments are a random sample. If this population is vague and unreal, the analysis is likely to be a waste of time, at least from the strictly practical point of view.

It seems appropriate to regard all the quantities in the equation as random variables except the general mean μ and the true effects τ_i of the treatments, because if we could tabulate all the values of say $(\pi\tau)_{ij}$ in the population, they would follow some frequency distribution from which the values in our data are a sample. For the full application of the analysis of variance we require the assumptions that the experimental errors and all interactions of treatments are normally and independently distributed, with variances that are denoted by use of the letters that enter into the interaction.

An unweighted analysis of variance of the treatment means is obtained by the standard procedure for factorial experiments. The important parts of his analysis are sketched in table 14.12.

TABLE 14.12 ANALYSIS OF VARIANCE OF TREATMENT MEANS WITH TIME AND PLACE VARIATIONS

Source of variation	d.f.	Expectation of mean square
Treatments	$(t-1)$	$\sigma_e^2 + \sigma_{tpy}^2 + p\sigma_{ty}^2 + y\sigma_{tp}^2 + py\dfrac{\sum(\tau_j - \bar{\tau})^2}{(t-1)}$
Treatments × places	$(t-1)(p-1)$	$\sigma_e^2 + \sigma_{tpy}^2 + y\sigma_{tp}^2$
Treatments × years	$(t-1)(y-1)$	$\sigma_e^2 + \sigma_{tpy}^2 + p\sigma_{ty}^2$
Treatments × places × years	$(t-1)(p-1)(y-1)$	$\sigma_e^2 + \sigma_{tpy}^2$
Pooled error		σ_e^2

In the most general case, σ_e^2 is the average value of σ_{ik}^2/r_{ik}, where σ_{ik}^2 is the error variance in the individual experiment at the ith place in the kth year, and r_{ik} is the number of replications in that experiment. Consequently, the pooled error should be estimated as the average of s_{ik}^2/r_{ik}.

The method to be followed in testing the significance of the successive terms is made clear by the expected values of the mean square. The treatments × places × years interactions are tested against the pooled error. Both treatments × years and treatments × places are tested against treatments × places × years.

The situation with regard to the test for treatments is interesting. It is evident that no other mean square in the analysis is suitable as a denominator of F, unless either the treatments × places or the treatments × years interactions appear to be negligible, so that the corresponding variance is assumed to be zero. If this happens, the other interaction m.s. is an appropriate denominator for F (subject to some uncertainty in case the assumption should not be correct).

When both two-factor interactions are present, the hypothesis that all τ_j are zero is equivalent to the hypothesis that

$$\theta_t + \theta_{tpy} = \theta_{tp} + \theta_{ty}$$

where the θ's stand for the expected mean squares in table 14.12. No exact test for this kind of relationship is at present known. By analogy with the F-test, one suggestion is to use the ratio

$$\frac{s_t^2 + s_{tpy}^2}{s_{tp}^2 + s_{ty}^2}$$

where the s^2 values are the mean squares in the analysis of variance. This ratio does not follow the F-distribution, but following an approximation suggested by Satterthwaite (14.7) and others, we might use the F-tables with n_1' and n_2' degrees of freedom, where

$$n_1' = \frac{(s_t^2 + s_{tpy}^2)^2}{\dfrac{s_t^4}{n_t} + \dfrac{s_{tpy}^4}{n_{tpy}}}$$

where the n's are the numbers of degrees of freedom in the corresponding mean squares. The analogous expression is used for n_2'.

After completion of the analysis, the next step is to examine the nature of the interactions and average effects of the treatments, bringing to bear any external knowledge that will throw light on the interpretation. Often, for the practical uses of the results, we would like to decide whether a single recommendation can be made for application in the whole population, or whether, on the other hand, it is necessary to have several recommendations for different parts of the population. These decisions are facilitated by calculating confidence limits for the differences among the means of the most likely candidates. For a discussion of methods for making this calculation, which can give rise to some complications, see reference (14.5).

REFERENCES

14.1 Bliss, C. I. The U.S.P. collaborative cat assay for digitalis. *Jour. Amer. Pharm. Assoc.* 33, 225–245, 1944.

14.2 Nelson, W. L., and Hawkins, A. Response of Irish potatoes to phosphorus and potassium. *Jour. Amer. Soc. Agron.* 39, 1053–1067, 1947.

14.3 Bartlett, M. S. Some examples of statistical methods of research in agriculture and applied biology. *Jour. Roy. Stat. Soc. Suppl.* 4, 121, 1937.

14.4 Fisher, R. A., and Yates, F. *Statistical tables for biological, agricultural and medical research.* Oliver and Boyd, Edinburgh, 3rd ed., 1948.

14.5 Yates, F., and Cochran, W. G. The analysis of groups of experiments. *Jour. Agr. Sci.* 28, 556–580, 1938.

14.6 Cochran, W. G. Problems arising in the analysis of a series of similar experiments. *Jour. Roy. Stat. Soc. Suppl.* 4, 102–118, 1937.

14.7 Satterthwaite, F. E. An approximate distribution of estimates of variance components. *Biom. Bull.* 2, 110–114, 1946.

14.1a James, G. S. The comparison of several groups of observations when the ratios of the population variances are unknown. *Biometrika* 38, 324–329, 1951.

14.2a Welch, B. L. On the comparison of several mean values: an alternative approach. *Biometrika* 38, 330–336, 1951.

ADDITIONAL READING

Cochran, W. G. Long-term agricultural experiments. *Jour. Roy. Stat. Soc. Suppl.* 6, 104–148, 1939.

Cochran, W. G. The combination of estimates from different experiments. *Biometrics* 10, 101–129, 1954.

Crowther, F., and Cochran, W. G. Rotation experiments with cotton in the Sudan Gezira. *Jour. Agr. Sci.* 32, 390–405, 1942.

Snedecor, G. W., and Haber, E. S. Statistical methods for an incomplete experiment on a perennial crop. *Biometrics* 2, 61–67, 1946.

Yates, F. The analysis of experiments containing different crop rotations. *Biometrics* 10, 324–346, 1954.

CHAPTER 15

Random Permutations of 9 and 16 Numbers

15.1 Use of the Random Permutations

In practically all the experimental designs, the randomization consists in arranging a set of objects, whether treatments, blocks, rows, or columns, in random order. If the number of objects does not exceed 9, a randomization is obtained at once from table 15.6, which contains 1000 random arrangements of the numbers from 1 to 9. For example, to arrange 7 treatments in random order, select a starting place in table 15.6 (without inspection of the numbers in the table). Suppose that the permutation chosen is

1 8 9 2 6 7 5 3 4

Omitting the digits 8 and 9, we obtain

1 2 6 7 5 3 4

for the desired arrangement. Similarly, table 15.7, which contains 1000 permutations of the numbers between 1 and 16, may be used for any number of objects between 2 and 16.

Occasionally the randomization involves dividing the objects into groups, as in the completely randomized and cross-over designs. Thus, with a cross-over design (section 4.4) having 2 treatments and 10 blocks, treatment A must appear in the first row in five of the blocks, chosen at random, and in the second row in the remaining five blocks. Select a random permutation of the numbers up to 16 from table 15.7, say

6 16 10 5 12 7 9 4 1 2 15 14 13 11 8 3

The numbers from 11 to 16 are ignored. The first five of the remaining numbers, 6, 10, 5, 7, and 9, are chosen as the blocks in which A appears in the first row.

If the number of objects to be randomized exceeds 16, see section 15.3.

15.2 Construction of the Random Permutations

The idea of using random permutations instead of the ordinary random digits was suggested by Professor George W. Snedecor. Before the writing of this book, 200 random permutations of each of the numbers from 3 to 13 had been made at the Statistical Laboratory at Iowa State College. These were obtained by first constructing a large number of ordinary random digits, which were then transformed into random permutations. The method of transformation was the usual one: it may be illustrated for the numbers from 1 to 11. To obtain a random permutation of the numbers 1 to 11, a series of pairs of random digits is taken, say

65, 04, 29, 82, 37, 52, 53, etc.

When we divide each number by 11 the remainders are

10, 4, 7, 5, 4 (omitted), 8, 9, etc.

This process is continued until 10 of the 11 numbers have appeared, repetitions being omitted.

Rather than present 200 permutations of each of a series of numbers, it was decided to concentrate on two numbers. This decision was made partly because any permutation of 9 numbers, say, can be used to give permutations of all numbers less than 9. Thus 1000 permutations of 9 numbers are more useful than 200 of each of the numbers 9, 8, 7, 6, and 5. Also, certain numbers, such as 11 and 13, are less often used in experiments than others. The numbers 9 and 16 were selected for presentation because of their occurrence in factorial experiments.

The 1000 permutations of 9 were obtained from the permutations already available for 9, 10, 11, 12, and 13, by omitting the numbers above 9. Of the permutations of 16, 800 were obtained from random digits. With as many as 16 numbers to permute, the method of division as illustrated above is rather slow. A more rapid alternative, suggested by Mr. Paul Peach, was used. In this method 16 pairs of random digits are taken, for example,

75	10	17	28	91	85	74	56	71	06	26	01	06	21	07	60
14	5	6	9	16	15	13	10	12	3	8	1	2	7	4	11

The permutation is produced by ranking the pairs in order of size, as shown below each pair. In the event of a tie, as happens in this example with the two 06's, the two ranks in question may be allotted from an additional random digit. Ties occur frequently (the probability that all 16 pairs are distinct is about .28) but cause little delay. Although the method has the additional merit of using fewer random digits than the method of division, with 32 random digits per permutation a large supply of random digits is required. The remaining 200 permutations were obtained by drawing numbered marbles from an urn.

15.3 Randomization of More than 16 Numbers

The most suitable method depends on the facilities available and on the frequency with which randomizations have to be made. Navy beans, drawn from a box, have been found quite expeditious and convenient, though the numbers on the beans rub off with repeated usage. Alternatively one of the standard sets of random digits, references (15.1–15.4), may be used. For example, with 25 numbers to be placed in random order, choose a starting place in the table and select the next 25 sets of *three-digit* random numbers. These numbers are then placed in increasing order by the method used in the previous section for the permutations of 16. Three-digit random numbers are taken instead of two-digit numbers in order to avoid ties. With three-digit numbers the frequency of ties is negligible even if as many as 100 numbers are to be randomized. If punched card machines are available, the numbers 1 to 25 may be punched in columns 1 and 2 and the random digits in columns 3, 4, and 5 of the card. The permutation may then be produced and printed by sorting on columns 3 to 5. By filling all the columns from 3 onwards with random digits, a number of permutations can be obtained by sorting on different sets of columns, though the number is limited if the permutations are to be kept independent of each other.

15.4 Tests of Randomness

Many tests of different aspects of the randomness of the permutations could be made. So far as their use in experimental design is concerned, the test most immediately relevant is a test of the null hypothesis that

TABLE 15.1 Number of occurrences of 1, 2, ···, 9 in 1st, 2nd, ···, 9th position in 1000 permutations of 9

Number	Position									Totals
	1	2	3	4	5	6	7	8	9	
1	130	111	111	108	94	121	114	106	105	1000
2	120	118	104	102	102	133	98	108	115	1000
3	92	100	116	111	113	110	116	123	119	1000
4	121	98	97	108	117	112	124	102	121	1000
5	95	103	120	140	122	84	113	113	110	1000
6	107	118	118	88	113	120	124	111	101	1000
7	108	119	121	124	110	94	95	113	116	1000
8	115	116	110	98	107	109	117	121	107	1000
9	112	117	103	121	122	117	99	103	106	1000
Totals	1000	1000	1000	1000	1000	1000	1000	1000	1000	1000

TABLE 15.2 Number of occurrences of 1, 2, ···, 16 in 1st, 2nd, ···, 16th position in 1000 permutations of 16

Number	Position																Totals
	1	2	3	4	5	6	7	8	9	10	11	12	13	14	15	16	
1	52	70	50	82	57	55	71	58	75	63	59	59	61	55	59	74	1000
2	74	60	71	58	52	64	65	73	70	54	55	59	66	53	58	68	1000
3	66	56	48	59	57	71	70	65	73	59	61	78	66	54	66	51	1000
4	64	57	66	63	65	70	64	58	62	68	56	67	55	59	61	65	1000
5	67	54	71	55	82	63	53	57	63	60	63	57	71	68	62	54	1000
6	72	60	62	68	68	63	64	59	67	62	51	63	56	55	74	56	1000
7	61	50	57	58	68	65	64	62	56	74	70	54	61	79	65	56	1000
8	64	53	58	65	71	67	73	67	46	58	66	60	66	55	58	73	1000
9	52	58	64	64	75	63	68	57	49	63	71	58	78	66	53	61	1000
10	54	69	58	62	63	60	63	57	59	62	65	81	67	51	58	71	1000
11	58	62	62	74	58	68	51	63	49	53	65	72	58	72	65	70	1000
12	77	56	66	56	61	59	58	67	55	72	77	66	55	72	56	47	1000
13	57	78	69	62	57	54	59	65	79	66	57	55	55	57	64	66	1000
14	62	71	60	49	61	66	56	69	71	68	59	58	59	74	67	50	1000
15	54	75	69	57	51	52	70	69	56	64	60	53	64	69	65	72	1000
16	66	71	69	68	54	60	51	54	70	54	65	60	62	61	69	66	1000
Totals	1000	1000	1000	1000	1000	1000	1000	1000	1000	1000	1000	1000	1000	1000	1000	1000	1000

the numbers 1, 2, $\cdots$, 9 have an equal probability of appearing in the 1st, 2nd, $\cdots$, 9th positions. This test is made by counting the number of times that each number has occurred in each position. The results are shown in two-way presentations in tables 15.1 and 15.2.

For the permutations of 9, the expectation in each cell is 1000/9, or 111.1. Each row and each column in table 15.1 adds to 1000. Taking the values in a single column, we may test by χ^2 whether the 9 numbers are equally represented in the corresponding position. If the x's are the observed entries in the column,

$$\chi^2 = \sum \frac{(x - m)^2}{m} = \frac{9}{1000} \sum \left(x - \frac{1000}{9}\right)^2$$

For computational purposes, this simplifies to

$$\chi^2 = \tfrac{9}{1000} \sum (x^2) - 1000$$

Thus, for the first position,

$$\chi^2 = \tfrac{9}{1000}(130^2 + 120^2 + \cdots + 112^2) - 1000 = 10.81$$

with 8 d.f. Similarly, from the rows we may test whether any given number has appeared equally often, apart from sampling fluctuations, in each position. Although the two sets of tests are not independent, both are of interest. The values of χ^2 are shown in table 15.3.

TABLE 15.3 VALUES OF χ^2 FROM 1000 PERMUTATIONS OF 9

Position test	χ^2	Number test	χ^2
1	10.81	1	7.46
2	5.19	2	9.17
3	5.08	3	6.70
4	17.16	4	7.75
5	6.06	5	18.90
6	15.52	6	8.97
7	8.82	7	8.25
8	3.88	8	3.45
9	3.44	9	5.32
Totals	75.96		75.97

In the *position* tests, the individual probability values range from .90 ($\chi^2 = 3.44$) to .028 ($\chi^2 = 17.16$). With 8 d.f., the probability that the largest of 9 independent values of χ^2 should exceed 17.16 works out at .23. Actually, owing to the fact that the cell numbers add to 1000 in any row or column, the 9 values for the position χ^2 have a small positive correlation which reduces the probability value for the largest χ^2 to about .20. The individual probability values for the *number* tests all lie between .90 ($\chi^2 = 3.45$) and .015 ($\chi^2 = 18.90$). The probability that the largest χ^2 should exceed 18.90 is about 0.11. Thus neither series of tests indicates any marked departure from randomness.

The total, 75.97, which is $\sum (x - m)^2/m$ taken over all cells in table 15.1, supplies a composite test of the complete table. On account of the positive correlation referred to in the previous paragraph, the value 75.97 does not itself follow a χ^2 distribution. In fact, it may be shown to be distributed approximately as $(9/8)\chi^2$, where χ^2 has 64 d.f. Consequently we may test $\frac{(8)(75.97)}{9}$, or 67.52, as a χ^2 value with 64 d.f. As would be expected from the results of the previous tests, the probability value, .36, provides no evidence for the rejection of the null hypothesis.

The corresponding χ^2 values for permutations of 16 are given in table 15.4.

TABLE 15.4 VALUES OF χ^2 FROM 1000 PERMUTATIONS OF 16

Position test	χ^2	Number test	χ^2
1	14.40	1	20.25
2	17.70	2	12.96
3	11.87	3	16.58
4	15.14	4	4.80
5	17.70	5	14.94
6	7.42	6	9.57
7	12.29	7	13.98
8	7.42	8	13.63
9	24.54	9	15.55
10	9.15	10	12.45
11	10.62	11	14.05
12	16.51	12	18.56
13	9.92	13	14.50
14	18.85	14	13.06
15	6.98	15	14.78
16	19.36	16	10.21
Totals	219.87		219.87

The only figure in the group which has an unusual probability is the low value 4.80 for number 4. This has an individual probability of .9932 (15 d.f.). When account is taken of the fact that this is the smallest of the 16 values for the number tests, the probability is found to be .10. The total, 219.87, is distributed as $(16/15)\ \chi^2$ with 225 d.f. and is found to be slightly but not abnormally below its expectation.

A further test of randomness was made by counting the number of *inversions* for each permutation. The inversions are counted as follows. For each number, record how many numbers to the left are greater than the number. Thus for the permutation

$$3\quad 4\quad 8\quad 1\quad 9\quad 7\quad 2\quad 6\quad 5$$

we record

$$0\quad 0\quad 0\quad 3\quad 0\quad 2\quad 5\quad 3\quad 4$$

The total $(3 + 2 + 5 + 3 + 4 = 17)$ gives the number of inversions. The extreme values for the number of inversions are zero (for the permutation 1 2 3 4 5 6 7 8 9) and 36 (for the permutation 9 8 7 6 5 4 3 2 1).

TABLE 15.5 OBSERVED AND THEORETICAL FREQUENCIES OF NUMBERS OF INVERSIONS FOR PERMUTATIONS OF 9

Number of inversions	Observed frequency	Theoretical frequency	Goodness of fit, χ^2
0–6	8	6.3	0.46
7	4	6.0	0.67
8	11	9.9	0.12
9	10	15.3	1.84
10	13	22.1	3.75
11	36	30.4	1.03
12	38	39.7	0.07
13	50	49.5	0.01
14	71	59.1	2.40
15	65	67.8	0.12
16	80	74.6	0.39
17	71	79.0	0.81
18	90	80.6	1.10
19	75	79.0	0.20
20	80	74.6	0.39
21	57	67.8	1.72
22	58	59.1	0.02
23	45	49.5	0.41
24	40	39.7	0.00
25	27	30.4	0.38
26	27	22.1	1.09
27	18	15.3	0.48
28	7	9.9	0.85
29	11	6.0	4.17
30–36	8	6.3	0.46
Totals	1000	1000.0	22.94

In intermediate cases the number of inversions serves as a measure of the extent to which the order is inverted from the natural order towards the reverse order. The frequency distribution of the number of inversions has been studied by Rosander (15.5) and Kendall (15.6), who derived from the number a measure of the rank correlation between two series of figures. For random permutations of 1 to 10 numbers, Kendall also tabulated the exact distribution of the number of inversions. His distribution was extended to 16 figures for the test given here.

For permutations of 9 the observed and theoretical distributions are shown in table 15.5.

The total χ^2, 22.94, has 24 d.f. and agrees closely with expectation. The corresponding value for permutations of 16 is $\chi^2 = 47.52$ with 54 d.f. and a probability value of .72.

REFERENCES

15.1 Tippett, L. H. C. Random sampling numbers (10,400 digits). *Tracts for computers*, XV. Cambridge University Press, 1927.

15.2 Kendall, M. G., and Babington Smith, B. Tables of random sampling numbers (100,000 digits). *Tracts for computers*, XXIV. Cambridge University Press, 1939.

15.3 Fisher, R. A., and Yates, F. *Statistical tables for biological, agricultural and medical research* (15,000 digits). Oliver and Boyd, Edinburgh, 2nd ed., 1943.

15.4 Snedecor, G. W. *Statistical methods* (10,000 digits). Iowa State College Press, Ames, Iowa, 4th ed., table 1.2, 1946.

15.5 Rosander, A. C. The use of inversions as a test of random order. *Jour. Amer. Stat. Assoc.* 37, 352–358, 1942.

15.6 Kendall, M. G. A new measure of rank correlation. *Biometrika* 30, 81–93, 1938.

15.5 Tables of Random Permutations

TABLE 15.6 Permutations of 9

5 5 6 7 1	4 3 3 7 3	8 7 4 6 3	9 7 4 9 4	9 2 2 8 8	2 7 9 3 5	8 3 1 9 4
4 1 2 8 2	7 1 1 2 9	9 5 7 8 2	8 9 3 6 6	1 7 7 2 4	4 8 5 7 3	3 7 4 5 6
9 3 3 2 9	8 8 8 4 5	2 4 6 1 6	3 6 7 7 8	7 4 4 7 1	7 3 2 8 6	6 1 2 2 2
7 9 7 4 3	5 5 2 9 2	1 6 5 3 5	7 8 5 1 9	5 1 9 1 3	6 5 1 4 9	2 9 8 7 8
1 6 9 6 5	6 9 4 3 6	4 3 9 2 9	5 1 8 2 3	8 3 3 3 2	8 9 6 1 2	4 5 7 6 9
6 4 4 3 6	2 4 6 8 1	7 9 3 4 1	6 2 6 4 2	2 9 8 5 9	9 2 4 2 8	9 6 9 8 1
8 7 8 1 7	1 2 5 6 8	3 1 2 9 8	4 4 1 8 7	6 5 1 6 7	5 4 3 5 1	1 4 3 1 7
3 2 1 9 4	3 6 7 5 7	6 8 8 7 7	2 5 9 5 1	3 8 5 4 6	3 6 7 9 4	5 2 5 4 5
2 8 5 5 8	9 7 9 1 4	5 2 1 5 4	1 3 2 3 5	4 6 6 9 5	1 1 8 6 7	7 8 6 3 3
7 4 6 1 5	9 2 2 2 9	2 8 1 7 3	2 4 2 1 9	2 4 8 3 1	2 6 5 4 8	8 4 9 4 2
9 3 8 3 2	1 1 1 9 8	9 4 9 5 4	8 8 8 8 6	7 7 5 4 6	5 3 2 7 6	9 3 8 2 1
1 6 3 4 7	6 5 8 4 5	6 1 7 1 9	5 2 5 6 3	8 5 7 5 5	6 9 9 8 1	3 6 7 9 7
6 8 2 8 4	4 8 7 8 6	5 7 5 1 5	9 6 7 5 8	5 9 9 7 7	8 5 3 3 5	6 9 4 6 9
4 1 4 7 8	2 3 9 3 4	4 2 2 3 6	4 7 4 2 5	6 3 3 6 9	1 7 8 5 4	4 5 2 1 4
2 9 1 9 3	7 9 6 6 2	1 6 4 6 1	7 9 9 7 4	1 8 4 1 8	9 2 7 9 3	1 8 3 5 5
5 5 5 5 1	3 7 4 7 7	8 5 8 9 2	1 5 1 3 2	9 6 2 8 4	3 8 1 1 9	5 7 1 3 3
8 2 9 2 9	8 6 5 5 3	7 9 6 8 8	3 1 6 9 7	4 1 6 9 3	4 4 6 6 2	7 2 6 8 8
3 7 7 6 6	5 4 3 1 1	3 3 3 2 7	6 3 3 4 1	3 2 1 2 2	7 1 4 2 7	2 1 5 7 6
9 7 7 5 5	9 9 9 3 8	9 8 6 1 7	5 8 6 1 2	1 9 8 3 3	3 1 7 7 3	7 6 6 5 5
3 8 1 7 2	6 2 7 1 6	4 1 3 4 2	3 6 2 4 3	2 6 1 2 8	8 8 6 2 7	8 9 7 4 7
4 3 4 2 7	7 3 1 7 2	1 5 4 8 6	6 2 1 6 1	7 8 5 1 7	5 9 1 3 6	3 1 2 3 1
5 9 2 8 3	3 7 5 8 9	2 9 1 7 1	2 3 8 3 4	3 5 9 9 9	7 2 3 4 1	5 7 1 7 8
1 6 5 1 1	5 6 4 4 1	7 3 7 2 3	4 7 3 8 8	9 3 2 5 6	6 6 9 5 9	9 8 9 1 2
6 2 8 3 6	8 4 6 2 5	5 2 2 6 8	9 1 7 5 6	4 7 4 6 4	1 7 4 6 4	1 2 8 8 6
2 4 9 6 4	1 8 3 5 4	3 6 5 9 4	8 5 9 7 9	8 1 6 8 1	4 5 5 9 5	2 4 5 9 4
8 5 6 9 9	2 5 2 6 7	8 7 8 3 9	1 9 4 2 5	6 4 7 4 5	2 3 2 8 2	6 3 3 2 3
7 1 3 4 8	4 1 8 9 3	6 4 9 5 5	7 4 5 9 7	5 2 3 7 2	9 4 8 1 8	4 5 4 6 9
7 4 9 8 7	9 7 1 7 1	9 2 3 8 7	7 8 5 3 5	5 1 6 4 9	7 8 6 1 8	2 9 7 3 4
5 6 1 1 2	6 4 6 1 4	5 9 1 2 8	2 4 6 8 7	7 3 7 6 1	5 1 7 4 1	9 3 4 7 7
4 9 3 5 6	1 1 8 4 8	3 5 4 9 3	3 6 1 2 3	2 6 8 7 7	4 5 3 8 5	8 5 9 5 1
3 3 2 2 8	5 2 3 2 2	7 3 8 6 9	4 1 8 6 1	1 9 2 3 6	3 9 5 7 7	1 2 8 1 2
2 1 4 9 4	4 6 2 8 3	2 7 6 5 1	5 7 3 1 2	9 8 4 1 3	6 3 1 2 9	6 1 5 8 8
9 7 5 4 5	3 9 7 9 9	1 4 2 3 4	6 9 7 4 4	3 2 5 2 2	8 4 2 6 3	5 6 3 6 3
6 2 6 3 9	8 8 5 5 5	8 6 7 7 2	9 3 4 5 8	8 7 9 9 4	9 2 4 9 4	4 8 1 2 9
8 5 8 7 1	2 3 9 3 7	4 1 5 1 5	8 5 9 7 6	4 5 3 5 8	1 6 8 5 2	3 4 6 4 5
1 8 7 6 3	7 5 4 6 6	6 8 9 4 6	1 2 2 9 9	6 4 1 8 5	2 7 9 3 6	7 7 2 9 6
8 4 6 8 6	2 1 9 9 7	2 2 1 8 9	5 1 9 2 4	5 2 6 2 8	1 6 8 8 3	8 1 9 4 1
9 9 4 5 8	4 4 8 7 8	8 7 5 9 7	3 6 4 7 7	3 8 5 3 6	4 4 6 7 7	6 6 8 7 8
6 6 3 1 1	6 8 3 1 9	7 5 7 5 5	6 5 1 8 5	2 4 3 8 2	5 1 4 3 6	4 9 7 8 6
7 3 7 7 2	7 3 6 2 2	3 8 9 4 6	4 7 2 6 9	7 9 7 4 1	3 8 2 6 5	3 5 3 1 4
2 8 9 3 4	1 5 5 5 1	5 4 3 6 4	7 8 7 5 3	9 5 8 6 5	8 2 7 9 2	5 3 4 3 5
3 7 2 6 9	8 6 4 6 3	4 1 8 2 1	1 9 6 4 8	4 7 2 1 3	6 3 5 5 1	2 2 6 9 9
5 1 8 4 5	9 9 1 8 4	1 9 4 3 2	8 2 8 9 6	6 3 4 9 9	2 7 1 2 4	9 8 2 6 2
4 5 5 2 7	3 2 7 3 6	9 3 2 1 8	9 3 5 1 2	1 6 9 7 7	9 5 9 1 8	7 7 1 5 7
1 2 1 9 3	5 7 2 4 5	6 6 6 7 3	2 4 3 3 1	8 1 1 5 4	7 9 3 4 9	1 4 5 2 3

TABLE 15.6 PERMUTATIONS OF 9 (*Continued*)

86224 54583 96522 49829 86656 35665 11391
22339 81837 65138 36916 37435 93252 58135
75695 42661 74496 13375 94888 54187 22882
54453 75224 59984 58594 61277 11549 44579
68182 96145 47869 74282 25544 48813 97668
19911 39419 33375 65141 48911 29938 86257
91848 17976 21743 82737 79169 66426 65724
33566 28752 82617 27463 53792 87791 33413
47777 63398 18251 91658 12323 72374 79946

65348 95382 16446 84991 43379 93459 32561
23759 61754 84591 37185 71913 69815 64492
74982 47219 67287 55662 88437 25731 55325
52296 72595 32618 62439 65621 34563 93244
47517 29868 59925 93228 29248 42642 79756
89433 36127 28869 48774 12594 76127 28673
11621 84641 91132 21857 37862 58296 47137
38164 13933 75354 19316 56155 87384 11918
96875 58476 43773 76543 94786 11978 86889

74242 12262 86522 61847 12183 97745 46147
85987 83923 11659 77293 54478 34117 65816
96459 31375 74288 99458 29617 46336 98378
17763 66748 97167 15785 65246 62598 14592
22875 59159 55946 23364 76534 25674 71954
59121 48516 69394 34929 91921 79261 33229
63616 77894 38413 56511 88892 53452 57785
41598 25437 43835 82636 47769 88923 22461
38334 94681 22771 48172 33355 11889 89633

97597 32358 11761 22663 73827 41767 89437
25775 26672 27813 48859 81989 79388 92192
72339 51744 53276 73232 38665 83826 56281
44852 49519 82455 99945 14346 95693 78849
53988 15296 64594 55577 96291 68442 37768
66146 93127 48942 67128 47773 54939 11615
39464 74935 99628 31784 62554 16174 44926
88221 67863 76187 86491 29418 37551 23374
11613 88481 35339 14316 55132 22215 65553

25665 42832 48626 42413 98197 47496 13414
81157 65718 77278 31168 37265 85672 41959
49239 36666 34347 58235 81671 32824 32873
78481 94493 56835 19396 62832 53169 54765
54572 23945 99482 93549 15344 64948 25341
32396 18177 65169 75624 76788 11583 67136
16923 57329 23711 67771 53556 98331 76698
63718 89284 11993 24957 24929 26217 98527
97844 71551 82554 86882 49413 79755 89282

TABLE 15.6 PERMUTATIONS OF 9 (*Continued*)

73653	32896	31468	84686	63712	58535	18475
55426	91537	22274	66364	78368	94979	72957
48815	49759	16912	29199	46293	87428	94523
37348	55121	88559	51522	37574	61114	57848
89597	14912	64687	33971	59155	13646	33639
26981	28363	47821	48413	14437	79287	49261
64232	77685	53393	72235	82946	32362	65194
91179	63448	95745	15858	21829	26751	26382
12764	86274	79136	97747	95681	45893	81716

47712	44429	87455	71191	22417	88271	47698
94224	21664	52337	43589	97793	26848	89181
31836	53988	43178	84268	75838	45792	74752
18347	87231	95823	25374	39956	53134	53523
53998	96596	28244	17842	58684	62566	22914
62571	68847	31696	96927	86549	94955	95269
26189	35353	74712	38636	44161	79319	68475
79655	72112	69561	59413	61375	37687	16336
85463	19775	16989	62755	13222	11423	31847

74785	35957	67413	22284	42225	44417	11517
17546	54176	88385	51956	37751	58694	85133
26119	88833	26596	95845	64683	19721	43264
63627	77415	93131	33171	95137	95953	69856
35498	26669	42957	67737	56392	63272	58791
48962	99782	39779	16599	21569	81168	37625
91251	41241	74644	48413	13446	22536	22488
52834	62328	51222	84662	78874	37349	94379
89373	13594	15868	79328	89918	76885	76942

11289	72161	91142	71644	61991	79225	61949
67153	49644	84687	92426	95222	63359	55261
73975	56786	35875	59962	49676	46161	44852
39512	31259	66731	86553	74559	57787	37475
54491	28532	23224	14185	36733	34816	29198
85836	15977	52519	38718	83818	15444	93336
92768	83828	19398	45297	58164	21593	82723
46344	64313	47453	63339	27345	88938	76587
28627	97495	78966	27871	12487	92672	18614

37415	52698	87644	17966	71833	33716	89525
78962	21365	96222	71459	93211	27565	62949
24649	45486	63316	95111	29765	19348	55117
46536	19533	32767	24348	16149	76893	93281
12771	34951	75971	56775	54627	45979	21872
59897	73117	59185	33882	68954	88124	18498
65323	97772	48438	62537	37478	62657	76666
91184	88829	24599	48223	82396	51432	47333
83258	66244	11853	89694	45582	94281	34754

TABLE 15.6 PERMUTATIONS OF 9 (*Continued*)

9 8 1 1 9	4 7 6 3 4	6 2 1 2 8	7 4 8 2 4	2 6 3 1 6	6 9 9 6 7	9 9 2 4 2
4 2 2 9 3	6 2 7 8 1	3 9 6 3 7	5 6 9 4 5	9 3 6 6 1	3 5 1 5 3	2 6 8 3 7
7 1 9 2 6	1 9 5 6 3	5 8 8 7 3	4 1 6 1 1	1 2 1 9 4	2 4 2 2 8	1 7 7 9 8
1 7 4 5 5	5 8 8 5 7	1 1 7 6 4	1 9 4 5 2	5 7 9 7 5	4 7 8 1 5	5 2 5 2 3
6 6 8 3 4	2 5 2 4 5	2 7 2 8 5	2 5 2 9 9	7 1 7 8 2	8 8 6 7 9	3 4 1 1 4
2 9 6 6 2	8 3 1 9 6	9 3 5 1 6	3 2 7 7 7	6 4 8 4 3	9 2 5 8 1	7 3 3 7 5
3 5 3 4 1	7 4 9 1 8	4 4 9 4 9	9 3 1 8 8	8 5 4 2 8	7 1 7 4 2	6 8 4 8 6
8 4 7 8 7	9 6 4 7 9	7 6 4 9 1	6 8 5 6 3	3 8 2 5 9	1 6 3 9 6	8 1 6 5 9
5 3 5 7 8	3 1 3 2 2	8 5 3 5 2	8 7 3 3 6	4 9 5 3 7	5 3 4 3 4	4 5 9 6 1
2 4 8 1 4	9 9 9 5 2	5 6 3 7 8	3 2 3 8 1	2 1 1 4 8	9 7 2 9 7	7 2 8 4 8
5 2 4 9 8	8 7 3 8 3	2 2 1 3 1	6 9 9 1 9	1 5 4 6 6	3 6 8 6 6	9 8 7 9 4
7 1 6 7 5	7 8 5 3 6	7 3 4 9 5	2 7 7 7 8	7 7 6 2 2	5 5 1 5 9	5 1 2 6 3
9 8 5 8 1	3 3 1 6 4	4 9 7 6 9	8 6 2 5 7	8 8 2 5 5	7 2 9 2 8	8 5 4 2 6
3 7 1 3 7	4 5 7 7 5	9 7 9 1 3	5 1 4 3 5	9 3 9 1 3	1 4 3 1 2	1 6 9 7 5
6 3 2 2 6	1 6 6 9 1	3 8 5 8 6	9 3 1 2 2	5 4 8 7 4	2 9 5 8 1	4 4 1 1 2
8 6 3 4 9	6 4 8 4 9	6 1 6 2 2	1 5 6 4 4	6 2 3 3 1	6 3 6 4 5	3 9 6 8 1
1 9 7 5 2	2 2 2 1 7	8 5 8 4 7	4 8 8 9 3	4 6 5 9 7	4 8 7 7 4	6 3 5 5 7
4 5 9 6 3	5 1 4 2 8	1 4 2 5 4	7 4 5 6 6	3 9 7 8 9	8 1 4 3 3	2 7 3 3 9
3 5 9 5 1	4 1 8 8 6	6 5 4 5 5	3 9 8 6 3	8 6 8 9 1	2 6 5 3 1	1 2 7 7 2
6 2 7 3 7	2 8 6 2 2	9 7 7 7 4	9 4 3 9 9	6 5 6 1 5	7 1 3 8 5	3 6 3 9 6
9 3 2 8 9	5 3 4 9 1	2 1 3 9 7	1 6 2 1 2	9 8 5 3 2	1 2 4 6 3	8 3 5 2 7
5 7 6 9 3	7 5 7 4 3	4 9 6 6 1	7 7 7 4 7	5 4 3 5 8	4 4 1 2 4	5 9 2 1 3
7 4 8 6 2	8 7 5 1 5	5 4 2 4 6	4 3 5 8 5	1 9 7 7 9	8 7 8 5 9	6 4 9 5 9
1 9 1 7 4	3 9 2 5 9	3 3 1 1 3	8 2 1 5 1	4 7 9 2 4	5 9 6 4 2	7 5 8 3 4
8 6 5 1 8	6 6 1 6 8	1 2 8 3 8	2 5 9 7 4	3 1 1 4 7	9 3 9 9 8	4 1 1 4 1
4 8 4 2 6	1 4 9 3 7	8 8 5 2 2	6 8 6 2 8	7 3 4 8 3	3 8 2 1 7	9 7 4 6 8
2 1 3 4 5	9 2 3 7 4	7 6 9 8 9	5 1 4 3 6	2 2 2 6 6	6 5 7 7 6	2 8 6 8 5
4 6 6 2 2	2 2 5 6 5	5 7 5 6 4	6 2 7 1 6	4 8 3 4 6	2 2 5 7 5	7 6 3 5 6
3 2 2 3 4	7 7 9 7 9	6 1 2 9 1	1 9 1 4 3	1 9 2 2 2	9 8 3 1 3	5 4 1 2 7
2 5 5 5 3	6 6 1 2 7	8 9 6 5 6	4 7 9 3 1	8 4 5 7 7	4 6 7 8 1	4 3 7 1 8
1 4 8 7 1	3 3 8 1 3	4 8 8 7 9	3 3 2 2 9	3 7 7 6 9	7 9 6 9 8	3 8 5 9 9
9 1 4 8 5	9 8 3 8 4	2 3 4 4 8	7 5 5 8 2	9 5 1 1 8	6 7 8 6 2	8 2 8 6 5
5 7 9 9 7	1 5 6 5 6	3 6 1 8 5	5 6 8 6 4	2 1 8 8 4	5 3 9 4 6	6 1 4 8 4
7 8 3 6 6	8 4 2 9 2	7 4 7 2 2	2 8 6 5 7	7 6 6 9 3	8 4 1 3 7	9 9 6 3 2
6 3 1 4 8	4 9 4 3 8	1 5 9 3 7	8 1 4 9 8	5 2 4 5 5	3 1 2 2 9	1 7 2 7 3
8 9 7 1 9	5 1 7 4 1	9 2 3 1 3	9 4 3 7 5	6 3 9 3 1	1 5 4 5 4	2 5 9 4 1
8 5 3 9 2	1 7 9 9 6	5 8 8 8 5	3 8 2 4 7	8 4 1 3 8	7 1 1 6 5	4 4 7 2 2
7 2 5 7 5	9 9 4 7 7	9 1 1 1 7	9 3 8 5 6	7 7 3 4 7	8 2 8 7 2	2 9 1 4 7
5 1 4 5 7	7 2 3 4 1	7 2 3 9 4	4 7 9 1 9	6 2 5 1 9	3 4 7 3 1	8 2 8 9 8
9 6 7 2 4	4 6 8 1 5	2 3 9 3 1	7 5 7 8 5	9 5 7 9 4	1 5 9 2 3	5 7 2 1 3
4 8 6 8 3	2 8 6 2 4	4 6 5 7 8	5 2 1 6 8	1 1 9 8 3	9 9 4 8 8	6 1 5 8 6
6 4 9 6 8	5 1 1 8 3	6 4 7 6 3	1 9 3 3 2	3 3 6 2 2	2 7 2 9 9	7 3 3 5 5
2 7 8 4 6	6 4 5 6 9	8 5 2 5 6	8 1 4 7 1	4 9 4 6 1	5 8 6 1 7	9 5 6 3 4
1 9 2 1 1	3 5 2 3 2	1 9 4 4 9	2 6 6 2 4	5 8 2 5 6	6 6 3 5 6	1 8 4 6 1
3 3 1 3 9	8 3 7 5 8	3 7 6 2 2	6 4 5 9 3	2 6 8 7 5	4 3 5 4 4	3 6 9 7 9

TABLE 15.6 Permutations of 9 (*Continued*)

4 2 6 5 9	1 4 9 7 8	7 4 6 4 3	2 1 2 2 4	3 3 6 8 1	4 7 1 6 4	9 9 3 2 3
6 8 1 3 1	9 6 4 4 2	2 1 8 3 9	5 9 6 5 9	2 2 7 1 8	7 9 8 9 5	2 4 2 5 4
3 6 4 7 8	7 5 1 8 4	9 2 2 7 8	7 4 4 7 8	1 9 9 2 4	6 3 7 4 9	6 1 5 6 6
1 4 8 2 4	3 7 5 5 6	3 5 9 8 2	6 3 7 3 7	4 5 5 3 9	5 6 2 5 2	4 6 1 3 2
7 9 9 6 6	6 1 7 1 3	5 7 5 6 1	8 5 3 9 3	5 4 4 9 5	3 8 9 7 8	7 3 7 7 1
9 5 3 1 3	4 3 3 2 7	1 6 4 1 5	4 2 8 8 1	9 7 2 4 2	8 4 5 8 6	5 8 4 8 8
2 1 5 8 5	8 2 6 3 5	4 3 1 5 4	1 6 5 4 5	8 6 3 6 3	9 1 3 3 7	8 2 6 9 5
5 3 7 9 7	2 8 8 9 9	6 9 7 2 7	3 8 1 1 2	6 1 1 5 7	1 2 6 1 1	3 7 8 4 7
8 7 2 4 2	5 9 2 6 1	8 8 3 9 6	9 7 9 6 6	7 8 8 7 6	2 5 4 2 3	1 5 9 1 9
1 6 3 8 3	7 2 6 7 9	7 8 1 6 5	1 1 4 4 8	1 2 7 8 1	8 9 7 6 9	7 5 8 1 7
2 2 8 9 8	2 5 5 2 6	3 4 8 5 1	4 8 7 2 1	6 5 1 2 2	9 5 1 4 2	3 9 4 3 8
8 7 7 5 1	9 7 9 5 3	4 3 7 7 9	5 5 2 9 6	2 7 9 5 6	5 3 8 9 4	6 2 6 8 3
3 5 6 1 4	3 6 8 9 1	1 5 4 1 4	8 7 1 7 5	8 8 8 6 7	7 2 9 7 8	8 4 7 7 4
9 3 2 7 5	4 1 7 3 7	9 1 9 3 7	2 4 9 3 4	7 9 4 7 8	3 6 2 5 6	1 7 9 9 1
4 4 9 4 6	1 8 2 1 2	2 2 6 8 2	7 9 3 6 3	4 6 6 1 5	2 4 4 2 5	2 6 2 5 6
7 8 1 6 7	5 4 3 0 8	6 9 3 2 6	3 6 5 8 7	3 4 3 4 9	6 8 6 1 7	9 1 3 6 5
6 1 4 3 9	8 9 4 4 5	5 6 2 9 3	9 2 6 1 2	9 1 5 9 4	4 1 5 8 1	5 8 1 2 9
5 9 5 2 2	6 3 1 8 4	8 7 5 4 8	6 3 8 5 9	5 3 2 3 3	1 7 3 3 3	4 3 5 4 2
4 9 6 6 1	1 1 8 3 1	3 7 5 4 9	9 7 4 9 9	9 4 8 8 3	3 2 5 1 3	9 5 6 8 8
5 3 1 9 6	6 8 2 5 9	6 5 4 9 2	2 8 5 6 3	2 1 9 4 2	8 6 4 2 6	8 1 7 9 6
8 6 8 5 7	7 5 1 1 3	7 3 9 2 7	6 9 7 3 6	8 6 5 1 1	9 5 9 9 8	7 3 8 1 1
3 7 4 4 2	2 2 9 4 5	9 1 3 3 8	1 2 1 1 7	3 9 6 2 9	4 8 2 5 4	1 2 3 7 7
9 8 3 3 9	4 9 6 7 2	8 6 7 8 3	8 1 9 2 8	1 7 3 5 6	5 3 3 3 1	2 9 9 6 9
2 4 7 1 4	3 3 3 8 6	2 9 1 1 4	3 6 3 7 1	4 2 1 3 4	6 9 8 7 5	5 4 2 5 5
1 5 9 2 8	9 6 5 6 8	1 4 6 6 1	5 5 2 5 2	7 5 4 6 7	1 1 1 8 9	4 7 4 3 2
7 1 5 8 3	5 4 7 9 4	5 8 8 7 5	7 4 8 8 5	5 3 7 9 5	2 7 7 6 7	3 8 5 4 4
6 2 2 7 5	8 7 4 2 7	4 2 2 5 6	4 3 6 4 4	6 8 2 7 8	7 4 6 4 2	6 6 1 2 3
3 1 7 2 7	5 4 3 6 3	9 8 6 4 4	8 6 6 9 6	5 8 1 2 6	5 4 1 1 1	1 2 1 7 3
2 2 5 8 8	9 6 5 5 5	3 1 4 8 8	3 9 3 1 7	7 3 7 5 7	6 7 4 4 9	3 7 3 3 4
1 5 8 6 9	2 2 1 2 4	4 9 9 9 1	1 3 4 6 8	8 4 6 7 4	2 8 3 9 2	8 9 5 9 2
6 3 2 7 6	8 5 8 8 1	7 5 7 2 2	4 5 2 5 1	1 2 5 6 5	7 2 9 7 6	4 4 2 4 7
9 8 4 1 4	7 3 7 3 8	6 4 5 3 9	5 7 7 2 9	3 6 2 9 9	4 6 5 2 7	7 6 4 8 1
5 7 6 3 3	4 1 2 7 9	5 2 2 7 7	9 4 1 4 4	2 1 3 3 1	1 9 2 6 3	2 3 8 5 6
4 6 1 5 5	1 7 4 4 6	1 3 1 1 5	6 8 9 8 3	6 7 4 4 8	3 3 8 5 5	9 8 6 6 8
8 4 9 9 1	3 9 6 9 2	8 6 8 5 3	2 1 5 7 5	4 5 9 1 2	8 5 7 3 8	5 1 7 1 9
7 9 3 4 2	6 8 9 1 7	2 7 3 6 6	7 2 8 3 2	9 9 8 8 3	9 1 6 8 4	6 5 9 2 5
9 2 4 9 5	4 8 4 4 8	1 9 4 8 5	2 7 9 6 5	9 8 7 3 4	3 8 2 1 3	3 5 3 2 6
1 1 8 1 3	8 6 5 9 9	2 7 6 7 7	6 8 6 9 8	2 2 2 2 9	1 4 8 6 2	2 8 9 8 4
3 9 5 5 7	2 4 9 3 3	8 1 9 2 3	7 6 5 7 7	6 7 8 6 7	2 5 9 5 7	1 4 1 1 8
8 6 6 7 2	6 9 1 5 6	9 6 5 3 1	1 1 7 5 1	8 3 4 5 8	9 3 4 2 8	5 1 6 7 2
7 5 3 3 9	7 3 6 8 7	6 8 2 5 4	3 4 1 4 6	5 9 5 9 2	6 2 5 7 5	6 9 7 3 7
2 4 9 2 1	9 7 2 2 4	4 2 7 4 8	8 3 4 3 2	4 6 9 7 1	7 7 6 9 6	7 2 2 6 1
4 3 7 8 4	5 1 8 1 2	7 3 3 9 9	9 9 2 1 9	3 1 1 1 3	8 9 7 3 4	4 3 5 4 9
6 8 1 6 6	3 5 7 7 1	5 5 1 1 6	5 2 3 8 3	1 5 6 8 6	4 6 3 8 9	8 6 4 9 5
5 7 2 4 8	1 2 3 6 5	3 4 8 6 2	4 5 8 2 4	7 4 3 4 5	5 1 1 4 1	9 7 8 5 3

TABLE 15.6 PERMUTATIONS OF 9 (*Continued*)

2 3 1 7 1	4 2 4 3 1	9 2 1 1 1	4 6 9 3 5	3 6 2 5 5	1 1 4 1 2	7 1 4 5 3
5 9 6 3 3	3 4 6 8 6	4 1 8 6 6	8 1 2 7 9	1 5 1 2 8	6 6 3 4 5	5 6 1 4 6
8 4 5 5 8	6 1 9 1 5	5 6 9 4 7	6 3 5 4 7	5 1 7 7 2	3 8 1 5 4	3 2 2 8 7
1 6 8 1 7	2 9 2 5 3	8 9 4 7 9	7 4 7 2 2	9 4 9 9 3	9 5 9 2 1	9 8 6 7 4
4 1 2 4 9	5 5 3 6 9	7 8 6 3 8	1 8 4 9 6	6 2 8 1 7	4 7 8 3 7	8 3 3 9 8
6 2 7 8 6	8 3 7 4 7	2 5 7 5 3	3 2 6 8 8	2 8 5 3 9	2 2 2 9 6	1 7 8 2 2
3 5 4 9 2	1 7 8 7 4	3 4 2 8 5	5 9 8 6 4	8 3 3 8 4	5 4 6 8 3	4 9 5 6 9
9 8 3 6 5	7 6 5 2 2	1 7 5 9 2	2 7 1 5 3	7 9 4 6 6	8 3 7 7 8	6 4 9 3 1
7 7 9 2 4	9 8 1 9 8	6 3 3 2 4	9 5 3 1 1	4 7 6 4 1	7 9 5 6 9	2 5 7 1 5

9 9 6 8 1	9 5 1 6 2	8 9 5 8 9	7 7 4 9 9	4 9 5 4 6	5 9 5 5 2	6 8 9 7 1
1 7 9 5 6	1 8 7 5 1	1 1 7 7 8	2 4 7 4 7	9 5 1 2 2	1 6 1 9 8	8 6 6 1 8
7 3 2 7 9	4 1 2 4 3	3 6 1 3 3	3 3 8 6 1	5 6 3 5 5	4 4 4 2 9	5 3 1 3 7
8 5 7 4 4	5 9 4 1 4	7 2 3 4 6	5 9 2 2 6	3 8 8 1 1	7 8 7 1 4	4 1 8 9 9
4 8 8 6 8	3 2 3 9 7	6 7 2 5 1	6 6 3 8 5	2 1 4 3 3	3 7 6 7 7	1 9 7 6 2
6 1 4 9 7	8 3 9 8 6	5 8 6 1 2	8 2 5 1 2	1 4 2 9 8	9 2 2 8 3	2 7 2 4 5
2 6 1 3 2	6 7 5 3 5	2 5 4 6 7	1 1 6 5 3	6 7 7 8 9	8 3 3 6 5	3 5 3 8 3
3 2 5 1 3	7 6 6 2 8	9 3 8 2 5	9 8 1 3 8	8 3 6 7 4	6 1 9 3 6	7 4 4 2 6
5 4 3 2 5	2 4 8 7 9	4 4 9 9 4	4 5 9 7 4	7 2 9 6 7	2 5 8 4 1	9 2 5 5 4

9 1 7 8 7	9 8 6 1 4	2 9 1 7 7	4 8 4 8 6	8 2 3 8 2	8 2 3 2 6	7 3 8 7 5
4 8 9 6 3	3 7 9 2 1	9 5 9 4 4	9 4 3 4 3	7 3 5 7 6	5 9 4 6 8	5 8 4 5 6
7 4 2 2 2	6 3 3 6 9	3 1 5 2 1	6 9 9 9 7	5 6 2 6 9	2 6 6 8 1	8 6 5 1 8
5 9 1 3 4	1 9 7 5 3	6 2 8 6 5	1 7 8 7 9	1 4 4 9 8	9 1 7 3 9	1 9 6 6 2
3 3 3 7 5	8 4 2 7 6	1 6 3 5 6	3 6 1 6 1	9 1 1 5 5	6 7 9 4 4	3 1 3 8 7
2 2 5 1 1	7 5 8 8 8	5 7 2 8 3	8 3 6 3 8	4 7 6 4 3	1 3 5 1 3	9 2 1 4 9
1 6 8 5 6	2 2 5 3 5	8 3 4 9 8	2 5 2 5 4	2 5 7 3 7	4 5 1 5 2	6 4 2 2 1
8 5 6 4 9	4 6 4 4 7	7 8 6 3 2	5 1 5 1 5	6 8 9 1 4	3 4 2 9 5	4 7 7 9 3
6 7 4 9 8	5 1 1 9 2	4 4 7 1 9	7 2 7 2 2	3 9 8 2 1	7 8 8 7 7	2 5 9 3 4

6 3 8 7 9	5 1 9 8 5	5 1 8 5 4	1 1 2 2 4
5 4 6 4 8	7 4 2 4 7	9 3 6 8 9	5 2 1 6 8
9 8 5 3 7	6 7 8 5 6	6 9 4 1 1	7 5 5 3 3
1 2 1 9 4	8 6 1 1 8	7 5 3 7 8	2 9 9 5 9
3 7 2 6 1	1 8 6 6 3	4 8 9 4 5	8 4 8 4 5
4 9 3 1 6	3 5 4 2 2	8 7 7 6 2	9 7 7 8 2
8 5 4 5 5	9 9 7 9 9	3 4 5 2 7	3 8 3 1 7
7 6 7 2 3	2 3 3 7 1	2 6 1 3 3	4 3 6 7 1
2 1 9 8 2	4 2 5 3 4	1 2 2 9 6	6 6 4 9 6

TABLE 15.7 PERMUTATIONS OF 16

7	12	15	15	1	2	7	16	10	2	14	15	7	13	13	10	6	1	8	10
13	3	8	16	7	10	11	10	13	5	11	7	13	16	7	7	5	13	2	14
3	1	4	5	14	13	3	14	9	13	13	2	9	15	6	2	8	4	5	8
11	8	16	14	15	6	2	6	2	16	8	5	12	3	9	13	4	3	10	4
14	9	1	6	3	9	14	13	8	6	5	8	14	7	3	15	13	11	4	7
2	16	10	13	5	5	13	2	11	7	3	12	5	14	12	16	2	2	9	15
4	6	13	7	2	15	1	9	1	4	7	10	6	9	11	9	7	6	16	11
6	14	6	10	4	14	4	15	3	3	4	16	2	6	5	1	12	10	6	9
10	15	2	1	13	12	16	3	4	8	10	1	15	5	14	12	14	12	3	2
12	10	7	12	9	11	9	8	12	14	15	4	11	8	16	8	9	14	14	1
15	7	5	2	10	7	8	12	6	15	6	13	16	12	15	4	11	8	12	6
16	2	11	8	8	8	15	5	16	1	1	9	8	1	8	14	16	5	13	5
9	13	14	3	6	4	10	11	5	12	9	3	10	4	4	3	10	9	1	3
8	11	0	4	11	0	12	7	7	10	12	14	3	10	1	6	15	16	15	12
1	5	12	11	16	16	5	4	14	9	16	11	1	2	10	5	1	15	7	13
5	4	3	9	12	1	6	1	15	11	2	6	4	11	2	11	3	7	11	16

11	8	16	5	5	13	1	13	2	16	14	12	9	8	7	5	13	3	13	3
2	2	8	8	14	16	4	3	8	11	10	14	15	1	2	11	4	5	15	9
6	13	2	13	6	5	9	15	11	10	12	6	16	15	16	9	10	12	16	15
14	12	4	16	16	11	14	10	5	12	3	3	12	14	15	13	6	4	1	16
8	6	3	9	4	10	6	4	16	2	2	9	8	16	4	6	5	15	7	8
9	15	12	10	3	2	12	6	1	15	4	13	7	7	9	12	14	8	8	11
2	10	11	12	13	12	5	11	7	8	9	5	14	11	10	1	3	13	3	5
6	1	13	14	8	14	15	5	3	7	11	15	6	12	5	7	11	1	14	4
1	14	14	2	9	15	16	14	6	14	7	8	3	13	11	8	7	7	12	7
4	4	6	4	12	3	11	8	15	9	8	1	13	6	3	3	15	9	9	12
15	5	1	11	10	6	3	7	10	5	5	11	10	10	12	15	16	14	5	2
5	3	5	6	7	7	13	2	14	3	16	4	5	5	13	4	9	16	2	6
12	7	15	15	15	9	8	12	12	13	15	10	1	4	6	16	2	6	11	1
10	11	10	3	2	4	2	1	4	6	6	7	11	9	14	10	8	11	4	13
7	9	7	7	11	1	7	16	13	1	13	2	4	2	1	2	12	2	10	14
13	16	9	1	1	8	10	9	9	4	1	16	2	3	8	14	1	10	6	10

1	6	7	4	8	6	5	2	8	15	4	6	6	1	4	5	7	13	2	10
9	15	11	3	11	15	9	10	1	3	8	2	15	7	9	8	16	1	14	3
10	16	4	5	12	9	16	11	7	1	7	16	11	8	3	3	12	2	3	4
4	14	1	9	5	5	4	13	6	8	15	5	12	5	7	16	5	11	8	1
7	3	13	14	15	2	1	14	16	5	14	9	2	16	1	12	6	14	4	13
16	11	2	1	14	16	6	9	3	4	16	14	3	15	11	11	3	9	12	5
3	10	16	16	13	7	13	1	11	14	9	10	16	2	10	2	10	7	10	16
11	13	9	13	4	13	8	3	5	13	10	12	5	12	5	14	13	16	5	6
15	2	3	12	9	12	2	4	13	10	3	13	14	4	2	1	14	8	6	12
14	1	14	6	10	1	3	12	4	2	2	4	13	3	16	9	9	3	7	14
13	12	5	11	3	11	15	8	2	7	11	7	8	14	6	4	4	4	15	11
12	5	10	7	2	14	7	15	14	16	13	1	9	10	12	10	11	10	9	8
8	9	8	10	6	4	11	7	10	11	6	8	4	9	8	15	8	6	11	9
2	7	0	2	1	8	10	6	15	12	1	11	7	11	13	6	1	15	13	15
6	4	15	8	16	10	14	16	9	6	12	3	10	6	14	7	2	12	16	7
5	8	12	15	7	3	12	5	12	9	5	15	1	13	15	13	15	5	1	2

13	4	10	4	16	13	16	13	5	3	6	14	1	16	8	7	2	3	3	12
5	14	4	6	8	2	15	1	13	14	16	4	15	4	3	12	12	1	4	7
2	2	2	15	14	16	9	12	16	6	10	15	14	9	10	1	14	8	8	16
7	12	15	8	12	3	5	14	7	12	5	13	16	1	7	5	11	2	9	3
6	9	7	14	9	14	10	11	15	11	12	1	12	12	14	16	3	11	11	8
14	5	16	7	10	8	11	8	14	13	7	11	6	3	11	4	4	6	6	9
15	11	8	9	7	12	8	7	1	15	9	3	3	7	13	11	10	4	5	1
11	6	6	1	4	1	3	16	12	5	4	9	13	13	6	8	15	9	1	14
4	10	3	16	2	11	7	9	6	9	1	8	4	11	5	2	16	10	12	4
1	8	1	13	1	15	4	4	11	4	2	16	5	8	1	9	5	12	16	6
9	7	14	2	6	4	14	10	9	8	15	10	7	10	9	10	6	14	10	11
12	1	9	10	15	5	2	15	10	2	14	2	8	2	4	13	8	5	15	5
3	3	12	11	5	9	6	6	3	10	13	12	9	6	2	15	7	15	7	13
10	15	11	5	13	7	12	5	2	7	11	5	10	15	12	3	1	13	13	10
8	13	13	3	3	10	13	2	4	1	8	6	11	14	15	6	9	16	2	2
16	16	5	12	11	6	1	3	8	16	3	7	2	5	16	14	13	7	14	15

TABLE 15.7 Permutations of 16 (*Continued*)

9	16	15	12	2	11	4	16	11	10	2	5	5	14	11	2	14	13	16	6
11	3	2	6	15	13	10	1	4	13	11	8	16	16	4	3	5	15	5	15
14	14	8	16	11	15	5	14	14	11	1	14	15	15	13	5	7	11	11	16
4	13	1	3	5	7	6	2	16	1	14	9	14	3	3	1	6	16	6	10
6	6	10	7	13	10	16	7	2	12	6	12	6	13	8	9	15	9	1	11
2	10	14	9	12	3	3	10	5	6	5	16	12	10	15	10	11	4	9	8
5	15	11	14	10	4	14	13	6	4	12	4	11	5	10	14	16	5	7	9
16	5	13	10	3	9	12	6	3	7	3	7	3	11	14	7	3	14	4	12
8	12	7	11	7	8	13	15	13	9	4	3	8	1	12	6	9	8	15	14
1	8	3	2	1	5	15	9	9	3	10	11	13	8	5	13	12	3	3	5
13	9	9	1	6	2	11	3	8	8	15	1	7	9	7	8	8	6	2	3
15	1	5	5	9	6	9	4	10	5	8	13	10	7	9	15	2	10	8	4
7	4	12	13	16	1	2	11	12	2	16	15	2	4	2	11	1	7	13	1
10	2	4	15	4	16	1	12	7	15	9	10	9	12	16	4	13	2	10	13
3	7	6	8	8	14	7	5	1	14	13	2	4	2	1	16	4	1	12	7
12	11	16	4	14	12	8	8	15	16	7	6	1	6	6	12	10	12	14	2

12	6	13	4	5	7	2	1	9	2	5	1	15	2	14	13	13	11	2	13
6	11	4	15	12	12	6	15	6	15	6	3	12	5	15	11	16	9	8	1
13	5	1	6	7	6	13	5	7	8	15	6	4	15	1	14	5	14	10	4
11	1	11	7	8	15	8	4	12	13	16	9	3	10	7	2	12	3	9	8
3	7	3	14	15	4	12	11	4	10	8	12	1	4	16	6	2	2	16	7
10	12	15	11	4	13	5	10	3	14	11	2	9	11	2	9	9	12	12	11
15	9	16	16	9	2	16	2	15	6	7	15	8	1	8	12	4	13	6	9
14	15	2	13	3	16	10	14	13	9	10	7	14	9	6	5	6	4	11	12
1	2	12	9	1	8	15	3	8	11	2	5	10	3	3	10	10	7	13	10
5	10	5	3	13	9	9	13	10	1	3	8	7	8	9	4	15	15	7	15
7	14	9	2	11	14	11	6	14	12	9	10	16	12	13	3	7	5	4	14
9	8	10	1	6	3	3	8	5	5	14	16	2	7	12	16	14	10	15	5
2	3	7	5	10	1	1	12	2	7	1	4	6	16	10	8	8	1	5	16
16	13	14	10	2	5	7	16	1	16	13	11	11	6	5	1	11	16	3	3
4	4	6	8	14	10	14	7	11	3	4	13	13	13	11	15	3	6	14	6
8	16	8	12	16	11	4	9	16	4	12	14	5	14	4	7	1	8	1	2

3	14	11	8	9	14	14	2	13	1	8	4	15	16	7	6	15	13	13	13
12	9	6	9	8	10	12	13	14	5	11	10	10	12	9	10	5	16	6	3
11	11	7	1	11	13	11	4	2	7	16	5	8	3	11	12	6	12	5	11
1	16	9	3	1	7	8	15	5	4	3	7	16	8	12	15	7	5	9	4
13	3	1	2	13	5	4	9	7	6	5	15	4	6	4	1	10	6	1	14
7	12	10	10	5	15	5	8	16	2	12	3	5	13	14	13	13	2	3	7
10	15	15	4	14	1	16	16	12	11	9	16	1	2	10	11	8	7	16	8
15	7	4	14	7	4	7	10	6	10	1	1	2	11	3	16	2	4	2	1
9	5	2	7	3	3	13	14	15	15	6	12	9	15	15	9	16	15	15	10
8	6	16	5	15	8	2	12	1	3	10	8	3	14	13	2	1	10	8	12
2	10	5	11	4	9	3	6	11	12	15	9	7	5	2	8	14	1	4	5
5	4	3	15	2	2	15	11	10	14	7	14	14	7	6	3	11	11	10	2
4	1	12	12	16	6	1	3	4	16	13	11	11	4	1	7	12	3	7	9
6	13	14	6	12	16	9	1	8	8	4	13	12	10	5	5	4	9	12	16
16	8	8	13	10	11	10	5	9	13	14	2	6	9	8	14	9	14	11	15
14	2	13	16	6	12	6	7	3	9	2	6	13	1	16	4	3	8	14	6

1	2	14	12	4	4	3	6	12	7	11	11	9	13	13	7	4	10	16	9
9	3	10	13	3	5	5	13	15	9	14	13	14	9	9	4	8	4	15	2
13	6	15	10	11	3	15	12	4	5	5	4	3	6	4	5	12	14	14	3
8	5	5	15	8	9	8	8	2	3	1	12	8	3	11	2	9	16	10	12
11	12	9	14	16	11	4	15	1	4	3	15	5	15	7	11	16	15	7	1
10	4	13	6	1	13	12	9	8	6	7	8	15	7	3	8	13	9	8	10
16	11	11	16	7	15	9	5	7	2	6	10	16	10	6	1	3	6	1	13
5	7	4	3	2	1	14	2	10	13	16	1	6	4	15	6	15	12	11	16
3	15	12	2	14	8	11	16	14	16	9	7	13	8	2	16	2	11	2	15
7	9	7	9	13	6	2	4	13	14	15	6	10	11	8	12	10	3	3	8
2	10	8	8	15	14	6	3	5	1	4	3	7	2	14	15	14	2	6	4
15	8	3	1	6	2	10	7	3	10	10	2	4	1	5	3	7	13	13	14
6	13	2	5	5	7	13	10	16	15	12	16	1	14	16	14	6	1	9	7
12	16	16	4	10	10	7	1	9	8	2	5	12	16	12	9	11	7	4	5
4	1	1	7	9	12	16	11	6	11	8	14	2	5	1	10	1	5	12	6
14	14	6	11	12	16	1	14	11	12	13	9	11	12	10	13	5	8	5	11

TABLE 15.7 PERMUTATIONS OF 16 (*Continued*)

16	2	15	13	13	2	11	14	4	3	12	11	11	11	7	4	13	7	16	16
7	10	8	2	9	5	14	15	5	16	5	9	9	3	1	2	5	6	3	5
9	16	3	8	4	8	7	2	11	1	13	15	2	5	6	8	2	11	6	10
8	11	14	11	14	7	5	7	6	9	16	4	5	16	9	9	16	4	8	15
3	15	11	14	8	10	10	10	12	10	2	2	13	1	11	11	12	13	5	3
12	7	13	15	12	3	12	8	8	11	10	12	10	7	14	15	14	10	12	4
15	13	9	3	3	14	6	16	2	13	8	14	15	14	12	7	1	1	7	13
10	4	1	4	16	6	1	12	16	15	3	16	3	6	16	6	8	3	9	11
6	1	6	5	6	1	15	11	7	8	1	1	8	9	2	16	10	14	14	2
11	8	16	16	5	13	13	6	15	2	15	5	12	8	3	10	11	8	2	14
13	9	10	1	15	15	9	9	3	14	6	6	1	10	13	13	9	5	15	1
14	5	4	6	1	12	3	4	14	6	4	10	4	4	15	3	15	2	1	12
5	12	7	12	10	11	8	3	1	4	7	8	7	15	10	5	3	9	11	7
2	3	2	10	7	4	16	5	10	5	14	13	6	12	5	12	4	16	4	9
4	14	12	7	11	16	4	13	13	12	11	3	14	2	4	14	6	15	10	8
1	6	5	9	2	9	2	1	9	7	9	7	16	13	8	1	7	12	13	6

13	4	3	11	16	8	7	16	3	16	1	10	1	15	16	10	11	4	6	4
4	1	6	16	6	11	11	14	4	8	3	3	5	3	6	7	7	8	5	12
2	7	15	5	7	7	16	3	2	7	10	8	11	11	7	16	12	15	9	9
3	8	12	10	8	6	4	4	11	10	9	2	4	7	1	3	15	10	14	11
6	15	4	6	3	10	2	12	15	6	6	16	9	4	8	12	8	2	8	6
5	6	10	2	13	15	8	7	16	3	2	12	10	9	12	14	9	5	13	5
12	11	1	13	4	5	3	15	5	12	12	1	15	5	15	6	4	12	4	14
9	9	11	8	2	3	12	9	14	13	13	6	8	6	13	11	3	1	10	7
7	13	5	7	1	14	6	13	8	11	14	5	3	16	4	13	10	13	15	3
10	2	14	14	15	16	1	2	1	9	8	15	2	14	9	1	1	7	3	13
14	16	2	1	14	9	13	5	6	4	11	13	16	12	10	15	2	9	2	16
16	3	8	9	5	1	14	6	10	1	16	9	6	10	14	5	6	6	7	15
11	10	16	12	12	12	10	1	12	14	7	14	12	8	3	4	5	16	12	8
15	5	9	15	9	13	9	11	9	5	4	4	13	2	5	8	14	3	1	1
1	12	7	3	10	2	5	10	13	15	5	7	14	1	2	2	16	11	16	2
8	14	13	4	11	4	15	8	7	2	15	11	7	13	11	9	13	14	11	10

1	4	10	3	8	16	4	10	6	10	12	13	16	6	12	10	5	12	1	10
3	10	12	5	7	11	3	3	7	1	6	1	13	2	7	12	16	10	11	13
13	15	5	1	1	12	15	12	14	4	10	14	11	12	4	15	13	16	9	3
10	5	2	9	13	2	16	9	4	2	7	7	4	10	10	6	1	9	5	9
7	6	15	14	15	15	2	7	9	15	2	16	10	8	11	14	4	3	8	15
11	1	11	8	9	7	12	11	11	5	11	3	6	4	15	1	12	4	6	1
12	3	4	2	3	3	1	13	13	12	9	11	1	3	6	2	15	14	7	4
14	9	7	16	14	1	10	16	16	6	8	15	3	16	14	16	14	2	2	7
5	12	13	4	4	14	14	4	5	7	16	5	14	14	8	5	7	15	10	12
4	13	14	13	6	6	8	14	8	9	13	10	7	7	9	3	3	13	12	6
8	14	16	12	12	13	7	2	10	13	4	12	12	5	16	9	11	6	3	16
16	7	3	10	11	9	13	5	12	16	5	9	8	9	3	13	8	8	14	11
6	2	9	15	5	8	9	8	1	11	15	6	9	13	5	11	2	11	13	5
15	8	6	6	16	4	11	1	2	8	3	4	15	15	13	7	6	5	4	14
9	11	1	7	10	5	6	6	15	14	1	2	5	11	1	8	10	1	15	8
2	16	8	11	2	10	5	15	3	3	14	8	2	1	2	4	9	7	16	2

6	12	2	3	9	14	1	16	11	14	5	7	15	11	10	14	10	16	4	16
1	2	15	9	13	15	5	12	12	9	8	6	1	2	5	1	11	13	13	15
9	7	8	13	12	7	11	13	10	5	6	13	6	5	13	9	15	4	9	10
11	1	16	1	1	5	12	4	2	10	16	4	3	16	2	8	9	8	7	7
7	14	11	5	3	4	2	15	5	12	14	9	5	10	15	5	13	9	3	4
8	11	10	10	15	12	8	11	16	13	10	10	10	3	4	12	1	3	2	11
2	4	5	6	16	16	3	2	8	4	7	11	4	7	3	2	7	14	11	1
5	9	7	8	5	6	14	7	14	1	1	2	2	15	7	4	14	15	8	8
16	5	1	11	4	8	15	5	13	8	15	16	8	9	9	11	16	12	12	3
13	10	6	12	6	11	10	6	15	15	2	3	13	1	8	10	5	10	5	12
10	3	14	4	11	3	9	1	3	11	3	14	14	6	12	3	8	1	15	5
12	15	3	14	8	9	6	14	1	6	9	15	11	14	6	7	6	2	6	6
3	16	9	15	2	1	4	3	9	16	13	1	7	13	1	16	3	11	16	13
4	8	4	2	10	10	13	9	6	7	12	8	16	8	11	15	2	6	14	2
15	13	13	16	14	2	16	8	7	3	4	5	9	4	14	13	12	5	1	14
14	6	12	7	7	13	7	10	4	2	11	12	12	12	16	6	4	7	10	9

TABLE 15.7 PERMUTATIONS OF 16 (*Continued*)

14	15	12	8	7	4	13	14	3	10	13	14	6	3	14	8	10	13	11	14
7	1	8	11	15	11	7	2	12	15	5	8	13	5	6	13	14	2	10	13
13	9	2	15	2	13	6	5	16	3	16	7	5	15	4	7	2	12	4	10
11	3	4	6	11	16	16	4	9	7	11	1	14	9	9	12	3	10	6	2
3	16	3	13	5	1	4	6	14	6	4	3	16	8	15	11	6	1	1	15
5	12	6	9	13	2	12	10	6	1	14	16	11	12	16	3	1	7	12	1
16	5	10	10	12	5	14	15	7	12	15	2	2	10	5	9	7	6	15	7
2	4	16	12	14	3	11	1	13	14	10	6	10	11	13	14	5	15	13	4
6	6	9	3	16	6	2	12	10	13	12	5	4	14	10	15	12	5	3	8
4	7	15	14	6	14	10	9	11	8	8	15	8	13	11	2	9	8	14	12
15	14	5	7	8	10	15	8	4	2	3	11	1	6	2	16	4	4	5	5
9	13	14	4	4	9	5	3	8	9	2	4	3	4	12	5	11	16	8	6
1	2	1	2	1	7	9	11	15	16	1	13	15	2	1	10	15	3	2	9
12	11	13	1	9	8	3	7	2	11	9	9	9	7	3	1	13	11	7	11
8	8	11	16	3	12	1	13	1	5	6	12	12	16	8	6	16	14	16	16
10	10	7	5	10	15	8	16	5	4	7	10	7	1	7	4	8	9	9	3

15	1	3	2	7	5	4	11	1	11	16	13	13	12	15	8	2	7	6	11
13	16	13	7	4	16	7	9	5	1	11	11	5	14	13	5	15	15	4	12
5	14	12	11	14	4	8	7	15	13	6	16	2	1	9	3	9	14	5	6
3	15	9	3	9	11	9	8	12	6	9	14	4	2	7	11	16	5	3	15
6	11	10	9	16	14	2	10	7	14	2	6	16	3	2	4	6	11	8	1
9	2	1	16	13	6	5	5	4	5	5	4	15	16	11	12	4	2	10	7
4	7	2	12	8	1	12	12	3	9	3	15	9	5	4	14	13	8	9	9
14	3	7	4	2	13	16	2	9	7	14	10	14	13	10	2	8	12	13	5
8	12	4	10	5	2	6	1	16	10	1	12	12	9	12	15	10	4	15	2
7	4	16	8	10	9	3	15	2	3	7	2	8	7	1	7	1	1	12	10
10	9	8	13	12	3	11	6	10	8	13	1	11	6	16	10	12	13	7	4
2	13	5	1	3	15	15	3	8	2	10	5	3	15	6	13	7	3	14	3
16	10	11	15	15	7	1	16	13	12	4	9	6	8	8	9	5	9	1	13
12	5	6	14	1	12	10	14	6	15	8	7	10	10	5	16	3	10	16	16
11	6	15	6	11	10	14	4	11	16	12	8	7	11	3	1	14	16	2	8
1	8	14	5	6	8	13	13	14	4	15	3	1	4	14	6	11	6	11	14

9	16	5	1	10	2	8	1	13	11	4	14	6	7	12	15	2	8	8	1
13	7	12	16	7	9	16	12	1	14	1	9	8	10	3	16	9	11	13	13
12	8	16	2	8	8	14	4	3	2	14	7	1	16	4	3	5	13	10	11
11	6	8	7	15	13	12	11	4	10	7	2	5	3	10	8	4	2	4	3
3	11	9	3	1	1	11	14	15	9	3	1	10	8	1	9	14	12	5	8
7	12	10	8	3	15	9	7	10	8	12	8	4	11	2	2	7	6	11	6
5	2	14	14	13	14	4	13	9	7	6	5	14	2	9	7	11	15	6	9
4	5	4	15	9	4	13	6	16	4	10	15	2	5	6	10	12	16	16	7
14	1	11	5	11	5	7	5	14	12	11	10	12	15	7	4	10	5	15	15
15	4	13	4	4	6	6	3	6	5	5	4	3	1	13	12	3	10	2	2
1	13	2	6	2	3	3	2	8	6	13	12	7	9	5	11	13	3	1	10
6	10	7	12	14	11	5	10	5	13	2	11	9	14	16	14	6	14	12	4
16	9	1	10	16	7	1	9	2	16	8	6	15	12	11	13	15	9	9	5
2	14	6	13	12	12	15	8	11	3	16	16	13	6	14	5	1	1	3	12
8	3	15	9	5	16	2	15	7	15	9	13	16	4	15	1	8	4	14	14
10	15	3	11	6	10	10	16	12	1	15	3	11	13	8	6	16	7	7	16

10	7	12	15	9	16	10	8	2	2	2	12	10	4	15	3	2	6	9	5
15	5	13	8	14	12	12	15	7	9	7	2	3	16	7	16	7	8	1	14
12	10	8	3	8	5	6	14	4	4	15	3	7	7	16	6	9	9	4	15
9	15	14	2	5	3	8	10	12	8	3	7	5	5	12	8	6	13	10	10
7	2	3	4	3	14	3	3	11	11	4	4	4	6	4	2	10	2	7	3
13	13	16	10	7	1	1	7	8	13	10	1	8	14	10	1	3	12	15	12
1	3	1	6	4	15	7	6	5	15	12	5	6	2	2	10	15	10	12	2
5	14	2	12	13	2	14	13	6	5	9	10	12	3	11	12	8	1	3	9
3	8	5	1	12	9	15	1	13	10	14	16	9	13	6	7	5	5	11	7
2	1	9	5	2	8	4	4	16	7	11	9	13	15	1	15	13	11	2	11
6	9	4	7	6	10	13	5	10	14	8	15	2	8	13	5	16	4	16	16
8	16	10	9	1	6	5	2	1	3	6	8	15	11	9	9	11	7	8	4
11	6	15	16	16	13	9	11	9	12	16	6	11	9	8	13	14	16	14	1
4	4	11	11	15	11	2	16	3	16	13	11	14	12	5	11	4	14	5	13
14	12	7	13	10	4	16	12	14	6	5	13	16	1	14	4	1	15	13	6
16	11	6	14	11	7	11	9	15	1	1	14	1	10	3	14	12	3	6	8

TABLE 15.7 Permutations of 16 (*Continued*)

1	10	14	16	12	14	3	10	16	15	15	5	2	6	9	8	4	12	6	2
10	16	15	3	4	11	16	6	3	6	1	15	10	2	4	3	12	11	12	7
11	8	7	14	14	16	2	3	6	3	4	14	7	14	3	10	16	7	4	9
14	3	13	9	11	8	13	16	8	9	6	13	6	15	14	5	15	14	3	14
4	12	16	15	5	2	12	7	10	5	14	12	5	11	7	2	7	5	2	4
2	2	9	5	9	15	11	2	4	11	9	8	12	1	16	16	6	1	9	15
7	15	5	7	10	13	6	8	15	7	12	7	16	13	8	12	9	15	7	10
12	6	10	6	6	1	1	15	1	8	2	9	3	12	15	11	14	9	10	3
15	4	8	13	13	10	5	9	9	16	11	11	11	3	12	1	11	13	5	6
16	14	2	11	2	9	14	13	13	4	16	1	4	7	13	4	3	4	1	8
3	1	3	1	8	4	7	11	5	1	8	2	8	5	1	13	10	6	13	1
9	13	4	12	3	5	8	1	14	14	13	6	9	10	10	7	5	10	8	13
13	7	6	4	15	12	9	5	11	10	3	16	1	9	11	14	8	2	14	12
5	11	1	2	7	6	4	4	12	2	7	10	13	8	5	6	1	8	16	16
6	9	11	10	1	7	10	14	7	13	10	4	15	4	6	15	13	16	11	5
8	5	12	8	16	3	15	12	2	12	5	3	14	16	2	9	2	3	15	11

5	1	3	6	16	14	16	1	12	5	8	8	4	5	13	3	8	6	10	1
7	11	10	4	11	13	8	8	6	14	3	4	16	14	14	7	6	10	9	16
15	8	16	10	12	6	15	6	3	10	13	14	12	7	6	14	15	3	14	6
9	13	1	13	7	9	9	10	2	3	16	2	1	10	8	10	3	7	2	14
16	5	6	15	5	4	10	11	15	13	14	16	10	11	5	5	16	13	6	15
4	2	11	16	6	3	14	15	9	1	11	5	8	16	7	6	1	9	4	5
3	6	14	3	8	12	12	5	1	6	6	13	5	9	4	11	9	4	5	4
6	9	2	1	15	15	7	7	13	8	2	1	9	13	15	4	10	16	16	10
14	7	8	2	9	8	11	2	14	16	7	7	14	8	3	13	12	15	11	13
12	16	5	11	1	7	3	3	11	15	1	9	7	2	10	9	14	8	12	9
2	14	13	12	2	2	4	13	5	9	10	12	15	3	1	1	2	5	3	11
8	4	7	7	4	1	13	4	4	4	12	11	6	6	9	8	13	1	8	7
13	15	12	5	3	11	5	9	16	7	4	15	2	1	11	16	5	12	13	2
11	12	4	9	10	5	1	12	7	2	9	6	3	4	16	2	7	2	7	12
1	10	15	14	13	10	6	16	10	12	15	10	13	12	2	15	4	11	1	8
10	3	9	8	14	16	2	14	8	11	5	3	11	15	12	12	11	14	15	3

8	8	15	3	2	15	16	15	7	1	8	11	1	5	9	1	9	8	11	13
10	1	13	16	13	16	11	6	4	11	4	1	14	6	15	9	6	5	16	1
11	6	5	1	8	8	6	16	10	2	5	3	2	13	8	14	12	7	14	5
1	13	10	14	3	11	13	1	6	3	6	2	12	2	7	3	8	10	10	14
7	11	2	6	5	3	12	2	14	9	16	7	10	3	10	12	11	4	12	12
13	5	6	4	12	4	9	7	16	4	9	14	3	1	4	4	1	12	4	15
5	12	1	10	10	12	10	3	1	10	11	10	16	15	13	10	2	3	5	11
9	14	3	13	1	1	15	13	11	14	14	12	13	9	12	16	3	2	9	10
15	4	14	9	9	5	5	5	12	6	15	8	4	4	14	7	14	15	7	2
16	9	16	12	7	14	1	14	15	13	2	15	5	10	3	8	5	13	13	16
4	2	8	8	6	7	7	10	5	12	1	4	8	11	16	13	15	9	1	8
6	15	11	7	4	10	4	4	3	16	10	13	11	16	1	5	13	16	8	6
14	10	12	11	11	6	2	8	9	7	7	5	7	7	2	15	7	1	3	7
3	16	9	5	15	9	8	9	2	8	12	6	9	8	5	2	10	14	15	3
12	3	4	15	16	13	3	11	8	15	13	16	15	14	11	6	4	6	6	9
2	7	7	2	14	2	14	12	13	5	3	9	6	12	6	11	16	11	2	4

13	1	1	9	7	6	1	2	13	2	4	9	1	4	13	2	2	3	16	11
10	14	3	13	4	5	13	9	10	8	6	10	14	16	11	13	8	11	6	9
4	15	14	11	12	10	11	8	15	12	9	11	9	6	14	14	10	14	11	3
15	12	10	4	14	11	4	12	7	3	7	8	2	3	7	4	14	8	1	15
2	9	16	16	16	2	6	4	9	10	3	2	15	10	2	3	11	13	10	1
7	10	8	8	15	14	15	10	12	13	5	5	6	8	8	9	9	1	9	2
9	8	15	3	13	7	16	11	6	4	10	6	4	12	3	1	15	5	13	13
6	5	7	10	1	3	3	16	14	7	16	12	16	7	5	6	5	12	15	5
16	6	11	7	9	8	14	6	2	11	13	14	5	13	12	11	12	9	12	4
5	4	12	15	5	1	2	1	8	5	15	1	8	5	10	8	1	7	7	7
11	2	4	6	3	9	10	14	11	9	2	7	12	14	6	10	16	16	8	10
14	7	13	1	11	4	7	3	16	1	11	15	11	11	1	7	7	15	14	14
12	3	2	12	6	16	5	13	3	15	8	16	3	1	9	5	4	2	3	6
3	16	9	2	10	12	9	15	1	6	12	4	7	9	4	12	13	6	2	8
1	11	6	5	2	13	12	7	4	14	14	13	10	2	16	15	6	10	5	16
8	13	5	14	8	15	8	5	5	16	1	3	13	15	15	16	3	4	4	12

TABLE 15.7 PERMUTATIONS OF 16 (*Continued*)

9	2	4	7	11	8	9	12	11	3	10	8	11	14	16	2	13	2	2	15
13	1	7	8	2	3	5	5	12	7	9	7	4	2	5	11	1	10	10	3
8	5	16	15	16	15	16	13	15	11	5	12	14	7	15	1	12	16	16	12
5	6	15	3	6	2	10	11	13	13	3	6	9	16	4	13	6	15	3	5
12	8	9	2	3	9	6	3	3	1	4	9	3	4	13	7	10	4	7	6
1	12	14	14	1	14	4	7	5	9	16	14	1	15	7	16	5	5	1	7
14	7	5	11	10	16	1	6	4	6	2	10	2	8	1	6	3	13	12	8
16	13	6	13	15	1	15	14	8	15	13	3	6	5	3	8	14	3	13	1
6	3	11	4	8	12	8	10	2	4	6	15	16	11	11	4	7	1	14	2
7	4	10	10	7	7	7	15	14	16	7	1	13	9	6	15	16	12	6	4
3	16	1	16	12	10	12	8	7	12	11	11	7	12	14	14	11	14	9	16
4	11	2	12	5	4	13	4	1	5	8	5	12	6	9	9	9	11	4	13
10	10	3	6	9	11	14	1	9	2	1	4	8	3	2	3	8	8	8	14
11	14	8	9	13	13	3	2	6	8	14	16	15	13	12	12	2	9	11	11
15	9	13	1	14	6	2	9	10	14	15	13	10	10	10	5	4	7	15	10
2	15	12	5	4	5	11	16	16	10	12	2	5	1	8	10	15	6	5	9

13	15	5	7	2	14	16	4	9	6	5	10	7	15	6	6	2	9	15	9
2	9	9	8	15	11	10	9	2	3	13	11	10	16	7	2	9	2	12	6
11	1	12	11	13	16	14	3	7	7	8	1	14	7	11	5	8	10	9	16
12	3	4	5	14	10	11	15	15	14	2	2	2	9	2	16	6	8	5	1
6	5	13	1	3	7	12	12	16	8	10	8	6	12	4	15	7	11	3	5
7	13	10	2	8	5	15	14	3	9	14	15	9	3	15	9	11	16	6	12
15	12	6	6	9	15	3	5	13	1	12	14	4	6	16	14	10	12	1	4
5	8	1	16	16	2	7	11	10	4	9	4	3	14	9	12	4	15	2	2
9	10	8	15	6	13	9	8	11	15	1	6	5	4	12	1	12	14	14	3
1	2	16	4	5	3	2	16	4	2	4	3	11	5	14	13	1	5	16	15
3	6	3	14	11	8	5	13	5	5	15	16	13	1	8	8	5	4	13	10
8	4	2	12	12	4	1	7	8	16	11	9	8	2	13	11	16	7	4	13
16	16	11	10	10	12	6	2	6	12	3	5	1	10	1	4	14	1	11	14
4	11	7	13	1	9	13	6	12	13	6	13	15	13	3	7	15	3	7	8
10	14	15	3	7	1	4	1	14	10	7	7	16	8	5	3	3	13	8	7
14	7	14	9	4	6	8	10	1	11	16	12	12	11	10	10	13	6	10	11

16	4	6	16	9	2	5	11	10	5	1	14	2	14	8	8	13	6	5	13
6	12	12	2	15	16	3	14	9	11	8	5	8	9	9	14	8	5	7	1
13	5	14	7	12	3	1	4	6	1	13	6	9	5	1	9	1	11	14	6
8	2	16	6	13	6	12	7	2	12	16	9	14	1	13	5	16	13	1	3
5	9	9	12	16	1	6	16	13	3	11	16	3	10	6	13	15	4	12	14
10	15	10	11	1	5	4	8	1	15	12	15	11	4	7	7	10	7	3	7
14	8	8	15	8	9	9	6	8	16	15	13	10	8	11	11	3	3	10	11
1	10	11	13	2	4	7	2	7	13	6	10	1	7	2	16	6	15	6	4
3	13	3	3	5	8	16	15	14	7	2	12	12	6	5	4	7	16	4	15
4	16	5	10	7	13	8	1	3	6	14	3	13	15	12	15	14	9	13	16
9	1	1	5	14	14	15	3	15	10	3	8	16	2	10	10	11	1	15	8
7	6	4	14	11	11	11	13	5	8	5	1	7	16	15	6	12	2	11	9
11	11	7	9	3	7	10	10	12	4	10	2	5	11	3	2	5	8	2	2
15	14	13	8	4	10	2	5	16	2	9	7	4	12	16	3	2	14	16	10
12	3	2	1	10	12	13	9	11	9	4	4	15	13	14	1	9	10	9	5
2	7	15	4	6	15	14	12	4	14	7	11	6	3	4	12	4	12	8	12

15	9	15	13	16	4	10	5	15	4	6	10	6	4	7	9	6	8	14	14
7	4	11	4	2	3	11	1	14	11	3	5	9	5	6	13	4	12	9	11
12	10	1	11	14	1	7	11	5	10	13	14	14	11	2	7	14	16	12	1
16	14	9	8	1	11	8	2	16	15	4	6	13	7	12	14	7	6	15	13
10	3	10	9	13	14	1	9	12	3	15	8	11	10	15	15	5	4	11	9
2	15	8	6	10	15	5	3	7	12	1	3	2	16	1	5	13	14	5	15
8	13	4	1	3	10	12	16	10	1	2	4	7	14	11	4	11	10	4	8
14	12	14	10	7	8	4	10	1	14	16	1	12	1	16	8	8	13	10	6
11	11	13	15	6	16	14	13	13	7	14	13	3	15	5	2	15	2	16	12
5	6	3	7	9	12	9	12	9	13	12	11	1	13	4	1	10	15	1	7
13	7	2	2	12	13	6	6	2	6	11	9	8	2	3	6	3	5	3	5
9	8	7	5	15	6	16	7	4	9	5	2	10	12	13	11	2	9	6	4
1	2	12	3	5	7	2	15	11	16	10	7	16	3	14	12	16	3	13	10
6	1	6	16	11	5	3	14	6	8	9	16	15	9	8	16	1	1	2	2
3	5	16	14	8	9	13	4	8	5	8	15	4	8	9	3	9	11	7	3
4	16	5	12	4	2	15	8	3	2	7	12	5	6	10	10	12	7	8	16

TABLE 15.7 PERMUTATIONS OF 16 (*Continued*)

2	8	5	14	1	8	4	8	14	9	2	9	5	13	8	15	10	15	13	3
14	1	16	7	3	14	16	14	9	6	8	14	4	8	14	6	6	10	5	16
7	9	12	8	16	15	2	10	8	16	7	2	12	6	2	13	2	13	15	10
9	11	9	10	9	16	9	15	12	14	12	10	13	10	1	12	14	8	7	5
15	10	14	9	5	13	1	9	11	13	4	11	3	11	10	8	8	9	12	4
8	15	7	1	11	3	8	1	16	10	9	15	14	7	11	14	9	4	8	15
1	7	10	5	13	7	10	7	6	12	15	4	11	16	6	4	11	7	14	6
4	3	3	4	14	6	14	11	13	2	6	1	16	15	9	16	5	1	4	11
16	4	15	15	2	9	13	16	3	1	10	16	2	3	13	10	7	2	1	12
11	2	8	12	12	2	15	12	4	7	13	7	15	14	4	1	4	14	10	7
6	16	4	3	8	11	12	6	7	8	1	6	7	4	7	9	1	12	16	1
5	12	1	2	6	12	6	4	2	4	3	13	9	12	5	5	3	16	6	8
13	6	6	11	10	10	7	3	15	15	16	8	6	1	15	11	15	11	11	14
3	10	11	13	7	5	3	5	1	5	14	3	1	9	3	2	12	3	3	2
10	5	2	16	4	4	11	2	5	11	11	12	8	2	16	7	13	5	9	13
12	14	13	6	15	1	5	13	10	3	5	5	10	5	12	3	16	6	2	9
4	2	6	6	15	8	10	15	13	9	8	9	12	13	10	3	16	8	9	8
6	9	10	4	4	4	14	16	6	5	14	8	11	14	2	10	6	11	6	6
1	10	5	1	6	16	1	2	4	1	6	3	16	12	14	1	12	16	11	2
9	13	9	3	5	14	16	8	11	16	11	13	4	1	6	7	15	12	14	11
16	14	7	15	3	7	2	9	9	6	9	12	10	7	13	8	2	1	8	9
5	7	12	9	14	2	13	7	1	3	15	6	6	3	9	2	3	9	1	13
11	1	8	16	12	5	7	13	3	11	16	10	3	8	7	9	1	13	7	12
12	15	13	12	16	10	11	11	12	14	7	5	2	10	16	15	14	5	5	3
13	3	11	14	2	9	6	6	10	2	2	11	15	16	8	4	10	2	10	4
15	4	2	10	8	6	8	10	16	12	3	14	5	11	15	6	9	6	3	15
2	12	16	11	11	15	3	3	8	4	5	16	7	4	4	5	13	15	12	16
7	8	3	5	10	11	5	4	5	7	12	15	14	9	12	11	8	7	15	5
14	16	1	7	1	1	15	14	15	10	10	4	8	15	5	13	4	10	13	7
3	6	14	8	9	13	9	1	2	15	4	7	13	5	11	12	5	4	2	1
10	11	15	13	13	12	12	12	14	13	1	2	1	2	3	14	7	14	16	14
8	5	4	2	7	3	4	5	7	8	13	1	9	6	1	16	11	3	4	10
5	16	3	12	13	13	1	2	7	3	9	6	15	15	1	13	1	1	16	2
4	1	13	7	3	2	12	1	9	5	4	10	6	3	10	16	7	10	12	12
1	11	7	16	7	11	15	5	4	15	3	12	11	2	15	3	16	16	6	7
9	2	2	11	5	12	11	10	14	14	1	16	10	13	9	1	5	8	15	14
11	5	12	1	8	15	5	11	12	1	6	8	4	6	4	10	6	7	1	16
16	10	15	2	6	16	8	7	13	12	16	7	9	10	5	12	2	4	14	5
10	9	14	5	14	7	14	12	2	16	15	1	13	7	12	9	9	13	9	8
6	6	11	13	10	4	3	4	5	8	10	2	2	14	13	8	4	6	13	13
14	15	8	14	9	14	16	6	1	10	14	5	7	12	2	6	10	2	11	1
8	4	6	3	11	5	4	9	3	9	11	4	3	8	11	4	11	9	10	15
7	7	16	4	15	1	7	8	10	2	7	14	14	11	16	2	14	15	4	11
15	14	1	10	12	10	10	16	6	4	8	11	1	9	3	14	8	3	7	3
3	3	9	15	16	6	9	3	11	11	12	9	5	1	6	15	3	5	3	10
2	8	10	6	4	8	2	15	8	7	13	13	12	5	14	11	15	14	8	9
13	13	5	8	2	3	6	13	16	6	5	15	16	4	8	7	13	11	5	6
12	12	4	9	1	9	13	14	15	13	2	3	8	16	7	5	12	12	2	4
10	4	5	3	3	5	11	12	15	12	14	12	6	15	10	7	7	9	8	10
6	6	12	11	12	11	3	13	3	3	2	10	2	2	15	15	8	11	5	12
14	3	8	15	16	9	7	4	14	13	8	15	9	3	9	5	14	10	13	8
16	7	6	1	10	16	2	6	16	9	1	1	5	13	12	8	2	7	10	6
13	16	13	13	11	12	9	7	7	4	12	5	12	10	14	12	1	1	2	5
4	11	11	8	13	3	6	10	10	5	16	16	8	4	7	1	11	15	1	3
11	15	15	14	15	1	8	15	9	11	3	3	1	7	1	10	10	8	7	14
15	12	16	10	6	10	12	1	8	7	4	13	10	9	2	14	6	16	4	7
1	13	7	7	8	2	1	16	6	6	13	11	7	16	6	3	16	2	11	1
7	2	10	9	1	8	13	3	1	10	7	6	4	8	8	11	12	6	9	16
9	8	2	4	7	14	15	11	13	8	11	14	3	12	3	16	13	5	16	2
5	14	1	12	2	6	14	2	12	1	15	2	14	1	11	4	5	3	3	9
8	5	9	16	9	4	10	14	4	14	6	8	16	6	16	2	4	12	6	11
2	10	14	5	5	15	4	8	11	2	9	7	11	5	5	6	9	14	12	15
12	9	4	2	14	7	5	9	2	15	10	4	15	11	13	9	15	13	15	13
3	1	3	6	4	13	16	5	5	16	5	9	13	14	4	13	3	4	14	4

TABLE 15.7 Permutations of 16 (*Continued*)

5	3	6	7	2	8	9	13	4	7	10	14	8	9	9	9	11	16	16	13
3	6	10	13	4	4	14	1	12	10	11	3	4	8	8	6	12	7	5	6
9	13	3	12	16	10	5	7	1	1	9	5	16	1	7	10	7	11	1	14
15	10	14	2	13	2	11	2	15	15	2	7	13	3	4	16	16	9	9	7
10	2	8	6	12	13	15	11	16	8	3	6	3	11	15	3	9	12	8	10
2	12	2	5	10	6	10	8	11	14	6	13	10	7	5	14	13	2	13	8
1	8	13	3	15	15	2	9	9	11	12	4	5	6	3	12	14	8	10	5
14	5	5	8	5	12	8	10	13	9	16	2	6	16	2	11	15	5	3	1
13	7	11	11	1	16	3	5	8	13	4	1	2	12	16	2	3	14	11	2
6	14	1	15	7	1	6	4	10	6	15	16	11	15	14	13	10	6	15	12
7	4	4	14	9	5	16	6	3	12	14	10	12	2	10	7	4	3	2	3
8	9	16	9	6	11	12	3	7	16	8	9	9	4	13	8	8	10	7	9
11	1	15	1	11	14	1	14	14	2	7	8	14	10	1	5	1	13	12	11
16	11	9	10	14	3	7	12	2	5	13	15	7	13	12	15	5	4	4	15
12	15	12	16	8	7	13	16	6	3	1	11	15	5	11	4	6	1	14	16
4	16	7	4	3	9	4	15	5	4	5	12	1	14	6	1	2	15	6	4

2	11	6	14	1	9	10	9	5	5	8	8	6	7	15	14	15	7	8	7
11	15	14	7	7	15	13	3	4	2	11	13	7	9	1	3	12	5	11	11
9	14	12	11	10	12	7	12	16	14	7	15	1	8	14	4	6	8	14	15
8	13	3	15	2	8	12	10	9	12	14	9	4	5	12	5	13	16	4	1
3	1	10	3	8	10	1	2	14	4	10	10	13	1	10	6	2	12	16	10
14	3	2	9	11	7	4	14	13	15	13	7	16	15	2	12	3	11	12	16
12	4	11	2	15	6	14	7	8	6	16	1	15	13	16	8	7	10	13	6
1	7	1	10	5	16	16	16	10	16	2	3	8	2	11	13	1	13	15	13
10	12	16	4	9	14	6	1	12	8	5	11	5	16	9	2	4	9	7	4
16	2	7	1	6	13	2	5	11	11	3	16	12	12	8	16	9	14	6	8
13	10	9	6	4	5	9	6	15	13	4	12	3	4	5	1	16	4	5	3
7	16	15	16	12	4	8	8	2	7	6	4	2	10	3	10	10	1	9	5
4	5	4	5	3	11	3	13	1	1	1	14	14	11	4	9	11	2	10	2
15	8	8	13	16	3	15	11	7	9	15	2	10	6	7	7	14	6	2	14
5	9	13	8	14	1	5	15	3	3	9	6	9	3	13	15	5	3	1	12
6	6	5	12	13	2	11	4	6	10	12	5	11	14	6	11	8	15	3	9

5	16	1	8	13	8	15	4	10	8	15	4	15	16	8	3	8	14	16	1
14	15	8	7	4	6	8	14	13	9	6	14	14	15	16	16	6	13	14	14
4	13	14	4	10	13	7	2	8	4	10	11	11	11	6	4	3	16	10	12
11	10	9	15	16	3	5	5	12	16	16	12	13	6	5	6	16	15	13	2
3	6	10	11	7	11	11	12	16	13	7	1	7	9	13	10	14	7	3	15
8	9	16	3	6	1	13	7	5	3	2	9	1	10	4	14	4	11	8	7
1	7	7	12	8	4	12	13	6	15	14	2	5	2	3	15	15	12	1	8
13	8	4	2	3	2	6	10	4	7	12	3	3	1	10	1	13	3	9	10
16	12	11	13	2	15	2	1	7	1	1	7	8	13	2	11	1	6	7	11
7	14	6	6	1	12	3	11	14	6	11	5	16	14	7	2	5	10	2	6
6	4	2	1	15	5	1	16	2	5	9	8	10	3	14	12	7	4	5	3
2	1	3	14	11	10	9	15	11	11	4	15	2	12	12	9	10	1	15	13
10	2	5	16	14	7	10	9	1	12	8	6	12	8	1	5	2	5	11	9
12	11	12	5	5	16	4	6	3	14	5	13	6	7	11	13	12	9	4	16
9	3	13	9	9	14	14	3	15	2	3	10	9	4	9	8	9	8	12	5
15	5	15	10	12	9	16	8	9	10	13	16	4	5	15	7	11	2	6	4

1	5	8	15	4	9	7	6	8	8	13	11	8	6	2	9	11	15	12	14
15	6	13	6	15	3	1	15	3	6	10	4	15	9	9	13	2	5	15	16
8	7	7	12	16	2	6	5	7	9	2	9	12	2	13	4	16	7	13	8
3	9	1	5	1	5	11	3	10	1	11	10	1	1	14	6	15	10	14	6
7	13	5	8	13	11	2	1	14	7	16	8	3	13	11	10	7	8	11	10
5	2	14	9	2	10	5	11	11	12	5	14	16	16	6	12	1	9	10	11
9	12	9	1	9	14	4	16	16	4	14	3	4	14	10	2	4	12	7	4
16	11	2	2	11	13	8	9	1	5	6	2	7	5	3	1	12	3	16	7
11	14	12	3	6	6	3	7	5	14	1	12	2	3	8	15	13	13	2	3
14	8	11	4	14	4	12	2	2	15	9	13	6	8	1	3	3	11	3	13
10	10	4	11	3	12	9	10	13	11	3	7	13	7	7	7	10	14	6	2
13	1	16	7	10	16	16	8	15	10	4	5	5	15	4	16	14	1	9	1
6	4	6	16	12	15	14	4	6	3	8	6	9	10	12	11	8	4	1	15
2	15	15	13	7	1	15	14	12	2	7	16	10	12	15	8	5	16	5	5
12	16	10	10	5	8	13	13	4	13	15	1	11	4	16	5	6	6	8	9
4	3	3	14	8	7	10	12	9	16	12	15	14	11	5	14	9	2	4	12

TABLE 15.7 Permutations of 16 (*Continued*)

15	2	11	8	9	2	11	15	3	1	15	16	1	2	1	3	12	1	4	3
3	5	3	2	7	3	4	9	8	4	4	13	3	4	5	6	10	16	1	8
4	6	9	4	12	11	9	6	12	13	7	2	9	15	15	14	11	8	6	5
5	13	5	11	11	10	3	14	7	16	1	15	4	12	9	9	8	12	15	2
10	8	1	10	16	14	10	10	16	2	9	11	7	1	13	4	16	9	13	12
12	16	4	7	14	13	16	11	14	7	16	3	14	7	6	10	1	13	7	16
9	9	8	12	1	15	2	4	1	12	3	12	13	16	16	1	14	15	2	11
14	10	7	1	10	9	1	7	10	14	5	4	10	13	7	7	9	7	3	10
8	3	6	13	15	8	7	2	13	6	13	9	5	14	4	5	15	10	12	15
7	14	16	15	13	12	6	3	6	8	8	5	15	3	8	2	7	5	8	7
2	11	2	9	6	5	13	13	4	3	12	6	2	8	11	13	2	2	10	14
11	7	14	3	8	16	12	1	11	11	6	1	8	6	2	16	5	11	16	6
1	4	13	5	3	6	5	5	9	9	10	7	16	5	10	15	6	14	9	13
6	12	15	14	4	4	14	8	5	5	11	8	6	11	14	12	13	3	5	9
13	1	10	16	2	7	8	16	15	15	14	14	11	10	12	11	4	4	14	1
16	15	12	6	5	1	15	12	2	10	2	10	12	9	3	8	3	6	11	4

7	1	3	9	5	12	2	1	5	6	12	12	10	2	14	15	3	10	9	2
9	14	16	14	16	10	13	4	6	1	7	15	8	15	10	4	16	12	13	14
12	7	4	5	12	2	1	12	9	11	9	9	11	6	2	9	8	1	15	4
13	3	15	16	10	7	12	9	2	5	4	2	16	4	8	8	15	13	10	8
8	4	11	12	9	15	6	5	3	13	3	4	3	14	12	2	5	6	14	10
10	13	6	8	2	5	8	3	13	7	16	10	15	8	3	10	14	2	7	3
16	8	9	13	13	4	5	2	1	9	8	5	2	7	7	16	11	14	16	13
6	11	1	11	14	16	7	13	11	3	2	6	4	16	9	5	2	11	2	12
5	16	2	2	7	8	11	15	4	10	15	14	13	11	6	3	12	3	4	7
15	15	14	1	3	1	14	10	16	15	6	13	9	9	1	11	6	15	11	5
3	5	7	3	15	6	16	11	10	8	11	8	1	13	5	1	1	5	12	9
2	9	5	7	4	13	10	16	8	16	10	3	12	12	15	7	13	8	3	16
4	10	12	6	1	3	3	14	12	12	1	7	14	10	4	13	4	16	1	15
14	12	8	15	6	11	9	6	7	14	14	1	6	5	13	14	10	4	5	1
1	2	10	10	8	14	15	7	15	2	5	16	5	1	16	6	9	7	6	6
11	6	13	4	11	9	4	8	14	4	13	11	7	3	11	12	7	9	8	11

9	13	10	7	11	11	11	7	11	11	16	11	15	9	2	12	4	10	10	1
6	8	11	16	8	14	6	16	13	2	14	12	14	1	7	16	7	12	8	6
15	3	12	11	14	7	15	4	12	14	8	10	13	6	5	5	9	5	1	7
7	7	5	9	15	1	2	10	4	8	5	6	2	8	3	7	13	15	4	14
2	15	1	8	16	9	8	12	16	12	13	8	8	12	10	4	2	6	14	11
13	12	2	6	7	3	13	9	9	13	2	5	1	3	9	2	1	16	2	5
16	2	13	14	6	2	9	13	5	7	6	4	10	5	14	8	8	8	6	13
8	11	9	13	4	13	14	8	6	16	3	7	11	14	6	1	16	1	3	4
4	10	15	4	13	12	3	2	3	10	7	3	16	16	16	6	6	11	9	12
3	9	7	3	9	5	12	5	15	4	4	14	3	7	1	11	10	3	5	2
14	14	4	15	1	16	10	11	8	9	11	13	12	10	4	3	11	2	7	15
12	1	3	10	12	10	7	1	2	15	15	15	4	11	12	15	5	14	12	10
5	4	6	1	5	6	5	3	7	5	1	9	7	15	11	9	14	4	15	16
10	6	16	2	2	15	4	14	14	3	9	1	5	13	15	10	12	13	11	8
11	5	8	12	3	8	1	15	10	6	12	16	6	2	13	13	15	9	16	3
1	16	14	5	10	4	16	6	1	1	10	2	9	4	8	14	3	7	13	9

3	4	12	11	10	13	11	10	1	4	12	11	5	11	16	7	14	5	4	7
6	10	13	4	1	11	1	2	6	8	10	5	11	12	7	4	11	3	5	8
2	13	11	1	16	1	10	8	14	5	7	9	12	7	1	12	5	6	9	15
11	1	6	10	2	3	4	4	5	2	3	6	1	8	14	9	7	8	8	9
10	16	14	3	6	10	6	3	11	3	8	2	10	14	8	16	8	12	12	13
16	14	15	15	8	6	9	5	16	9	13	15	15	15	6	14	16	14	10	3
1	7	7	7	3	15	15	6	9	1	14	13	2	13	9	3	1	2	14	12
13	8	2	16	12	5	14	11	3	7	6	1	4	1	13	1	13	16	2	5
9	9	1	6	13	7	5	12	7	10	9	8	8	9	5	10	6	15	15	11
5	5	8	9	9	12	16	1	10	16	15	12	6	3	3	5	4	13	11	14
15	6	9	14	11	2	2	16	4	11	10	14	9	5	2	8	3	7	7	1
4	15	4	13	5	9	7	15	12	6	5	16	14	4	12	2	9	9	3	6
8	3	10	5	4	8	8	9	8	15	4	3	7	6	10	11	15	4	1	16
7	11	3	8	14	4	12	7	13	13	2	4	16	16	15	13	2	10	13	10
14	12	16	2	15	14	13	14	15	14	1	10	3	2	11	15	10	1	16	2
12	2	5	12	7	16	3	13	2	12	11	7	13	10	4	6	12	11	6	4

TABLE 15.7 PERMUTATIONS OF 16 (*Continued*)

10	9	10	9	4	4	7	2	5	16	6	13	15	11	16	16	10	16	12	15
12	5	5	7	5	1	11	5	13	10	2	6	16	4	15	14	8	12	13	11
4	4	7	4	9	9	15	13	7	13	3	8	14	5	8	11	7	7	14	12
9	6	15	6	10	8	6	14	1	5	15	9	4	8	11	9	12	11	6	9
5	10	16	13	8	2	3	7	4	9	11	14	5	2	10	3	11	15	3	2
13	2	6	12	2	14	16	9	3	4	1	2	2	15	7	13	14	14	15	13
11	7	2	5	6	5	2	8	14	7	10	3	1	7	2	12	2	5	7	14
15	14	14	10	16	15	8	3	12	6	4	15	10	1	14	5	15	1	1	6
7	8	12	1	3	6	4	11	6	15	5	11	3	6	1	10	4	3	10	16
6	3	11	11	11	12	9	4	15	3	8	10	7	12	13	4	9	6	11	7
1	15	3	16	15	3	5	12	16	2	12	16	6	14	6	8	13	10	9	8
3	11	13	2	12	7	1	6	9	14	9	5	13	16	4	7	3	13	8	1
16	13	8	15	1	10	12	15	2	8	16	4	8	3	3	6	5	4	16	4
14	12	9	14	14	11	10	16	10	1	7	1	11	9	12	15	1	9	5	5
2	16	1	8	13	13	13	1	11	11	14	12	12	13	9	1	16	2	2	10
8	1	4	3	7	16	14	10	8	12	13	7	9	10	5	2	6	8	4	3

6	5	3	12	16	7	7	13	8	4	6	12	15	5	1	9	5	16	10	15
16	1	8	13	14	15	4	12	13	7	16	14	7	14	7	7	16	15	14	2
10	12	11	7	7	5	1	15	3	9	5	6	9	15	4	11	8	14	12	13
14	16	2	16	15	13	12	3	10	1	14	8	10	11	12	5	15	1	11	5
3	10	10	6	3	12	8	4	5	3	1	5	3	1	11	13	7	3	4	10
5	14	9	8	8	4	2	16	14	13	12	10	4	10	16	1	14	12	8	11
15	13	16	10	9	8	9	6	12	14	15	1	6	16	3	4	6	13	1	16
1	3	13	15	4	1	16	5	1	8	7	13	12	8	9	15	13	4	7	4
11	8	6	2	10	11	5	9	15	5	8	2	16	6	5	16	4	7	5	7
12	4	15	14	1	2	3	2	6	11	9	9	5	3	8	10	2	5	15	14
7	6	4	9	13	10	13	8	11	12	11	4	8	4	2	8	9	8	9	3
4	9	14	11	2	3	15	7	9	6	10	15	11	9	6	3	12	6	2	12
2	7	7	4	11	9	11	10	4	2	3	11	1	13	13	2	10	2	3	8
13	15	1	1	12	14	10	11	16	10	4	16	14	2	15	6	1	9	16	6
9	11	12	3	5	6	14	14	7	16	2	3	13	12	14	14	3	10	13	1
8	2	5	5	6	16	6	1	2	15	13	7	2	7	10	12	11	11	6	9

13	5	4	11	11	12	9	9	16	12	4	8	2	2	5	1	9	3	8	10
8	14	5	2	10	4	5	3	2	2	7	4	13	3	2	15	15	11	4	6
1	13	12	14	12	15	7	15	13	3	8	9	14	10	16	5	4	12	2	13
4	15	11	3	13	5	6	7	6	5	2	13	3	9	7	12	7	10	7	7
3	12	13	6	6	14	4	8	14	14	12	3	7	16	1	14	5	16	10	9
16	10	9	10	8	6	2	11	4	16	1	15	11	14	3	6	6	6	12	3
11	6	7	4	5	2	15	10	3	4	3	14	15	1	14	16	10	2	3	14
9	16	10	5	2	13	14	12	15	11	14	7	16	11	9	9	3	13	11	1
5	4	3	7	9	3	8	6	9	7	15	12	9	13	6	8	14	5	13	2
12	7	1	12	1	8	12	14	11	10	16	16	1	12	11	13	11	14	9	12
7	1	16	1	16	7	10	4	8	9	6	10	12	5	15	4	12	1	14	11
14	8	14	15	7	10	16	13	12	15	5	11	8	7	13	11	16	8	6	16
15	11	6	8	3	1	1	16	1	13	9	1	10	8	8	2	2	9	1	4
10	3	8	9	14	9	13	2	5	6	10	2	6	15	10	10	13	15	5	8
6	9	2	13	15	11	11	1	10	1	11	5	4	4	4	3	1	4	15	5
2	2	15	16	4	16	3	5	7	8	13	6	5	6	12	7	8	7	16	15

9	12	16	8	10	7	4	15	5	14	11	3	8	15	13	16	14	1	1	15
16	13	1	12	12	14	3	16	6	10	16	11	2	1	15	11	2	14	3	14
2	7	8	9	5	6	13	13	4	2	5	13	13	16	7	2	9	9	15	5
7	15	9	5	13	2	2	1	11	7	10	15	3	5	10	4	13	8	16	3
10	2	12	14	3	11	5	8	13	8	7	12	1	11	11	10	5	13	4	1
11	16	6	4	1	16	12	14	16	13	1	5	7	8	3	12	12	10	7	2
14	10	5	13	8	10	7	6	12	9	9	6	12	4	12	3	7	4	8	13
1	3	3	7	4	15	15	11	2	11	12	2	15	2	16	7	8	12	10	12
4	9	2	6	11	12	11	2	9	6	6	1	4	9	2	9	16	7	14	8
8	4	7	10	6	8	10	4	8	12	8	14	6	3	9	14	1	5	6	16
15	5	15	16	15	5	16	3	7	1	3	10	5	14	6	15	3	15	11	9
13	8	10	11	9	13	14	12	1	5	14	9	10	10	8	6	6	6	9	11
6	1	14	3	14	3	6	5	14	16	13	7	9	6	4	13	4	11	13	4
3	6	4	15	7	9	1	9	15	4	15	16	11	7	14	5	15	16	2	7
12	11	11	1	16	1	8	10	10	3	4	4	16	13	1	8	10	3	12	6
5	14	13	2	2	4	9	7	3	15	2	8	14	12	5	1	11	2	5	10

TABLE 15.7 PERMUTATIONS OF 16 (*Continued*)

10	11	7	9	8	10	12	11	4	10	16	6	11	4	11	10	6	5	10	16
1	6	2	16	6	1	3	1	8	8	6	3	1	12	16	11	3	4	11	5
6	10	11	15	16	14	15	10	11	9	5	15	15	8	6	14	5	12	15	3
11	3	16	1	12	15	4	15	15	6	4	2	10	2	9	4	11	15	16	10
7	16	15	10	10	11	13	13	7	4	10	10	12	9	3	1	1	10	2	2
4	5	1	6	5	6	11	16	12	11	11	1	9	11	4	16	2	8	5	4
14	4	8	5	3	4	5	14	1	12	12	11	4	10	12	7	12	14	14	14
3	9	14	11	15	9	16	7	3	16	15	7	2	16	7	13	14	3	3	12
2	14	5	14	14	7	1	3	5	14	3	12	13	3	14	2	15	1	4	1
9	2	3	3	13	3	14	2	9	7	9	4	7	7	2	12	16	7	13	7
12	13	13	7	7	8	6	9	14	3	7	16	6	5	13	9	8	9	12	11
16	1	9	2	11	13	10	8	10	15	14	13	3	6	10	5	13	11	7	8
15	12	6	4	4	5	7	12	16	2	8	5	5	13	5	8	7	10	0	6
8	15	12	8	2	16	9	5	2	13	2	9	14	1	1	6	4	2	9	15
5	8	4	13	1	12	2	4	13	1	13	14	16	15	15	3	10	13	1	9
13	7	10	12	9	2	8	6	6	5	1	8	8	14	8	15	9	6	8	13
11	7	16	10	15	3	1	14	1	2	15	4	3	14	8	15	1	5	16	6
1	6	1	6	7	7	6	6	15	5	10	12	6	11	7	4	14	15	2	10
3	3	12	5	6	11	8	11	2	12	16	14	2	15	9	11	12	14	13	7
10	1	10	11	3	9	7	2	11	16	3	7	15	5	13	7	13	12	9	13
12	8	7	16	12	2	3	1	16	3	1	8	10	1	16	5	4	13	4	4
6	15	4	8	8	5	9	13	9	10	6	1	11	12	3	10	7	8	5	1
9	11	6	15	1	8	2	4	6	13	5	3	4	13	6	8	3	9	1	16
13	16	3	4	11	14	4	12	5	11	7	5	9	7	11	3	10	6	6	15
14	4	11	14	4	10	5	10	12	8	8	16	7	2	4	12	2	11	7	2
5	5	2	9	9	13	10	9	13	15	2	15	1	10	15	2	15	3	10	5
4	14	8	3	16	4	16	16	7	7	4	2	14	4	5	16	8	2	3	12
7	12	5	12	5	16	15	5	14	6	9	6	16	9	14	13	9	4	15	14
8	2	9	2	14	1	12	8	4	4	13	11	13	16	1	6	11	16	11	8
16	10	15	13	10	15	14	7	10	14	11	10	8	8	2	14	5	7	8	11
15	9	14	1	2	12	13	15	3	1	12	9	12	6	10	9	16	10	12	9
2	13	13	7	13	6	11	3	8	9	14	13	5	3	12	1	6	1	14	3
7	3	11	1	15	5	13	11	14	9	1	2	11	6	5	11	1	10	15	7
4	4	6	11	12	8	12	16	11	3	7	11	13	13	10	15	15	1	5	14
12	15	13	4	13	4	3	14	5	4	14	1	1	11	16	14	6	7	10	15
9	8	16	9	11	1	14	5	1	11	8	15	2	8	8	9	14	5	6	12
10	2	7	2	10	14	6	8	2	7	9	14	5	15	9	3	9	9	1	5
15	13	12	12	5	9	7	7	10	1	11	9	9	12	1	12	5	8	3	3
8	1	2	13	14	12	1	12	12	8	6	4	16	1	4	13	12	16	2	4
5	9	4	14	6	2	15	13	15	5	2	12	6	5	3	2	2	13	9	2
14	10	5	3	9	13	8	6	16	2	13	13	7	7	15	8	13	11	4	1
3	16	1	15	2	15	4	4	13	6	10	7	4	16	14	1	4	3	14	9
11	14	8	7	3	6	9	9	9	12	12	3	12	9	12	5	8	2	16	16
6	7	9	10	1	7	11	2	7	15	16	6	3	4	11	10	3	6	7	11
13	6	14	16	16	10	2	10	6	13	15	16	14	14	2	0	7	4	8	13
1	5	15	5	4	16	16	3	8	14	5	5	8	2	6	4	11	14	13	6
16	11	3	8	7	3	5	15	4	10	4	10	10	10	13	7	16	12	11	10
2	12	10	6	8	11	10	1	3	16	3	8	15	3	7	16	10	15	12	8
8	2	1	14	10	6	1	1	8	3	7	7	1	2	12	2	12	15	3	4
6	10	2	2	6	13	9	13	7	12	11	1	13	3	16	4	6	3	6	8
11	9	3	4	3	10	4	7	4	4	15	8	4	9	8	7	4	10	7	2
5	3	14	10	9	8	6	10	2	15	10	14	2	6	1	9	5	2	15	10
10	15	7	16	13	12	12	15	11	6	4	11	6	12	11	8	11	11	16	3
12	16	8	11	12	5	11	5	12	11	8	9	5	7	9	13	8	13	14	9
16	12	13	5	7	4	8	14	16	13	3	6	11	15	10	14	1	16	9	11
15	5	4	12	1	3	2	12	10	8	2	4	9	8	5	15	3	4	12	13
2	6	10	1	16	14	3	8	1	9	13	2	14	10	7	5	16	8	10	1
3	4	6	8	5	16	13	2	6	7	5	5	15	13	14	6	10	7	2	12
9	11	10	13	8	1	15	4	14	1	14	10	16	5	6	10	2	1	11	15
1	13	9	7	4	9	10	16	5	16	1	12	3	1	3	3	7	5	8	6
4	1	11	15	15	11	14	6	9	5	16	13	10	14	13	11	15	14	4	14
13	8	12	9	2	2	16	9	13	10	9	15	8	16	4	12	13	6	13	5
14	14	5	3	11	7	5	11	15	2	6	16	12	11	15	1	9	12	5	7
7	7	15	6	14	15	7	3	3	14	12	3	7	4	2	16	14	9	1	16

TABLE 15.7 PERMUTATIONS OF 16 (*Continued*)

13	1	9	7	7	13	6	15	1	12	11	10	11	7	5	3	9	11	10	3
2	12	15	4	16	10	2	16	16	3	13	12	1	3	15	1	15	1	7	5
3	9	2	9	9	1	5	10	9	13	7	15	13	12	4	11	16	12	11	10
5	2	4	10	12	8	13	6	2	6	9	9	15	13	1	5	12	4	5	1
4	6	7	8	13	9	15	4	5	5	14	5	4	4	2	8	10	14	8	14
14	4	8	1	11	7	12	14	4	7	3	16	12	15	10	4	5	3	4	16
1	14	13	12	5	11	10	5	14	2	10	3	5	2	7	12	13	15	15	8
16	11	6	5	3	14	8	1	12	8	5	14	8	5	3	15	1	10	14	13
15	15	12	15	10	16	3	11	10	1	15	8	14	8	12	10	11	13	3	9
8	7	10	16	8	12	4	2	7	10	4	2	7	6	16	2	4	6	2	11
7	16	5	14	15	6	1	3	15	14	2	11	3	11	6	7	6	8	13	12
6	10	14	13	14	5	14	12	13	9	12	7	9	14	13	9	8	5	1	6
11	8	11	11	4	15	7	13	3	16	8	4	2	1	9	6	14	7	9	7
12	13	3	2	1	3	16	7	8	4	6	13	16	9	14	14	3	2	12	4
9	5	16	6	2	2	9	8	11	15	1	6	6	16	11	16	7	9	16	2
10	3	1	3	6	4	11	9	6	11	16	1	10	10	8	13	2	16	6	15

10	4	1	16	14	11	1	11	8	11	16	15	9	4	2	3	10	10	16	5
13	2	5	8	6	10	3	15	12	14	10	3	16	1	9	16	8	1	3	9
9	16	15	3	11	7	11	1	14	9	12	7	1	16	1	2	1	7	2	3
8	14	11	6	15	12	6	10	5	1	3	1	11	13	4	5	7	14	7	10
3	5	8	11	2	9	15	4	10	5	8	13	14	12	11	11	3	16	13	13
11	3	4	12	8	13	9	9	1	16	7	16	3	8	15	8	4	8	15	16
14	11	12	13	13	14	16	12	3	4	4	9	5	2	5	13	9	9	11	14
5	9	7	1	5	1	13	14	2	15	2	2	2	11	16	6	6	15	5	12
1	8	6	2	4	2	10	6	4	8	15	12	15	7	13	15	11	3	9	15
6	7	9	4	10	3	8	8	15	7	14	11	6	15	14	4	2	6	6	8
4	6	2	5	7	8	4	7	13	13	9	14	10	10	7	9	15	12	1	4
12	12	14	10	9	5	7	5	7	10	5	10	8	9	6	7	5	2	12	1
15	1	10	7	16	16	5	13	9	6	1	6	7	5	3	10	13	13	8	6
7	10	3	9	3	4	2	2	16	12	6	4	4	14	10	14	14	5	10	11
16	15	13	15	1	6	12	16	11	3	11	8	12	3	8	1	12	11	4	7
2	13	16	14	12	15	14	3	6	2	13	5	13	6	12	12	16	4	14	2

15	15	8	2	4	10	16	12	12	9	3	6	11	10	16	8	7	4	9	1
5	6	1	4	15	15	9	2	1	16	10	11	13	4	9	4	15	1	16	16
12	8	3	1	1	2	8	3	6	4	12	15	15	6	5	7	6	16	2	2
1	12	15	16	2	14	10	7	15	3	7	1	4	15	14	3	9	5	14	13
2	16	5	12	11	4	12	13	10	14	16	7	2	9	6	1	11	6	6	14
16	11	7	10	12	16	6	9	14	13	5	9	3	5	13	9	4	7	4	6
8	2	9	6	8	3	4	16	7	7	8	14	14	11	12	5	10	12	15	10
6	4	13	5	14	5	11	11	5	6	13	12	1	3	3	2	1	3	10	15
13	14	14	8	9	6	3	8	2	10	1	2	6	14	1	12	13	15	3	4
10	1	2	9	10	1	15	1	13	11	6	8	16	1	2	16	2	10	1	5
7	9	6	3	5	9	14	4	11	1	14	4	12	13	11	14	12	9	13	7
9	7	16	7	6	13	5	5	3	12	2	5	8	8	4	13	16	14	5	12
14	3	12	13	13	7	13	10	8	8	11	3	10	2	8	11	8	11	8	8
4	5	11	15	16	11	2	15	16	5	4	16	9	12	10	15	5	8	7	11
11	10	10	11	3	8	7	14	9	15	15	10	5	7	7	10	14	13	12	3
3	13	4	14	7	12	1	6	4	2	9	13	7	16	15	6	3	2	11	9

11	1	3	7	8	10	11	8	15	9	15	8	13	8	4	8	4	3	6	15
7	8	2	9	14	4	7	10	3	7	16	7	2	10	7	5	2	16	14	8
1	15	11	3	7	9	12	13	4	13	4	3	10	14	5	2	3	5	11	11
15	13	14	16	4	14	15	16	10	14	13	6	8	13	11	12	7	10	4	7
12	11	1	8	5	11	9	4	5	16	14	1	9	1	13	1	14	11	2	9
4	9	4	15	15	7	3	6	8	11	12	11	15	16	12	3	12	9	13	1
9	6	12	5	13	13	4	11	12	15	3	15	3	4	10	6	5	1	7	4
6	14	6	11	12	5	6	1	14	1	1	12	12	2	3	7	8	13	16	16
10	12	7	4	16	1	13	9	1	12	8	9	1	11	2	9	15	14	3	12
13	16	16	10	1	15	16	7	11	10	10	13	14	15	15	10	10	8	9	2
14	5	9	12	3	16	8	15	2	5	7	4	16	12	14	4	11	4	12	14
3	2	8	2	11	8	1	12	9	3	5	2	11	5	16	15	9	12	5	13
16	7	13	13	10	6	10	14	6	4	9	14	7	6	9	13	1	2	1	6
8	4	5	1	6	3	2	5	16	8	6	16	4	7	1	16	16	7	15	3
2	10	15	14	9	12	14	3	7	2	2	10	6	3	6	11	6	6	8	5
5	3	10	6	2	2	5	2	13	6	11	5	5	9	[illegible]	14	13	15	10	10

TABLE 15.7 Permutations of 16 (*Continued*)

10	14	13	1	6	6	13	1	8	4	9	9	1	2	9	3	1	15	12	8
7	12	2	5	2	16	2	14	16	16	14	10	15	3	14	15	9	8	15	10
11	6	10	10	10	8	1	9	3	14	12	11	7	5	1	5	10	12	11	5
14	11	16	16	15	12	3	12	9	1	16	8	11	11	6	7	5	3	10	15
5	16	5	8	11	13	16	7	5	10	11	1	3	14	4	1	15	1	4	16
15	15	8	14	1	1	4	8	7	8	3	15	14	6	2	14	6	11	7	4
13	7	14	9	14	11	14	2	15	5	10	7	4	8	7	10	4	6	5	9
4	2	1	11	13	14	15	16	4	13	1	2	9	13	5	4	13	9	9	13
3	4	3	12	8	10	11	3	10	11	5	12	10	7	3	16	7	13	1	14
12	9	6	13	12	5	9	15	12	6	2	16	5	10	15	2	11	2	8	7
16	3	11	6	4	9	7	6	11	3	15	4	8	1	16	11	14	4	2	6
8	13	12	4	3	7	12	13	13	7	8	6	6	4	11	13	12	10	14	2
1	10	4	15	5	2	6	11	2	9	6	13	13	16	8	8	8	14	16	12
6	1	9	7	7	4	8	5	1	12	4	3	12	12	13	12	16	16	13	11
9	5	15	3	9	15	10	4	6	15	7	5	2	9	10	6	2	7	6	3
2	8	7	2	16	3	5	10	14	2	13	14	16	15	12	9	3	5	3	1

7	11	1	15	10	2	14	4	14	14	6	4	15	4	6	11	11	5	13	10
2	6	14	8	12	16	6	10	11	7	5	12	1	7	14	9	15	12	7	14
12	8	5	14	1	7	11	16	5	12	16	10	7	9	9	3	2	6	5	7
11	2	9	4	2	6	5	1	9	11	7	5	9	2	8	13	1	10	14	5
6	4	6	3	5	5	10	15	3	13	9	15	12	10	2	10	12	15	16	16
16	14	7	13	9	1	16	13	12	10	3	9	10	14	7	7	3	7	2	12
5	7	15	12	3	3	4	5	6	9	11	13	8	13	16	2	6	14	8	11
8	15	4	2	16	8	8	3	10	8	1	2	13	12	5	8	10	3	3	13
9	5	10	11	8	13	13	14	4	16	13	1	3	6	3	4	14	9	4	2
1	16	12	10	6	9	3	7	16	1	12	8	14	1	10	1	4	8	11	3
3	9	8	1	4	14	1	12	7	4	10	3	5	11	11	14	13	16	10	9
14	1	11	7	7	4	9	8	8	3	4	14	4	5	12	16	5	1	6	1
10	12	2	5	11	12	15	11	13	2	14	11	16	8	1	12	7	13	12	4
4	10	13	16	14	11	7	6	2	6	15	6	11	15	15	6	9	11	15	8
13	3	3	9	13	15	2	9	15	15	8	16	6	3	4	15	16	4	1	15
15	13	16	6	15	10	12	2	1	5	2	7	2	16	13	5	8	2	9	6

Selected Bibliography

BOOKS

ANDERSON, R. L., and BANCROFT, T. A. *Statistical theory in research.* McGraw-Hill, New York, 1952.

BROWNLEE, K. A. *Industrial experimentation.* 2nd ed., Chemical Publishing Co., Brooklyn, 1948. 4th ed., His Majesty's Stationery Office, London, 1949.

DAVIES, O. L. *Design and analysis of industrial experiments.* 1st ed., Oliver & Boyd, London, 1954.

FEDERER, W. T. *Experimental design, theory and application.* 1st ed., The Macmillan Co., New York, 1955.

FINNEY, D. J. *Experimental design and its statistical basis.* Univ. Chicago Press, Chicago, 1955.

KEMPTHORNE, O. *The design and analysis of experiments.* John Wiley and Sons, New York, 1952.

MANN, H. B. *Analysis and design of experiments.* Dover Publications, New York, 1949.

PEARCE, S. C. Field experimentation with fruit trees and other perennial plants. *Commonwealth Bur. Hort. Tech. Comm.* 23, 1953.

RAO, C. R. *Advanced statistical methods in biometric research.* John Wiley and Sons, New York, 1952.

WISHART, J. *Field trials: their layout and statistical analysis.* Imp. Bur. Plant Breed. and Genetics, Cambridge, 1940.

APPLICATIONS

BRUNK, M. E., and FEDERER, W. T. Experimental designs and probability sampling in marketing research. *Jour. Am. Stat. Assoc.* 33, 440–452, 1948.

COMSTOCK, R. E., and WINTERS, L. M. Design of experimental comparisons between lines of breeding in livestock. *Jour. Agr. Res.* 64, 523–532, 1942.

COX, G. M. Statistics as a tool for research. *Jour. Home Econ.* 36, 575–580, 1944.

COX, G. M., and COCHRAN, W. G. Designs of greenhouse experiments for statistical analysis. *Soil Sci.* 62, 87–98, 1946.

DICKERSON, G. E. Experimental design for testing inbred lines of swine. *Jour. Animal Sci.* 1, 326–341, 1942.

STEARMAN, ROEBERT L. Statistical concepts in microbiology. *Bacteriological Rev.* 19, 160–215, 1955.

WISHART, J. Statistical treatment of animal experiments. *Jour. Roy. Stat. Soc. Suppl.* 6, 1–22, 1939.

WOOD, E. C., and FINNEY, D. J. The design and statistical analysis of microbiological assays. *Quart. Jour. Pharm. and Pharmacol.* 19, 112–127, 1946.

List of Author References

Index

Table of t

Degrees of freedom	P = 0.9	0.8	0.7	0.6	0.5	0.4	0.3	0.2	0.1	0.05	0.02	0.01
1	0.158	0.325	0.510	0.727	1.000	1.376	1.963	3.078	6.314	12.706	31.821	63.657
2	0.142	0.289	0.445	0.617	0.816	1.061	1.386	1.886	2.920	4.303	6.965	9.925
3	0.137	0.277	0.424	0.584	0.765	0.978	1.250	1.638	2.353	3.182	4.541	5.841
4	0.134	0.271	0.414	0.569	0.741	0.941	1.190	1.533	2.132	2.776	3.747	4.604
5	0.132	0.267	0.408	0.559	0.727	0.920	1.156	1.476	2.015	2.571	3.365	4.032
6	0.131	0.265	0.404	0.553	0.718	0.906	1.134	1.440	1.943	2.447	3.143	3.707
7	0.130	0.263	0.402	0.549	0.711	0.896	1.119	1.415	1.895	2.365	2.998	3.499
8	0.130	0.262	0.399	0.546	0.706	0.889	1.108	1.397	1.860	2.306	2.896	3.355
9	0.129	0.261	0.398	0.543	0.703	0.883	1.100	1.383	1.833	2.262	2.821	3.250
10	0.129	0.260	0.397	0.542	0.700	0.879	1.093	1.372	1.812	2.228	2.764	3.169
11	0.129	0.260	0.396	0.540	0.697	0.876	1.088	1.363	1.796	2.201	2.718	3.106
12	0.128	0.259	0.395	0.539	0.695	0.873	1.083	1.356	1.782	2.179	2.681	3.055
13	0.128	0.259	0.394	0.538	0.694	0.870	1.079	1.350	1.771	2.160	2.650	3.012
14	0.128	0.258	0.393	0.537	0.692	0.868	1.076	1.345	1.761	2.145	2.624	2.977
15	0.128	0.258	0.393	0.536	0.691	0.866	1.074	1.341	1.753	2.131	2.602	2.947
16	0.128	0.258	0.392	0.535	0.690	0.865	1.071	1.337	1.746	2.120	2.583	2.921
17	0.128	0.257	0.392	0.534	0.689	0.863	1.069	1.333	1.740	2.110	2.567	2.898
18	0.127	0.257	0.392	0.534	0.688	0.862	1.067	1.330	1.734	2.101	2.552	2.878
19	0.127	0.257	0.391	0.533	0.688	0.861	1.066	1.328	1.729	2.093	2.539	2.861
20	0.127	0.257	0.391	0.533	0.687	0.860	1.064	1.325	1.725	2.086	2.528	2.845
21	0.127	0.257	0.391	0.532	0.686	0.859	1.063	1.323	1.721	2.080	2.518	2.831
22	0.127	0.256	0.390	0.532	0.686	0.858	1.061	1.321	1.717	2.074	2.508	2.819
23	0.127	0.256	0.390	0.532	0.685	0.858	1.060	1.319	1.714	2.069	2.500	2.807
24	0.127	0.256	0.390	0.531	0.685	0.857	1.059	1.318	1.711	2.064	2.492	2.797
25	0.127	0.256	0.390	0.531	0.684	0.856	1.058	1.316	1.708	2.060	2.485	2.787
26	0.127	0.256	0.390	0.531	0.684	0.856	1.058	1.315	1.706	2.056	2.479	2.779
27	0.127	0.256	0.389	0.531	0.684	0.855	1.057	1.314	1.703	2.052	2.473	2.771
28	0.127	0.256	0.389	0.530	0.683	0.855	1.056	1.313	1.701	2.048	2.467	2.763
29	0.127	0.256	0.389	0.530	0.683	0.854	1.055	1.311	1.699	2.045	2.462	2.756
30	0.127	0.256	0.389	0.530	0.683	0.854	1.055	1.310	1.697	2.042	2.457	2.750
∞	0.12566	0.25335	0.38532	0.52440	0.67449	0.84162	1.03643	1.28155	1.64485	1.95996	2.32634	2.57582

Reproduced from *Statistical Methods for Research Workers*, 10th ed., with the permission of the author, R. A. Fisher, and his publisher, Oliver and Boyd, Edinburgh.

Table of F

5% (Roman Type) and 1% (Bold-Face Type) Points for the Distribution of F

Degrees of freedom n_2	Degrees of freedom n_1: 1	2	3	4	5	6	7	8	9	10	11	12	14	16	20	24	30	40	50	75	100	200	500	∞
1	161 **4052**	200 **4999**	216 **5403**	225 **5625**	230 **5764**	234 **5859**	237 **5928**	239 **5981**	241 **6022**	242 **6056**	243 **6082**	244 **6106**	245 **6142**	246 **6169**	248 **6208**	249 **6234**	250 **6258**	251 **6286**	252 **6302**	253 **6323**	253 **6334**	254 **6352**	254 **6361**	254 **6366**
2	18.51 **98.49**	19.00 **99.00**	19.16 **99.17**	19.25 **99.25**	19.30 **99.30**	19.33 **99.33**	19.36 **99.34**	19.37 **99.36**	19.38 **99.38**	19.39 **99.40**	19.40 **99.41**	19.41 **99.42**	19.42 **99.43**	19.43 **99.44**	19.44 **99.45**	19.45 **99.46**	19.46 **99.47**	19.47 **99.48**	19.47 **99.48**	19.48 **99.49**	19.49 **99.49**	19.49 **99.49**	19.50 **99.50**	19.50 **99.50**
3	10.13 **34.12**	9.55 **30.82**	9.28 **29.46**	9.12 **28.71**	9.01 **28.24**	8.94 **27.91**	8.88 **27.67**	8.84 **27.49**	8.81 **27.34**	8.78 **27.23**	8.76 **27.13**	8.74 **27.05**	8.71 **26.92**	8.69 **26.83**	8.66 **26.69**	8.64 **26.60**	8.62 **26.50**	8.60 **26.41**	8.58 **26.35**	8.57 **26.27**	8.56 **26.23**	8.54 **26.18**	8.54 **26.14**	8.53 **26.12**
4	7.71 **21.20**	6.94 **18.00**	6.59 **16.69**	6.39 **15.98**	6.26 **15.52**	6.16 **15.21**	6.09 **14.98**	6.04 **14.80**	6.00 **14.66**	5.96 **14.54**	5.93 **14.45**	5.91 **14.37**	5.87 **14.24**	5.84 **14.15**	5.80 **14.02**	5.77 **13.93**	5.74 **13.83**	5.71 **13.74**	5.70 **13.69**	5.68 **13.61**	5.66 **13.57**	5.65 **13.52**	5.64 **13.48**	5.63 **13.46**
5	6.61 **16.26**	5.79 **13.27**	5.41 **12.06**	5.19 **11.39**	5.05 **10.97**	4.95 **10.67**	4.88 **10.45**	4.82 **10.27**	4.78 **10.15**	4.74 **10.05**	4.70 **9.96**	4.68 **9.89**	4.64 **9.77**	4.60 **9.68**	4.56 **9.55**	4.53 **9.47**	4.50 **9.38**	4.46 **9.29**	4.44 **9.24**	4.42 **9.17**	4.40 **9.13**	4.38 **9.07**	4.37 **9.04**	4.36 **9.02**
6	5.99 **13.74**	5.14 **10.92**	4.76 **9.78**	4.53 **9.15**	4.39 **8.75**	4.28 **8.47**	4.21 **8.26**	4.15 **8.10**	4.10 **7.98**	4.06 **7.87**	4.03 **7.79**	4.00 **7.72**	3.96 **7.60**	3.92 **7.52**	3.87 **7.39**	3.84 **7.31**	3.81 **7.23**	3.77 **7.14**	3.75 **7.09**	3.72 **7.02**	3.71 **6.99**	3.69 **6.94**	3.68 **6.90**	3.67 **6.88**
7	5.59 **12.25**	4.74 **9.55**	4.35 **8.45**	4.12 **7.85**	3.97 **7.46**	3.87 **7.19**	3.79 **7.00**	3.73 **6.84**	3.68 **6.71**	3.63 **6.62**	3.60 **6.54**	3.57 **6.47**	3.52 **6.35**	3.49 **6.27**	3.44 **6.15**	3.41 **6.07**	3.38 **5.98**	3.34 **5.90**	3.32 **5.85**	3.29 **5.78**	3.28 **5.75**	3.25 **5.70**	3.24 **5.67**	3.23 **5.65**
8	5.32 **11.26**	4.46 **8.65**	4.07 **7.59**	3.84 **7.01**	3.69 **6.63**	3.58 **6.37**	3.50 **6.19**	3.44 **6.03**	3.39 **5.91**	3.34 **5.82**	3.31 **5.74**	3.28 **5.67**	3.23 **5.56**	3.20 **5.48**	3.15 **5.36**	3.12 **5.28**	3.08 **5.20**	3.05 **5.11**	3.03 **5.06**	3.00 **5.00**	2.98 **4.96**	2.96 **4.91**	2.94 **4.88**	2.93 **4.86**
9	5.12 **10.56**	4.26 **8.02**	3.86 **6.99**	3.63 **6.42**	3.48 **6.06**	3.37 **5.80**	3.29 **5.62**	3.23 **5.47**	3.18 **5.35**	3.13 **5.26**	3.10 **5.18**	3.07 **5.11**	3.02 **5.00**	2.98 **4.92**	2.93 **4.80**	2.90 **4.73**	2.86 **4.64**	2.82 **4.56**	2.80 **4.51**	2.77 **4.45**	2.76 **4.41**	2.73 **4.36**	2.72 **4.33**	2.71 **4.31**

10	4.96 **10.04**	4.10 **7.56**	3.71 **6.55**	3.48 **5.99**	3.33 **5.64**	3.22 **5.39**	3.14 **5.21**	3.07 **5.06**	3.02 **4.95**	2.97 **4.85**	2.94 **4.78**	2.91 **4.71**	2.86 **4.60**	2.82 **4.52**	2.77 **4.41**	2.74 **4.33**	2.70 **4.25**	2.67 **4.17**	2.64 **4.12**	2.61 **4.05**	2.59 **4.01**	2.56 **3.96**	2.55 **3.93**	2.54 **3.91**
11	4.84 **9.65**	3.98 **7.20**	3.59 **6.22**	3.36 **5.67**	3.20 **5.32**	3.09 **5.07**	3.01 **4.88**	2.95 **4.74**	2.90 **4.63**	2.86 **4.54**	2.82 **4.46**	2.79 **4.40**	2.74 **4.29**	2.70 **4.21**	2.65 **4.10**	2.61 **4.02**	2.57 **3.94**	2.53 **3.86**	2.50 **3.80**	2.47 **3.74**	2.45 **3.70**	2.42 **3.66**	2.41 **3.62**	2.40 **3.60**
12	4.75 **9.33**	3.88 **6.93**	3.49 **5.95**	3.26 **5.41**	3.11 **5.06**	3.00 **4.82**	2.92 **4.65**	2.85 **4.50**	2.80 **4.39**	2.76 **4.30**	2.72 **4.22**	2.69 **4.16**	2.64 **4.05**	2.60 **3.98**	2.54 **3.86**	2.50 **3.78**	2.46 **3.70**	2.42 **3.61**	2.40 **3.56**	2.36 **3.49**	2.35 **3.46**	2.32 **3.41**	2.31 **3.38**	2.30 **3.36**
13	4.67 **9.07**	3.80 **6.70**	3.41 **5.74**	3.18 **5.20**	3.02 **4.86**	2.92 **4.62**	2.84 **4.44**	2.77 **4.30**	2.72 **4.19**	2.67 **4.10**	2.63 **4.02**	2.60 **3.96**	2.55 **3.85**	2.51 **3.78**	2.46 **3.67**	2.42 **3.59**	2.38 **3.51**	2.34 **3.42**	2.32 **3.37**	2.28 **3.30**	2.26 **3.27**	2.24 **3.21**	2.22 **3.18**	2.21 **3.16**
14	4.60 **8.86**	3.74 **6.51**	3.34 **5.56**	3.11 **5.03**	2.96 **4.69**	2.85 **4.46**	2.77 **4.28**	2.70 **4.14**	2.65 **4.03**	2.60 **3.94**	2.56 **3.86**	2.53 **3.80**	2.48 **3.70**	2.44 **3.62**	2.39 **3.51**	2.35 **3.43**	2.31 **3.34**	2.27 **3.26**	2.24 **3.21**	2.21 **3.14**	2.19 **3.11**	2.16 **3.06**	2.14 **3.02**	2.13 **3.00**
15	4.54 **8.68**	3.68 **6.36**	3.29 **5.42**	3.06 **4.89**	2.90 **4.56**	2.79 **4.32**	2.70 **4.14**	2.64 **4.00**	2.59 **3.89**	2.55 **3.80**	2.51 **3.73**	2.48 **3.67**	2.43 **3.56**	2.39 **3.48**	2.33 **3.36**	2.29 **3.29**	2.25 **3.20**	2.21 **3.12**	2.18 **3.07**	2.15 **3.00**	2.12 **2.97**	2.10 **2.92**	2.08 **2.89**	2.07 **2.87**
16	4.49 **8.53**	3.63 **6.23**	3.24 **5.29**	3.01 **4.77**	2.85 **4.44**	2.74 **4.20**	2.66 **4.03**	2.59 **3.89**	2.54 **3.78**	2.49 **3.69**	2.45 **3.61**	2.42 **3.55**	2.37 **3.45**	2.33 **3.37**	2.28 **3.25**	2.24 **3.18**	2.20 **3.10**	2.16 **3.01**	2.13 **2.96**	2.09 **2.89**	2.07 **2.86**	2.04 **2.80**	2.02 **2.77**	2.01 **2.75**
17	4.45 **8.40**	3.59 **6.11**	3.20 **5.18**	2.96 **4.67**	2.81 **4.34**	2.70 **4.10**	2.62 **3.93**	2.55 **3.79**	2.50 **3.68**	2.45 **3.59**	2.41 **3.52**	2.38 **3.45**	2.33 **3.35**	2.29 **3.27**	2.23 **3.16**	2.19 **3.08**	2.15 **3.00**	2.11 **2.92**	2.08 **2.86**	2.04 **2.79**	2.02 **2.76**	1.99 **2.70**	1.97 **2.67**	1.96 **2.65**
18	4.41 **8.28**	3.55 **6.01**	3.16 **5.09**	2.93 **4.58**	2.77 **4.25**	2.66 **4.01**	2.58 **3.85**	2.51 **3.71**	2.46 **3.60**	2.41 **3.51**	2.37 **3.44**	2.34 **3.37**	2.29 **3.27**	2.25 **3.19**	2.19 **3.07**	2.15 **3.00**	2.11 **2.91**	2.07 **2.83**	2.04 **2.78**	2.00 **2.71**	1.98 **2.68**	1.95 **2.62**	1.93 **2.59**	1.92 **2.57**
19	4.38 **8.18**	3.52 **5.93**	3.13 **5.01**	2.90 **4.50**	2.74 **4.17**	2.63 **3.94**	2.55 **3.77**	2.48 **3.63**	2.43 **3.52**	2.38 **3.43**	2.34 **3.36**	2.31 **3.30**	2.26 **3.19**	2.21 **3.12**	2.15 **3.00**	2.11 **2.92**	2.07 **2.84**	2.02 **2.76**	2.00 **2.70**	1.96 **2.63**	1.94 **2.60**	1.91 **2.54**	1.90 **2.51**	1.88 **2.49**
20	4.35 **8.10**	3.49 **5.85**	3.10 **4.94**	2.87 **4.43**	2.71 **4.10**	2.60 **3.87**	2.52 **3.71**	2.45 **3.56**	2.40 **3.45**	2.35 **3.37**	2.31 **3.30**	2.28 **3.23**	2.23 **3.13**	2.18 **3.05**	2.12 **2.94**	2.08 **2.86**	2.04 **2.77**	1.99 **2.69**	1.96 **2.63**	1.92 **2.56**	1.90 **2.53**	1.87 **2.47**	1.85 **2.44**	1.84 **2.42**
21	4.32 **8.02**	3.47 **5.78**	3.07 **4.87**	2.84 **4.37**	2.68 **4.04**	2.57 **3.81**	2.49 **3.65**	2.42 **3.51**	2.37 **3.40**	2.32 **3.31**	2.28 **3.24**	2.25 **3.17**	2.20 **3.07**	2.15 **2.99**	2.09 **2.88**	2.05 **2.80**	2.00 **2.72**	1.96 **2.63**	1.93 **2.58**	1.89 **2.51**	1.87 **2.47**	1.84 **2.42**	1.82 **2.38**	1.81 **2.36**
22	4.30 **7.94**	3.44 **5.72**	3.05 **4.82**	2.82 **4.31**	2.66 **3.99**	2.55 **3.76**	2.47 **3.59**	2.40 **3.45**	2.35 **3.35**	2.30 **3.26**	2.26 **3.18**	2.23 **3.12**	2.18 **3.02**	2.13 **2.94**	2.07 **2.83**	2.03 **2.75**	1.98 **2.67**	1.93 **2.58**	1.91 **2.53**	1.87 **2.46**	1.84 **2.42**	1.81 **2.37**	1.80 **2.33**	1.78 **2.31**
23	4.28 **7.88**	3.42 **5.66**	3.03 **4.76**	2.80 **4.26**	2.64 **3.94**	2.53 **3.71**	2.45 **3.54**	2.38 **3.41**	2.32 **3.30**	2.28 **3.21**	2.24 **3.14**	2.20 **3.07**	2.14 **2.97**	2.10 **2.89**	2.04 **2.78**	2.00 **2.70**	1.96 **2.62**	1.91 **2.53**	1.88 **2.48**	1.84 **2.41**	1.82 **2.37**	1.79 **2.32**	1.77 **2.28**	1.76 **2.26**
24	4.26 **7.82**	3.40 **5.61**	3.01 **4.72**	2.78 **4.22**	2.62 **3.90**	2.51 **3.67**	2.43 **3.50**	2.36 **3.36**	2.30 **3.25**	2.26 **3.17**	2.22 **3.09**	2.18 **3.03**	2.13 **2.93**	2.09 **2.85**	2.02 **2.74**	1.98 **2.66**	1.94 **2.58**	1.89 **2.49**	1.86 **2.44**	1.82 **2.36**	1.80 **2.33**	1.76 **2.27**	1.74 **2.23**	1.73 **2.21**
25	4.24 **7.77**	3.38 **5.57**	2.99 **4.68**	2.76 **4.18**	2.60 **3.86**	2.49 **3.63**	2.41 **3.46**	2.34 **3.32**	2.28 **3.21**	2.24 **3.13**	2.20 **3.05**	2.16 **2.99**	2.11 **2.89**	2.06 **2.81**	2.00 **2.70**	1.96 **2.62**	1.92 **2.54**	1.87 **2.45**	1.84 **2.40**	1.80 **2.32**	1.77 **2.29**	1.74 **2.23**	1.72 **2.19**	1.71 **2.17**
26	4.22 **7.72**	3.37 **5.53**	2.98 **4.64**	2.74 **4.14**	2.59 **3.82**	2.47 **3.59**	2.39 **3.42**	2.32 **3.29**	2.27 **3.17**	2.22 **3.09**	2.18 **3.02**	2.15 **2.96**	2.10 **2.86**	2.05 **2.77**	1.99 **2.66**	1.95 **2.58**	1.90 **2.50**	1.85 **2.41**	1.82 **2.36**	1.78 **2.28**	1.76 **2.25**	1.72 **2.19**	1.70 **2.15**	1.69 **2.13**

5% (Roman Type) and 1% (Bold-Face Type) Points for the Distribution of F

Degrees of freedom n_2	Degrees of freedom n_1: 1	2	3	4	5	6	7	8	9	10	11	12	14	16	20	24	30	40	50	75	100	200	500	∞
27	4.21	3.35	2.96	2.73	2.57	2.46	2.37	2.30	2.25	2.20	2.16	2.13	2.08	2.03	1.97	1.93	1.88	1.84	1.80	1.76	1.74	1.71	1.68	1.67
	7.68	**5.49**	**4.60**	**4.11**	**3.79**	**3.56**	**3.39**	**3.26**	**3.14**	**3.06**	**2.98**	**2.93**	**2.83**	**2.74**	**2.63**	**2.55**	**2.47**	**2.38**	**2.33**	**2.25**	**2.21**	**2.16**	**2.12**	**2.10**
28	4.20	3.34	2.95	2.71	2.56	2.44	2.36	2.29	2.24	2.19	2.15	2.12	2.06	2.02	1.96	1.91	1.87	1.81	1.78	1.75	1.72	1.69	1.67	1.65
	7.64	**5.45**	**4.57**	**4.07**	**3.76**	**3.53**	**3.36**	**3.23**	**3.11**	**3.03**	**2.95**	**2.90**	**2.80**	**2.71**	**2.60**	**2.52**	**2.44**	**2.35**	**2.30**	**2.22**	**2.18**	**2.13**	**2.09**	**2.06**
29	4.18	3.33	2.93	2.70	2.54	2.43	2.35	2.28	2.22	2.18	2.14	2.10	2.05	2.00	1.94	1.90	1.85	1.80	1.77	1.73	1.71	1.68	1.65	1.64
	7.60	**5.42**	**4.54**	**4.04**	**3.73**	**3.50**	**3.33**	**3.20**	**3.08**	**3.00**	**2.92**	**2.87**	**2.77**	**2.68**	**2.57**	**2.49**	**2.41**	**2.32**	**2.27**	**2.19**	**2.15**	**2.10**	**2.06**	**2.03**
30	4.17	3.32	2.92	2.69	2.53	2.42	2.34	2.27	2.21	2.16	2.12	2.09	2.04	1.99	1.93	1.89	1.84	1.79	1.76	1.72	1.69	1.66	1.64	1.62
	7.56	**5.39**	**4.51**	**4.02**	**3.70**	**3.47**	**3.30**	**3.17**	**3.06**	**2.98**	**2.90**	**2.84**	**2.74**	**2.66**	**2.55**	**2.47**	**2.38**	**2.29**	**2.24**	**2.16**	**2.13**	**2.07**	**2.03**	**2.01**
32	4.15	3.30	2.90	2.67	2.51	2.40	2.32	2.25	2.19	2.14	2.10	2.07	2.02	1.97	1.91	1.86	1.82	1.76	1.74	1.69	1.67	1.64	1.61	1.59
	7.50	**5.34**	**4.46**	**3.97**	**3.66**	**3.42**	**3.25**	**3.12**	**3.01**	**2.94**	**2.86**	**2.80**	**2.70**	**2.62**	**2.51**	**2.42**	**2.34**	**2.25**	**2.20**	**2.12**	**2.08**	**2.02**	**1.98**	**1.96**
34	4.13	3.28	2.88	2.65	2.49	2.38	2.30	2.23	2.17	2.12	2.08	2.05	2.00	1.95	1.89	1.84	1.80	1.74	1.71	1.67	1.64	1.61	1.59	1.57
	7.44	**5.29**	**4.42**	**3.93**	**3.61**	**3.38**	**3.21**	**3.08**	**2.97**	**2.89**	**2.82**	**2.76**	**2.66**	**2.58**	**2.47**	**2.38**	**2.30**	**2.21**	**2.15**	**2.08**	**2.04**	**1.98**	**1.94**	**1.91**
36	4.11	3.26	2.86	2.63	2.48	2.36	2.28	2.21	2.15	2.10	2.06	2.03	1.98	1.93	1.87	1.82	1.78	1.72	1.69	1.65	1.62	1.59	1.56	1.55
	7.39	**5.25**	**4.38**	**3.89**	**3.58**	**3.35**	**3.18**	**3.04**	**2.94**	**2.86**	**2.78**	**2.72**	**2.62**	**2.54**	**2.43**	**2.35**	**2.26**	**2.17**	**2.12**	**2.04**	**2.00**	**1.94**	**1.90**	**1.87**
38	4.10	3.25	2.85	2.62	2.46	2.35	2.26	2.19	2.14	2.09	2.05	2.02	1.96	1.92	1.85	1.80	1.76	1.71	1.67	1.63	1.60	1.57	1.54	1.53
	7.35	**5.21**	**4.34**	**3.86**	**3.54**	**3.32**	**3.15**	**3.02**	**2.91**	**2.82**	**2.75**	**2.69**	**2.59**	**2.51**	**2.40**	**2.32**	**2.22**	**2.14**	**2.08**	**2.00**	**1.97**	**1.90**	**1.86**	**1.84**
40	4.08	3.23	2.84	2.61	2.45	2.34	2.25	2.18	2.12	2.07	2.04	2.00	1.95	1.90	1.84	1.79	1.74	1.69	1.66	1.61	1.59	1.55	1.53	1.51
	7.31	**5.18**	**4.31**	**3.83**	**3.51**	**3.29**	**3.12**	**2.99**	**2.88**	**2.80**	**2.73**	**2.66**	**2.56**	**2.49**	**2.37**	**2.29**	**2.20**	**2.11**	**2.05**	**1.97**	**1.94**	**1.88**	**1.84**	**1.81**
42	4.07	3.22	2.83	2.59	2.44	2.32	2.24	2.17	2.11	2.06	2.02	1.99	1.94	1.89	1.82	1.78	1.73	1.68	1.64	1.60	1.57	1.54	1.51	1.49
	7.27	**5.15**	**4.29**	**3.80**	**3.49**	**3.26**	**3.10**	**2.96**	**2.86**	**2.77**	**2.70**	**2.64**	**2.54**	**2.46**	**2.35**	**2.26**	**2.17**	**2.08**	**2.02**	**1.94**	**1.91**	**1.85**	**1.80**	**1.78**
44	4.06	3.21	2.82	2.58	2.43	2.31	2.23	2.16	2.10	2.05	2.01	1.98	1.92	1.88	1.81	1.76	1.72	1.66	1.63	1.58	1.56	1.52	1.50	1.48
	7.24	**5.12**	**4.26**	**3.78**	**3.46**	**3.24**	**3.07**	**2.94**	**2.84**	**2.75**	**2.68**	**2.62**	**2.52**	**2.44**	**2.32**	**2.24**	**2.15**	**2.06**	**2.00**	**1.92**	**1.88**	**1.82**	**1.78**	**1.75**
46	4.05	3.20	2.81	2.57	2.42	2.30	2.22	2.14	2.09	2.04	2.00	1.97	1.91	1.87	1.80	1.75	1.71	1.65	1.62	1.57	1.54	1.51	1.48	1.46
	7.21	**5.10**	**4.24**	**3.76**	**3.44**	**3.22**	**3.05**	**2.92**	**2.82**	**2.73**	**2.66**	**2.60**	**2.50**	**2.42**	**2.30**	**2.22**	**2.13**	**2.04**	**1.98**	**1.90**	**1.86**	**1.80**	**1.76**	**1.72**
48	4.04	3.19	2.80	2.56	2.41	2.30	2.21	2.14	2.08	2.03	1.99	1.96	1.90	1.86	1.79	1.74	1.70	1.64	1.61	1.56	1.53	1.50	1.47	1.45
	7.19	**5.08**	**4.22**	**3.74**	**3.42**	**3.20**	**3.04**	**2.90**	**2.80**	**2.71**	**2.64**	**2.58**	**2.48**	**2.40**	**2.28**	**2.20**	**2.11**	**2.02**	**1.96**	**1.88**	**1.84**	**1.78**	**1.73**	**1.70**

50	4.03 **7.17**	3.18 **5.06**	2.79 **4.20**	2.56 **3.72**	2.40 **3.41**	2.29 **3.18**	2.20 **3.02**	2.13 **2.88**	2.07 **2.78**	2.02 **2.70**	1.98 **2.62**	1.95 **2.56**	1.90 **2.46**	1.85 **2.39**	1.78 **2.26**	1.74 **2.18**	1.69 **2.10**	1.63 **2.00**	1.60 **1.94**	1.55 **1.86**	1.52 **1.82**	1.48 **1.76**	1.46 **1.71**	**1.44** **1.68**
55	4.02 **7.12**	3.17 **5.01**	2.78 **4.16**	2.54 **3.68**	2.38 **3.37**	2.27 **3.15**	2.18 **2.98**	2.11 **2.85**	2.05 **2.75**	2.00 **2.66**	1.97 **2.59**	1.93 **2.53**	1.88 **2.43**	1.83 **2.35**	1.76 **2.23**	1.72 **2.15**	1.67 **2.06**	1.61 **1.96**	1.58 **1.90**	1.52 **1.82**	1.50 **1.78**	1.46 **1.71**	1.43 **1.66**	1.41 **1.64**
60	4.00 **7.08**	3.15 **4.98**	2.76 **4.13**	2.52 **3.65**	2.37 **3.34**	2.25 **3.12**	2.17 **2.95**	2.10 **2.82**	2.04 **2.72**	1.99 **2.63**	1.95 **2.56**	1.92 **2.50**	1.86 **2.40**	1.81 **2.32**	1.75 **2.20**	1.70 **2.12**	1.65 **2.03**	1.59 **1.93**	1.56 **1.87**	1.50 **1.79**	1.48 **1.74**	1.44 **1.68**	1.41 **1.63**	1.39 **1.60**
65	3.99 **7.04**	3.14 **4.95**	2.75 **4.10**	2.51 **3.62**	2.36 **3.31**	2.24 **3.09**	2.15 **2.93**	2.08 **2.79**	2.02 **2.70**	1.98 **2.61**	1.94 **2.54**	1.90 **2.47**	1.85 **2.37**	1.80 **2.30**	1.73 **2.18**	1.68 **2.09**	1.63 **2.00**	1.57 **1.90**	1.54 **1.84**	1.49 **1.76**	1.46 **1.71**	1.42 **1.64**	1.39 **1.60**	1.37 **1.56**
70	3.98 **7.01**	3.13 **4.92**	2.74 **4.08**	2.50 **3.60**	2.35 **3.29**	2.23 **3.07**	2.14 **2.91**	2.07 **2.77**	2.01 **2.67**	1.97 **2.59**	1.93 **2.51**	1.89 **2.45**	1.84 **2.35**	1.79 **2.28**	1.72 **2.15**	1.67 **2.07**	1.62 **1.98**	1.56 **1.88**	1.53 **1.82**	1.47 **1.74**	1.45 **1.69**	1.40 **1.62**	1.37 **1.56**	1.35 **1.53**
80	3.96 **6.96**	3.11 **4.88**	2.72 **4.04**	2.48 **3.56**	2.33 **3.25**	2.21 **3.04**	2.12 **2.87**	2.05 **2.74**	1.99 **2.64**	1.95 **2.55**	1.91 **2.48**	1.88 **2.41**	1.82 **2.32**	1.77 **2.24**	1.70 **2.11**	1.65 **2.03**	1.60 **1.94**	1.54 **1.34**	1.51 **1.78**	1.45 **1.70**	1.42 **1.65**	1.38 **1.57**	1.35 **1.52**	1.32 **1.49**
100	3.94 **6.90**	3.09 **4.82**	2.70 **3.98**	2.46 **3.51**	2.30 **3.20**	2.19 **2.99**	2.10 **2.82**	2.03 **2.69**	1.97 **2.59**	1.92 **2.51**	1.88 **2.43**	1.85 **2.36**	1.79 **2.26**	1.75 **2.19**	1.68 **2.06**	1.63 **1.98**	1.57 **1.89**	1.51 **1.79**	1.48 **1.73**	1.42 **1.64**	1.39 **1.59**	1.34 **1.51**	1.30 **1.46**	1.28 **1.43**
125	3.92 **6.84**	3.07 **4.78**	2.68 **3.94**	2.44 **3.47**	2.29 **3.17**	2.17 **2.95**	2.08 **2.79**	2.01 **2.65**	1.95 **2.56**	1.90 **2.47**	1.86 **2.40**	1.83 **2.33**	1.77 **2.23**	1.72 **2.15**	1.65 **2.03**	1.60 **1.94**	1.55 **1.85**	1.49 **1.75**	1.45 **1.68**	1.39 **1.59**	1.36 **1.54**	1.31 **1.46**	1.27 **1.40**	1.25 **1.37**
150	3.91 **6.81**	3.06 **4.75**	2.67 **3.91**	2.43 **3.44**	2.27 **3.14**	2.16 **2.92**	2.07 **2.76**	2.00 **2.62**	1.94 **2.53**	1.89 **2.44**	1.85 **2.37**	1.82 **2.30**	1.76 **2.20**	1.71 **2.12**	1.64 **2.00**	1.59 **1.91**	1.54 **1.83**	1.47 **1.72**	1.44 **1.66**	1.37 **1.56**	1.34 **1.51**	1.29 **1.43**	1.25 **1.37**	1.22 **1.33**
200	3.89 **6.76**	3.04 **4.71**	2.65 **3.88**	2.41 **3.41**	2.26 **3.11**	2.14 **2.90**	2.05 **2.73**	1.98 **2.60**	1.92 **2.50**	1.87 **2.41**	1.83 **2.34**	1.80 **2.28**	1.74 **2.17**	1.69 **2.09**	1.62 **1.97**	1.57 **1.88**	1.52 **1.79**	1.45 **1.69**	1.42 **1.62**	1.35 **1.53**	1.32 **1.48**	1.26 **1.39**	1.22 **1.33**	1.19 **1.28**
400	3.86 **6.70**	3.02 **4.66**	2.62 **3.83**	2.39 **3.36**	2.23 **3.06**	2.12 **2.85**	2.03 **2.69**	1.96 **2.55**	1.90 **2.46**	1.85 **2.37**	1.81 **2.29**	1.78 **2.23**	1.72 **2.12**	1.67 **2.04**	1.60 **1.92**	1.54 **1.84**	1.49 **1.74**	1.42 **1.64**	1.38 **1.57**	1.32 **1.47**	1.28 **1.42**	1.22 **1.32**	1.16 **1.24**	1.13 **1.19**
1000	3.85 **6.66**	3.00 **4.62**	2.61 **3.80**	2.38 **3.34**	2.22 **3.04**	2.10 **2.82**	2.02 **2.66**	1.95 **2.53**	1.89 **2.43**	1.84 **2.34**	1.80 **2.26**	1.76 **2.20**	1.70 **2.09**	1.65 **2.01**	1.58 **1.89**	1.53 **1.81**	1.47 **1.71**	1.41 **1.61**	1.36 **1.54**	1.30 **1.44**	1.26 **1.38**	1.19 **1.28**	1.13 **1.19**	1.08 **1.11**
∞	3.84 **6.64**	2.99 **4.60**	2.60 **3.78**	2.37 **3.32**	2.21 **3.02**	2.09 **2.80**	2.01 **2.64**	1.94 **2.51**	1.88 **2.41**	1.83 **2.32**	1.79 **2.24**	1.75 **2.18**	1.69 **2.07**	1.64 **1.99**	1.57 **1.87**	1.52 **1.79**	1.46 **1.69**	1.40 **1.59**	1.35 **1.52**	1.28 **1.41**	1.24 **1.36**	1.17 **1.25**	1.11 **1.15**	1.00 **1.00**

Applied Probability and Statistics (Continued)

JOHNSON and KOTZ · Distributions in Statistics
Discrete Distributions
Continuous Univariate Distributions-1
Continuous Univariate Distributions-2
Continuous Multivariate Distributions

JOHNSON and LEONE · Statistics and Experimental Design: In Engineering and the Physical Sciences, Volumes I and II, *Second Edition*

KEENEY and RAIFFA · Decisions with Multiple Objectives

LANCASTER · The Chi Squared Distribution

LANCASTER · An Introduction to Medical Statistics

LEWIS · Stochastic Point Processes

MANN, SCHAFER and SINGPURWALLA · Methods for Statistical Analysis of Reliability and Life Data

MEYER · Data Analysis for Scientists and Engineers

OTNES and ENOCHSON · Digital Time Series Analysis

PRENTER · Splines and Variational Methods

RAO and MITRA · Generalized Inverse of Matrices and Its Applications

SARD and WEINTRAUB · A Book of Splines

SEAL · Stochastic Theory of a Risk Business

SEARLE · Linear Models

THOMAS · An Introduction to Applied Probability and Random Processes

WHITTLE · Optimization under Constraints

WONNACOTT and WONNACOTT · Econometrics

YOUDEN · Statistical Methods for Chemists

ZELLNER · An Introduction to Bayesian Inference in Econometrics

Tracts on Probability and Statistics

BHATTACHARYA and RAO · Normal Approximation and Asymptotic Expansions

BILLINGSLEY · Convergence of Probability Measures

CRAMER and LEADBETTER · Stationary and Related Stochastic Processes

JARDINE and SIBSON · Mathematical Taxonomy

RIORDAN · Combinatorial Identities